Challenges and Advances in Computational Chemistry and Physics

Volume 27

Series editor

Jerzy Leszczynski
Department of Chemistry and Biochemistry
Jackson State University, Jackson, MS, USA

This book series provides reviews on the most recent developments in computational chemistry and physics. It covers both the method developments and their applications. Each volume consists of chapters devoted to the one research area. The series highlights the most notable advances in applications of the computational methods. The volumes include nanotechnology, material sciences, molecular biology, structures and bonding in molecular complexes, and atmospheric chemistry. The authors are recruited from among the most prominent researchers in their research areas. As computational chemistry and physics is one of the most rapidly advancing scientific areas such timely overviews are desired by chemists, physicists, molecular biologists and material scientists. The books are intended for graduate students and researchers.

All contributions to edited volumes should undergo standard peer review to ensure high scientific quality, while monographs should be reviewed by at least two experts in the field. Submitted manuscripts will be reviewed and decided by the series editor, Prof. Jerzy Leszczynski.

More information about this series at http://www.springer.com/series/6918

C. Gopi Mohan
Editor

Structural Bioinformatics: Applications in Preclinical Drug Discovery Process

Springer

Editor
C. Gopi Mohan
Amrita Centre for Nanosciences
and Molecular Medicine
Amrita Institute of Medical Sciences
and Research Centre
Kochi, India

ISSN 2542-4491 ISSN 2542-4483 (electronic)
Challenges and Advances in Computational Chemistry and Physics
ISBN 978-3-030-05281-2 ISBN 978-3-030-05282-9 (eBook)
https://doi.org/10.1007/978-3-030-05282-9

Library of Congress Control Number: 2018962784

This Springer imprint is published by the registered company Springer Nature Switzerland AG
The registered company address is: Gewerbestrasse 11, 6330 Cham, Switzerland

Preface

Human society has immense faith in the potential of drugs. Our belief towards therapeutically safer drugs to alleviate the symptoms of different types of diseases is accelerating nowadays. The twenty-first century witnessed tremendous progress in the scientific and technical aspects in several therapeutic domains, such as viral, bacterial, cancer and other metabolic and infectious diseases. Further, bioinformatics and computational biology disciplines are integrated into all levels of medicine and health care. Future breakthroughs will depend on the strong collaborations between experimental and computational biologists. Areas such as building predictive models of the cell, organelles, and organs, understanding ageing, designing enzymes, and improving drug design and target validation are becoming crucial for the drug discovery programme.

The main concept of the present book includes computer-aided molecular modelling and protein/enzyme design in preclinical discovery towards understanding the molecular mechanisms of different diseases. This technique can be successfully employed in different areas of medical research, including rare and neglected diseases. Different case studies integrated with the experimental research as well the future plan for clinical aspects are described effectively. The present 12 chapters of the book have been contributed by leading and internationally recognized scientists. It addresses computer simulation techniques for studying biological phenomena from the perspective of both methodology and applications. The chapters are organized on the methodology of molecular simulations and its applications, chemoinformatics methods and its use of experimental information in computational simulations. Selected applications of structural biology and structure-based drug design, focussing towards druggable targets, and its physiological molecular mechanisms of actions are critically addressed.

The first five chapters are devoted to theories and methodologies, which form the backbone of the structure-based drug design concepts as well as different molecular modeling techniques in computer-aided drug design. Chapter "Structure-Based Drug Design of *Pf*DHODH Inhibitors as Antimalarial Agents" describes the latest theories and computational methodologies in structure-based drug design for the development of inhibitors against key druggable target *Plasmodium falciparum* dihydroorotate dehydrogenase. Chapter "Recent Advancements in Computing Reliable Binding Free Energies in Drug Discovery Projects" is dedicated to understanding the protein–ligand binding affinities and different concepts and methods towards free energy calculations for the drug discovery projects. Next chapter (Chapter "Integrated Chemoin–formatics Approaches Towards Epigenetic Drug Discovery") addresses the epigenetics molecular mechanism and its key targets involved in different diseases by efficiently employing different chemoinformatics strategies. Chapter "Structure-Based Drug Design with a Special Emphasis on Herbal Extracts" directly deals with the natural products, a component of Ayurinformatics, and its emphasis on the application of structure-based drug design. Chapter "Impact of Target-Based Drug Design in Anti-bacterial Drug Discovery for the Treatment of Tuberculosis" is devoted completely towards tuberculosis drug discovery and the role of three-dimensional druggable targets in the structure-based anti-tuberculosis design. The role of big data and high-performance computing is prevalent nowadays in different fields, and the concept and algorithms presented in Chapter "Turbo Analytics: Applications of Big Data and HPC in Drug Discovery" directly address its importance and application towards the preclinical drug discovery aspects. Finally, Chapter "Single-Particle cryo-EM as a Pipeline for Obtaining Atomic Resolution Structures of Druggable Targets in Preclinial Structure-Based Drug Design" is devoted towards the latest technique in structural biology, i.e. single-particle cryo-EM to solve the atomic structures of single and multi-protein druggable targets and which is key to the structure-based drug design studies.

> In the future, Computers will design, discover, people will verify—John Rumble
>
> Science knows no country, because knowledge belongs to humanity, and is the torch which illuminates the world—Louis Pasteur
>
> Science is beautiful when it makes simple explanations of phenomena or connections between different observations. Examples include the double helix in biology and the fundamental equations of physics—Stephen Hawking

The purpose of this book is to explore the theoretical strategies involved in drug discovery and development by proper integration with the experimental concepts as well. Further, the book is intended to deliver the reader with an overview of multifaceted, challenging and rapidly evolving field. We feel that the scientific material covered herein will provide the reader with an excellent overview in preclinical drug discovery programme.

Ämrita Vishwa Vidyapeetham, Kochi, India
October 2018

C. Gopi Mohan

Contents

Editor and Contributors

About the Editor

Dr. C. Gopi Mohan is Incharge, Bioinformatics and Computational Biology Laboratory, Center for Nanosciences and Molecular Medicine, Amrita Vishwa Vidyapeetham, Kochi. He graduated with a Ph.D. degree from Banaras Hindu University, Varanasi. He gained experience as Postdoctoral Fellow from Molecular Biophysics Unit at IISc, Bangalore, and as Research Officer from Department of Biology and Biochemistry, University of Bath, UK. Further, he worked as Associate Researcher of CNRS, University Henri Poincare, Nancy, France. During his research career, he visited different countries which include UK, Canada, France, Finland, and USA.

He has supervised many Ph.D. and postgraduate students and completed different research and industrial consultancy projects. He has published more than 80 peer-reviewed research papers and chapters and is Active Reviewer of international/national journals, thesis, and grants. His research interests are Computational Biology & Structural Bioinformatics, Structure-Based Drug Design, Protein Crystallography, and Nanoinformatics. He is Member of the Indian Biophysical Society and the American Chemical Society. He was Invited Speaker for different international conferences and recently was awarded ICMR-Senior Biomedical Scientist International Fellowship.

Contributors

Avinash Bayaskar High Performance Computing-Medical & Bioinformatics Applications Group, Centre for Development of Advanced Computing (C-DAC), Savitribai Phule Pune University Campus, Pune, India

Shweta Bhagat Department of Medicinal Chemistry, National Institute of Pharmaceutical Education and Research (NIPER), S.A.S. Nagar, Punjab, India

Prasad V. Bharatam Department of Medicinal Chemistry, National Institute of Pharmaceutical Education and Research (NIPER), S.A.S. Nagar, Punjab, India

Atanu Bhattacharjee Department of Biotechnology & Bioinformatics, North-Eastern Hill University, Shillong, India

Raja Biswas Center for Nanosciences and Molecular Medicine, Amrita Institute of Medical Sciences and Research Centre, Amrita Vishwa Vidyapeetham, Kochi, Kerala, India

Chinmayee Choudhury Center for Molecular Modelling, Indian Institute of Chemical Technology, Hyderabad, India; Department of Biochemistry, All India Institute of Medical Sciences, Basni, Jodhpur, Rajasthan, India

Evans C. Coutinho Molecular Simulations Group, Department of Pharmaceutical Chemistry, Bombay College of Pharmacy, Mumbai, India

Anuj Gahlawat Department of Pharmacoinformatics, National Institute of Pharmaceutical Education and Research (NIPER), S.A.S. Nagar, Punjab, India

Vivek Gavane High Performance Computing-Medical & Bioinformatics Applications Group, Centre for Development of Advanced Computing (C-DAC), Savitribai Phule Pune University Campus, Pune, India

Indira Ghosh School of Computational and Integrative Sciences (SCIS), Jawaharlal Nehru University, New Delhi, India

Pawan Gupta Advanced Protein Science Building, Institute of Microbial Technology, Chandigarh, India

Vinod Jani High Performance Computing-Medical & Bioinformatics Applications Group, Centre for Development of Advanced Computing (C-DAC), Savitribai Phule Pune University Campus, Pune, India

Rajendra R. Joshi High Performance Computing-Medical & Bioinformatics Applications Group, Centre for Development of Advanced Computing (C-DAC), Savitribai Phule Pune University Campus, Pune, India

Vikrant Karmahapatra Advanced Protein Science Building, Institute of Microbial Technology, Chandigarh, India

Shruti Koulgi High Performance Computing-Medical & Bioinformatics Applications Group, Centre for Development of Advanced Computing (C-DAC), Savitribai Phule Pune University Campus, Pune, India

N. H. V. Kutumbarao CAS in Crystallography and Biophysics, University of Madras, Chennai, Tamil Nadu, India

Saurabh Loharch Advanced Protein Science Building, Institute of Microbial Technology, Chandigarh, India

Rethi Madathil Advanced Protein Science Building, Institute of Microbial Technology, Chandigarh, India

Rashmi Mahajan High Performance Computing-Medical & Bioinformatics Applications Group, Centre for Development of Advanced Computing (C-DAC), Savitribai Phule Pune University Campus, Pune, India

Sandeep Malviya High Performance Computing-Medical & Bioinformatics Applications Group, Centre for Development of Advanced Computing (C-DAC), Savitribai Phule Pune University Campus, Pune, India

Elvis A. F. Martis Molecular Simulations Group, Department of Pharmaceutical Chemistry, Bombay College of Pharmacy, Mumbai, India

C. Gopi Mohan Center for Nanosciences and Molecular Medicine, Amrita Institute of Medical Sciences and Research Centre, Amrita Vishwa Vidyapeetham, Kochi, Kerala, India

N. Arul Murugan Department of Theoretical Chemistry and Biology, School of Engineering Sciences in Chemistry, Biotechnology and Health, Royal Institute of Technology, Stockholm, Sweden

G. Narahari Sastry Center for Molecular Modelling, Indian Institute of Chemical Technology, Hyderabad, India

Ramanathan Natesh School of Biology, Indian Institute of Science Education and Research Thiruvananthapuram (IISER-TVM), Trivandrum, Kerala, India

Debnath Pal Department of Computational and Data Sciences, Indian Institute of Science, Bangalore, India

Shailesh Kumar Panday School of Computational and Integrative Sciences (SCIS), Jawaharlal Nehru University, New Delhi, India

Sudhir Pandey High Performance Computing-Medical & Bioinformatics Applications Group, Centre for Development of Advanced Computing (C-DAC), Savitribai Phule Pune University Campus, Pune, India

Raman Parkesh Advanced Protein Science Building, Institute of Microbial Technology, Chandigarh, India

Vasanthanathan Poongavanam Department of Physics, Chemistry, Pharmacy, University of Southern Denmark, Odense M, Denmark

U. Deva Priyakumar CCNSB, International Institute of Information Technology, Gachibowli, Hyderabad, India

Anju Choorakottayil Pushkaran Center for Nanosciences and Molecular Medicine, Amrita Institute of Medical Sciences and Research Centre, Amrita Vishwa Vidyapeetham, Kochi, Kerala, India

E. P. Ramakrishnan High Performance Computing-Medical & Bioinformatics Applications Group, Centre for Development of Advanced Computing (C-DAC), Savitribai Phule Pune University Campus, Pune, India

Chandan Raychaudhury Department of Computational and Data Sciences, Indian Institute of Science, Bangalore, India

Md. Imbesat Hassan Rizvi Department of Computational and Data Sciences, Indian Institute of Science, Bangalore, India

Amit Saxena High Performance Computing-Medical & Bioinformatics Applications Group, Centre for Development of Advanced Computing (C-DAC), Savitribai Phule Pune University Campus, Pune, India

Neeru Sharma High Performance Computing-Medical & Bioinformatics Applications Group, Centre for Development of Advanced Computing (C-DAC), Savitribai Phule Pune University Campus, Pune, India

Dmitrii Shcherbinin Laboratory of Structural Bioinformatics, Institute of Biomedical Chemistry, Moscow, Russia

Uddhavesh Sonavane High Performance Computing-Medical & Bioinformatics Applications Group, Centre for Development of Advanced Computing (C-DAC), Savitribai Phule Pune University Campus, Pune, India

Mallikarjunachari Uppuladinne High Performance Computing-Medical & Bioinformatics Applications Group, Centre for Development of Advanced Computing (C-DAC), Savitribai Phule Pune University Campus, Pune, India

D. Velmurugan CAS in Crystallography and Biophysics, University of Madras, Chennai, Tamil Nadu, India

Alexander Veselovsky Laboratory of Structural Bioinformatics, Institute of Biomedical Chemistry, Moscow, Russia

V. Viswanathan CAS in Crystallography and Biophysics, University of Madras, Chennai, Tamil Nadu, India

Free Energy-Based Methods to Understand Drug Resistance Mutations

Elvis A. F. Martis and Evans C. Coutinho

Abstract In this chapter, we present an overview of various computational methods, particularly, those that are used to compute the free energy of binding to understand target site mutations that will enable us to foresee mutations that could significantly affect drug binding. We begin by looking at the driving forces that lead to drug resistance and throw some light on the various mechanisms by which drugs can be rendered ineffective. Next, we studied molecular dynamic simulations and its use to understand the thermodynamics of protein–ligand interactions. Building on these fundamentals, we discuss various methods that are available to compute the free energy binding, their mathematical formulations, the practical aspects of each these methods and finally their use in understanding drug resistance.

Keywords Molecular dynamics · Drug resistance · MM-PB(GB)-SA Free energy perturbation · Linear interaction energy · Computational mutational scanning · Thermodynamic integration

1 Drug Resistance Problem

Every organism attempts to survive in hostile conditions by making minor modifications in its life cycle. Though these modifications are observed phenotypically, genetic reshuffling and alterations are the underlying cause of these changes. Although we are unable to accurately explain this phenomenon and its initiation, we have been able to use this observed knowledge and empirically derive explanations for such modifications. However, it may not always be necessary to know all the details regarding genetic modifications, so long as we can correctly, at least empirically, understand such observations, and put it to effective use to predict and understand the drug resistance problem. Often the enzymes in the biochemical

E. A. F. Martis · E. C. Coutinho (✉)
Molecular Simulations Group, Department of Pharmaceutical Chemistry,
Bombay College of Pharmacy, Kalina, Santacruz [E], Mumbai 400098, India
e-mail: evans.coutinho@bcp.edu.in

C. G. Mohan (ed.), *Structural Bioinformatics: Applications in Preclinical Drug Discovery Process*, Challenges and Advances in Computational Chemistry and Physics 27, https://doi.org/10.1007/978-3-030-05282-9_1

pathways undergo mutations to improve the survival rate of the organism by either improving the protein function or catalytic efficiency and stability to escape the inhibitory action of the drug. In the latter case, the motive for modifying the drug target is to ensure that drug binding is weakened. Moreover, the mutations are such that substrate binding is unaffected or minimally affected. Most of the computational methods employed to study the mechanism of drug resistance, attempt to understand the differences in the binding patterns of the substrate and the drug molecule, i.e. understanding the "**substrate-envelope hypothesis**". Here, we present an overview of those computational methods that employ free energy of binding as a tool to gauge the differences in the binding of the substrate and the drug molecule before and after mutation.

In the Sect. 1, we discuss the driving force for resistant mutations and throw some light on the different mechanisms by which drug resistance can occur. In Sect. 2, we present a brief overview of molecular dynamics, thermodynamics of protein–ligand binding, and various methods for computing the free energy of binding. The last section, Sect. 3, has a detailed discussion on various free energy-based methods used to understand and predict the target site mutations leading to loss in drug binding.

1.1 Overview of the Mechanisms of Drug Resistance

The drug-induced selection pressure [1–4] is the major driving force for infectious organisms to try to evade the effects of drugs. One of the primary moves that any organism will adopt is to disrupt the action of drug molecules by one or more possible mechanisms. To show its effect, the drug must enter the cells and find its target protein. As a primary defence mechanism against drugs, the organism may down regulate the expression of influx channels that enable the entry of the drug, resulting in a decreased concentration build-up within the cell. Another strategy that hinders the build-up of the drug inside the cell is the upregulation of the expression of efflux channels/pumps that facilitate the egress of the drug molecules. These strategies are often very difficult to understand owing to the complicated pathways involved in the upregulation or downregulation of various proteins associated in the regulation of traffic to and from the cell. This attribute is difficult to study using computational techniques that use free energy-based methods. Target site mutations [5–8] that lead to disruption in the drug binding without significant loss of the protein function [9, 10] is another mechanism of drug resistance. Such mutations can be studied using computer simulations that enable us to estimate the free energy difference between the drug binding to the mutant and the wild-type protein. An essential factor to consider while understanding target site mutation is the fitness cost associated with the mutational change. This can be estimated by the change in the free energy of binding of the natural ligands/substrates; for example, a drop in their binding energy indicates that substrate binding is impeded, which this leads to increased fitness cost. This means the enzyme now must expend more energy to carry out the same reaction. Hence, we can assume that such mutations are seldom seen, and if at all they occur, a compensatory mutation(s) will be seen to counter the detrimental effects of those mutations [11, 12]. Another strategy adopted by organisms is to increase the production of drug-metabolizing enzymes that modify the drugs to their inactive form eventually leading to their elimination. A classic example of this is the inactivation of penicillin by the enzyme β-lactamase.

1.2 Overview of Computational Methods to Study Drug Resistance

Broadly, computer-assisted methods used to study drug resistance can be classified into two categories based on the information they require and the output they return. The first category of methods requires only 1D sequence data as input and the output is generally a classification type, i.e. the test sequence is classified as a resistant or a non-resistant sequence. Thus, the methods grouped under this class are collectively called as "sequence-based" methods [13]. The workflow of these methods is akin to machine learning or QSAR type classification methods. In a nutshell, sequence-based methods require sequences with the corresponding biological activity data (K_i or IC_{50} or any other suitable numerical value) for the drug under study. Such data can be curated from databases like HIVDB (for HIV resistance, curated and maintained by Stanford University; [14, 15]) CancerDR (for cancer resistance, curated by CSIR Institute of Microbial Technology and OSDD, India; [16]), tuberculosis resistance mutation database (curated and maintained by various departments and schools with Harvard University; [17], and many other such databases. The data is then split into training and test sets to develop and validate the predictive models. The advantage of such methods is that it is not necessary to know the tertiary structure of the protein or the drug-receptor interactions. Therefore, sequence-based methods are computationally inexpensive and large amount of data can be trained to obtain decent quality predictive models in a short time. However, they suffer from two major drawbacks; (1) a lot of a priori information on drug-resistant mutations is needed to train/develop predictive models and (2) no mechanistic insights or atomistic details can be obtained.

The drawbacks seen in the sequence-based methods are efficiently overcome by structure-based methods [13, 18, 19]. Further, structure-based methods are the methods of choice when atomistic details are desired. However, these additional details come at an added computational cost and require high-resolution protein structures to be able to make accurate and reliable predictions. However, unlike the sequence-based methods, they do not require large a priori information on mutations; on the contrary, they can be applied to systems where no data on mutation is available. To assess the binding stability which is the basis for predictions, these methods employ either empirical scoring functions that implicitly try to reflect the free energy of binding or use techniques that compute the free energy of binding per se. Molecular docking-based methods use empirical scoring functions to find the best docking conformations, and these methods are computationally less expensive. Therefore, they can be applied to assess many protein–ligand complexes. The ligand can be docked to various mutant proteins to predict their binding strength before and after mutations, and this will allow one to understand the effect of the mutation on the binding strength. The accuracy of docking-based methods relies on the accuracy of the scoring function, and they are best suited for rank ordering of compounds rather than computing the absolute free energy of binding. The major issue with docking-based methods is that most docking programs treat proteins as rigid entities, and therefore, mutations in highly flexible protein–ligand systems are poorly understood [19]. However, in recent times there have been several attempts to incorporate protein flexibility in molecular docking [20]. This has largely improved the enrichment scores. Due to the limited scope of this chapter, such docking methods will not be discussed here and have been treated elsewhere [21–25]. Molecular dynamics-based methods can incorporate flexibility in the protein–ligand complexes, and in most cases, are the methods of choice as a conformational sampling tool to explore the phase space accessible to the system under study. The conformations sampled are used to compute the free energy change. However, the drawback of MD-based methods is the computational cost, which is several magnitudes higher compared to docking-based methods.

Another critical issue that must be addressed about the structure-based methods is, how fast predictions can be made, in addition to how reliable are the predictions. These methods find application in drug discovery programs, wherein additional filters can be placed to weed out molecules likely to encounter a high level of resistance or assist in suitably modifying leads to inhibit the mutant proteins. Drug discovery itself is an extremely lengthy and expensive process, and an additional filter like resistance should be economical in terms of time as well as money. Moreover, such methods should also assist medicinal chemists during lead optimization stages to identify potential groups that will help evade drug resistance and avoid late-stage failures that lead to huge financial losses.

2 Molecular Dynamics Simulations and Free Energy Calculations

2.1 Overview of MD and Conformational Sampling Methods

Computer simulations are very useful in predicting changes in molecular properties brought about by alterations in an atom or a group of atoms, particularly, amino acid residues. Therefore, they find good application in predicting the effect of mutations on drug binding at the active site or elsewhere. Protein design experiments clarify the effect of a mutation on drug or substrate binding, thereby facilitating prediction of drug-resistant mutations. This way the program can be used to select all mutations wherein drug binding is hampered and substrate binding is either improved or [26].

In case of free energy calculations, molecular dynamics (MD) simulations are the most commonly used technique to generate conformational ensembles. Hence, it is rightly called as one of the main toolkits for theoretically studying biological molecules (Hansson et al. [27], Binder et al. [28]. MD calculates the time-dependent behaviour of particles or atoms, by numerical integration of Newton's second law of motion and predicts the future positions and momenta. MD simulations have provided detailed information on the fluctuations and conformational changes of proteins and nucleic acids upon drug/substrate binding. As a result, it is now routinely used to investigate the structure, dynamics and thermodynamics of biological molecules and their complexes. MD simulations have an advantage in that, starting from an X-ray or NMR solved structure, it can provide insights into the dynamic nature of biomolecules that are inaccessible to experiments. To accurately simulate the behaviour of molecules, one must be able to account for the thermal fluctuations and the environment-mediated interactions arising in diverse and complex systems (e.g., a protein-binding site or bulk solution). This depends on how accurately the force fields represent the atoms and treats the non-bonded interactions. A complete account of force fields can be found in the review by Pissurlenkar et al. [29]. However, most of the biological events occur at timescales that are not routinely reachable by classical MD simulations, for example, protein folding occurs in the timescale of few seconds, whereas drug binding and unbinding occur in the timescale of few microseconds to milliseconds. The routine timescale that is feasible using high-end servers equipped with graphic processing units [30–32] and distributed grid computing [33, 34], is few tens of microseconds, that is nearly 1/100th of the timescale required to study protein folding. Conventional MD suffers from the severe limitation that it is extremely difficult to sample high-energy regions and surmount energy barriers, leading to inaccuracies in free energy calculations.

The limitations of classical MD simulations have motivated the development of new conformational sampling algorithms that facilitate the sampling of conformational space that is inaccessible to classical MD simulation. The simplest way to encourage the system to sample the high-energy regions on the phase space is to increase the target temperature [35]. This leads to increased kinetic energy of the system that enables it to surmount these barriers. However, it has been argued by many, that such elevated temperatures ($\sim$400 K and above) lead to physiologically unrealistic states that may severely distort the results; however, such methods have been found to be advantageous in improving the sampling efficiency during MD simulations. Another method that uses elevated temperature to enhance the sampling is the replica-exchange molecular dynamics (parallel tempering, [36, 37]). In this approach, several replicas are simulated in parallel at different temperatures. At appropriate intervals, the replicas switch temperatures with the nearest replica, and this exchange is governed by the Metropolis acceptance criteria. However, all these methods do not prohibit the system from revisiting the same conformational space. This problem was resolved by adding the memory concept in molecular dynamics (local elevation method [38] Metadynamics [39]) uses Gaussian potentials that discourage the system from sampling the same conformational space. These are few of the most commonly used methods to tackle sampling problems in molecular dynamics, a complete account on enhanced sampling algorithms can be found elsewhere [40–44].

2.2 An Overview of Thermodynamics of Protein–Ligand Binding

Molecular interactions, between the ligand and receptor, are primarily non-covalent in nature and governed by attractive and repulsive forces. In drug design experiments, the goal is always to optimize the attractive interactions and reduce the repulsive ones [45–47]. Moreover, these associations are temporary, and the lifespan of such complexes are governed by the off rates (K_{off}) or the dissociation constant (K_{d}), both of which indicate the binding strength of a ligand to its protein counterpart. In the realm of thermodynamics, binding is governed by enthalpic and entropic components [48] given by Eq. 1.

$$\Delta G = \Delta H - T\Delta S \tag{1}$$

where ΔG is the binding free energy; ΔH is enthalpy; ΔS is entropy and T is the temperature in Kelvin.

The association is favourable, i.e. spontaneous when the $\Delta \text{G}_{\text{Gibbs}}$ is negative and unfavourable otherwise. All the binding and pre-binding (recognition and pre-organization) events in biomolecular associations are either enthalpy (ΔH) driven or entropy (ΔS) driven. The enthalpic component represents several types of non-covalent interactions like electrostatic, van der Waals, ionic, hydrogen bonds and halogen bonds, while the entropic components reflect the contribution to binding due the dynamics or flexibility of the system. Computing the enthalpic component of binding has reached far heights, in terms of methods available for calculating the aforementioned type of interactions. However, till date, calculation of the entropic component is extremely difficult, and the algorithms are computationally very demanding.

The Gibbs equation is more relevant in biochemistry for calculating the free energy and is given by Eq. 2:

$$\Delta G_{\text{Gibbs}} = -RT \ln K_d \tag{2}$$

where ΔG_{Gibbs} is Gibbs free energy, R is universal gas constant, T is the temperature in Kelvin, K_{d} is the dissociation constant. Equations 1 and 2, along with the Born–Haber cycle [46] (Fig. 1) form the basis for the development of the methods used to compute the free energy binding. The two main methods are Free energy perturbation (FEP) and Thermodynamics Integration (TI), both of which will be

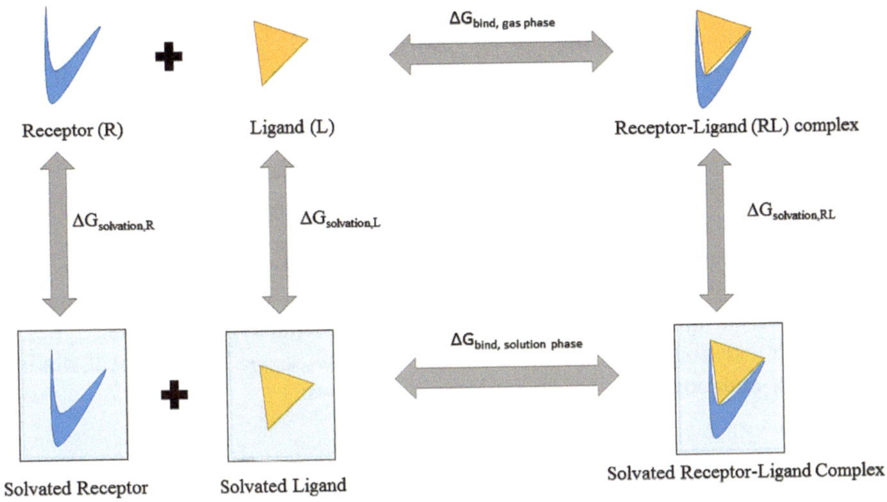

Fig. 1 Thermodynamic or Born–Haber cycle for the receptor-ligand binding

dealt with in the subsequent Sect. 2.3.2. However, measuring the dissociation constants from simulations is a daunting task; nevertheless, computing the partition functions from the molecular simulations is relatively easy. Hence, the ratios of the partition functions can be used to estimate the free energy of binding, which is given by Eq. 2a,

$$\Delta G = -k_{\mathrm{B}}T \ln \frac{Q_{\mathrm{PL}}}{Q_{\mathrm{P}}Q_{\mathrm{L}}} \tag{2a}$$

where k_{B} is the Boltzmann constant, T is the temperature in Kelvin, Q is the partition function with subscripts PL, P and L indicating protein–ligand complex, protein, and ligand, respectively. This section presents a summary of thermodynamics, which is imperative for understanding the application and methods developed to compute binding free energy. More elaborate discussions on the thermodynamics of protein–ligand binding can be found in the reviews by Bronowska [48], and Homans [46].

2.3 *Methods to Compute Free Energy Binding*

Free energy is a quantity that can be measured for systems such as liquids or flexible macromolecules with several minimum energy configurations separated by high-energy barriers. However, its computation is far from trivial and the associated quantities such as entropy and chemical potential are also difficult to calculate. More so, the free energy cannot be accurately determined from classical molecular dynamics or Monte Carlo simulations due to their inability to sample adequately from the high-energy regions of the phase space, which also make important contributions to the free energy. However, the free energy differences ($\Delta\Delta G$) are rather simple to compute. The free energy binding for the non-covalent association of two molecules (protein and ligand in this case) may be written as follows:

$$\Delta G_{\text{bind}} = G_{\text{complex}} - \left(G_{\text{protein}} + G_{\text{ligand}}\right) \tag{3}$$

The binding event is an additive interaction of many events [49–52], for example solvation energy (G_{sol}), conformational energy (G_{conf}), energy due to interaction with residues in the vicinity (G_{int}), and energy associated with different types of motions (translational, rotational and vibrational, G_{motion}). The classical binding free energy equation now can be rewritten as follows:

$$\Delta G_{\text{bind}} = G_{\text{sol}} + G_{\text{conf}} + G_{\text{int}} + G_{\text{motion}} \tag{4}$$

Directly computing the free energy from an MD or MC simulation is not trivial; hence, the following methods have been formulated. Broadly, the methods used for computing free energy are classified as partitioning-based methods or end-state free energy methods and non-partitioning-based methods. The partitioning-based methods partition the binding energy into various components as shown in Eq. 4; however, this method has been highly criticized [53] stating that it is physically unreal to partition the free energy into components.

2.3.1 End-State Free Energy Methods or Partitioning-Based Methods

The human body majorly comprises of water; hence, it is imperative to carefully include the solvation effects while computing the free energy of binding. More importantly, water plays a crucial role in ligand recognition and in the binding phenomenon. In computational chemistry, the methods for incorporation of solvent are divided into three groups: (i) continuum electrostatic methods/implicit solvent, (ii) explicit solvent models with microscopic detail and (iii) hybrid approaches. Historically, the continuum electrostatic methods were among the first to consider the solvent effect, and they still represent very popular approaches to evaluate solvation free energies, especially in quantum chemistry. Polarizable continuum model (PCM, [54]), COnductor-like Screening MOdel (COSMO, [55]) and SMD solvation model [56] are few popular models for treating solvent effects implicitly in quantum chemistry. Continuum solvation methods are computationally economical; however, the frictional drag of the solvent is highly underestimated and as a consequence may drive the system to non-physical states. Moreover, solvent–solvent and solute–solvent interactions are inadequately treated, posing a danger of underestimating the effects of such interactions. The explicit treatment of solvent enables one to consider the solvent–solvent and solute–solvent interactions. This prohibits the systems from visiting non-physical states due to the inclusion of the dampening effect shown by the solvent atoms. The principal drawback of explicit solvent models is the number of atoms to be considered in the system leading to increased computational cost. However, with the help of GPU-based acceleration, this drawback, now, is hardly any cause for worry.

The end-state free energy methods use the conformations extracted from an MD or MC simulation, wherein the system is simulated by explicitly defining the solvent. However, while solving the GB or PB equation, the solvent is implicitly treated by defining the external dielectric constant for water (for most drug design cases) and a suitable internal dielectric constant [57–61].

Molecular Mechanics-Poisson Boltzmann/Generalized Born Surface Area (MM-PB/GB-SA)

The MM-GBSA [62–65] approach employs molecular mechanics-based energy calculations and the generalized Born model to account for the solvation effects in the calculation of the free energy. Similarly, the MM-PBSA [66–68] approach solves the linear or nonlinear Poisson–Boltzmann equation [69–71], to account for the solvation electrostatics, whereas the MM part is calculated as in MM-GBSA from the derivative of the force field equations. Both these approaches are parameterized such that they partition the energy components into various terms, and the net free energy change is the sum of these individual terms (Coulomb, vdW, solvation, etc.). MM-PBSA has gained considerable attention for estimating the binding free energies of molecular complexes due to its exhaustive nature of computing the solvation electrostatics by iteratively solving the PB equation, whereas the GB method does not involve any rigorous and iterative procedure and hence is faster. However, this does not necessarily guarantee that the MM-PBSA method always outperforms MM-GBSA method. In MM-PB(GB)SA methods, MD- or MC-derived conformational ensembles are used to compute the "**average**" free energy of a state and this is approximated as follows:

$$\langle G \rangle = \langle E_{\text{MM}} \rangle + \left\langle G_{\text{PBSA/GBSA}} \right\rangle - T\langle S_{\text{MM}} \rangle \tag{5}$$

where the angular bracket <> indicates average over the MD/MC conformations, E_{MM} is the molecular mechanics energy that typically includes bond, angle, torsion, van der Waals, and electrostatic terms (see Eqs. 7c and 7d) and is evaluated with no or extremely large (virtually infinite) non-bonded cut-off limit. The second term is solved as mentioned in the preceding stanza and it forms the crux of this method. The last term $T\langle S_{\text{MM}} \rangle$, is the solute entropy, which is estimated by quasi-harmonic analysis [72, 73] of the trajectory or by normal mode analysis [74–76].

The following equation (Eq. 6) shows how the binding free energy is computed from the energies of the ligand, protein, and its complex over all the MD or MC snapshots. However, the snapshots can be obtained in two possible ways—one is called the single trajectory approach and other is the multiple trajectory approach. In the single trajectory approach, only the protein–ligand complex is simulated, and the snapshots for the protein, ligand and the complex are extracted by defining appropriate atom numbers from the parameter and coordinate file. However, in the multiple trajectory approach, three separate simulations are performed, one each for the protein, ligand and protein–ligand complex.

$$\langle \Delta G_{\text{bind}} \rangle = \langle G_{\text{complex}} \rangle - (\langle G_{\text{protein}} \rangle - \langle G_{\text{ligand}} \rangle) \tag{6}$$

Furthermore, Eq. 1 is modified to accommodate solvation electrostatics and hydrophobic terms as shown in Eq. 5. Here, Eqs. 7a–7d give the computation of the individual terms,

$$\Delta G_{\text{bind}} = \Delta E_{\text{MM}} + \Delta G_{\text{sol}} - T\Delta S \tag{7a}$$

$$\Delta G_{\text{sol}} = \Delta G_{\text{sol-elect}} + \Delta G_{\text{nonpolar}} \tag{7b}$$

$$\Delta E_{\text{MM}} = \Delta E_{\text{int}} + \Delta E_{\text{elect}} + \Delta E_{\text{vdW}} \tag{7c}$$

$$\Delta E_{\text{int}} = \Delta E_{\text{bond}} + \Delta E_{\text{angle}} + \Delta E_{\text{torsion}} \tag{7d}$$

Here, ΔE_{MM} is computed in the gas phase using classical force fields, ΔG_{sol} is computed using PBSA or GBSA method, $\Delta G_{\text{sol-elect}}$ is computed using PB or the GB method, and the $\Delta G_{\text{nonpolar}}$ is computed by the solvent accessible surface area (SA). While employing the single trajectory approach, Eq. 7d generally cancels out and hence makes negligible contribution to the binding energy.

Linear Interaction Energy (LIE)

Linear interaction energy [77–79] is similar to the MM-PB/GB-SA method with regard to the partitioning of the electrostatic and van der Waals terms (polar and non-polar contribution, respectively,); however, the use of the weighting parameter for electrostatic and van der Waals interactions, is unique to this method. LIE measures the binding energy by estimating the difference in the interaction energies of the ligand in the solvent (unbound state) and in the protein environment (bound state). Hence, to obtain these interactions, two separate MD simulations are performed. In one simulation, only the ligand is placed in the solvent (mostly water) and in the other, the protein–ligand complex is placed in the solvent. The formulation of this method is based on deriving the linear response approximation from converged ensemble interactions, most often extracted from well-equilibrated trajectories from the MD simulation of the ligand with its surroundings (solvent or protein).

The mathematical formula for computing free energies using LIE method is given in Eq. 8

$$\Delta G_{\text{bind}} = \alpha\left[\langle E_{\text{coul}}^{\text{L-S}} \rangle_{\text{PL}} - \langle E_{\text{coul}}^{\text{L-S}} \rangle_{\text{L}}\right] + \beta\left[\langle E_{\text{vdW}}^{\text{L-S}} \rangle_{\text{PL}} - \langle E_{\text{vdW}}^{\text{L-S}} \rangle_{\text{L}}\right] \tag{8}$$

where the angular bracket <> indicates ensemble over the MD trajectory, $E_{\text{coul}}^{\text{L-S}}$ and $E_{\text{vdW}}^{\text{L-S}}$ are electrostatic and van der Waals interactions between the ligand and its medium in the vicinity (PL—protein–ligand complex; L—ligand in solvent), and α is the weighting parameter for electrostatic interactions, which is most often set to 0.5 [78]. This value is assumed due to the linear response of the surroundings to the electrostatic field and was validated using more extensive computations on the ions (Na^+ and Ca^{2+}) in water [80]. β is the weighting parameter for van der Waals interactions and is set to 0.16–0.18 [81], which is a subject of much debate owing to the difficulty in estimating the vdW's contribution to the free energy of binding. However, these values are obtained by empirical fitting the experimental binding free energies. Moreover, the linear response of the vdW term is assumed by observing the linear trend in the interaction of the hydrocarbons with the solvent (water) that depends on the number of carbons in a hydrocarbon.

2.3.2 Non-partitioning-Based Methods

In non-partitioning methods, there is no partitioning of the free energy into various components. Statistical mechanics plays a crucial role in deriving the relationship between the free energy of a system and the ensemble average of the Hamiltonian that describes the system. These methods are far more accurate than the previously mentioned end-state free energy methods, but at the same time, are computationally very demanding. Hence, while dealing with a large dataset of molecules against a particular protein target, it is worthwhile to screen the molecules using a fast method like high-throughput virtual screening [82, 83], followed by a flexible docking-based screening, then use an end-state free energy method, and finally employ the non-partitioning methods to study few tens of molecules. Here, we will present a brief discussion on FEP and TI methods along with their mathematical treatment, and then move on to explain the idea behind alchemical free energy predictions.

Free Energy Perturbation (FEP) and Thermodynamic Integration (TI)

Most of the methods for free energy calculations are generally formulated in terms of estimating the relative free energy differences, ΔG, between two equilibrium states, or binding of two similar ligands to a common target. The free energy difference between the two states I and II can be formally obtained by Zwanzig's formula [84, 85].

$$\Delta G = G_{\text{II}} - G_{\text{I}} = \beta^{-1} \ln e_{\text{I}}^{(-\beta \Delta \text{V})} \tag{9}$$

Here, $\beta = (k_B T)^{-1}$

This represents a sampling of the differences in potentials (ΔV) of the two states using Monte Carlo or molecular dynamics simulation over the potential of state I. To ensure the convergence of these calculations, it is recommended that the potentials of the two systems should thermodynamically overlap. For satisfying this condition, correct conformations must be selected, which is a daunting task, and hence, to achieve this, a multistep process is usually implemented. A path between the states I and II is defined by introducing a set of intermediate potential energy functions that are constructed as linear combinations of the initial (I) and final (II) state potentials and these intermediate states are non-physical states (Eq. 10).

$$V_m = (1 - \lambda_m)V_{\mathrm{I}} - \lambda_m V_{\mathrm{II}} \tag{10}$$

where the transition from one state to another is discretized into many points (m = 1,…,n), each represented by a separate potential energy function that corresponds to a given value of λ, such that λ_m varies from 0 to 1. Here, zero indicates the pure initial state of the system and one indicates pure final state of the system. The total free energy, thus, can be obtained by summing over the intermediate states along the λ variable.

$$\Delta G = G_{\mathrm{II}} - G_{\mathrm{I}} = -\beta^{-1} \sum_{m=1}^{n-1} \ln \langle^{[-\beta(V_{m+1} - V_m)]} \rangle_m \tag{11}$$

This approach is known as free energy perturbation (FEP) where $\Delta\lambda_m = \lambda_{m-1} - \lambda_m$; hence, it can be written as

$$\Delta G = -\beta^{-1} \sum_{m=1}^{n-1} \ln \langle e^{[-\beta \Delta V \Delta\lambda_m)]} \rangle_m \tag{12}$$

Since the potential difference can also be described as the derivative of the potential with respect to λ_{m}, Eq. 12 can also be written as,

$$\Delta G = -\beta^{-1} \sum_{m=1}^{n-1} \ln \langle e^{\left[-\beta \frac{\partial V_m}{\partial \lambda_m} \Delta\lambda_m)\right]} \rangle_m \tag{13}$$

Now, expansion of the Eq. 13 by the Taylor expansion series gives Eq. 14,

$$\Delta G = \sum_{m=1}^{n-1} \langle e^{\left[-\beta \frac{\partial V_m}{\partial \lambda_m} \Delta\lambda_m)\right]} \rangle_m \tag{14}$$

wherein $0 \rightarrow \lambda$ can instead be written as an integral over λ

$$\Delta G = \int_0^1 \langle \beta \frac{\partial V(\lambda)}{\partial \lambda} \rangle_\lambda \mathrm{d}\lambda \tag{15}$$

Equation 15 is usually referred to as the thermodynamic integration (TI) method for calculating the free energy change [86, 87]. In the early days of free energy simulations, the TI approach was synonymous with the slow-growth method [88]. In the slow-growth method, the value of λ is changed at each time step during the MD simulation. While this method was claimed to be more efficient than the discrete FEP formulation, nowadays, a "non-continuous" change in λ is a better choice (50–100 discrete points are usually recommended). This facilitates equilibration at each point, the addition of extra points at any time, and use of any pattern of spacing between the λ-points, to optimize the efficiency.

Alchemical Free Energy Perturbation

Here, the free energy is computed by transforming a molecule from one state (bound-solvated) to another state (unbound-solvated) through several physically unrealistic states, that are called as alchemical states, hence the name "Alchemical Free energy" [89, 90]. This method is regarded as one of the apt methods to study

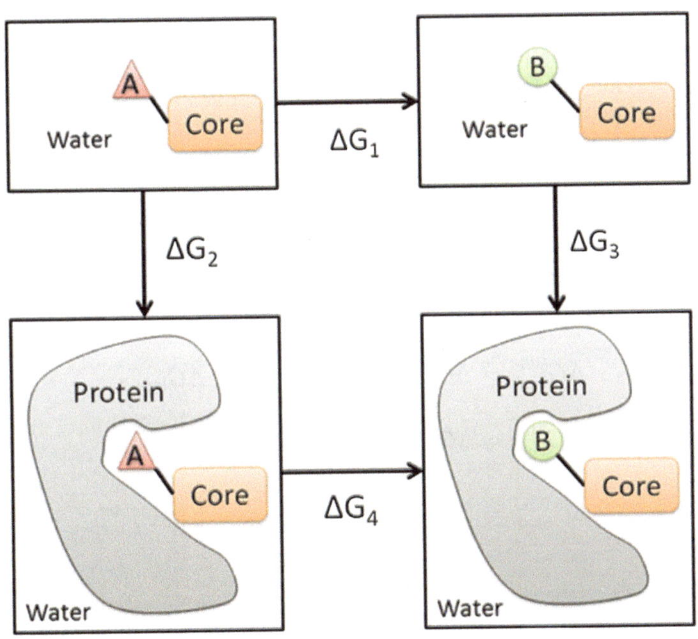

Fig. 2 Thermodynamics cycle for computing alchemical free energy binding. Image reproduced from Wang et al. [91] [open-access article distributed under the terms of the Creative Commons Attribution License (CC BY)]

the effect of mutations on the drug binding affinity (Fig. 2). The total free energy change in a thermodynamic cycle in any alchemical transformation is equal to zero.

$$\Delta G_1 - \Delta G_4 - (\Delta G_2 - \Delta G_3) = 0 \tag{16a}$$

$$\Delta G_1 - \Delta G_4 = \Delta G_2 - \Delta G_3 \tag{16b}$$

3 Application of Computational Methods to Understand Drug-Resistant Mutations

3.1 Computational Mutation Scanning

Computational mutation scanning [92] is a useful method to explore the sensitivity to changes in the composition of the amino acid in a protein-binding site (Fig. 3). In computational mutation scanning, the wild-type amino acid residue is mutated to another amino acid in the binding pocket or elsewhere. However, the most widely practised method is to mutate any amino acid residue to an alanine, since it is the simplest amino acid with a side chain (not glycine because it is devoid of a side chain). Hence, this method is equivalent to the experimental "alanine-scanning mutagenesis", which is a powerful tool to investigate and confirm the important interactions in the protein–protein interface and protein–ligand interactions. In computational alanine scanning, all atoms from the C_β carbon atom of the amino acid under study are replaced by three hydrogen atoms to convert it to an alanine. After the mutation, the change in the binding energy is estimated either using docking with an appropriate scoring function or by MM-PBSA or MM-GBSA to compute $\Delta\Delta G$ (Eq. 17c). By scanning with alanine at various positions in the binding cavity, important residues can be identified, as mutating an important amino acid will drastically decrease the binding energy.

$$\Delta G_{\text{bind}}^{\text{Wild}} = \Delta G_{\text{complex}}^{\text{Wild}} - \Delta G_{\text{receptor}}^{\text{Wild}} - \Delta G_{\text{ligand}} \tag{17a}$$

$$\Delta G_{\text{bind}}^{\text{Mut}} = \Delta G_{\text{complex}}^{\text{Mut}} - \Delta G_{\text{receptor}}^{\text{Mut}} - \Delta G_{\text{ligand}} \tag{17b}$$

$$\Delta\Delta G = \Delta G_{\text{bind}}^{\text{Mut}} - \Delta G_{\text{bind}}^{\text{Wild}} = \left[\Delta G_{\text{complex}}^{\text{Mut}} - \Delta G_{\text{complex}}^{\text{Wild}}\right] - \left[\Delta G_{\text{receptor}}^{\text{Mut}} - \Delta G_{\text{receptor}}^{\text{Wild}}\right] \tag{17c}$$

In the context of predicting drug-resistant mutations, one must perform alanine scanning in the binding site on two complexes, i.e. with the substrate bound complex and the inhibitor-bound complex. The change in the binding energy after mutation is computed for both the systems, viz., for inhibitor and the substrate. A decrease in the binding affinity for the inhibitor with negligible or no change in the binding affinity for the substrate indicates a hotspot amenable to resistant mutation, these spots are

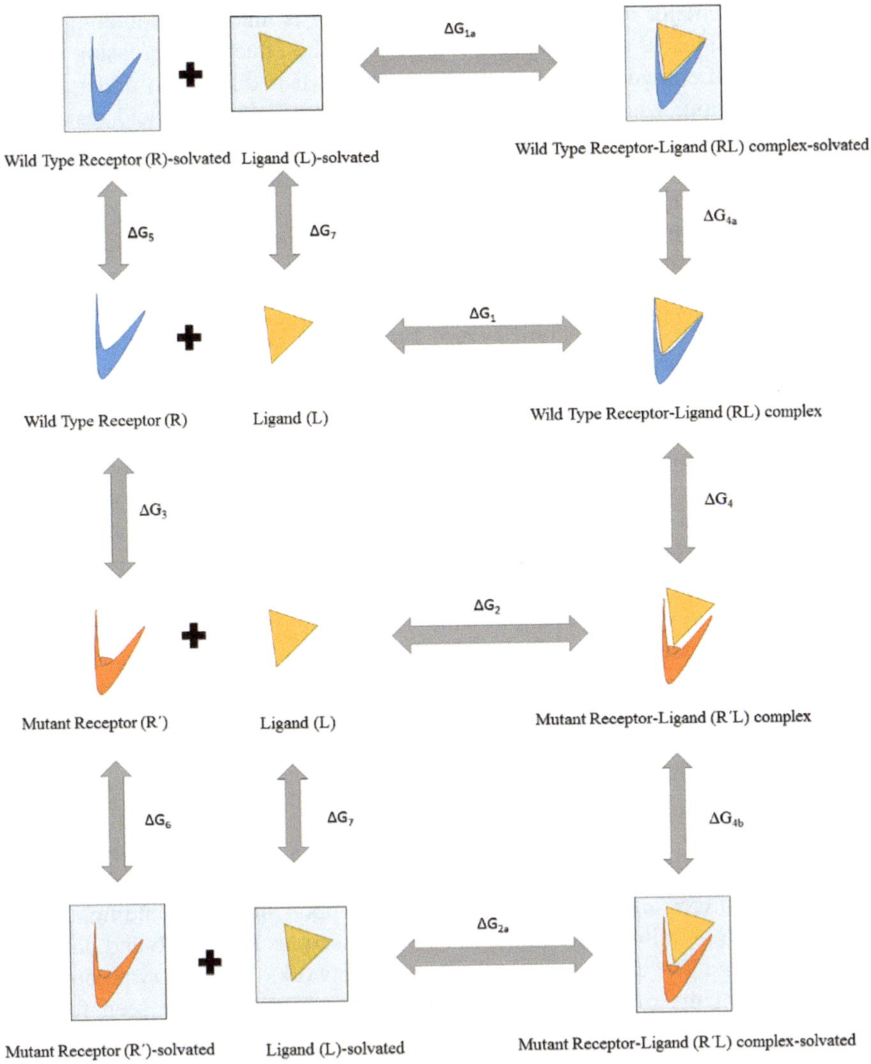

Fig. 3 Thermodynamic cycle for computing free energy change between mutated and wild-type protein

termed as "mutational hotspots". The method follows the substrate-envelope hypothesis [93–95], which states that there is a large fitness cost that needs to be paid if one mutates an amino acid residue that is involved in substrate binding. Mutating such amino acids could lead to impaired enzyme function resulting in the death of an organism. This can be put to appropriate use by developing inhibitors that completely overlap in the substrate binding region, leading to a lower predisposition towards developing drug resistance [96–99].

However, a major drawback in alanine scanning is that when mutating a large amino acid residue to alanine one can only study the effect of decreasing the side chain or loss of charged groups in the binding site. It is difficult to understand the resistant mutation, wherein there is a change in charged amino acid residue, for example, arginine replacing aspartate or a large amino acid replaces a small amino acid residue. Nevertheless, computational alanine scanning has been successfully used to predict mutational hotspots.

Hao et al. [100] reported a modification of computational alanine scanning (CAS), named computational mutation scanning (CMS) to study drug resistance in six HIV-1 protease inhibitors. This protocol is an improvised version of the classical CAS that enables a geometry optimization step and incorporates entropy calculations by means of normal model analysis. Using a single trajectory approach and modifying the standard MM-PBSA protocol, to allow for mutations with other amino acid residues, they computed the change in the binding affinities ($\Delta\Delta G$) of 77 drug-mutant combinations (includes single and double mutants). They obtained promising results with $\sim$83% consistency with the experimental observations, demonstrating that the prowess of the method lies in identifying the binding hotspots. However, Hao et al., do not report the change in the binding affinity for various substrates, from which they could have investigated the substrate-envelope hypothesis for the HIV-1 protease. This could have led to interesting findings facilitating our understanding about those mutations that would lead to a decrease in the enzyme function, either leading to the death of the organism or compelling a compensatory mutation to counter the lethal effects of any mutation. This information can be used to unravel the role and need for double, triple or even multiple mutations.

Tse and Verkhivker [101] used CAS along with residue interaction network to elucidate the effects of inhibitor binding on the network of residues in ABL kinase. They showed the utility of this combination in deducing the critical networks of amino acid residues and the changes that follow upon inhibitor binding, using a selective kinase inhibitor (nilotinib) and two promiscuous (bosutinib and dasatinib) kinase inhibitors. The changes in the interaction networks in the enzyme holds key hints to unravel the mystery of how drug-resistant mutations are seen for ABL kinase inhibitors. Moreover, the mutations that occur far from the binding site can also be explained, since a mutation far off from the site can affect drug binding through a cascade of events that eventually percolate into the binding site through the changes in the residue interaction network. CAS followed by MM-PBSA added the energetic component to locate the hotspots that could lead to drug resistance in the kinase inhibitors

3.2 *MM-PB(GB)-SA*

MM-GBSA or MM-PBSA are two widely used free energy methods employed to understand the effects of mutations on the drug binding affinity, moreover, these methods are successful in predicting likely mutations leading to drug resistance. These methods are able to predict due to their amenability to decompose the free energy into its components at the residue level that leads to better understanding of the effect of mutations on drug binding. Lethal effects of the V82F/I84V double mutation in HIV-1 protease on amprenavir were demonstrated using MM-PBSA approach on snapshots obtained from the well-equilibrated protein–ligand complex [102]. It was reported that amprenavir lost its binding affinity due to distortions in the binding site, hence weakening many favourable interactions ($\Delta\Delta G = 3.73$ kcal/mol). Such a distortion of the binding site was previously observed and attributed to the rapid flap movements seen in this double mutant which is absent in the wild-type HIV-1 protease [103]. Furthermore, newer inhibitors, that are very close structural analogues of amprenavir, like TMC126 ($\Delta\Delta G = 2.01$ kcal/mol) and TMC114 (darunavir, $\Delta\Delta G = 3.45$ kcal/mol) were also seen to be affected by these mutations, though to a lesser extent than amprenavir. Despite structural distortions in the binding site, it had no effect on the substrate binding, and hence, the catalytic process was unhindered.

Hou et al. [104] combined MM-GBSA with the positional variability approach, to modify Kollman's FV value [105] to give a new scoring function also called FV (Free energy/Variability) score. Using the FV score, they evaluated the binding of six substrates that are hydrolysed by HIV-1 protease and confirmed Kollman's [105] observation that drug-resistant mutations are more likely to occur at less conserved regions. The FV score reported by Hou et al. comprises two components, one that reflects the binding energetics at the per-residue level, obtained by MM-GBSA, and the second component is the sequence variability that represents the conservation of amino acids at each position. Using this score, one can identify amino acid residues that are crucial for substrate and inhibitor binding, and thus classify the residues that are exclusively involved in substrate binding and those that are exclusive for inhibitor binding. Such a classification when coupled with the positional variability of amino acid residues can extract those positions with low conservation and exclusivity for inhibitor binding; such positions are highly amenable to mutations leading to drug resistance. Employing this method Hou et al. confirmed their previous observation [102] that the V82F/I84V double mutations are lethal for many FDA approved HIV-1 protease inhibitors, whereas TMC126 is still active against this mutant.

3.3 *Vitality Analysis*

One of the primary drawbacks of the aforementioned methods to predict drug-resistant mutations is their inability to accurately estimate the binding affinity for the substrate molecule(s). The fitness cost of the mutation can be estimated by gauging the change in the binding affinity of the substrate to its enzyme target; any perturbation in the substrate binding is likely to affect the function of the enzyme. Therefore, computing the catalytic efficiency of the enzyme before and after mutation will enable us to understand the fitness cost. Pioneering work in this line was done by Gulnik et al. [1]. In this work, they have determined the catalytic efficiency (Eq. 18a) of HIV-1 protease following few active and non-active site mutations. This principle was incorporated in terms of free energy change by Warshel et al., and they employed this method (Eq. 18b) to computationally predict the likely mutations that could potentially abolish drug binding leading to drug resistance. This method is aptly named as "Vitality approach" wherein higher vitality values indicate that the resistance is more likely as there is little chance of increase in the catalytic efficiency of the enzyme. The basic workflow adopted by Warshel et al. [106, 107] is to estimate the change in the drug binding before and after mutation, depicted in the first part of Eq. 18b and then estimate the catalytic efficiency by determining the binding of the substrate by modelling the transition state (TS) conformation of the enzyme, depicted in the second part of Eq. 18b. However, the challenge of employing this method to predict likely mutations is that a thorough knowledge of the catalytic mechanism of the enzyme is essential. Nonetheless, this method is far more accurate and truly predictive in nature. This is exemplified by the fact that Warshel et al. successfully used this method on six clinical agents active against HIV-1 protease.

$$\text{Vitality value} = \frac{\left(\frac{K_i k_{\text{cat}}}{K_m}\right)_{\text{mutant}}}{\left(\frac{K_i k_{\text{cat}}}{K_m}\right)_{\text{WT}}} \tag{18a}$$

$$\ln \frac{\gamma_M}{\gamma_N} \cong \frac{1}{\text{RT}} \left(\Delta\Delta G_{\text{bind}}^{N \to M}(\text{drug}) - \Delta\Delta G_{\text{bind}}^{N \to M}(\text{TS}) \right) \tag{18b}$$

where K_i = inhibition constant; k_{cat} = constant that defines the turnover rate of an enzyme-substrate complex to the product; K_{m} = Michaelis constant.

4 Concluding Remarks

This chapter describes important computational methods that have been proven extremely helpful in gaining insights into mutations leading to drug resistance. We have attempted to introduce methods used to compute the free energy of binding along with their mathematical formulations, practical implementation and pros and cons of such methods. Finally, we have discussed a few applications of such methods to study drug resistance.

Acknowledgements E. A. F. Martis and E. C. Coutinho are grateful to Ian R. Craig, Ph.D. (BASF, Ludwigshafen) for his critical comments and feedback on this chapter. The authors are grateful to Department of Science and Technology (DST), Department of Biotechnology (DBT) and Council of Scientific and Industrial Research (CSIR) for their financial support to build the High-Performance Computing system at the Department of Pharmaceutical Chemistry,

Bombay College of Pharmacy. E. A. F. Martis and E. C. Coutinho are also thankful to nVIDIA Corporation for their hardware support grant. E. A. F. Martis is indebted to BASF, Ludwigshafen, Germany for the Ph.D. fellowship and the MCBR4 (2015) consortium (Prof. Dr. P. Comba, University of Heidelberg; Prof. Dr H. Zipse LMU, Munich and Prof. Dr. G. N. Sastry, IICT, Hyderabad for MCBR visiting fellowship to Heine-Heinrich University of Düsseldorf, Germany). E. A. F. Martis would also like to thank Prof. Dr. Holger Gohlke, Heine-Heinrich University of Düsseldorf for his guidance during the sabbatical in his CPCLab. Gratitude is expressed to Sandhya Subash, Ph.D. (Bristol-Meyer-Squibb, India), for her assistance in preparing and proofreading the drafts of this manuscript.

References

1. Gulnik SV, Suvorov LI, Liu B, Yu B, Anderson B, Mitsuya H, Erickson JW (1995) Kinetic characterization and cross-resistance patterns of HIV-1 protease mutants selected under drug pressure. Biochem 34(29):9282–9287
2. Schliekelman P, Garner C, Slatkin M (2001) Natural selection and resistance to HIV. Nature 411(6837):545–546
3. Toprak E, Veres A, Michel J-B, Chait R, Hartl DL, Kishony R (2012) Evolutionary paths to antibiotic resistance under dynamically sustained drug selection. Nature Genet 44(1):101–105
4. Yang Z, Nielsen R, Goldman N, Pedersen A-MK (2000) Codon-substitution models for heterogeneous selection pressure at amino acid sites. Genetics 155(1):431–449
5. Blanchard JS (1996) Molecular mechanisms of drug resistance in Mycobacterium tuberculosis. Annu Rev Biochem 65(1):215–239
6. Borst P, Ouellette M (1995) New mechanisms of drug resistance in parasitic protozoa. Annu Rev Microbiol 49(1):427–460
7. Longley D, Johnston P (2005) Molecular mechanisms of drug resistance. J Pathol 205 (2):275–292
8. Walsh C (2000) Molecular mechanisms that confer antibacterial drug resistance. Nature 406:775–781
9. Andersson DI, Levin BR (1999) The biological cost of antibiotic resistance. Curr Opin Microbiol 2(5):489–493
10. Gagneux S, Long CD, Small PM, Van T, Schoolnik GK, Bohannan BJ (2006) The competitive cost of antibiotic resistance in Mycobacterium tuberculosis. Science 312:1944–1946
11. Böttger EC, Springer B, Pletschette M, Sander P (1998) Fitness of antibiotic-resistant microorganisms and compensatory mutations. Nature Med 4(12):1343–1344
12. Sander P, Springer B, Prammananan T, Sturmfels A, Kappler M, Pletschette M, Böttger EC (2002) Fitness cost of chromosomal drug resistance-conferring mutations. Antimicrob Agents Chemother 46(5):1204–1211
13. Cao ZW, Han LY, Zheng CJ, Ji ZL, Chen X, Lin HH, Chen YZ (2005) Computer prediction of drug resistance mutations in proteins. Drug Discov Today 10(7):521–529
14. Rhee S-Y, Gonzales MJ, Kantor R, Betts BJ, Ravela J, Shafer RW (2003) Human immunodeficiency virus reverse transcriptase and protease sequence database. Nucleic Acids Res 31(1):298–303
15. Shafer RW (2006) Rationale and uses of a public HIV drug-resistance database. J Infect Dis 194(Supplement 1):S51–S58
16. Kumar R, Chaudhary K, Gupta S, Singh H, Kumar S, Gautam A, Kapoor P, Raghava GP (2013) CancerDR: cancer drug resistance database. Sci Rep 3:1445
17. Sandgren A, Strong M, Muthukrishnan P, Weiner BK, Church GM, Murray MB (2009) Tuberculosis drug resistance mutation database. PLoS Med 6(2):e1000002

18. Carbonell P, Trosset J-Y (2014) Overcoming drug resistance through in silico prediction. Drug Discov Today Technol 11:101–107
19. Hao G-F, Yang G-F, Zhan C-G (2012) Structure-based methods for predicting target mutation-induced drug resistance and rational drug design to overcome the problem. Drug Discov Today 17(19):1121–1126
20. Martis EAF, Joseph B, Gupta SP, Coutinho EC, Hdoufane I, Bjij I, Cherqaoui D (2017) Flexibility of important HIV-1 targets and in silico design of anti-HIV drugs. Curr Chem Biol 12(1):23–39
21. Chandrika B-R, Subramanian J, Sharma SD (2009) Managing protein flexibility in docking and its applications. Drug Discov Today 14(7):394–400
22. Coupez B, Lewis R (2006) Docking and scoring-theoretically easy, practically impossible? Curr Med Chem 13(25):2995–3003
23. Davis IW, Baker D (2009) RosettaLigand docking with full ligand and receptor flexibility. J Mol Biol 385(2):381–392
24. Lin J-H (2011) Accommodating protein flexibility for structure-based drug design. Curr Top Med Chem 11(2):171–178
25. Mohan V, Gibbs AC, Cummings MD, Jaeger EP, DesJarlais RL (2005) Docking: successes and challenges. Curr Pharm Des 11(3):323–333
26. van Gunsteren WF (1988) The role of computer simulation techniques in protein engineering. Protein Eng 2(1):5–13
27. Hansson T, Oostenbrink C, van Gunsteren WF (2002) Molecular dynamics simulations. Curr Opin Struct Biol 12(2):190–196
28. Binder K, Horbach J, Kob W, Paul W, Varnik F (2004) Molecular dynamics simulations. J Phys Condens Matter 16:S429
29. Pissurlenkar RR, Shaikh MS, Iyer RP, Coutinho EC (2009) Molecular mechanics force fields and their applications in drug design. AntiInfect Agents Med Chem 8(2):128–150
30. Anderson JA, Lorenz CD, Travesset A (2008) General purpose molecular dynamics simulations fully implemented on graphics processing units. J Comput Phys 227(10):5342–5359
31. Götz AW, Williamson MJ, Xu D, Poole D, Le Grand S, Walker RC (2012) Routine microsecond molecular dynamics simulations with AMBER on GPUs. 1. generalized born. J Chem Theory Comput 8(5):1542–1555
32. Salomon-Ferrer R, Götz AW, Poole D, Le Grand S, Walker RC (2013) Routine microsecond molecular dynamics simulations with AMBER on GPUs. 2. explicit solvent particle mesh Ewald. J Chem Theory Comput 9(9):3878–3888
33. Beberg AL, Ensign DL, Jayachandran G, Khaliq S, Pande VS (2009) Folding@ home: lessons from eight years of volunteer distributed computing. In: IEEE international symposium on parallel & distributed processing, 2009. IPDPS 2009. IEEE
34. Larson SM, Snow CD, Shirts M, Pande VS (2009) Folding@ Home and Genome@ Home: Using distributed computing to tackle previously intractable problems in computational biology. DOI: arXiv preprint arXiv:0901.0866
35. Bruccoleri RE, Karplus M (1990) Conformational sampling using high-temperature molecular dynamics. Biopolymers 29(14):1847–1862
36. Earl DJ, Deem MW (2005) Parallel tempering: theory, applications, and new perspectives. Phys Chem Chem Phys 7(23):3910–3916
37. Sugita Y, Okamoto Y (1999) Replica-exchange molecular dynamics method for protein folding. Chem Phys Lett 314(1):141–151
38. Huber T, Torda AE, van Gunsteren WF (1994) Local elevation: a method for improving the searching properties of molecular dynamics simulation. J Comput Aided Mol Des 8(6):695–708
39. Laio A, Parrinello M (2002) Escaping free-energy minima. Proc Natl Acad Sci USA 99 (20):12562–12566

40. Belubbi AV, Martis EAF (2017) Advanced techniques in bimolecular simulations. In: Bharati SK (ed) Handbook of research on medicinal chemistry, Apple Academic Press (in Press)
41. Berne BJ, Straub JE (1997) Novel methods of sampling phase space in the simulation of biological systems. Curr Opin Struct Biol 7(2):181–189
42. Hamelberg D, Mongan J, McCammon JA (2004) Accelerated molecular dynamics: a promising and efficient simulation method for biomolecules. J Chem Phys 120(24):11919–11929
43. Lei H, Duan Y (2007) Improved sampling methods for molecular simulation. Curr Opin Struct Biol 17(2):187–191
44. Zuckerman DM (2011) Equilibrium sampling in biomolecular simulation. Annu Rev Biophys 40:41–62
45. Böhm HJ, Klebe G (1996) What can we learn from molecular recognition in protein-ligand complexes for the design of new drugs? Angew Chem Int Ed 35(22):2588–2614
46. Homans S (2007) Dynamics and thermodynamics of ligand–protein interactions. In: Peters T (ed) Bioactive Conformation I. Springer, Berlin, Heidelberg, pp 51–82
47. Whitesides GM, Krishnamurthy VM (2005) Designing ligands to bind proteins. Q Rev Biophys 38(4):385–396
48. Bronowska, A. K. (2011). Thermodynamics of ligand-protein interactions: implications for molecular design. In: Moreno-Pirajan JC (ed) Thermodynamics—Interaction Studies—Solids, Liquids and Gases. INTECH Open Access Publisher, Croatia, pp 1–48
49. Datar PA, Khedkar SA, Malde AK, Coutinho EC (2006) Comparative residue interaction analysis (CoRIA): a 3D-QSAR approach to explore the binding contributions of active site residues with ligands. J Comput Aided Mol Des 20(6):343–360
50. Martis EA, Chandarana RC, Shaikh MS, Ambre PK, D'Souza JS, Iyer KR, Coutinho EC, Nandan SR, Pissurlenkar RR (2015) Quantifying ligand–receptor interactions for gorge-spanning acetylcholinesterase inhibitors for the treatment of Alzheimer's disease. J Biomol Struct Dyn 33(5):1107–1125
51. Verma J, Khedkar VM, Prabhu AS, Khedkar SA, Malde AK, Coutinho EC (2008) A comprehensive analysis of the thermodynamic events involved in ligand–receptor binding using CoRIA and its variants. J Comput Aided Mol Des 22(2):91–104
52. Wang T, Wade RC (2001) Comparative binding energy (COMBINE) analysis of influenza neuraminidase—inhibitor complexes. J Med Chem 44(6):961–971
53. van Gunsteren WF (1993) Molecular dynamics studies of proteins. Curr Opin Struct Biol 3 (2):277–281
54. Mennucci B (2012) Polarizable continuum model. Wiley Interdisc Rev Comput Mol Sci 2 (3):386–404
55. Klamt A, Schuurmann G (1993) COSMO: a new approach to dielectric screening in solvents with explicit expressions for the screening energy and its gradient. J Chem Soc Perkin Trans 2(5):799–805
56. Marenich AV, Cramer CJ, Truhlar DG (2009) Universal solvation model based on solute electron density and on a continuum model of the solvent defined by the bulk dielectric constant and atomic surface tensions. J Phy Chem 113(18):6378–6396
57. Hou T, Wang J, Li Y, Wang W (2010) Assessing the performance of the MM/PBSA and MM/GBSA methods. 1. the accuracy of binding free energy calculations based on molecular dynamics simulations. J Chem Inf Model 51(1):69–82
58. Hou T, Wang J, Li Y, Wang W (2011) Assessing the performance of the molecular mechanics/Poisson Boltzmann surface area and molecular mechanics/generalized Born surface area methods. II. The accuracy of ranking poses generated from docking. J Comput Chem 32(5):866–877
59. Sun H, Li Y, Shen M, Tian S, Xu L, Pan P, Guan Y, Hou T (2014) Assessing the performance of MM/PBSA and MM/GBSA methods. 5. improved docking performance using high solute dielectric constant MM/GBSA and MM/PBSA rescoring. Phys Chem Chem Phys 16(40):22035–22045

60. Sun H, Li Y, Tian S, Xu L, Hou T (2014) Assessing the performance of MM/PBSA and MM/GBSA methods. 4. accuracies of MM/PBSA and MM/GBSA methodologies evaluated by various simulation protocols using PDBbind data set. Phys Chem Chem Phys 16 (31):16719–16729
61. Xu L, Sun H, Li Y, Wang J, Hou T (2013) Assessing the performance of MM/PBSA and MM/GBSA methods. 3. the impact of force fields and ligand charge models. J Phy Chem B 117(28):8408–8421
62. Dominy BN, Brooks CL (1999) Development of a generalized born model parametrization for proteins and nucleic acids. J Phy Chem B 103(18):3765–3773
63. Jayaram B, Sprous D, Beveridge D (1998) Solvation free energy of biomacromolecules: parameters for a modified generalized born model consistent with the AMBER force field. J Phy Chem B 102(47):9571–9576
64. Onufriev A, Bashford D, Case DA (2000) Modification of the generalized Born model suitable for macromolecules. J Phy Chem B 104(15):3712–3720
65. Onufriev A, Bashford D, Case DA (2004) Exploring protein native states and large-scale conformational changes with a modified generalized born model. Proteins Struct Funct Bioinf 55(2):383–394
66. Homeyer N, Gohlke H (2012) Free energy calculations by the molecular mechanics Poisson —Boltzmann surface area method. Mol Inform 31(2):114–122
67. Kollman PA, Massova I, Reyes C, Kuhn B, Huo S, Chong L, Lee M, Lee T, Duan Y, Wang W (2000) Calculating structures and free energies of complex molecules: combining molecular mechanics and continuum models. Acc Chem Res 33(12):889–897
68. Srinivasan J, Cheatham TE, Cieplak P, Kollman PA, Case DA (1998) Continuum solvent studies of the stability of DNA, RNA, and phosphoramidate-DNA helices. J Am Chem Soc 120(37):9401–9409
69. Edinger SR, Cortis C, Shenkin PS, Friesner RA (1997) Solvation free energies of peptides: comparison of approximate continuum solvation models with accurate solution of the Poisson-Boltzmann equation. J Phy Chem B 101(7):1190–1197
70. Gilson MK, Davis ME, Luty BA, McCammon JA (1993) Computation of electrostatic forces on solvated molecules using the Poisson-Boltzmann equation. J Phy Chem 97(14):3591–3600
71. Im W, Beglov D, Roux B (1998) Continuum solvation model: computation of electrostatic forces from numerical solutions to the Poisson-Boltzmann equation. Comput Phys Commun 111(1):59–75
72. Baron R, van Gunsteren WF, Hünenberger PH (2006) Estimating the configurational entropy from molecular dynamics simulations: anharmonicity and correlation corrections to the quasi-harmonic approximation. Trends Phys Chem 11:87–122
73. Harris S, Laughton C (2007) A simple physical description of DNA dynamics: quasi-harmonic analysis as a route to the configurational entropy. J Phys: Condens Matter 19(7):076103
74. Case DA (1994) Normal mode analysis of protein dynamics. Curr Opin Struct Biol 4 (2):285–290
75. Karplus M, Kushick JN (1981) Method for estimating the configurational entropy of macromolecules. Macromolecules 14(2):325–332
76. Tidor B, Karplus M (1993) The contribution of cross-links to protein stability: a normal mode analysis of the configurational entropy of the native state. Proteins Struct Funct Bioinf 15(1):71–79
77. Aqvist J, Marelius J (2001) The linear interaction energy method for predicting ligand binding free energies. Comb Chem High Throughput Screen 4(8):613–626
78. Åqvist J, Medina C, Samuelsson J-E (1994) A new method for predicting binding affinity in computer-aided drug design. Protein Eng 7(3):385–391
79. Hansson T, Marelius J, Åqvist J (1998) Ligand binding affinity prediction by linear interaction energy methods. J Comput Aided Mol Des 12(1):27–35

80. Åqvist J (1990) Ion-water interaction potentials derived from free energy perturbation simulations. J Phys Chem 94(21):8021–8024
81. Wang W, Wang J, Kollman PA (1999) What determines the van der waals coefficient β in the LIE (linear interaction energy) method to estimate binding free energies using molecular dynamics simulations? Proteins Struct Funct Bioinf 34(3):395–402
82. Bajorath J (2002) Integration of virtual and high-throughput screening. Nat Rev Drug Discov 1(11):882–894
83. Shoichet BK (2004) Virtual screening of chemical libraries. Nature 432(7019):862–865
84. Zwanzig RW (1954) High-temperature equation of state by a perturbation method. i. nonpolar gases. J Chem Phy 22(8):1420–1426
85. Zwanzig RW (1955) High-temperature equation of state by a perturbation method. II. polar gases. J Chem Phy 23(10):1915–1922
86. van Gunsteren WF (1989) Methods for calculation of free energies and binding constants: successes and problems. In: van Gunsteren WF, Weiner PK (eds) Computer simulation of biomolecular systems: theoretical and experimental applications. Escom, Leiden, pp 27–59
87. van Gunsteren WF, Berendsen HJ (1987) Thermodynamic cycle integration by computer simulation as a tool for obtaining free energy differences in molecular chemistry. J Comput Aided Mol Des 1(2):171–176
88. Kollman P (1993) Free energy calculations: applications to chemical and biochemical phenomena. Chem Rev 93(7):2395–2417
89. Chodera JD, Mobley DL, Shirts MR, Dixon RW, Branson K, Pande VS (2011) Alchemical free energy methods for drug discovery: progress and challenges. Curr Opin Struct Biol 21 (2):150–160
90. Shirts MR, Mobley DL, Chodera JD (2007) Alchemical free energy calculations: ready for prime time? Annu Rep Comput Chem D A Dixon 3:41–59
91. Wang Q, Edupuganti R, Tavares CD, Dalby KN, Ren P (2015) Using docking and alchemical free energy approach to determine the binding mechanism of eEF2K inhibitors and prioritizing the compound synthesis. Front Mol Biosci 2:9
92. Massova I, Kollman PA (1999) Computational alanine scanning to probe protein-protein interactions: a novel approach to evaluate binding free energies. J Am Chem Soc 121 (36):8133–8143
93. Chellappan S, Kairys V, Fernandes MX, Schiffer C, Gilson MK (2007) Evaluation of the substrate envelope hypothesis for inhibitors of HIV-1 protease. Proteins Struct Funct Bioinf 68(2):561–567
94. Nalam MN, Ali A, Altman MD, Reddy GKK, Chellappan S, Kairys V, Özen A, Cao H, Gilson MK, Tidor B (2010) Evaluating the substrate-envelope hypothesis: structural analysis of novel HIV-1 protease inhibitors designed to be robust against drug resistance. J Virol 84 (10):5368–5378
95. Shen Y, Altman MD, Ali A, Nalam MN, Cao H, Rana TM, Schiffer CA, Tidor B (2013) Testing the substrate-envelope hypothesis with designed pairs of compounds. ACS Chem Biol 8(11):2433–2441
96. Chellappan S, Kiran Kumar Reddy G, Ali A, Nalam MN, Anjum SG, Cao H, Kairys V, Fernandes MX, Altman MD, Tidor B (2007). Design of mutation-resistant HIV protease inhibitors with the substrate envelope hypothesis. Chem Biol Drug Des 69(5): 298–313
97. Kairys V, Gilson MK, Lather V, Schiffer CA, Fernandes MX (2009) Toward the design of mutation-resistant enzyme inhibitors: further evaluation of the substrate envelope hypothesis. Chem Biol Drug Des 74(3):234–245
98. Nalam MN, Ali A, Reddy GKK, Cao H, Anjum SG, Altman MD, Yilmaz NK, Tidor B, Rana TM, Schiffer CA (2013) Substrate envelope-designed potent HIV-1 protease inhibitors to avoid drug resistance. Chem Biol 20(9):1116–1124
99. Nalam MN, Schiffer CA (2008) New approaches to HIV protease inhibitor drug design II: testing the substrate envelope hypothesis to avoid drug resistance and discover robust inhibitors. Curr Opin HIV AIDS 3(6):642

100. Hao G-F, Yang G-F, Zhan C-G (2010) Computational mutation scanning and drug resistance mechanisms of HIV-1 protease inhibitors. J Phy Chem B 114(29):9663–9676
101. Tse A, Verkhivker GM (2015) Molecular determinants underlying binding specificities of the ABL kinase inhibitors: combining alanine scanning of binding hot spots with network analysis of residue interactions and coevolution. PLoS ONE 10(6):e0130203
102. Hou T, Yu R (2007) Molecular dynamics and free energy studies on the wild-type and double mutant HIV-1 protease complexed with amprenavir and two amprenavir-related inhibitors: mechanism for binding and drug resistance. J Med Chem 50(6):1177–1188
103. Perryman AL, Lin JH, McCammon JA (2004) HIV-1 protease molecular dynamics of a wild-type and of the V82F/I84V mutant: possible contributions to drug resistance and a potential new target site for drugs. Protein Sci 13(4):1108–1123
104. Hou T, McLaughlin WA, Wang W (2008) Evaluating the potency of HIV-1 protease drugs to combat resistance. Proteins: Struct, Funct, Bioinf 71(3):1163–1174
105. Wang W, Kollman PA (2001) Computational study of protein specificity: the molecular basis of HIV-1 protease drug resistance. Proc Natl Acad Sci USA 98(26):14937–14942
106. Ishikita H, Warshel A (2008) Predicting drug-resistant mutations of HIV protease. Angew Chem Int Ed 47(4):697–700
107. Singh N, Frushicheva MP, Warshel A (2012) Validating the vitality strategy for fighting drug resistance. Proteins Struct Funct Bioinf 80(4):1110–1122

Pharmacophore Modelling and Screening: Concepts, Recent Developments and Applications in Rational Drug Design

Chinmayee Choudhury and G. Narahari Sastry

Abstract Computational design of molecules with desired properties has become indispensable in many areas of research, particularly in the pharmaceutical industry and academia. Pharmacophore is one of the essential state-of-the-art techniques widely used in various ways in the computer-aided drug design projects. The pharmacophore modelling approaches have been an important part of many drug discovery strategies due to its simple yet diverse usage. It has been extensively applied for virtual screening, lead optimization, target identification, toxicity prediction and de novo lead design and has a huge scope for application in fragment-based drug design and lead design targeting protein–protein interaction interfaces and target-based classification of chemical space. In this chapter, we have briefly discussed the basic concepts and methods of generation of pharmacophore models. The diverse applications of the pharmacophore approaches have been discussed using number of case studies. We conclude with the limitations of the approaches and its wide scope for the future application depending on the research problem and the type of initial data available.

Keywords Computer-aided drug design · Pharmacophore mapping
Receptor-based pharmacophore · Ligand-based pharmacophore
Pharmacophore features · Pharmacophore fingerprints · Virtual screening
Pharmacophore searching · Docking · QSAR · De novo design

Abbreviations

ADMET	Absorption, distribution, metabolism, excretion, toxicity
CADD	Computer-aided drug design
CmaA1	Mycobacterial cyclopropane synthase

C. Choudhury · G. Narahari Sastry (✉)
Center for Molecular Modelling, Indian Institute of Chemical Technology,
Hyderabad, India
e-mail: gnsastry@gmail.com

C. Choudhury
Department of Biochemistry, All India Institute of Medical Sciences,
Basni, Jodhpur, Rajasthan, India

C. G. Mohan (ed.), *Structural Bioinformatics: Applications in Preclinical Drug Discovery Process*, Challenges and Advances in Computational Chemistry and Physics 27, https://doi.org/10.1007/978-3-030-05282-9_2

HHCPF	Hexadecahydro-1H-Cyclopenta[a]Phenanthrene Framework
HTS	High-throughput screening
MD	Molecular dynamics
Mtb	Mycobacterium tuberculosis
QSAR	Quantitative structure-activity relationship
TB	Tuberculosis

1 Introduction

Rational drug discovery is highly interdisciplinary and is one of the outstanding challenges, besides being highly arduous and expensive. The process of designing new medications requires investment of roughly 14 years [1] of time and cost as high as 1 billion USD [2]. Along with rapidly evolving HTS [3] and combinatorial chemistry technologies, computer-aided drug design (CADD) strategies are also effectively contributing to accelerate and economize the process of drug development [4–6]. A broad range of CADD applications are employed at almost all early stages of the drug discovery pipelines, starting from target identification, target structure prediction, screening of initial hits to prioritization and optimization of leads and understanding their structure-property relationships [7, 8]. We have been working in state-of-the-art CADD techniques such as homology modelling [9], molecular dynamics simulations [10–12], QSAR [13–15], molecular docking [16], pharmacophore modelling [17], virtual screening [18, 19] and cheminformatics [20] since more than a decade. One of the fundamental applications of cheminformatics is to develop programmes that store, manage and retrieve molecular structures in various formats, their calculated/experimental properties and bioactivities. Cheminformatics also involves computing molecular fingerprints and descriptors based on the molecular structures that label a physicochemical property and can be used as screening filters [21, 22]. These molecular descriptors of known active molecules can also be used to develop quantitative structure-activity/property relationship (QSAR/QSPR) models to predict the inhibitory activity or toxicity of novel compounds and preliminarily profile them in silico without performing expensive in vitro and in vivo assays [23–26]. Docking and simulations predict the three-dimensional binding mode of a given molecule in the binding site of a macromolecular receptor (protein/DNA), and their affinity is quantitatively assessed by a docking score. This technique has not only been proved enormously useful to study receptor–ligand interactions but also is used as a popular tool to virtually screen compound libraries to obtain a hit or to identify the target for a molecule by reverse engineering [27–29]. A large number of studies from our group have focused on application of these techniques to a plethora of drug targets such as phosphodiesterases [14], kinases [12, 30], HIV proteases [10, 13] and reverse transcriptase [31] and Mtb cyclopropane synthases [11, 17, 18]. We have also initiated development of a disease (tuberculosis) specific Web portal, integrating all these techniques, which will be of tremendous help for researches working in the field of Mtb drug discovery [32].

Pharmacophore modelling is one of the enormously useful sub-areas of CADD with diverse structure and ligand-based applications [33, 34]. Like docking, one of the basic applications of pharmacophore models is virtual screening, but at a much faster speed as compared to docking [33]. This approach can also be implemented complementarily with docking and QSAR studies [18, 20]. Many studies use pharmacophore models for target/off-target identification as well [35, 36]. In this chapter, we basically focus on the in silico representation of the concept and the varieties of ways of application of pharmacophore models in drug discovery projects.

2 The Concept of Pharmacophore

The term 'pharmacophore' has gained immense popularity in the field of medicinal chemistry paralleled with computer-aided structure-activity relationship studies. In 1909, Ehrlich gave an introductory definition of pharmacophore [37, 38], by combining the words 'phoros' meaning carrying and 'pharmacon' meaning drug. Hence, a pharmacophore is 'the molecular framework carrying the crucial features accountable for a drug's biological activity'. Since then, many groups have attributed various definitions and meanings to this term based on their scientific background and research view. IUPAC has officially defined a pharmacophore model as [39]

> An ensemble of steric and electronic features that is necessary to ensure the optimal supramolecular interactions with a specific biological target and to trigger (or block) its biological response.

However, a century's research and development has expanded its circumstantial meaning and application considerably. Due to their simple way of capturing and representing the chemical features of compounds, pharmacophore models have drawn the attention of the medicinal chemistry community in last few years as a tool to screen the cig (chemistry) data [40]. Upon administration, when a drug/small molecule enters the human body, it comes across thousands of proteins (receptors, transporters, carriers, plasma proteins, etc.) to potentially interact with. But it chooses to bind to only those proteins (targets) where the protein's active site and drug have compatible shape/size and the protein–drug interactions are energetically favourable. Similarly, size/volume/shape and the chemical features of the residues lining the binding pocket determine which type of small molecules it is able to bind. Hence, the right size, correct shape and complementary chemical features are the key factors for the protein–drug recognition to instigate a biological effect. The central concept of pharmacophore is based on the perception that the molecular interaction pattern of a group of compounds with their biological target can be credited to a small set of common features complementary to the chemical features present in the target's binding pocket. The general features include hydrogen-bond (HB) donors, HB acceptors, charged groups (positive and negative), hydrophobic sites and aromatic rings, which are used as chemical features in pharmacophore models by most of the programmes. Some programmes define a few additional features such as 'exclusion volumes' representing steric constraints. These features generally replicate the steric environment of the binding pocket to avoid clashes of the mapped of compounds with the protein surface. Pharmacophore models comprises distinct spatial arrangement of these features that denotes the chemical functionalities of active small molecules. Instead of real atoms/functional groups, a pharmacophore model emphasizes the chemical features of ligands/protein–ligand complexes, making it a better and fast tool to recognize molecular similarities.

3 A Typical Pharmacophore Model: Representation of Pharmacophoric Features

According to the definition, a pharmacophore model represents the binding patterns of bioactive molecules with the target binding site, by virtue of a distinct 3D arrangement of abstract interaction features accounting for different types of non-covalent interactions. These interaction types can be HB formation, columbic interactions, metal interactions, hydrophobic contacts, aromatic stacking or charge transfer interactions. Overall, a pharmacophore model characterizes a common binding mode of diverse ligands with a specific target. In pharmacophore modelling, the molecules are first segregated into a set of features, each representing a certain type of interaction with the binding site residues. Then, each feature is represented by points to be used for superimposition (least-squares fitting) of molecules with each other. Here we will be discussing features employed by most of the popular programmes [41–45].

HB donor (D): Hydroxyl groups, hydrogens bound to nitrogen, acetylenic CH groups and thiols (SH) are normally denoted as donors. However, the –CH and –SH groups are considered relatively weaker donors. Sometimes, along with acetylenes, other types of –CH such as the ones in nitrogen heterocycles of some kinase inhibitors are considered as donors. Keeping protonation in mind, basic amines such as $RCH_2N(Me)_2$ are considered as donors. Tautomeric and ionized states severely influence pharmacophore feature definition because they may amend the characteristic of a feature. Hence, molecules should be presented to the pharmacophore elucidation programmes in all possible protonation/ionization states.

HB acceptor (A): Generally, atoms with available lone pairs of electrons such as N, O, S are treated as acceptors. However, some programmes do not consider oxygen atoms present in furan/oxazole rings, as they are very weak acceptors according to theoretical and crystallographic evidence.

Along with defining the HB features, it is very essential to fix the positions of the complementary feature points to be overlapped in the resulting pharmacophore. That is why the pharmacophore modelling programmes link donor and acceptor features with the equivalent ligand atoms as well as the supposed locations of the corresponding complementary receptor atoms involved in the interaction.

Positive and negative features (P and N): In the molecules, atoms bearing formal charges are considered as positive or negative features provided they are not part of a dipole. Groups possessing net formal charges are also considered as positive/negative features. Centroid of the heteroatoms of a group is the region, where the positive/negative charged features are generally placed. Sometimes the positive and negative features are emphasized specifically based on their ionizability. For example, $R–NH_3^+$ is measured as positively ionizable feature, but $R–N(Me)_3^+$ is not as the interactions made by these two groups are significantly different.

Hydrophobic features (H): Choosing atoms/groups that should be measured as hydrophobic is neither easy nor straightforward. The most commonly used algorithm developed by Greene et al. [42] first allot a hydrophobicity score to each atom based on a set of empirical rules defined from medicinal chemists' perceptions and then atoms with amply large hydrophobicity values are grouped into clusters. Then a hydrophobic feature point is placed at the centroid of each such cluster. The order of hydrophobicity score is roughly rings/ring atoms > groups like $–CF_3$ > alkyl chains. Some simple algorithms [44] consider all non-donors/non-acceptor/non-charged atoms as steric groups (equivalent of hydrophobic groups), which also yield a depiction of molecular shape.

Aromatic rings (R): Aromatic rings are treated as a special type of hydrophobic feature represented by vectors instead of points so as to mimic the directionality of interactions like π–π stacking and cation–π interactions. Figure 1 shows an example of a typical pharmacophore model.

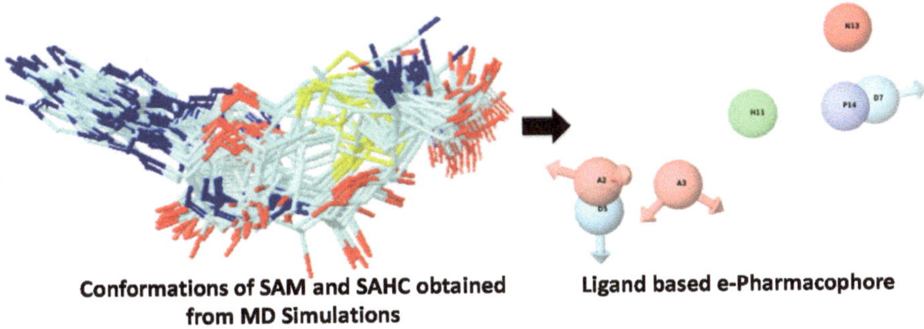

Fig. 1 An example of a pharmacophore model, generated from the conformations of S-adenosyl methionine (SAM) and S-adenosyl homocysteine (SAHC) [17] with Phase programme. Colour codes for the pharmacophoric features are as follows. Cyan: D, pink: A, red: N, blue: P, green: H and orange: R

Most recent pharmacophore modelling programmes define additional steric constraints features. These are called exclusion volumes (XVols), representing the steric effect of the binding pocket [46]. These features are required to avoid the clashes of the molecule with the protein surface while mapping. Feature generation not only facilitates the molecules to be aligned in an easy and rational way, but also can be used in scoring. The root mean square deviation (RMSD) between matched features gives quantitative account of the extent of overlay, which is often used as a fitness score [40]. Hence, the placement of feature points should be accurate, and one needs to be careful while deciding whether to consider all possible features or to choose few of them giving adequate information about the spatial orientation of a group of molecules. For example, sometimes there are huge number of hydrophobic features as compared to other features, which may bias the alignment and give a model with good score, but the model will be useless due to lack of specificity.

4 Evolution of the 'Pharmacophore' Concept: Historical Perspective

Paul Ehrlich first used the concept of pharmacophore in the end of nineteenth century, when he revealed the selective binding of methylene blue to nerve fibres. This realization ushered the beginning of pharmacophore concept as '*a molecular framework that carries (phoros) the essential features responsible for a drug's (pharmacon) biological activity*' [37, 38]. Based on this idea, Ehrlich improved the chemical structure of several compounds to yield efficacious drugs against syphilis (under the trade name Salvarsan), trypanosome and spirochete infections [37, 38], which made him win the Nobel prize in 1908 sharing with Ilya Metchnikoff. Although Ehrlich's early definition of pharmacophore is almost unchanged for over a century, Schueler proposed the first modern definition in his book 'Chemobiodynamics and Drug Design' in 1960 [47], where the 'chemical groups' were replaced by patterns of 'abstract features'. Beckett and co-workers [48] proposed the first pharmacophore model of muscarinic agents in 1963 that identified distance ranges between abstract features, and later in 1967, Kier developed the first 'computed' pharmacophore model for muscarinic receptor inhibitor binding pattern [49–51]. Simple pharmacophores were in application as tools for designing new drug molecules much before the dawn of a well-defined field like computer-aided drug design. In the 1940s, preliminary structure-activity relationship models were computed based on simple two-dimensional model structures utilizing the accessible information of the van der Waals sizes and bond lengths [52]. Eventually, in the 1960s, three-dimensional models could be built with the convenience of X-ray and conformational analysis techniques. Medicinal chemists could classify some common molecular frameworks that attributed to high biological activity more often as compared to other structures by retrospectively analysing the chemical structures of the various drugs. Evans et al. [52] named such frameworks as 'privileged structures', which offer the basic scaffold and the substituents at different positions impart receptor specificity. Dihydropyridines [53], Arylethylamines, N-arylpiperazines, diphenylmethane derivatives, biphenyls and pyridazines [52, 53], tricyclic psychotropics and sulphonamides, benzodiazepines [54] are among some popular examples of the privileged structures. Woods and Fildes [55] found that p-aminobenzoic acid (PABA) and *p*-aminobenzenesulphonamide have similar critical distances; hence, bind to the PABA target with similar efficacy and inhibits the biosynthesis of tetrahydrofolic acid. This was one of the examples of the early two-dimensional pharmacophore models. An early 3D pharmacophoric approach was the 'three-point contact model' proposed by Easson and Stedman [56] and Beckett [48] in the case of (R)-(−)-adrenaline [= (R)-(−)-epinephrine]. These models are based on a concept that when a chiral centre is present in a compound, the substituents on this asymmetric atom make three-point contacts with the binding pocket of the receptor, which can only be obtained for one of the two isomers of epinephrine (the more active natural (R)-(−)-epinephrine). Similarly, another three-dimensional approach was developed in the early 1970s, characterizing the activity of clonidine on the central norepinephrine receptor [57]. It was observed that the natural ligand norepinephrine fits into the binding pocket of its target by three main interactions [57], viz. ionic bond between an anion (carboxylate, phosphate) of the binding pocket and the protonated –NH_2 functional group, a HB between the NH–CO group of the binding site and the secondary alcoholic hydroxyl and a π-stacking between the protonated imidazole of a histidine residue of the binding pocket and the aromatic ring of the drug. It was also recognized that the cationic head must be light and the phenolic –OH groups are not important for the biological activity. Pullmann et al. [58] in their 3D pharmacophore model of the norepinephrine receptor computed the critical intramolecular distances for the above key interactions which could successfully explain the pharmacophoric similarity between clonidine and norepinephrine, which in turn enables clonidine to make the same kind of interactions as norepinephrine.

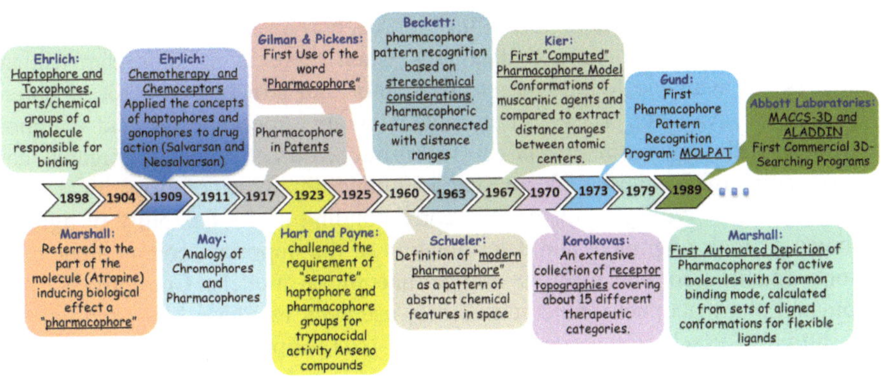

Fig. 2 Schematic presentation of timeline showing early developments in the field of pharmacophore modelling

These are some early efforts to explain pharmacophoric patterns that could act as key features for the design of new chemical entities. Figure 2 shows few early milestones in the field of emergence of pharmacophore modelling.

Nevertheless, in recent years, many effective pharmacophore modelling approaches and their contributions to drug discovery have been reported [59]. With the help of pharmacophoric insights and 3D searching tools, computer-aided drug design efforts are swiftly gaining efficiency since the 1990s. Still, this approach encounters many challenges that restrict its success. Pharmacophore approaches have been widely used in virtual screening, de novo ligand design, lead optimization and multi-target drug design. A range of automated pharmacophore modelling and screening tools have constantly appeared after the computational chemistry revolution witnessed in the past couple of decades [60]. Today, pharmacophore screening is one of the apt choices for researchers working in drug discovery and design.

5 Pharmacophore Model Generation

Pharmacophore models are typically generated either from a group of ligands, by aligning them and taking out the common interaction features indispensable for their biological activity. On the other hand, they can be constructed in a structure-based way, by probing probable interaction points in the receptor binding pocket, provided the 3D structure of the receptor is reported. The pharmacophore models can also be generated from a receptor–ligand complex by identifying the key interactions between the receptor and ligands.

5.1 Ligand-Based Pharmacophore Model Generation

Ligand-based pharmacophore modelling approach is used as a key strategy for facilitating screening compound databases when there is no three-dimensional structures are available for the target or receptor, but structure of a set of potent inhibitors are available. These active molecules are superimposed, and common pharmacophoric features representing crucial interactions between the ligands and the common target of these molecules are identified. Firstly, a conformational space of each of the active ligands is created corresponding to the flexibility of ligands, followed by their alignment and determination of the important common chemical features required for the creation of pharmacophore models. Currently, various automated pharmacophore generators are in use such as Phase [46] (Schrodinger Inc., http://www.schrodinger.com), HypoGen [61], HipHop [61] (Accelrys Inc., http://www.accelrys.com), GASP [62], DISCO [63], GALAHAD [64] (Tripos Inc., http://www.tripos.com) and MOE (Chemical Computing Group, http://www.chemcomp.com) [65]. Several academic programmes [40, 60, 66–68] are also popularly being used. The key differences among these tools are mostly in the algorithms that are implemented for conformational search and alignment. This chapter is about the general steps followed by most of the programmes to recognize a pharmacophore pattern from a group of molecules that interact with a common receptor and the diverse applications of the pharmacophore concept.

5.1.1 Picking the Right Set of Compounds and Their Initial Structures

As the resulting pharmacophore models are highly inclined by the type, size and structural diversity of the participating ligands, it is imperative to choose the set of ligands that take part in the process of pharmacophore model generation. Some programmes like RAPID [69], HipHop [61] and the Crandell Smith method [70] assume all the compounds in the set as active, some other methods consider the information on the inactive molecules to be important as they give an idea about the structural features responsible for reducing the activities and the ones essential for enhancing activity. For example, DISCO [62, 71] and CLEW [72] provide an option to include or exclude inactive molecules in generating a model so that the user can identify the distinguishing features, while HypoGen [61] provides an option for including activity ranges of the set of ligands. As far as size of the dataset is concerned, most of the programmes are capable of handling up to 100 ligands in a set. If the dataset contains large number of molecules, then it can be sorted and categorized based on the activity value ranges. However, some programmes like SCAMPI [73] can handle up to a few thousand molecules but compromising the quality of the models. The high structural diversity of the dataset also is important to identify features that are most essential for target binding and produce high-quality models. Correct compound structures with correct atomic valencies, bond orders and properly defined aromaticity and the appropriate stereochemical flags are crucial for model generation.

5.1.2 Conformational Search

Ligands being flexible may have multiple possible conformations, and each conformation may bind to the binding site of the target in a particular fashion. Thus, it is crucial to consider the flexibilities of each molecule during pharmacophore development. Conformational search is considered as a separate stage in most of the pharmacophore modelling programmes like HipHop, DISCO and RAPID, where a large number of conformations are generated for each ligand. Systematic search, Monte Carlo sampling and molecular dynamics are the methods of choice for most of the software for conformation generation. As, the number of all possible conformers for molecules (especially when they have complex structures with a large number of rotatable bonds) is too large to handle and incorporate in the pharmacophore model building, energy minimization and clustering methods are used to reduce the conformational space. The conformers with lowest energy or representatives from clusters of similar conformers are chosen to take part in model generation. In some other software, conformational search is parallelly performed along with pattern identification by retaining the conformers that possess certain features in a particular spatial arrangement. GASP [63] and GAMMA [74] use such an approach by the genetic algorithm (GA) techniques.

5.1.3 Feature Extraction and Representation

After conformational search, the molecules are subdivided into a set of features, each feature having the capability to form a particular type of non-covalent interaction with the receptor. There are three main levels of resolution for defining the features; (i) it may be atom based as implemented in MPHIL [75], GAMMA [74] and RAPID [69], where 3D atomic position related to the atom type is used as a feature; (ii) it can be atoms grouped into topological features such as a C = O group or a phenyl ring; or (iii) it may be function based, where the atoms are assembled into functional features describing the type of non-bonded interactions with the receptor. These features are HB acceptor (A), HB donor (D), base (+ve charge pH 7) (P), acid (−ve charge, pH 7) (N), aromatic moieties (rings) (R) and hydrophobic group (H). We have already discussed these features in Sect. 3 of this chapter. The third type of feature extraction method is immensely popular and is being used in many programmes like catalyst [43], Phase [46], HypoGen and HipHop [63]. Different topological features having the same chemical function can fall under same functional feature category. At the same time, the functional features are not assigned exclusively for any functional group. For instance, a –OH oxygen can act as both HB acceptor, a donor and at times may act as negatively charged feature. Commonly, the functional groups like a negatively/positively charged species, HB donor and acceptor are represented by their centres, which are nothing but the exact atom positions. Additionally, HB acceptors and donors are often represented by a vector that enforces a restriction of bond directionality between the feature on the binding site of the receptor and the complementary ligand feature. The centre of a hydrophobic site or an aromatic ring is defined as the centroid of the group.

After extracting the features, depiction of the whole molecule's structure is obtained by combining the selected features. These representations are generated mostly as: (i) 3D point set, where a ligand structure is represented as a group of categorized points in the 3D space, where each point is linked with a feature, (ii) a labelled graph, where nodes correspond to the features and the edges correspond to the relations, or (iii) a set of interpoint distances, where the ligand structure is represented as a collection of feature points, along with their interpoint distances. The third type of representation is commonly stored as a $n \times n$ distance matrix, n being the number of atoms.

5.1.4 Pattern Identification and Scoring

Once the features extracted for each ligand in the dataset, a pattern is identified as a set of relative positions in the 3D space, each linked to a feature. If a ligand holds a set of features in at least one of its conformations, the set of features can be aligned with the corresponding locations. Most of the methods are based on spatially overlaying conformations of various compounds with the pharmacophores points with minimal root mean square alignment errors. One can roughly classify the alignment methods as either point or property-based. In the first class of algorithms, pairs of pharmacophoric features are generally aligned using a least-squares fitting using clique detection methods [76, 77]. According to the graph-theoretical approach to molecular structures, a clique is a maximum completely connected sub-graph, which recognizes all imaginable combinations of atoms/functional groups to find out common substructures for the alignment. Property-based or field-based algorithms utilize grid or field descriptors, based on molecular properties such as volume, shape, charge distribution, electron density and electrostatic potentials of molecules. A 3D grid is generated about a ligand by computing the interaction energy components between the ligand and a probe placed at each grid point. Properties are calculated on a grid and later converted to a set of Gaussian representations. A number of either random or thoroughly sampled initial configurations are then generated followed by local optimizations with some similarity measure of the intermolecular overlap of the Gaussians.

After obtaining the pharmacophore candidates in the previous stages, they are generally scored and ranked. The basic obligation of a scoring scheme is implemented such that a high score implies higher chance of the ligands mapping to the pharmacophore model. Despite the great advances, molecular alignment handling ligand flexibility and proper selection of training set compounds are considered as the biggest challenges in ligand-based pharmacophore modelling.

5.2 Structure-Based Pharmacophore Model Generation

Structure-based pharmacophore modelling requires the 3D structure of the receptor or a receptor–ligand complex. The models are generated based on the spatial relationships of complementary interaction features of the binding pockets followed by selection and assembly of features to generate pharmacophore models.

5.2.1 Active Site Identification

The input for receptor-based pharmacophore modelling is the three-dimensional structure of a receptor usually in PDB format. The receptor binding pocket is identified using a spherical probe with customizable radius and location to include the binding site as well as the key interacting residues involved with ligands.

ThereĆare several programmes available for detection of clefts, crevices and binding pockets and to suggest possible active site locations based on the geometry of the surface [78, 79]. The key residues can be determined by user, deduced from studying the activity of the protein after mutation of a single residue. If mutation of a particular residue hampers function of the protein, then that residue may be part of the active site. Computational analyses such as multiple protein structural alignment techniques also help in identifying the active site of a protein by comparing it with a similar protein with known active site.

5.2.2 Complementary Image Construction

The receptor binding pocket is analysed to create an interaction map of features that the molecule is anticipated to satisfy for a reasonable interaction with the active site. In other words, a complement of the receptor binding site is created as the basis to create an input pharmacophore model. In particular, functional features like HB donors/acceptors and hydrophobic groups are identified in the binding site followed by rational placement of complementary features within the binding pockets in chemically acceptable positions [80, 81].

5.2.3 Generation of Queries, Searching and Hit Analysis

Once the active site is defined and chemically characterized, there is no straightforward single step to derive pharmacophore models from the binding site map. Since the receptor binding site has a potential to bind a variety of molecules in a variety of binding conformations, the interaction map often gives rise to huge number of features. To address this problem, adjacent features of the same type are clustered and the feature that lies nearest to the geometric centre of the cluster is retained as the cluster representative and all the other features are discarded. Sometimes, the number of the features is still very high even after the clustering, and all of them cannot be used as a single model because models possessing all such features would not be able to obtain any hits from the database. So, possible combinations of limited numbers of features are derived from the interaction map and multiple pharmacophore modes are composed. And then, these models are used by programmes like catalyst [43, 82] implemented in Accelrys Discovery Studio to search the compound database and test the validity of the models (also termed as pharmacophore 'queries' in catalyst) to screen or reject highly active compounds. It is always necessary to examine these models for how they interact with the binding site residues and how far the models extend within the binding pocket and if they fill specificity pockets and make the strongest interactions. Queries describing only the features present in an inhibitor might end up giving many false positive hits. At times, they screen compounds that are able to map to all the query features but also contain a bulky substituent causing steric hinderance and averting the compound from fitting into the binding site. That is why inclusion of some excluded volume features is often recommended which penalizes the molecules' score if some atoms or group are placed in positions where they are likely to collide with the active site atoms.

5.3 Generation of Pharmacophore Models from the Protein–Ligand Complexes

Protein–ligand complexes produced by X-ray crystallography provide a detailed picture of the interactions between the ligand and the receptor, showing which atoms of the ligand are in contact with the receptor along with the atomic coordinates of those atoms. Also, the type of interactions can also be delineated from the atom types, distances and orientations of the ligand and receptor atoms. The major interaction that occurs in the receptor–ligand interface is hydrogen bonding. But other non-covalent interactions such as π–π and cation–π interactions are also obviously essential for protein–ligand complex formation apart from the hydrogen bonding. We have extensively looked at the importance of these interactions and the cooperativity existing among themselves to maintain supramolecular structures [83–86]. This information is of immense importance to establish a pharmacophore model from the complex. However, one needs to give attention to the facts that alternative pharmacophore models are possible within a single binding pocket owing to the flexibilities of both the active site and the ligands which are capable of rearranging themselves to accommodate different ligands and also there is a possibility of more than one active sites for a particular receptor. The programmes like 'LigandScout' developed by Wolber and Langer [87] and Phase [46] module of Schrodinger suite generate structure-based pharmacophore models from the protein–ligand complexes given as an input. We will be discussing the steps of generation of pharmacophore models from the protein–ligand complexes by the LigandScout and Phase, where the former characterizes the pharmacophoric features using kekule's patterns and the latter prioritizes the features based on the XP docking energy components.

5.3.1 Pharmacophore Model Generation with LigandScout

With the LigandScout [87] programme, as a first step, the correct molecular topology of rings and of hybridization state are assigned to the ligands by analysing the neighbouring atoms followed by assignment of double bonds and Kekule's patterns for functional groups such as carboxylic acids and esters, nitro groups, sulphonyl groups, thio acids, thio acetic esters guanidine-like groups, acetamidine and phosphinoyl groups functional groups. Next, the pharmacophoric features based on the hydrogen bonds, electrostatic interactions, charge transfer or hydrophobic interactions between the ligand and the receptor are defined, and models are generated. Atoms belonging to nonacidic –OH groups (all –OHs excluding carboxylic, sulphinic, sulphonic, phosphonic or phosphinic acids), –SH groups, –C≡C– hydrogens and –NHs (barring trifluoromethyl sulphonamide hydrogens and tetrazoles) are recognized as HB donor atoms. When such an atom is found in the distance range of 2.5–3.8 Å from the heavy atom of a HB acceptor of the receptor molecule, a donor feature consisting of a donor point on the ligand side and a projected point on the macromolecule side is created. Atoms like –OH oxygen, –SH sulphur, –C≡C– carbon or –C≡N nitrogen are recognized as acceptor atoms, and an acceptor feature is placed with the initial point positioned on the acceptor atom and the projected point placed onto the heavy atom of the HB donor on the receptor within the distance range of 2.5–3.8 Å. The electrostatic interaction is represented as a vector resembling the definition of the H-bond acceptor. Hydrophobic areas are implemented in the form of spheres with a tolerance radius of 1.5 Å located in the centre of hydrophobic atom chains, branches or groups after testing a group of adjacent atoms to attain a sufficient overall hydrophobicity score.

5.3.2 e-pharmacophore Model Generation by Phase

The e-pharmacophores method of Phase module [46, 88] of Schrodinger suite is a new approach that utilizes the grid-based ligand docking with energetics (Glide) extra precision (XP) scoring function [89] to precisely quantify protein–ligand interactions. XP scoring function calculates enthalpic contribution of each interacting (pharmacophoric) site of a molecule towards the total score. Thus, each site gets a score based on the sum of enthalpic terms (such as HB, electrostatic, cation–π, π–π, hydrophobic and hydrophobically packed/associated HBs and other interactions) and is ranked. Then the e-pharmacophore models are generated from the top scoring features. The user can choose the number and type of features required to build a model. E-pharmacophores also include excluded volumes representing the regions of space occupied by the receptor where any portion of the ligand cannot be accommodated. E-pharmacophores have been shown to screen diverse set of bioactive molecules as compared to conventional structure-based methods, making it more useful.

5.4 Dynamic Pharmacophore Model Generation and Multicopy Simulations

The active sites of the drug targets being very flexible, structure-based pharmacophore models derived from a single conformational state of the protein may not satisfactorily account for all the possible potential drug–target interactions. In this situation, molecular dynamics simulation has been a very competent method to tackle the target flexibility issues in SBDD. Dynamic pharmacophore models recognize compounds, which complementarily bind to the protein considering flexibility of their binding pockets, theoretically reducing the entropic penalties experienced by the protein due to ligand binding. MD simulation trajectories would give rise to multiple conformations of a protein active site, describing the targets' intrinsic flexibilities. Multiple copy minimization is also a regularly used exercise in computational drug design. The technique first fills the active sites of the receptors with multiple copies of probe molecules those do not react among themselves. Then, molecular dynamics, Monte Carlo/steepest descent minimizations are performed to minimize all these probes parallelly to obtain local minima. When the probes are clustered in the various regions of the active site in different orientations, the relative preferences of the binding regions can be estimated from the number of probes or the interaction energies.

Highly ordered and smaller clusters represent highly crucial prerequisite for favourable interactions, while the haphazardly spread larger ones indicate highly flexible sites. The MUSIC algorithm [80, 90] implemented with the BOSS programme uses similar strategy. It is capable of performing Monte Carlo simulations for a wide range of biomolecular systems in solvent clusters and mixtures and periodic solvent boxes with multiple solutes. It is able to calculate the interaction energies between solvent–solvent, solvent–solute and solute–solute. Usually, the probe or solvent are small molecules. For example, hydroxyl groups, aromatic groups and carbonyl groups are represented by small probes like –CH_3OH, C_6H_6 (Benzene) and –CH_3CO (acetone), respectively. The probe molecules as well as the side chains of the receptor can be treated as rigid, partially/fully flexible or all-atom. The wide-ranging OPLS force field used in this programme is proven to be successfully handling the flexibilities of the receptor while generating pharmacophore models. Applications of the dynamic pharmacophore models will be discussed in the subsequent sections of the chapter.

6 Pharmacophore Finger Prints

The complex 3D structure of a molecule is reduced to an abstract collection of features in the pharmacophoric approach. Extending this concept, the structure of a molecule can be interpreted a as an exclusive data string by extracting all possible three-/four-point sets of pharmacophoric features. The inter-feature distances are assigned using distance binning or simply by bonds. These resulting unique strings describing the frequency of every possible combination at predefined loci of the string are known as pharmacophore fingerprints. Different types of molecular similarity analyses among libraries of molecules have been carried out using pharmacophore fingerprints [91, 92]. Also, the pharmacophoric fingerprint can be used to detect the common key features/groups contributing to the biological function of a group of active ligands.

7 Applications of Pharmacophore-Based Approaches

In this section, we discuss the diverse applications of the pharmacophore approaches under different scenarios.

7.1 Pharmacophore Approaches for Virtual Screening

Pharmacophore models being very simple by their definition can be used in a variety of ways depending on the research problem. This simplicity makes ‘pharmacophore based search’ a tool of choice for drug discovery scientists in the last decade [93]. When the structure of a set of molecules with similar or different scaffolds active on a particular target are known, then ligand-based pharmacophore models can be developed using their structures as described in Sect. 5.1. If the structures of some inactive derivatives are also known, then contribution of each feature towards the bioactivity can be compared between the positive and negative datasets to distinguish the wanted and unwanted features. The allowable steric arrangement of the ligands can also be mapped. When only the structure of the receptor or a receptor–ligand complex is available, then pharmacophore models are generated as described in Sects. 5.2 and 5.3 and can be utilized as queries to screen a database not only to screen compounds satisfying certain geometric and chemical restraints, but also to filter molecules with undesirable properties. For example, Voet and co-workers identified specific antagonists of human androgen receptor by applying two pharmacophoric filters back to back. One model is being generated from the available receptor-agonist complexes, while the other filter applied was a pharmacophore model generated from the receptor-antagonist complex. This approach enabled the authors to screen the compound that matches the antagonist-specific feature [94].

7.1.1 Dynamic E-pharmacophore Models: A Case Study with Mycobacterial CmaA1

We present here the summary of our recent work (Choudhury et al. [11, 17, 18]) on generation and application of dynamic structure and ligand-based pharmacophore models for screening a certain library against a mycobacterial target cyclopropane synthase (CmaA1). Mycolic acids are the characteristic constituents of Mtb cell wall which contribute towards the drug resistance, pathogenicity and persistence of the parasite. CmaA1 enzyme catalyses the cis-cyclopropanation of unsaturated mycolic acid chains at the distal position, which is an indispensable step in mycolic acid biosynthesis and maturation, thus making CmaA1 an important Mtb drug target. Five model systems of CmaA1 corresponding to different stages of cyclopropanation were studied using molecular dynamics (MD) simulations. A detailed picture of the structural changes in the two distinct binding sites, i.e. cofactor and acyl substrate binding sites of CmaA1 during the cyclopropanation process was obtained by analysing the MD simulations trajectories. The apo-state of CmaA1 was observed to have a closed conformation where the cofactor binding site is inaccessible. Upon cofactor binding, H-bond between Pro202 of loop10 (L10) and Asn11 of N-terminal α1 helix disrupts making the cofactor binding pocket accessible. Upon cofactor binding, the non-polar side chains of the substrate binding site position towards the inner side of the pocket forming a hydrophobic environment for the substrate. In order to exchange the methyl group from the cofactor to the substrate, both the ligands tend to come close to each other facilitated by the upliftment of loop10. These observations prompted to think that the protein can remain in diverse conformations at different stages of its catalytic function and considering only one conformation for drug design would not be sufficient. So multiple structures obtained from the MD trajectories were used to generate, validate and use structure and ligand-based pharmacophore models.

7.1.2 Generation of Dynamic Structure-Based Pharmacophore Models

The molecular dynamics simulations on CmaA1 revealed that the binding sites of the enzyme exhibit huge conformational diversity, when bound to different ligands at various stages of its function. To use this conformational diversity of the binding sites in structure-based drug design, representative structures (snapshots) were extracted from all the five MD trajectories at a regular interval of 5 ns, thus obtaining a total of forty conformations of CmaA1 bound to different ligands in the two binding sites. The crystal structure of CmaA1 reported in PDB was also added to this pool. Now these 41 protein–ligand complexes were used to obtain e-pharmacophore models as described in Sect. 5.3.2. The first step used was evaluating the Glide energy terms. Active site of each CmaA1 structure was defined as a cubical box of 12 * 12 * 12 $Å^3$ dimension, and the Glide [89] energy grids were generated. Glide scores with XP descriptor information were obtained for the already bound ligands keeping their original conformations unchanged (unlike a typical docking where protein is held rigid while ligands are kept flexible). This exercise calculated all the interaction energy components between the receptor–ligand complexes, which were then submitted to the Phase module of Schrodinger to develop energy-based e-pharmacophore [88, 95] models. Figure 3 depicts the steps of the e-pharmacophore model generation and selection of best ones as virtual screening filters.

7.1.3 Pharmacophore Model Validation

To examine the capabilities of the dynamics-based e-pharmacophore models to successfully distinguish inhibitors and non-inhibitors of CmaA1, a set of 23 reported CmaA1 inhibitors (MIC:0.0125–12.5 μg/mL) [96] were used as a positive

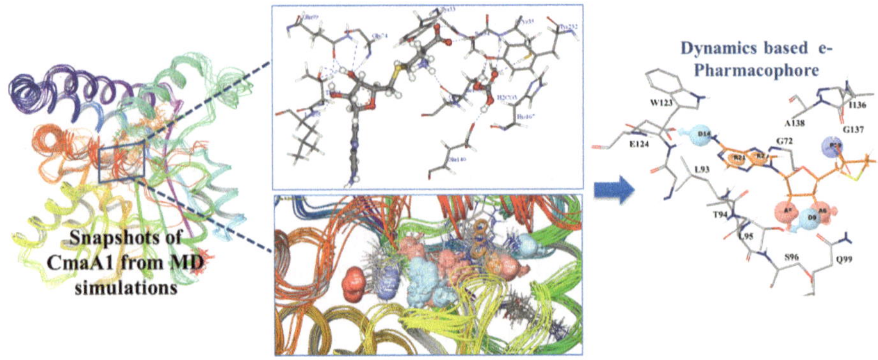

Fig. 3 Generation of dynamics-based e-pharmacophore models from the MD trajectory. The associated active site residues' interactions have been shown. The colour representations for the features are same as Fig. 1

dataset and 1398 Mtb inactive compounds reported in ChEMBL database (molecular weight ranging from 180 to 400, number of heavy atoms ranging from 12 to 27, similar to SAM/SAHC and the 23 inhibitors) were used as the negative dataset. Structures of these molecules were energy minimized and five lowest energy conformers were chosen for each of them. All these conformations were mapped to the 41 e-pharmacophore models using the 'advanced pharmacophore screening' option of Phase. Fast conformational sampling was used during pharmacophore screen, excluding molecules with >15 rotatable bonds. Molecules, which could be mapped to at least four pharmacophoric sites of each model were screened and among several conformers of a molecule the one with the best fitness score (*S*) given by the following equation [46] was retained for each compound. *S* is a measure of volume overlap and extent of match of chemical nature and directionalities of the pharmacophoric features with the corresponding complementary features of the molecules.

$$S = W_{\text{site}}\left(1 - S_{\text{align}}/C_{\text{align}}\right) + W_{\text{vec}}S_{\text{vec}} + W_{\text{vol}}S_{\text{vol}} + W_{\text{ivol}}S_{\text{ivol}}$$

where $W_{\text{site}} = \left(1 - S_{\text{align}}/C_{\text{align}}\right)$, S_{align} = alignment score, C_{align} = alignment cutoff, S_{vec} = vector score, W_{vec} = weight of vector score, S_{vol} $\left(V_{\text{common}}/V_{\text{total}}\right)$ = volume score, W_{vol} = weight of volume score, S_{ivol} = included volume score. Detailed explanations of the components of the fitness score are given in reference 47. Volumes were computed using van der Waals models of all atoms except non-polar hydrogens, and W_{ivol} is the weight of volume score. C_{align}, W_{site}, W_{vec}, W_{vol} and W_{ivol} are user-adjustable parameters, with default values of 1.20, 1.00, 1.00, 1.00 and 0.0, respectively.

Analysis of the hits obtained from these pharmacophore screening showed that most of the models developed from the CmaA1 complexes obtained from the MD trajectories were able to screen up to 17 reported inhibitors (out of 23), while the model developed from the crystal structure could screen only one inhibitor.

The fitness scores of the molecules with the dynamics-based models were also found to be higher. To further confirm our observation, a docking-based virtual screening was parallelly performed with the 41 CmaA1 snapshots and the reported inhibitors. Docking with the MD CmaA1 snapshots not only could bind the most active inhibitors as top scored hits, but also the docking scores were higher than the ones with the crystal structure. These results thus throw light on the effect of including multiple conformations of the targets on the screening abilities of the pharmacophore models. Five out of the 40 dynamic e-pharmacophore models were selected to be further used in our virtual screening study based on the consistency of docking and pharmacophore screening results.

7.1.4 Dynamic Ligand-Based Pharmacophore Models: Construction and Validation

Dynamic ligand-based pharmacophores were developed for the cofactors SAM and SAHC considering their conformational heterogeneity in CmaA1 binding sites as observed from MD trajectories of the respective model systems. Average structures of SAM/SAHC were created after superimposing the conformations obtained from each trajectory using uniform weighting method. Phase module of Schrodinger is used to build the ligand-based pharmacophore models, each comprising six types and 8–11 numbers of chemical features depending on the number and type of interactions with the CmaA1 binding sites. To verify the screening efficiencies of these models, a positive dataset of 23 CmaA1 inhibitors [96] and a negative dataset of 1398 non-inhibitors (the same dataset used to validate the structure-based models described in the previous section) were screened against each of the models. The ligand-based models created using multiple conformations of the cofactors obtained from the MD trajectories could screen up to 22 out of 23 CmaA1 active compounds when the condition for matching was minimum four features of a model. The fitness scores of the inhibitors matching the dynamic-ligand-based pharmacophore models were also higher as compared to the one developed from the conformation of SAHC bound to the crystal structure which was able to match to four CmaA1 inhibitors.

7.1.5 Pharmacophore-Based Virtual Screening

Once the best structure and ligand-based pharmacophore models were validated, they were employed as filters in a novel virtual screening workflow consisting of four different levels of screenings, viz. ligand-based pharmacophore mapping > structure-based pharmacophore mapping > docking > pharmacokinetic properties (ADMET) filters. A focused library of 18,239 molecules from three different sources was used for our virtual screening studies. As the first component of the dataset, 6583 drugs reported in DrugBank were chosen, targeting drug repurposing. The second component of the dataset was a set of 701 molecules which were already reported to be highly active (<1 μM activity) on Mtb cells/

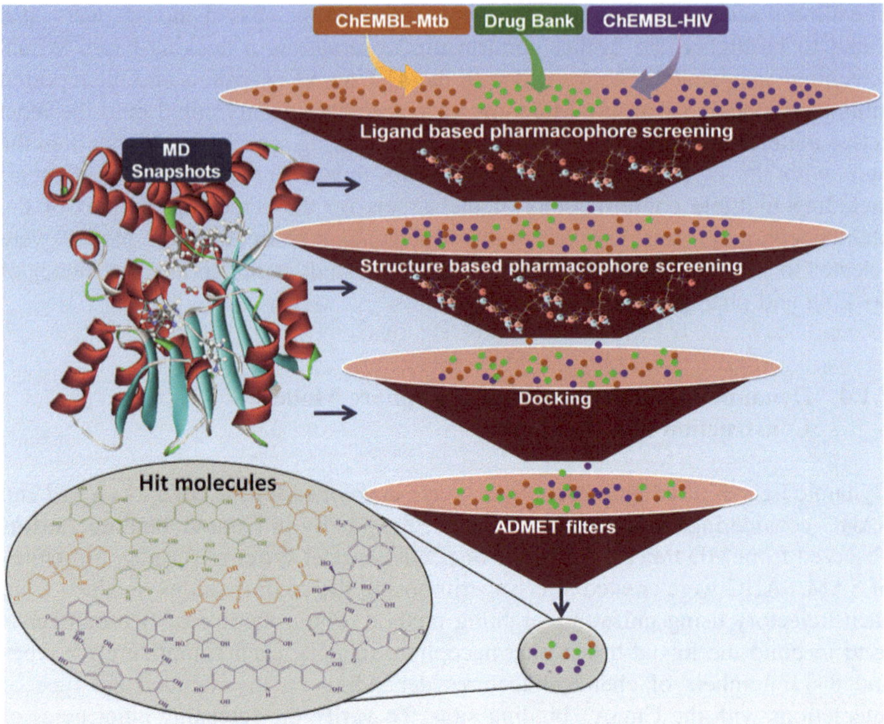

Fig. 4 Virtual screening workflow with structure and ligand-based pharmacophore models

targets and was considered to obtain molecules capable of acting on multiple Mtb targets including CmaA1. The third part of the dataset, i.e. a set of 11,089 highly active anti-HIV molecules (<1 μM activity on HIV cell lines/targets) was taken to screen molecules that can inhibit both Mtb-CmaA1 and HIV simultaneously. After subjecting these three subsets of molecules parallelly through the four screening filters, 12 compounds were obtained as potential anti-CmaA1 hits. As analysed from the Glide XP docking results, all of the identified hits made strong interactions with the important CmaA1 active site residues. Figure 4 shows virtual screening workflow with various levels of filters.

Virtual screening is usually a highly ordered approach combining diverse computational screening methods, where at each consecutive step, the filter criteria become more and more stringent, thus retaining the most promising compounds for experiments. As the steps proceed, the approaches used go on being more thorough and computationally expensive. So, being simple and fast by nature, pharmacophore models are usually implemented at the beginning of a hierarchical protocol to eliminate the compounds which do not even fulfil bare simple spatial and chemical requirements of the query, before subjecting the compound libraries to more complicated and computationally demanding docking calculations.

7.2 *Applications of Pharmacophores in Predicting Pharmacokinetic Properties*

Poor pharmacokinetic properties contribute majorly to failures of many drugs during development and clinical trials. Hence, these properties (also known as ADMET) must be profiled during the early drug discovery process so as to avoid failure at the later stages. Pharmacophore modelling approaches can be of great use for prediction of the ADMET properties. If one can identify the possible interactions made by a group of drug molecules having a well-defined ADMET profile with enzymes involved in drug metabolism, the common interacting features can be captured as pharmacophore models and equivalent features of the query molecules can be matched with the models. The cytochrome P450 (CYP) constitute the major group of enzymes involved in drug metabolism out of which isoenzymes 3A4, 2E1, 2D6, 2C19, 2C9 and 1A2 carry out 90% of the metabolism. Many recent studies report successful implementations [97, 98] of structure-based pharmacophore models trained from the known drugs CYP enzyme interactions to predict the suitability of query molecules to bind to a certain CYP. Also models to assess the probability of chemical alteration of the molecules by a CYP enzyme [99, 100] have been successfully developed and implemented. Inhibitors of the drug clearance enzymes such as the uridine 5′-diphospho-glucuronosyltransferases and transporters like P-glycoprotein/organic cation transporter have also been utilized to build pharmacophore models [101]. Pharmacophore models may also be employed to predict the possibilities of off-target binging of compounds accounting for the side effects, thereby helping design more target-specific compounds [102].

7.2.1 A Case Study with Hexadecahydro-1H-Cyclopenta[a] Phenanthrene Framework (HHCPF)

One of the recent studies from our group [20] reports implementation of ligand-based pharmacophore model features in combination with the QSAR techniques to establish a relationship between the number and type of pharmacophoric feature at a particular position of the core scaffold of a group of drugs with their drug-like properties and target binding affinities. A set of 110 FDA approved drugs containing the Hexadecahydro-1H-Cyclopenta[a]Phenanthrene Framework (HHCPF) (Fig. 5) was considered for the study to understand their structural and functional diversities and target specificities. Analyses of the target information collected from DrugBank, UniProt and PDB show the selectivity of the scaffolds for different targets and vice versa. The substituents present at 17 different positions of the scaffolds were classified as six pharmacophoric features, viz. H-bond donors, H-bond acceptors, aromatic rings, hydrophobic, charged and halogen groups. ADMET (human intestinal absorption, biodegradability, P-glycoprotein binding, carcinogenicity, Caco2 cell permeability, Ames test positivity, blood brain barrier permeability, hERG, CYP450 binding, Rat LD50, etc.)/physicochemical properties

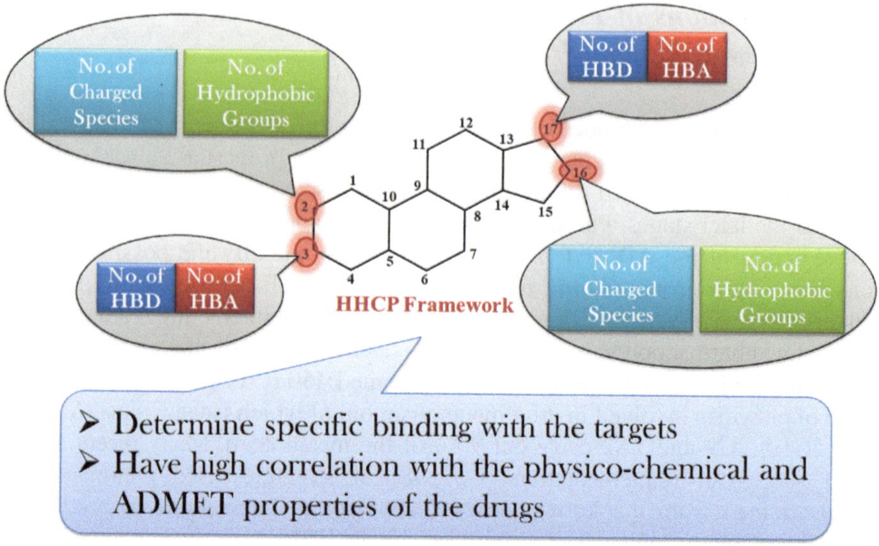

Fig. 5 Important substitution spots on the HHCPF, where number of different pharmacophoric features has a high correlation with target binding and ADMET properties

(polar surface area, polarizability, LogP, refractivity, etc., obtained from DrugBank) of the HHCPF drugs were observed to be highly correlated ($R > 0.8$) to the number and type of these pharmacophoric features at positions 3 and 17 of the framework. The chemical nature of the substitutions at different carbon atoms of the framework was observed to play extensive role in making specific interactions with the active site residues of their respective targets as revealed from analyses of the docking poses. The target binding was found to be greatly influenced by the presence/absence of aromatic rings, HB donors and HB acceptors as substitutions at different positions of the HHCPF scaffolds. Structure-based pharmacophore models were generated from the docked complexes of eight most important HHCPF drugs with their targets which can further be used to screen for new inhibitors. The general observation in the study was that the number and positions of double bonds in the framework regulate the preference of HHCPF drugs for a target class, and the substituents at particular carbon positions account for the target binding patterns and ADMET profiles.

7.3 Target Identification Using Pharmacophore Approaches

Pharmacophore models may also be employed to identify possible targets for active molecules, thereby facilitating the understanding of their mechanism of action. This approach is also proven to be helpful for studies that explore polypharmacology and drug repositioning [103–105]. Firstly, pharmacophore-based fingerprints can be employed to search for similar molecules, whose mechanisms of action are already understood. In the other way around, pharmacophore models can be generated from the active sites of a group of probable proteins involved in the particular disease pathway and then the active molecules can be mapped to them to find out the best fit. The structures of these groups of proteins may be obtained from PDB or models generated using various techniques. The active site pharmacophore mapped with high scores can be proposed as potential targets for the compounds. A study on a group of plant metabolites and pharmacophore models of their possible targets was carried out by Rollinger et al. The best mapping targets were later proven to be accurate by experimental testing, thus validating the usefulness of the pharmacophore mapping approach [106].

7.4 De Novo Ligand Design with Pharmacophores

Apart from acting as a query to screen molecules with features at desired spatial locations and thus possibly prompting a desired biological response, pharmacophore models can also be employed for de novo design, of compounds, satisfying a specific physicochemical constrains. For example, the NEWLEAD method is able to create novel molecules from distinct disconnected fragments (mostly derived from known active ligands) that are consistent with the features of a pharmacophore model by using linkers. The linkers are small connecting fragment may be few atoms, chains or sometimes ring moieties [107]. Software packages like LUDI [108] or BUILDER [109] can grow such novel molecules when the receptor structures are also known. Many other packages also perform such de novo ligand design from the receptor-based pharmacophore features [110, 111]. Thus, pharmacophore models have versatile ways of application for lead generation. De novo design is meant to create entirely novel compounds, while pharmacophore searching screens the available chemical space. However, pharmacophore searching is faster and easier.

8 Limitations of Pharmacophore-Based Approaches

Though the literature is flooded a plenty of successful and reliable applications of pharmacophore-based approaches in rational drug design, its limitations should be cautiously considered as with any method [33, 112]. A systematic or straightforward way of constructing pharmacophore models is not available. This is the case especially with the receptor-based pharmacophore models where many different combinations of features are possible and each model may screen completely different set of molecules [113]. Lack of accuracy in pharmacophore scoring/fitness functions is one of the limitations of pharmacophore searching. So, quality of mapping of a compound with a pharmacophore model which is often given by the RMSD between the feature of a model and atoms of the target molecule does not stand accurate as it does not take an account of similarity with the known active molecules [114]. Especially, the ligand-based pharmacophore models do not consider the overall compatibility with the receptor, thus sometimes end up with screening molecules those are very different from the other active compounds, with a completely different set of functional groups not complementary with the receptor. The pharmacophore-based searches against the compound databases lack fast conformation sampling as most of the programmes rely on conformer databases having only a limited number of energetically favourable conformations of molecules [115, 116]. There is a possibility of missing an active molecule if a suitable conformation is not available. So, it is desirable to generate as many low-energy conformers as possible for the database compounds, but again it would consume a lot of computational time. Especially for the rotatable bonds of small hydroxyl groups, it is difficult to sample all the different rotations.

9 Summary

Evolving from a simple concept to a well-validated and widely exploited method, the pharmacophore modelling approaches have been an essential part of many drug discovery strategies. The pharmacophore-based approaches are well known for their strength to propose a diverse set of molecules having diverse molecular frameworks but owing to a desired biological activity for one target. It has been extensively applied for virtual screening, lead optimization, target identification, toxicity prediction and de novo lead design, and it has ways to go [117]. Considering the strengths and limitations of the pharmacophore approaches, it can either be used alone to identify potential functional group substituents in molecules, design new molecules specific for a target by scaffold hopping keeping the substituents with certain pharmacophoric feature and orientation constant virtually screen for inhibitors, perform ADMET profiling of compounds, investigate possible off-targets or can be applied as a complementing approach along with other methods like docking and QSAR. The concept can be sensibly applied for fragment-based drug design, characterization of protein–protein interaction interfaces and target-based classification of chemical space. In this chapter, we touched upon the basic concepts and methods of generation of pharmacophore models. The diverse applications of the pharmacophore approaches exemplified though a number of case studies are believed to be useful for the readers. However, we believe that the choice and way of application of the method depends on the research problem and the type of initial data available.

Acknowledgements CC and GNS thank the Department of Science and Technology (DST), Government of India, for financial support in the forms of DST-INSPIRE Faculty Award [DST/INSPIRE/04/2016/000732] and JC Bose Fellowship, respectively.

References

1. Myers S, Baker A (2001) Drug discovery—an operating model for a new era. Nat Biotechnol 19:727–730
2. Moses H III, Dorsey ER, Matheson DH et al (2005) Financial anatomy of biomedical research. JAMA 294:1333–1342
3. Lahana R (1999) How many leads from HTS? Drug Discov Today 4:447–448
4. Veselovsky AV, Zharkova MS, Poroikov VV et al (2014) Computer-aided design and discovery of protein-protein interaction inhibitors as agents for anti-HIV therapy. SAR QSAR Environ Res 25:457–471
5. Song CM, Lim SJ, Tong JC (2009) Recent advances in computer-aided drug design. Brief Bioinform 10:579–591
6. Taft CA, Da Silva VB, Da Silva CH (2008) Current topics in computer-aided drug design. J Pharm Sci 97:1089–1098
7. Thiel KA (2004) Structure-aided drug design's next generation. Nat Biotechnol 22:513–519
8. Reddy AS, Amarnath HSD, Bapi RS et al (2008) Protein ligand intreraction database (PLID): datamining analysis of structure-function relationships. Comput Biol Chem 32:387–390
9. Reddy ChS, Vijayasarathy K, Srinivas E et al (2006) Homology modeling of membrane proteins: a critical assessment. Comput Biol Chem 30:120–126
10. Srivastava HK, Sastry GN (2012) A molecular dynamics investigation on a series of HIV protease inhibitors: assessing the performance of MM-PBSA and MM-GBSA approaches. J Chem Inf Model 52:3088–3098
11. Choudhury C, Priyakumar UD, Sastry GN (2014) Molecular dynamics investigation of the active site dynamics of mycobacterial cyclopropane synthase during various stages of the cyclopropanation process. J Struct Biol 187:38–48
12. Badrinarayan P, Sastry GN (2014) Specificity rendering 'hot-spots' for aurora kinase inhibitor design: the role of non-covalent interactions and conformational transitions. PLoS ONE 9:e113773
13. Srivastava HK, Choudhury C, Sastry GN (2012) The efficacy of conceptual DFT descriptors and docking scores on the QSAR models of HIV protease inhibitors. Med Chem 8:811–825
14. Srivani P, Srinivas E, Raghu R et al (2007) Molecular modeling studies of pyridopurinone derivatives—potential phosphodiesterase 5 inhibitors. J Mol Graph Model 26:378–390
15. Janardhan S, RamVivek M, Sastry GN (2016) Modeling the permeability of drug-like molecules through the cell wall of mycobacterium tuberculosis: an analogue based approach. Mol Bio Sys 12:3377–3384
16. Bohari MH, Sastry GN (2012) FDA approved drugs complexed to their targets: evaluating pose prediction accuracy of docking protocols. J Mol Model 18:4263–4274
17. Choudhury C, Priyakumar UD, Sastry GN (2015) Dynamics based pharmacophore models for screening potential inhibitors of mycobacterial cyclopropane synthase. J Chem Inf Model 55:848–860
18. Choudhury C, Priyakumar UD, Sastry GN (2016) Dynamic ligand-based pharmacophore modeling and virtual screening to identify mycobacterial cyclopropane synthase inhibitors. J Chem Sci 128:719–732
19. Reddy AS, Pati SP, Kumar PP et al (2007) Virtual screening in drug discovery—a computational perspective. Curr Protein Pept Sci 8:329–351
20. Choudhury C, Priyakumar UD, Sastry GN (2016) Structural and functional diversities of the hexadecahydro-1H-cyclopenta[a] phenanthrene framework, a ubiquitous scaffold in steroidal hormones. Mol Inform 35:145–157
21. Agrafiotis DK, Bandyopadhyay D, Wegner JK et al (2007) Recent advances in chemoinformatics. J Chem Inf Model 47:1279–1293
22. Vogt M, Bajorath J (2012) Chemoinformatics: a view of the field and current trends in method development. Bioorg Med Chem 20:5317–5323

23. Kapetanovic IM (2008) Computer-aided drug discovery and development (CADDD): in silico-chemico-biological approach. Chem Biol Inter 171:165–176
24. Karelson M, Lobanov VS, Katritzky AR (1996) Quantum-chemical descriptors in QSAR/QSPR studies. Chem Rev 96:1027–1044
25. Gozalbes R, Doucet JP, Derouin F (2002) Application of topological descriptors in QSAR and drug design: history and new trends. Curr Drug Targets Infect Disord 2:93–102
26. Perkins R, Fang H, Tong W et al (2003) Quantitative structure-activity relationship methods: perspectives on drug discovery and toxicology. Environ Toxicol Chem 22:1666–1679
27. Kitchen DB, Decornez H, Furr JR (2004) Docking and scoring in virtual screening for drug discovery: methods and applications. Nat Rev Drug Discov 3:935–949
28. Paul N, Kellenberger E, Bret G et al (2004) Recovering the true targets of specific ligands by virtual screening of the protein data bank. Proteins 54:671–680
29. Kharkar PS, Warrier S, Gaud RS (2014) Reverse docking: a powerful tool for drug repositioning and drug rescue. Future Med Chem 6:333–342
30. Ravindra GK, Srivani P, Achaiah G et al (2007) Strategies to design pyrazolyl urea derivatives for p38 kinase inhibition: a molecular modeling study. J Comput Aided Mol Des 25:155–166
31. Srivastava HK, Bohari M, Sastry GN (2012) Modeling anti-HIV compounds: the role of analogue based approaches. Curr Comput Aided Drug Des 8:224–248
32. Gaur AS, Bhardwaj A, Sharma A et al (2017) Assessing therapeutic potential of molecules: molecular property diagnostic suite for tuberculosis ($MPDS^{TB}$). J Chem Sci 129:515
33. Yang SY (2010) Pharmacophore modeling and applications in drug discovery: challenges and recent advances. Drug Discov Today 15:444–450
34. Braga RC, Andrade CH (2013) Assessing the performance of 3D pharmacophore models in virtual screening: how good are they? Curr Topics Med Chem 13:1127–1138
35. Wolber G, Langer T (2005) LigandScout: 3-D pharmacophores derived from protein-bound ligands and their use as virtual screening filters. J Chem Inf Model 45:160–169
36. Koutsoukas A, Simms B, Kirchmair J et al (2011) From in silico target prediction to multi-target drug design: current databases, methods and applications. J Proteomics 74:2554–2574
37. Ehrlich P (1909) Ueber den jetzigen Stand der Chemotherapie. Ber Dtsch Chem Ges 42:17–47
38. Ehrlich P, Morgenroth J. Über Haemolysine (1900) Dritte Mitteilung. Berl Klin Wochnschr 37:453–457
39. Wermuth CG, Ganellin CR, Lindberg P et al (1998) Glossary of terms used in medicinal chemistry (IUPAC recommendations 1997). Annu Rep Med Chem 33:385–395
40. Güner OF (ed) (2000) Pharmacophore perception, development, and use in drug design, vol 2. Internat'l University Line
41. Bush B, Sheridan RJ (1993) PATTY: a programmable atom type and language for automatic classification of atoms in molecular databases. Chem Inf Comput Sci 33:756–762
42. Greene J, Kahn S, Savoj H et al (1994) Chemical function queries for 3D database search. J Chem Inf Comput Sci 34:1297–1308
43. Molecular Simulations Inc. (MSI), "Catalyst Software". http://www.accelrys.com/about/msi.html
44. Wang T, Zhou JJ (1998) 3DFS: a new 3D flexible searching system for use in drug design. Chem Inf Comput Sci 38:71–77
45. Pickett S, Mason J, McLay IJ (1996) Diversity profiling and design using 3D pharmacophores: pharmacophore-derived queries (PDQ). Chem Inf Comput Sci 36:1214–1223
46. Dixon SL, Smondyrev AM, Knoll EH et al (2006) PHASE: a new engine for pharmacophore perception, 3D QSAR model development, and 3D database screening: 1. Methodology and preliminary results. J Comput Aided Mol Des 20:647–671
47. Schueler FW (1946) Sex hormonal action and chemical constitution. Science 103:221–223
48. Beckett AH (1959) Stereochemical factors in biological activity. In: Fortschritte der Arzneimittel Forschung. Birkhäuser Verlag, Basel, pp 455–530

49. Kier LB (1967) Molecular orbital calculation of preferred conformations of acetylcholine, muscarine, and muscarone. Mol Pharmacol 3:487–494
50. Kier LB (1970) Receptor mapping using mo theory. In: Danielli JF, Moran JF, Triggle DJ (eds) Fundamental concepts in drug-receptor interactions, Academic Press: New York
51. Kier LB (ed) (1971) MO theory in drug research. Academic Press, New York, pp 164–169
52. Evans BE, Rittle KE, Bock MG et al (1988) Methods for drug discovery: development of potent, selective, orally effective cholecystokinin antagonists. J Med Chem 31:2235–2246
53. Thompson LA, Ellman JA (1966) Synthesis and applications of small molecule libraries. Chem Rev 96:555–600
54. Wermuth CG (1998) Search for new lead compounds: the example of the chemical and pharmacological dissection of aminopyridazines. J Heterocycl Chem 35:1091–1100
55. Woods DD, Fildes P (1940) The anti-sulphanilamide activity (in vitro) ofp-aminobenzoic acid and related compounds. Chem Ind 59:133–134
56. Easson LH, Stedman E (1933) Studies on the relationship between chemical constitution and physiological action. V. Molecular dissymmetry and physiological activity. Biochem J 27:1257–1266
57. Peroutka SJ, U'Prichard DC, Greenberg DA et al (1977) Neuroleptic drug interactions with norepinephrine alpha receptor binding sites in rat brain. Neuropharmacology 16:549–556
58. Pullmann B, Coubeils JL, Courrière P et al (1972) Quantum mechanical study of the conformational properties of phenethylamines of biochemical and medicinal interest. J Med Chem 15:17–23
59. Leach AR, Gillet VJ, Lewis RA (2010) Three-dimensional pharmacophore methods in drug discovery. J Med Chem 53:539–558
60. Güner OF (2002) History and evolution of the pharmacophore concept in computer-aided drug design. Curr Top Med Chem 2:1321–1332
61. Maynard AJ (2004) HypoGenRefine and HipHopRefine: pharmacophore refinement using steric information from inactive compounds. Presented at the ACS national meeting, Spring, 2004
62. Jones G, Willett P, Glen R (2000) GASP: genetic algorithm superposition program. In: Pharmacophore perception, development, and use in drug design, vol 2. International University Line, La Jolla, CA, USA, pp 85–106
63. Jones G, Willett P, Glen RC (1995) A genetic algorithm for flexible molecular overlay and pharmacophore elucidation. J Comput Aided Mol Des 9:532–549
64. GALAHAD. Tripos, St. Louis, MO. http://www.tripos.com/
65. Lin A, Overview of pharmacophore applications in MOE. http://www.chemcomp.com/journal/ph4.htm
66. Vlachakis D, Fakourelis P, Makris C, Kossida S (2015) DrugOn: a fully integrated pharmacophore modeling and structure optimization toolkit. PeerJ 3:e725
67. Khedkar SA, Malde AK, Coutinho EC et al (2007) Pharmacophore modeling in drug discovery and development: an overview. Med Chem 3:187–197
68. Langer T, Hoffmann RD (eds) (2006) Pharmacophores and pharmacophore searches, pharmacophores and pharmacophore searches. Wiley-VCH Verlag GmbH & Co. KGaA, Weinheim
69. Finn PW, Kavraki LE, Latombe JC et al (1997) Rapid: randomized pharmacophore identification for drug design. Comput Geom Theor Appl 10:263–272
70. Crandell C, Smith D (1983) Computer-assisted examination of compounds for common three-dimensional substructures. J Chem Inf Comp Sci 23:186–197
71. Martin YC (2000) DISCO: what we did right and what we missed. In: Pharmacophore perception, development, and use in drug design. International University Line, pp 49–68
72. Dolata D, Parrill A, Walters W (1998) CLEW: the generation of pharmacophore hypotheses through machine learning. SAR QSAR Environ Res 9:53–81
73. Chen X, Rusinko A III, Tropsha A et al (1999) Automated pharmacophore Identification for large chemical data sets. J Chem Inf Comput Sci 39:887–896

74. Handschuh S, Wagener M, Gasteiger JJ (1998) Superposition of three-dimensional chemical structures allowing for conformational flexibility by a hybrid method. J Chem Inf Comput Sci 38:220–232
75. Holliday J, Willet P (1997) Using a genetic algorithm to identify common structural features in sets of ligands. J Mol Graph Model 15:203–253
76. Bron C, Kerbosch J (1973) Algorithm 457: finding all cliques of an undirected graph. Commun ACM 16:575–577
77. Brint A, Willett P (1987) Algorithms for the identification of three-dimensional maximal common substructures. J Chem Inf Comp Sci 27:152–158
78. Guilloux VL, Schmidtke P, Tuffery P (2009) Fpocket: an open source platform for ligand pocket detection. BMC Bioinform 10:168–178
79. Schmidtke P, Bidon-Chanal A, Luque FJ et al (2011) MDpocket: open-source cavity detection and characterization on molecular dynamics trajectories. Bioinformatics 27:3276–3285
80. Carlson HA, Masukawa KM, Rubins K et al (2000) Developing a dynamic pharmacophore model for HIV-1 integrase. J Med Chem 43:2100–2114
81. Masukawa KM, Carlson HA, McCammon JA (2000) Technique for developing a pharmacophore model that accommodates inherent protein flexibility: an application to HIV-1 integrase. In: Guner OF (ed) Pharmacophore perception, development, and use in drug design. International University Line
82. Clark DE, Westhead DR, Sykes RA et al (1996) Active-site-directed 3D database searching: pharmacophore extraction and validation of hits. J Comput Aided Mol Des 10:397–416
83. Mahadevi AS, Sastry GN (2013) Cation-p interaction: its role and relevance in chemistry, biology and material science. Chem Rev 113:2100–2138
84. Chourasia M, Sastry GM, Sastry GN (2011) Aromatic—aromatic database, A2ID: an analysis of aromatic networks in proteins. Int J Biol Macromol 48:540–552
85. Saha S, Sastry GN (2015) Cooperative or anticooperative: how noncovalent interactions influence each other. J Phys Chem B 119:11121–11135
86. Badrinarayan P, Choudhury C, Sastry GN (2015) Molecular modeling. In: Dhar PK and Singh V (eds) Systems and synthetic biology (S2B2). Springer Press, pp 93–128
87. Wolber G, Langer T (2005) LigandScout: 3-D pharmacophores derived from protein-bound ligands and their use as virtual screening filters. J Chem Inf Model 45:160–169
88. Salam NK, Nuti R, Sherman W (2009) Novel method for generating structure-based pharmacophores using energetic analysis. J Chem Inf Model 49:2356–2368
89. Friesner R, Murphy RB, Repasky MP et al (2006) Extra precision glide: docking and scoring incorporating a model of hydrophobic enclosure for protein-ligand complexes. J Med Chem 49:6177–6196
90. Carlson HA, Masukawa KM, McCammon JA (1999) Method for including the dynamic fluctuations of a protein in computer-aided drug design. J Phys Chem A 103:10213–10219
91. McGregor MJ, Muskal SM (1999) Pharmacophore fingerprinting. 1. Application to QSAR and focused library design. J Chem Inf Comput Sci 39:569–574
92. McGregor MJ, Muskal SM (2000) Pharmacophore fingerprinting. 2. Application to primary library design. J Chem Inf Comput Sci 40:117–125
93. Voet AR, Kumar A, Berenger F et al (2014) Combining in silico and in cerebro approaches for virtual screening and pose prediction in SAMPL4. J Comput Aided Mol Des 28:363–373
94. Voet A, Helsen C, Zhang KY et al (2013) The discovery of novel human androgen receptor antagonist chemotypes using a combined pharmacophore screening procedure. Chem Med Chem 8:644–651
95. Loving K, Salam NK, Sherman W (2009) Energetic analysis of fragment docking and application to structure-based pharmacophore hypothesis generation. J Comput Aided Mol Des 23:541–554
96. Anuradha A, Trivelli X, Guérardel Y et al (2007) Thiacetazone, an antitubercular drug that inhibits cyclopropanation of cell wall mycolic acids in mycobacteria. PLoS ONE 12:e1343

97. Guner OF, Bowen JP (2013) Pharmacophore modeling for ADME. Curr Top Med Chem 13:1327–1342
98. Yamashita F, Hashida M (2004) In silico approaches for predicting ADME properties of drugs. Drug Metab Pharmacokinet 19:327–338
99. de Groot MJ, Ekins S (2002) Pharmacophore modeling of cytochromes P450. Adv Drug Deliv Rev 54:367–383
100. Ekins S, de Groot MJ, Jones JP (2001) Pharmacophore and three-dimensional quantitative structure activity relationship methods for modeling cytochrome p450 active sites. Drug Metab Dispos 29:936–944
101. Sorich MJ, Miners JO, McKinnon RA et al (2004) Multiple pharmacophores for the investigation of human UDP-glucuronosyltransferase isoform substrate selectivity. Mol Pharmacol 65:301–308
102. Hu Y, Bajorath J (2010) Polypharmacology directed compound data mining: identification of promiscuous chemotypes with different activity profiles and comparison to approved drugs. J Chem Inf Model 50:2112–2118
103. Keiser MJ, Roth BL, Armbruster BN (2007) Relating protein pharmacology by ligand chemistry. Nat Biotechnol 25:197–206
104. Koutsoukas A, Simms B, Kirchmair J et al (2011) From in silico target prediction to multi-target drug design: current databases, methods and applications. J Proteomics 74:2554–2574
105. Xu Y, Liu X, Li S (2013) Combinatorial pharmacophore modeling of organic cation transporter 2 (OCT2) inhibitors: insights into multiple inhibitory mechanisms. 10:4611–4619
106. Rollinger JM, Schuster D, Danzl B et al (2009) In silico target fishing for rationalized ligand discovery exemplified on constituents of ruta graveolens. Planta Med 75:195–204
107. Tschinke V, Cohen NJ (1993) The NEWLEAD program: a new method for the design of candidate structures from pharmacophoric hypotheses. Med Chem 36:3863–3870
108. Bohm HJ (1992) The computer program LUDI: a new method for the de novo design of enzyme inhibitors. J Comput Aided Mol Des 6:61–78
109. Roe D, Kuntz IJ (1995) BUILDER v.2: improving the chemistry of a de novo design strategy. J Comput Aided Mol Des 9:269–282
110. Joseph-McCarthy D (1999) Computational approaches to structure-based ligand design. Pharmacol Ther 84:179–191
111. Schneider G, Bohm HJ (2002) Virtual screening and fast automated docking methods. Drug Discov Today 7:64–70
112. Scior T, Bender A, Tresadern G et al (2012) Recognizing pitfalls in virtual screening: a critical review. J Chem Inf Model 52:867–881
113. Vancraenenbroeck R, De Raeymaecker J, Lobbestael E et al (2014) In silico, in vitro and cellular analysis with a kinome-wide inhibitor panel correlates cellular LRRK2 dephosphorylation to inhibitor activity on LRRK2. Front Mol Neurosci 7:51
114. Schomburg KT, Bietz S, Briem H et al (2014) Facing the challenges of structure-based target prediction by inverse virtual screening. J Chem Inf Model 54:1676–1686
115. Kirchmair J, Wolber G, Laggner C et al (2006) Comparative performance assessment of the conformational model generators omega and catalyst: a large-scale survey on the retrieval of protein-bound ligand conformations. J Chem Inf Model 46:1848–1861
116. Kirchmair J, Laggner C, Wolber G et al (2005) Comparative analysis of protein-bound ligand conformations with respect to catalyst's conformational space subsampling algorithms. J Chem Inf Model 45:422–430
117. Nagamani S, Gaur AS, Tanneeru K et al (2017) Molecular property diagnostic suite (MPDS): development of disease-specific open source web portals for drug discovery. SAR QSAR Environ Res https://doi.org/10.1080/1062936x.2017.1402819

Analysis of Protein Structures Using Residue Interaction Networks

Dmitrii Shcherbinin and Alexander Veselovsky

Abstract The network description is widely used to analyze the topology and the dynamics of complex systems. Residue interaction network (RIN) represents three-dimensional structure of protein as a set of nodes (residues) with their connections (edges). Calculated topological parameters from RIN correlate with various aspects of protein structure and function. Here, we reviewed the applications of RIN for the analysis and prediction of functionally important residues and ligand binding sites, protein–protein interactions, allosteric regulation, influence of point mutations on structure and dynamics of proteins.

Keywords Residue interaction network · RIN · Protein–protein interactions Allosteric regulation · Scoring function · Allosteric pathway

Abbreviations

CAPRI	Critical assessment of predicted interactions
DDN	Differential network
GPCR	G protein-coupled receptor
HPNCscore	Hydrophobic and polar networks combined scoring function
MD	Molecular dynamics simulation
NACEN	Node-weighted amino acid contact energy network
PPI	Protein–protein interaction
RIN	Residue interaction network
SVM	Support vector machine

D. Shcherbinin · A. Veselovsky (✉)
Laboratory of Structural Bioinformatics, Institute of Biomedical Chemistry,
Pogodinskaya Str., 10, Moscow, Russia
e-mail: veselov@ibmh.msk.su

C. G. Mohan (ed.), *Structural Bioinformatics: Applications in Preclinical Drug Discovery Process*, Challenges and Advances in Computational Chemistry and Physics 27, https://doi.org/10.1007/978-3-030-05282-9_3

1 Introduction

Proteins play a vital role in biological systems and have numerous functions such as catalysts, transporters, regulators of signal transduction. They are linear heteropolymers folded into three-dimensional structures. The amino acid residues interact through various covalent and non-covalent bonds in a specific manner to obtain a particular three-dimensional structure, which determines their functions. Knowledge of the relationship between protein structure and its function is important in drug design, molecular medicine, and biotechnology.

Different computational methods have been used for investigations of protein structures and their functions, finding functionally important residues, prediction protein–protein interactions, discovering new biological active compounds. In the most approaches, the protein structures have been viewed as linear sequences of amino acid residues packed into 3D globules. In the last decade, an alternative view of proteins structures has emerged that describe the protein spatial structure as network of amino acids residues interaction.

Network analysis has successfully used in different fields, such as social networks [1], Internet networks [2], road networks [3]. In biology, this method is widely used for analysis of networks of gene regulation, protein–protein interaction, metabolites flow, prediction of drug side effects, etc., [4–9]. Applying network methodology for polypharmacology was reviewed in [10].

A network method is based on the graph theory and includes a set of entities (nodes) and of the relationships (edges) occurring among them. These nodes and edges can have various attributes. Depending on the object of the study, nodes can represent genes, proteins, small compounds, and edges connecting these nodes represent the physical interactions, genetic regulatory, or other properties linking the nodes. Edges can have additional information, such as weights, directions.

According to the structure of protein, every amino acid residue in it is considered to be a "node" or "vertex," and the interaction of residues represents "edge" (Fig. 1). The existence of an edge between two nodes depends only on their spatial position in protein globule and has no relation to position in their primary sequence. The interaction can be represented as distance between C_α or any other atoms of amino acid residues, non-covalent interaction (electrostatic, hydrophobic, H-bonds) of the particular amino acids [11]. Additionally, in residue interaction network (RIN), the energy of interaction between residues can be used for weighting the edges [12, 13]. Proteins can be also modeled as subnetworks of amino acid residues having similar physiochemical properties. RIN method reduces spatial protein architectures to simple maps including nodes (residues) and edges (inter-residue interactions). Analysis of these graphs yields a characterization of the protein's topology and network characteristics.

There are several names of the resultant intraprotein amino acid residue interaction networks. They are called residue interaction graphs [14], protein structure graphs [15, 16], protein residue networks [17], protein contact networks [18], protein energy networks [13], amino acid networks [19], protein structure networks

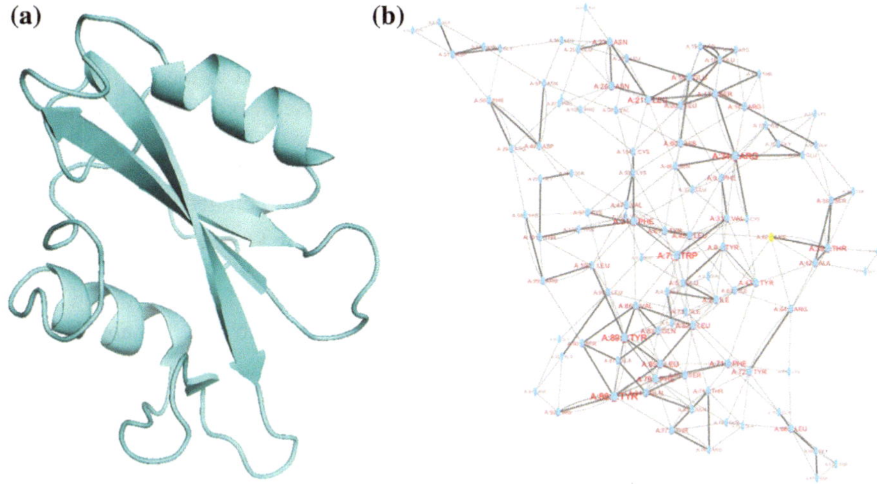

Fig. 1 Structure of SH2 domain of proto-oncogene tyrosine-protein kinase SRC (PDB ID 1o41) in cartoon (A) and RIN representation

[20], residue interaction networks [21]. In this review, we will use the residue interaction networks (RINs) to distinguish it from network of protein–protein interactions.

The application of RIN method in drug design is just at a beginning. RINs have been used to analyze protein stability and folding [22, 23], 3D structure modeling [19, 23], finding functionally important amino acid residues and sites [14, 24], analyzed protein–protein interactions [25], allosteric regulation [26], influence of amino acid mutations [27]. These studies showed that RIN method is valuable approaches allowed to improve the drug discovery process. Recently, several reviews on RINs have been published [28–31].

Herein, we aim to review the investigation of the construction, analysis, and application of RINs in fields related to drug design.

2 Graph Theory and Residue Interaction Network

Graph theory represents complex system as a set of elements (called *vertices* or *nodes*) with their connections (called *edges*). Each node can be connected to each other through multiple edges. Adding order of nodes in the graph, we get a *directed graph*, where edges are directed and usually represented as arrows. Introduction of the quantitative characteristics of the edges results in a *weighted graph*. Nodes with edges form a network. The network representation helps to analyze the interaction among individual elements and to characterize the whole system.

Residue interaction network is constructed on the base of the three-dimensional atomic coordinates of protein structure and consists of nodes and edges. Each node represents amino acid residue (or C_α atom) that is connected to the neighbor node. In the simplest variant, the edges are defined on the base of predefined cutoff of the distances in 3D structure between nodes. The values of distance may be varied based on nature of interactions (van der Waals, hydrophobic, electrostatic interactions, etc.). Frequently, the covalent backbones are included as edges in the networks. The edges can be weighed based on energy of interactions, knowledge-based potentials, or amino acid fluctuations in molecular dynamics simulation [30, 31]. The differential network (DDN) method was proposed where network formed by unique edges that are present only in one state but are absent in other ones [32].

Networks have several most common characteristics; some of them that frequently have been used for analysis of biological systems are listed below [28, 31, 33].

A degree of a node is a number of edges in a network that connect node with its neighbors. In a directed network, there might be two types of degrees, the in-degree, and the out-degree depending on the orientation of the edges. *An average degree* is the average number of connections that the nodes have in a network.

A connectivity represents a minimum number edges that need to be removed to make a disconnected graph. The connectivity structure and the degree of nodes analysis in RINs help to identify important residues, i.e., participating in ligand binding sites.

A shortest path is a path in which the two nodes are connected by the smallest number of intermediate nodes. A *characteristic path length* is defined as the number of edges in the shortest path between two nodes, averaged over all pairs of nodes. Residues with small shortest path lengths are often located in the active or ligand binding sites of proteins [17] and participate in allosteric pathways [34, 35].

A betweenness centrality of a node is the number of times that a node is included in the shortest path between each pair of nodes, normalized by the total number of pairs.

A closeness centrality of a node is the reciprocal of the average shortest path length.

The network concept is widely used to analyze and predict properties in different biological systems, from intramolecular interaction to whole cells and organisms. Biological networks are small worlds that means that two nodes are connected to each other via only a few other nodes [23, 30]. There are several network parameters for characterizing different aspects of biological networks.

A *hub* is defined as a node with a high degree or connectivity in a network. Hubs may play a structural role in proteins increasing the thermodynamic stability of proteins [14, 36].

A *cluster* is a set of nodes with the number of connections, which is higher than in the other nodes. Clusters often are equivalent to a domain of protein and participate in intramolecular interactions.

A *clique* is a set of nodes in which each node of graph is connected to every other node. Studies of cliques can help to understand ligand-induced population shift in protein [37].

There are several software packages, Web servers, and plug-ins available for construction and analyzing of RINs, such as Xpyder (http://xpyder.sourceforge.net/) [38], Network View [39], RING (http://protein.bio.unipd.it/ring/) [21, 40], RINalyzer (http://www.rinalyzer.de) [41], structureViz (http://www.cgl.ucsf.edu/cytoscape/structureViz/) [42].

Web server RING constructs physicochemically RINs from PDB files for subsequent visualization in the Cytoscape (software platform for the analysis and visualization of biological networks) (http://www.cytoscape.org) or Pymol (https://pymol.org/). Interactions (edges) are disulfide bonds, salt bridges, hydrogen bonds, aromatic interactions, and van der Waals contacts. Several features can be added to nodes and edges, such as secondary structure, solvent accessibility, energy score, sequence conservation. Subnetwork can be also constructed.

RINalyzer and structureViz are plug-ins for Cytoscape [43] that link Cytoscape with the molecular viewer UCSF Chimera (http://www.cgl.ucsf.edu/chimera/) [44]. They allow interactive structure analysis of RINs together with the corresponding 3D protein structure.

NetworkView plug-in for VMD (https://www.ks.uiuc.edu/Research/vmd/) allows to study allostery and signaling through network models. This plug-in can display the dynamical network representations.

3 RINs Application

3.1 Ligand Binding Sites

Identification of the ligand binding sites of proteins and functionally important residues is a crucial first step in drug design. However, it is a difficult task in the case of the absence of homologous proteins.

Several topological parameters of RINs may be used for the prediction of ligand binding sites. Several investigations showed that closeness and betweenness values of residues are correlated with ligand binding site residues [14, 34, 45–48]. The accuracy of prediction such residues may be improved by combining with such parameters as their solvent accessibility. So, Amitai et al. [14] could predict active site residues in 70% of the analyzed 178 enzymes proteins, using closeness centrality and solvent accessibility parameters. The similar result was obtained in [49]. The closeness centrality was used as parameter in machine learning methods for prediction of functionally important residues [50] or in score for docking [25].

However, for non-enzyme proteins correlation between closeness centrality and binding sites has not observed [34, 51]. In addition, global closeness centrality gave unsatisfactory result for non-globular and oligomer proteins. For such proteins, more tolerable prediction was obtained with local closeness [52]. It seems that the ligand binding sites in enzymes are correlated with centrality due to their typical location in cavities of the enzymes, whereas in oligomer proteins, the protein–protein interfaces are more flat [53], which reduces the centrality of their residues.

Coevolution residues networks, which include information about coevolved residues, were also used for predicting functionally important residues [54, 55]. RIN analysis was applied for prediction similarity of ligand binding sites in different proteins [56, 57].

The node-weighted RIN, called node-weighted amino acid contact energy network (NACEN) was developed for prediction hotspots, catalytic residues, and allosteric residues. Nodes were weighted based on structural, sequence, physicochemical and dynamic properties of the residues. SVM was used for design model to identify functionally important residues. The results revealed that parameters from node-weighted RIN have advantages over ones from unweighted network [58].

Poirrette et al. [56] designed RIN of the influenza sialidase binding site of Zanamivir and used it to predict proteins having the similar binding sites. Such an approach may be used for repurposing drugs or prediction of side effects.

3.2 Protein–Protein Interactions

Protein–protein interactions (PPIs) are crucial for many biological processes and functions; inhibition of PPIs with small molecules is a perspective way in drug design [53]. RIN method was used for analysis of protein–protein interfaces, prediction of hotspots, and selection of protein poses in the protein–protein docking.

Several investigations were done using RIN for analysis of protein–protein interfaces. They showed that hydrophobic and charged residues are predominant in the dimer interface and that arginine, histidine, glutamic acid, phenylalanine, and tyrosine are located in clusters at the interface [59, 60]. In those clusters, highly connected residues correlate with experimentally identified hotspots in the protein complexes [15, 16, 61, 62].

Correct prediction of protein–protein complexes using individual proteins by docking method is a big challenge, since the docking gives many false-positive solutions [63, 64]. Protein–protein complex formation may be viewed as combining of two RINs, where additional edges have appeared between nodes from different subunits. The interaction of residues occurs in accordance with their properties. Since native protein–protein complexes are far from random, the correct and incorrect poses have different topologies.

Chang et al. [65] designed hydrophobic and hydrophilic RINs of a protein–protein complex. Three terms based on these networks (degree, clustering coefficient, and characteristic path length) were calculated and used in network-based scoring function HPNet. Combining it with energy terms of RosettaDock [66] results in new combined scoring function HPNet-combine. It was found that HPNet-combine could improve the discrimination of the RosettaDock scoring function.

The similar methodology based on the construction of a hydrophobic and hydrophilic RINs of protein–protein complexes was used for the development NPPD scoring function [67]. Protein–protein docking, HoDock, and scoring function HPNCscore (hydrophobic, and polar network combined scoring function) were developed. It showed good results for several targets in Critical Assessment of PRedicted Interactions (CAPRI) rounds [68].

The weighed RINs were used for the development of Sn scoring function [69]. Two weighted parameters (strength and weighted average nearest neighbors' degree) were introduced to develop a scoring function. The testing of this scoring function for 42 protein–protein complexes had shown a satisfied performance.

The scoring function based on the local network patterns, iScore, was proposed [70]. It achieved 83.6% specificity with 82% sensitivity for training set of ~1800 two domain proteins, homo- and heterodimers.

3.3 Allosteric Regulation

Allosteric regulation is a common mechanism to control the protein activities. The perturbation at the allosteric site results in transmission of signal through the protein structure to other sites leading to modification of catalytic activity, oligomerization, etc. [71, 72].

Allosteric sites became attractive target for drug design at last decade. Allosteric drugs have several potential benefits over orthosteric drugs. They may be more specific due to less similarity of allosteric sites comparing to active site in homologous proteins; they can increase or decrease the activity of enzymes and receptors; partially inhibiting by allosteric drugs may cause less side effects [73, 72].

Using allosteric sites for drug design, it is required to predict allosteric sites, residues involved in signal transduction pathways to the active sites. The search of allosteric sites by RIN method is similar to the other sites described above.

Allosteric pathways show how the signal may be transmitted over a long distance from allosteric to active sites within the protein. RIN is accurate and not time-consuming method for prediction such pathways.

Once the RIN constructed, several algorithms can be used to find allosteric pathways within the RINs. The common method is to find the shortest paths connecting the allosteric and active sites [34, 35, 74]. The shortest path may be determined by Floyd–Warshall algorithm. It was shown that many proteins may be considered as a set of modules (subgraphs with many interconnections and with few connections to other subgraphs). The residues involved in the interaction of such modules can participate in allosteric pathways [75]. It is proposed that such residues are conservative that also may be used for their prediction [76–78]. Proteins can have multiple allosteric pathways, which may preexist without effector binding at allosteric site [79]. Various pathways may be involved depending on the different changes in allosteric site.

However, RINs constructed based on a single structure do not take into account the structural changes in protein globule. Therefore, the combination of molecular dynamics simulation (MD) followed by RINs design frequently has been used to detect and to analyze allosteric pathways. In these cases, the edges in RINs are defined using various parameters obtained from MD. The edges may reflect the correlation of displacements of the residues [74, 80], the fluctuation of distances [81], interaction energy [82], etc.

Aminoacyl-tRNA synthetases are convenient objects for analysis of allosteric communication. The combination of MD with RIN was used for discovering pathways from anticodon region to the aminoacylation region for methionyl-tRNA synthetase [74, 83], glutaminyl-tRNA synthetase [84], cysteinyl-tRNA synthetase [35], and tryptophanyl-tRNA synthetase [85, 86]. Particularly, analysis of tryptophanyl-tRNA synthetase showed changes of flexibility around the active site induced by allosteric ligands binding and allowed to explain the molecular mechanism of half-of-the-sites reactivity (tryptophanyl-tRNA synthetase is a homodimer).

Another popular object is G protein-coupled receptors (GPCRs) [87–89]. It is a large family of membrane receptors, which have ligand binding site on the extra-cellular side of membrane and activation domain on its internal side. Using RIN method, several conservative residues participating in the signal transduction were discovered for the lutropin receptor [76] and A_{2A} adenosine receptor [87] (Fig. 2).

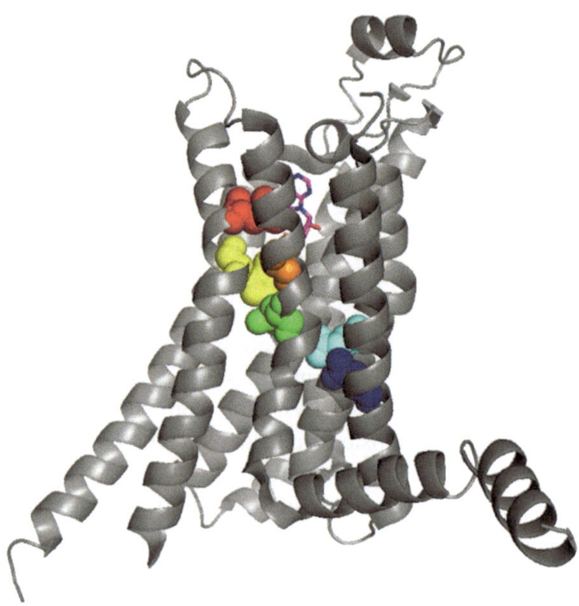

Fig. 2 Structure of A_{2A} adenosine receptor (PDB ID 2ydv). One of the predicted allosteric pathways is shown in rainbow color scheme. The synthetic agonist NECA is in stick

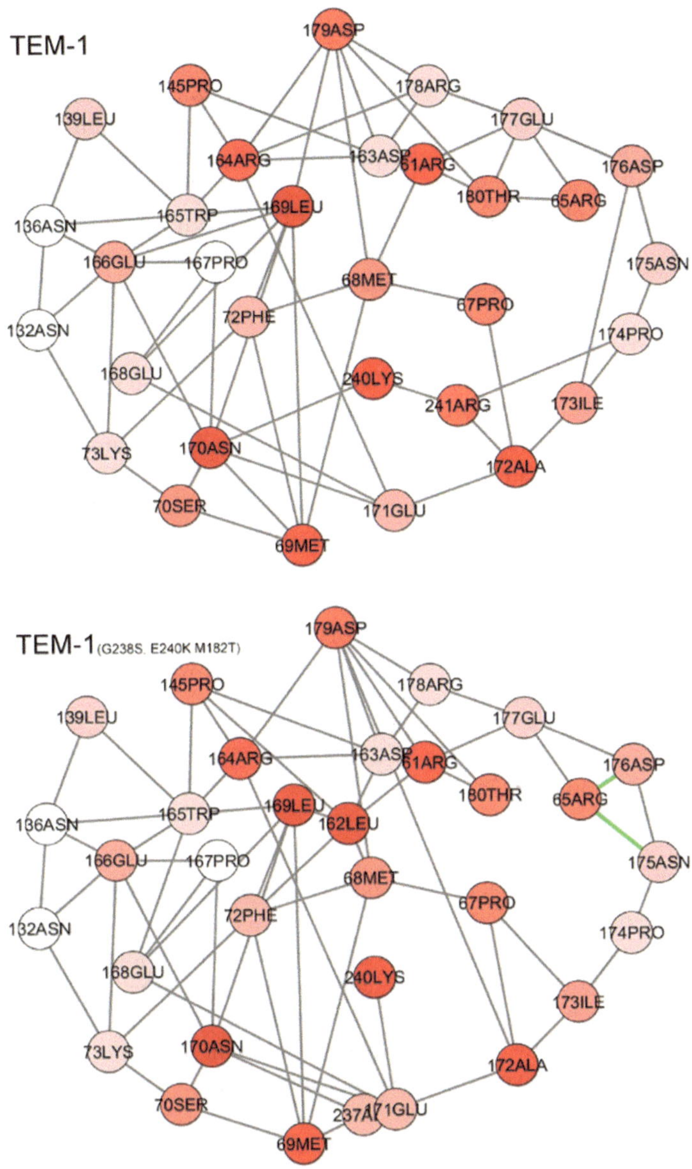

Fig. 3 Part of the networks near Ω-loop of β-lactamases TEM-1 and its triple mutant (G238S, E240K, M182T). The additional interactions appeared in the triple mutant that results in freeze of movement of Ω-loop are in green

3.4 Analyses of Mutations

RIN methods may be used for analysis and prediction of effects of amino acid mutation on protein properties, which may be useful for protein design, investigations of disease-associated single nucleotide polymorphisms, or mechanism of the drug resistance [27, 90–92].

Recently, we used RIN for investigation of the influence of several mutations on structure and flexibility of β-lactamase [93]. β-lactamases are class of enzymes responsible for bacteria resistant to β-lactam antibiotics. Besides, the key mutations, responsible for the extended spectrum β-lactamases or inhibitor resistance phenotype, secondary mutations, located far from active site and with a weak impact on the protein structure and enzyme activity, have been often appeared [94]. Analysis of MD trajectories showed that the secondary mutations, and the key mutations can exhibit opposite effect on the flexibility of the Ω-loop of β-lactamase that participate in antibiotic hydrolysis and transport in the active site [93]. Detailed analysis of RIN maps of proteins of consistent mutations from wild-type TEM-1 to TEM-72 (carrying two key mutations G238S and E240K and two secondary ones M182T and Q39K) showed that key mutations (responding for extended spectrum β-lactamases) lead to weakening interactions of the Ω-loop with protein globule. The appearance of secondary mutation M182T resulted in dramatic changing of conformation of R65, and this residue began to interact with the Ω-loop and fixed it near protein globule (manuscript submitted) (Fig. 3).

4 Conclusion

Herein, we have reviewed the development and current stage of RINs and their application for drug discovery.

RINs provide complex analysis of the proteins and their complexes. Residues are in tight contact with each other in protein globules, and RINs allowed to estimate their interdependence and to predict different properties and functionality of the individual residues and the whole proteins. In addition to topology, RINs allow to use chemicophysical properties of residues and energy of their interaction in RIN construction and analysis of proteins.

Besides, using RINs for investigation protein structure and functions, they may be applied in drug design in several ways.

Prediction of functionally important residues and sites can be helpful for understanding functions and regulation of uncharacterized proteins, finding active sites, allosteric and cryptic ligand binding sites. It may decrease the amount of "undruggable" protein, increasing field for drug design. On the other hand, many drug candidates fail in the late and costly stages of clinical trials [95]. Side effects are one of the main reasons for drug failure [96]. The detection of similarity in network topologies and interactions with ligands for several targets may indicate the promiscuity of drug candidates and possibly their side effects.

The development of inhibitors of protein–protein interactions is a perspective way in drug design, and RIN showed their applicability for this purpose. The analysis of networks may help to select correct poses in protein–protein docking that is important for the selection of inhibitor binding sites; incorporation of the terms from RINs may improve docking scoring functions.

Allosteric inhibitors are another mainstream in drug design in last decade. It is proposed that such inhibitors may regulate cellular processes more accurately. Allosteric regulation is the common property of protein, which may increase the number of druggable targets. RINs are convenient for finding allosteric sites, investigation of mechanism of intraprotein signal transmission. Prediction of the effect of amino acid mutations on protein structure and dynamics is crucial for the development drugs against diseases with a high probability of occurrence drug resistance, in particular antibacterial, antiviral, and anticancer drugs.

Nowadays, the application of RIN methods for drug discovery is at their early stage, but they already help to understand intimate properties of proteins and provide a new view for drug discovery.

References

1. Otte E, Rousseau R (2002) Social network analysis: a powerful strategy, also for the information sciences. J Inform Sci 28:441–453
2. Meusel R, Vigna S, Lehmberg O, Bizer C (2015) The graph structure in the web – analyzed on different aggregation levels. J Web Sci 1:33–47
3. Bottinelli A, Louf R, Gherardi M (2017) Balancing building and maintenance costs in growing transport networks. Phys Rev E 96:032316
4. Murakami Y, Tripathi LP, Prathipati P, Mizuguchi K (2017) Network analysis and in silico prediction of protein-protein interactions with applications in drug discovery. Curr Opin Struct Biol 44:134–142
5. Zhao B, Wang J, Wu FX (2017) Computational methods to predict protein functions from protein-protein interaction networks. Curr Protein Pept Sci 18:1120–1131
6. Miryala SK, Anbarasu A, Ramaiah S (2018) Discerning molecular interactions: a comprehensive review on biomolecular interaction databases and network analysis tools. Gene 642:84–94
7. Laddach A, Ng JC, Chung SS, Fraternali F (2018) Genetic variants and protein-protein interactions: a multidimensional network-centric view. Curr Opin Struct Biol 50:82–90
8. Yao V, Wong AK, Troyanskaya OG (2018) Enabling precision medicine through integrative network models. J Mol Biol 430(18 Pt A):2913–2923
9. Xie L, Li J, Xie L, Bourne PE (2009) Drug discovery using chemical systems biology: identification of the protein-ligand binding network to explain the side effects of CETP inhibitors. PLoS Comput Biol 5:e1000387
10. Li P, Fu Y, Wang Y (2015) Network based approach to drug discovery: a mini review. Mini-Rev Med Chem 15:687–695
11. Aftabuddin M, Kundu S (2007) Hydrophobic, hydrophilic, and charged amino acid networks within protein. Biophys J 93:225–231

12. Bhattacharyya M, Vishveshwara S (2011) Probing the allosteric mechanism in pyrrolysyl-tRNA synthetase using energy-weighted network formalism. Biochemistry 50:6225–6236
13. Vijayabaskar MS, Vishveshwara S (2010) Interaction energy based protein structure networks. Biophys J J99:3704–3715
14. Amitai G, Shemesh A, Sitbon E, Shklar M, Netanely D, Venger I, Pietrokovski S (2004) Network analysis of protein structures identifies functional residues. J Mol Biol 344:1135–1146
15. Brinda KV, Vishveshwara S (2005) A network representation of protein structures: implications for protein stability. Biophys J J89:4159–4170
16. Brinda KV, Vishveshwara S (2005) Oligomeric protein structure networks: insights into protein-protein interactions. BMC Bioinform 6:296
17. Atilgan AR, Akan P, Baysal C (2004) Small-world communication of residues and significance for protein dynamics. Biophys J J86:85–91
18. Bagler G, Sinha S (2007) Assortative mixing in protein contact networks and protein folding kinetics. Bioinformatics 23:1760–1767
19. Zhou J, Yan W, Hu G, Shen B (2014) Amino acid network for the discrimination of native protein structures from decoys. Curr Protein Pept Sci 15:522–528
20. Hu G, Zhou J, Yan W, Chen J, Shen B (2013) The topology and dynamics of protein complexes: insights from intra–molecular network theory. Curr Protein Pept Sci 14:121–132
21. Martin AJ, Vidotto M, Boscariol F, Di Domenico T, Walsh I, Tosatto SC (2011) RING: networking interacting residues, evolutionary information and energetics in protein structures. Bioinformatics 27:2003–2005
22. Rao F, Caflisch A (2004) The protein folding network. J Mol Biol 342:299–306
23. Grewal RK, Roy S (2015) Modeling proteins as residue interaction networks. Protein Pept Lett 22:923–933
24. Zhou J, Yan W, Hu G, Shen B (2016) Amino acid network for prediction of catalytic residues in enzymes: a comparison survey. Curr Protein Pept Sci 17:41–51
25. Pons C, Glaser F, Fernandez-Recio J (2011) Prediction of protein-binding areas by small-world residue networks and application to docking. BMC Bioinform 12:378
26. Schueler-Furman O, Wodak SJ (2016) Computational approaches to investigating allostery. Curr Opin Struct Biol 41:159–171
27. Cheng TMK, Lu Y-E, Vendruscolo M, Lio P, Blundell TL (2008) Prediction by graph theoretic measures of structural effects in proteins arising from non-synonymous single nucleotide polymorphisms. PLoS Comput Biol 4:e1000135
28. Di Paola L, De Ruvo M, Paci P, Santoni D, Giuliani A (2013) Protein contact networks: an emerging paradigm in chemistry. Chem Rev 113:1598–1613
29. Csermely P, Korcsmáros T, Kiss HJ, London G, Nussinov R (2013) Structure and dynamics of molecular networks: a novel paradigm of drug discovery: a comprehensive review. Pharmacol Ther 138:333–408
30. Yan W, Zhou J, Sun M, Chen J, Hu G, Shen B (2014) The construction of an amino acid network for understanding protein structure and function. Amino Acids 46:1419–1439
31. Bhattacharyya M, Ghosh S, Vishveshwara S (2016) Protein structure and function: looking through the network of side-chain interactions. Curr Protein Pept Sci 17:4–25
32. Grewal RK, Mitra D, Roy S (2015) Mapping networks of light-dark transition in LOV photoreceptors. Bioinformatics 31:3608–3616
33. Doncheva NT, Assenov Y, Domingues FS, Albrecht M (2012) Topological analysis and interactive visualization of biological networks and protein structures. Nat Protoc 7:670–685
34. del Sol A, Fujihashi H, Amoros D, Nussinov R (2006) Residues crucial for maintaining short paths in network communication mediate signaling in proteins. Mol Syst Biol 2:0019
35. Ghosh A, Sakaguchi R, Liu C, Vishveshwara S, Hou YM (2011) Allosteric communication in cysteinyl tRNA synthetase: a network of direct and indirect readout. J Biol Chem 286:37721–37731
36. Estrada E (2010) Universality in protein residue networks. Biophys J 98:890–900

37. Ghosh A, Vishveshwara S (2008) Variations in clique and community patterns in protein structures during allosteric communication: investigation of dynamically equilibrated structures of methionyl tRNA synthetase complexes. Biochemistry 47:11398–11407
38. Pasi M, Tiberti M, Arrigoni A, Papaleo E (2012) xPyder: a PyMOL plugin to analyze coupled residues and their networks in protein structures. J Chem Inf Model 52:1865–1874
39. Eargle J, Luthey-Schulten Z (2012) NetworkView: 3D display and analysis of protein· RNA interaction networks. Bioinformatics 28:3000–3001
40. Piovesan D, Minervini G, Tosatto SC (2016) The RING 2.0 web server for high quality residue interaction networks. Nucleic Acids Res 44(W1):W367–W374
41. Doncheva NT, Klein K, Domingues FS, Albrecht M (2011) Analyzing and visualizing residue networks of protein structures. Trends Biochem Sci 36:179–182
42. Morris JH, Huang CC, Babbitt PC, Ferrin TE (2007) structureViz: linking Cytoscape and UCSF Chimera. Bioinformatics 23:2345–2347
43. Shannon P, Markiel A, Ozier O, Baliga NS, Wang JT, Ramage D, Amin N, Schwikowski B, Ideker T (2003) Cytoscape: a software environment for integrated models of biomolecular interaction networks. Genome Res 13:2498–2504
44. Pettersen EF, Goddard TD, Huang CC, Couch GS, Greenblatt DM, Meng EC, Ferrin TE (2004) UCSF Chimera—a visualization system for exploratory research and analysis. J Comput Chem 25:1605–1612
45. Yan Y, Zhang SG, Wu FX (2011) Applications of graph theory in protein structure identification. Proteome Sci 9(Suppl 1):S17
46. Thibert B, Bredesen DE, del Rio G (2005) Improved prediction of critical residues for protein function based on network and phylogenetic analyses. BMC Bioinform 6:213
47. Emerson IA, Gothandam KM (2012) Residue centrality in alpha helical polytopic transmembrane protein structures. J Theor Biol 309:78–87
48. Tse A, Verkhivker GM (2015) Molecular dynamics simulations and structural network analysis of c-Abl and c-Src kinase core proteins: capturing allosteric mechanisms and communication pathways from residue centrality. J Chem Inf Model 55:1645–1662
49. Chea E, Livesay DR (2007) How accurate and statistically robust are catalytic site predictions based on closeness centrality? BMC Bioinform 8:153
50. Tang YR, Sheng ZY, Chen YZ, Zhang Z (2008) An improved prediction of catalytic residues in enzyme structures. Protein Eng Des Sel 21:295–302
51. Sheftel S, Muratore K, Black M, Costanzi S (2013) Graph analysis of β2 adrenergic receptor structures: a “social network” of GPCR residues. Silico Pharmacol 1:16
52. Slama P, Filippis I, Lappe M (2008) Detection of protein catalytic residues at high precision using local network properties. BMC Bioinform 9:517
53. Veselovsky AV, Archakov AI (2007) Inhibitors of protein-protein interactions as potential drugs. Curr Comput-Aided Drug Des 3:51–58
54. Marino Buslje C, Teppa E, Di Domenico T, Delfino JM, Nielsen M (2010) Networks of high mutual information define the structural proximity of catalytic sites: implications for catalytic residue identification. PLoS Comput Biol 6:e1000978
55. Aguilar D, Oliva B, Buslje CM (2012) Mapping the mutual information network of enzymatic families in the protein structure to unveil functional features. PLoS ONE 7:e41430
56. Poirrette AR, Artymiuk PJ, Grindley HM, Rice DW, Willett P (1994) Structural similarity between binding sites in influenza sialidase and isocitrate dehydrogenase: implications for an alternative approach to rational drug design. Protein Sci 3:1128–1130
57. Liu ZP, Wu LY, Wang Y, Zhang XS, Chen L (2008) Analysis of protein surface patterns by pocket similarity network. Prot Pept Lett 15:448–455
58. Yan W, Hu G, Liang Z, Zhou J, Yang Y, Chen J, Shen B (2018) Node-weighted amino acid network strategy for characterization and identification of protein functional residues. J Chem Inf Model. (in press). https://doi.org/10.1021/acs.jcim.8b00146
59. Brinda KV, Kannan N, Vishveshwara S (2002) Analysis of homodimeric protein interfaces by graph-spectral methods. Protein Eng 15:265–277

60. Reichmann D, Rahat O, Albeck S, Meged R, Dym O, Schreiber G (2005) The modular architecture of protein–protein binding interfaces. PNAS 102:57–62
61. Brinda KV, Surolia A, Vishveshwara S (2005) Insights into the quaternary association of proteins through structure graphs: a case study of lectins. Biochem J J391:1–15
62. Kannan N, Chander P, Ghosh P, Vishveshwara S, Chatterji D (2001) Stabilizing interactions in the dimer interface of alpha-subunit in Escherichia coli RNA polymerase: a graph spectral and point mutation study. Protein Sci 10:46–54
63. Soni N, Madhusudhan MS (2017) Computational modeling of protein assemblies. Curr Opin Struct Biol 44:179–189
64. Zhang Q, Feng T, Xu L, Sun H, Pan P, Li Y, Li D, Hou T (2016) Recent advances in protein-protein docking. Curr Drug Targets 17:1586–1594
65. Chang S, Jiao X, Li CH, Gong XQ, Chen WZ, Wang CX (2008) Amino acid network and its scoring application in protein–protein docking. Biophys Chem 134:111–118
66. Gray JJ, Moughon S, Wang C, Schueler-Furman O, Kuhlman B, Rohl CA, Baker D (2003) Protein–protein docking with simultaneous optimization of rigid-body displacement and side-chain conformations. J Mol Biol 33(1):281–299
67. Shih ESC, Hwang M-J (2015) NPPD: a protein-protein docking scoring function based on dyadic differences in networks of hydrophobic and hydrophilic amino acid residues. Biology 4:282–297
68. Gong X, Wang P, Yang F, Chang S, Liu B, He H, Cao L, Xu X, Li C, Chen W, Wang C (2010) Protein-protein docking with binding site patch prediction and network-based terms enhanced combinatorial scoring. Proteins 78:3150–3155
69. Jiao X, Chang S (2011) Scoring function based on weighted residue network. Int J Mol Sci 12:8773–8786
70. Luo Q, Hamer R, Reinert G, Deane CM (2013) Local network patterns in protein-protein interfaces. PLoS ONE 8:e57031
71. Greener JG, Sternberg MJ (2018) Structure-based prediction of protein allostery. Curr Opin Struct Biol 50:1–8
72. Nussinov R, Tsai CJ (2013) Allostery in disease and in drug discovery. Cell 153:293–305
73. Lu S, Li S, Zhang J (2014) Harnessing allostery: a novel approach to drug discovery. Med Res Rev 34:1242–1285
74. Ghosh A, Vishveshwara S (2007) A study of communication pathways in methionyl-tRNA synthetase by molecular dynamics simulations and structure network analysis. PNAS 104:15711–15716
75. del Sol A, Arauzo-Bravo MJ, Amoros D, Nussinov R (2007) Modular architecture of protein structures and allosteric communications: potential implications for signaling proteins and regulatory linkages. Genome Biol 8:R92
76. Angelova K, Felline A, Lee M, Patel M, Puett D, Fanelli F (2011) Conserved amino acids participate in the structure networks deputed to intramolecular communication in the lutropin receptor. Cell Mol Life Sci 68:1227–1239
77. Süel GM, Lockless SW, Wall MA, Ranganathan R (2003) Evolutionarily conserved networks of residues mediate allosteric communication in proteins. Nat Struct Biol 10:59–69
78. Tang S, Liao JC, Dunn AR, Altman RB, Spudich JA, Schmidt JP (2007) Predicting allosteric communication in myosin via a pathway of conserved residues. J Mol Biol 373:1361–1373
79. del Sol A, Tsai CJ, Ma B, Nussinov R (2009) The origin of allosteric functional modulation: multiple pre-existing pathways. Structure 17:1042–1050
80. Sethi A, Eargle J, Black AA, Luthey-Schulten Z (2009) Dynamical networks in tRNA: protein complexes. PNAS 106:6620–6625
81. Dixit A, Verkhivker GM (2011) Computational modeling of allosteric communication reveals organizing principles of mutation-induced signaling in ABL and EGFR kinases. PLoS Comput Biol 7:e1002179
82. Kong Y, Karplus M (2009) Signaling pathways of PDZ2 domain: a molecular dynamics interaction correlation analysis. Proteins 74:145–154

83. Vishveshwara S, Ghosh A, Hansia P (2009) Intra and inter-molecular communications through protein structure network. Curr Protein Pept Sci 10:146–160
84. Sathyapriya R, Vishveshwara S (2007) Structure networks of E-coli glutaminyl-tRNA synthetase: effects of ligand binding. Proteins 68:541–550
85. Bhattacharyya M, Ghosh A, Hansia P, Vishveshwara S (2010) Allostery and conformational free energy changes in human tryptophanyl-tRNA synthetase from essential dynamics and structure networks. Proteins 78:506–517
86. Hansia P, Ghosh A, Vishveshwara S (2009) Ligand dependent intra and inter subunit communication in human tryptophanyl tRNA synthetase as deduced from the dynamics of structure networks. Mol Bio Syst 5:1860–1872
87. Fanelli F, Felline A (2011) Dimerization and ligand binding affect the structure network of A_{2A} adenosine receptor. Biochim Biophys Acta 1808:1256–1266
88. Lee Y, Choi S, Hyeon C (2014) Mapping the intramolecular signal transduction of G-protein coupled receptors. Proteins 82:727–743
89. Miao Y, Nichols SE, Gasper PM, Metzger VT, McCammon JA (2013) Activation and dynamic network of the M2 muscarinic receptor. PNAS 110:10982–10987
90. Hu Z, Bowen D, Southerland WM, del Sol A, Pan Y, Nussinov R, Ma B (2007) Ligand binding and circular permutation modify residue interaction network in DHFR. PLoS Comput Biol 3:1097–1107
91. Li Y, Wen Z, Xiao J, Yin H, Yu L, Yang L, Li M (2011) Predicting disease-associated substitution of a single amino acid by analyzing residue interactions. BMC Bioinform 12:14
92. Tse A, Verkhivker GM (2015) Small-world networks of residue interactions in the Abl kinase complexes with cancer drugs: topology of allosteric communication pathways can determine drug resistance effects. Mol Biosyst 11:2082–2095
93. Shcherbinin DS, RubtsovaMYu Grigorenko VG, Uporov IV, Veselovsky AV, Egorov AM (2017) The study of the role of mutations M182T and Q39K in the TEM-72 β-lactamase structure by the molecular dynamics method. Biochem (Moscow), Suppl B: Biomed Chem 11:120–127
94. Grigorenko VG, RubtsovaMYu Uporov IV, Ishtubaev IV, Andreeva IP, Shcherbinin DS, Veselovsky AV, Egorov AM (2018) Bacterial TEM-type serine beta-lactamases: structure and analysis of mutations. biochemistry (Moscow). Suppl B: Biomed Chem 12:87–95
95. Nwaka S, Hudson A (2006) Innovative lead discovery strategies for tropical diseases. Nat Rev Drug Discov 5:941–955
96. Scheiber J, Chen B, Milik M, Sukuru SC, Bender A, Mikhailov D, Whitebread S, Hamon J, Azzaoui K, Urban L, Glick M, Davies JW, Jenkins JL (2009) Gaining insight into off-target mediated effects of drug candidates with a comprehensive systems chemical biology analysis. J Chem Inf Model 49:308–317

Combinatorial Drug Discovery from Activity-Related Substructure Identification

Md. Imbesat Hassan Rizvi, Chandan Raychaudhury and Debnath Pal

Abstract A newly developed drug discovery method composed of graph theoretical approaches for generating structures combinatorially from an activity-related root vertex, prediction of activity using topological distance-based vertex index and a rule-based algorithm and prioritization of putative active compounds using a newly defined Molecular Priority Score (MPS) has been described in this chapter. The rule-based method is also used for identifying suitable activity-related vertices (atoms) present in the active compounds of a data set, and identified vertex is used for combinatorial generation of structures. An algorithm has also been described for identifying suitable training set–test set splits (combinations) for a given data set since getting a suitable training set is of utmost importance for getting acceptable activity prediction. The method has also been used, to our knowledge for the first time, for matching and searching rooted trees and sub-trees in the compounds of a data set to discover novel drug candidates. The performance of different modules of the proposed method has been investigated by considering two different series of bioactive compounds: (1) convulsant and anticonvulsant barbiturates and (2) nucleoside analogues with their activities against HIV and a data set of 3779 potential antitubercular compounds. While activity prediction, compound prioritization and structure generation studies have been carried out for barbiturates and nucleoside analogues, activity-related tree–sub-tree searching in the said data set has been carried for screening potential antitubercular compounds. All the results show a high level of success rate. The possible relation of this work with scaffold hopping and inverse quantitative structure–activity relationship (iQSAR) problem has also been discussed. This newly developed method seems to hold promise for discovering novel therapeutic candidates.

Keywords Graph theory · Vertex index of molecular graph · Root vertex
Combinatorial molecular structure generation · Activity prediction
Compound prioritization and screening · Drug discovery

Md.I. H. Rizvi · C. Raychaudhury · D. Pal (✉)
Department of Computational and Data Sciences, Indian Institute of Science, Bangalore 560012, India
e-mail: dpal@iisc.ac.in

C. G. Mohan (ed.), *Structural Bioinformatics: Applications in Preclinical Drug Discovery Process*, Challenges and Advances in Computational Chemistry and Physics 27, https://doi.org/10.1007/978-3-030-05282-9_4

Abbreviations

QSAR	Quantitative structure–activity relationship
iQSAR	Inverse quantitative structure–activity relationship
vHTS	Virtual high-throughput screening
MIC	Minimum inhibitory concentration
Mtb	*Mycobacterium tuberculosis*
AAE	Acid alkyl ester
NA	Nucleoside analogue
HIV	Human immunodeficiency virus
MPS	Molecular Priority Score
ARL	Active range length
ARW	Active range weight
ARV	Active range value
MAI	Molecular activity index
IRL	Inactive range length
IRW	Inactive range weight
IRV	Inactive range value
MDI	Molecular de-activity index
SMILES	Simplified molecular-input line-entry system
MOL file	Molecular structural information file

1 Introduction

Exploring chemical space to discover a compound that elicits a desired pharmacologic response without undesired side effect is like searching a needle in a haystack problem. The problem arises because we seek to screen a limited subset that exists among many compounds that elicit a desired pharmacologic response. Different approaches have therefore evolved to make the problem tractable, namely effective use of macromolecular target information, if available, use synthesis tractability of the compounds as guidance, and most importantly, the pharmacological relevance of the compounds selected. While modern advances like targeted library search or chemogenomics have helped in bringing focus to the drug candidate search, the utility of drug candidate search using serendipity-based approaches has not diminished in face of increasing burden of drug resistance and adverse side effects. These problems may possibly be addressed by discovering novel compounds using new drug discovery methods. One of such a new line of thinking has been proposed by Ruddigkeit et al. [1] who have considered all possible compounds having 17 atoms taken from C, N, O, S and halogens to create a database of several billions of compounds. It is tempting to believe that such an effort of discovering novel drug molecules from such a huge collection of compounds can be useful. However, a method that enables searching of potential drug candidates from a relatively smaller set of compounds, quite exhaustive at the same time within given limits, activity linked and rationally guided too may help drug discovery more effectively.

Among the current drug discovery methods, data modelling and quantitative–qualitative prediction of activity [2–4], use of molecular docking methods and scoring functions for virtual high-throughput screening (vHTS) [5] and 3D quantitative structure–activity relationship (QSAR) studies [6] are some of the most used ones. At the same time, combinatorial generation of chemical compounds is also carried out since it increases the possibility of finding novel drug molecules from a large number of chemically diverse compounds generated particularly for the need of making scaffold hopping [7]. It also provides the opportunity to search for compounds having diverse structural characteristics which in turn may help decipher the role of molecular components which may be responsible for the biological activities of new drug molecules, particularly in situations where novel therapeutic candidates are sought for to handle the challenges arising out of drug resistance problem [8].

So far generating molecular structures are concerned, molecular topology-based approaches are in use for generating and designing molecular structures [9, 10] and graph theory [11] and graph theoretical methods [12] have been suitably used for doing that. However, in general these methods are used for generating structures combinatorially [10] with no connection to their biological activities and a separate method has to be used for the prediction of molecular properties and activities. It appears, therefore, that a method that generates a large number of compounds combinatorially and gets linked to their activities at the same time may be more efficient in designing and discovering novel drug molecules. In particular, topological molecular descriptors [2] can be useful in this regard. Moreover, if this is done using a single molecular (structural/substructural) descriptor, the process may also be looked upon from inverse QSAR (iQSAR) point of view [13] since the basic idea of doing iQSAR studies is to get molecular structures back from molecular descriptor which has been used for activity prediction. In this context, it seems reasonable to explore whether a method can be developed that is integrated in such a way that it can be used for generating structures combinatorially that would have molecules of diverse scaffold from a single molecular topological descriptor , can be used for predicting molecular properties/activities and can be used for compound prioritization and screening to help discover potential drug candidates.

So, the first question that may be asked in developing such an integrated method is: Can we have a method such that structures can be generated combinatorially from structural or substructural information that is already related to activity? In this regard, there are two primary aspects in designing potential bioactive compounds from activity-related substructural information—(1) identification of activity-related vertices using a suitable method; (2) a method that can be used for structure generation using topological information associated with such vertices. One of the most useful activity-related substructure identification method was proposed by Klopman [14] where molecular fragments of different length are identified from active and inactive compounds, and the fragments are weighed on the basis of the number of fragments obtained from active and inactive compounds using a suitable measure to assess their usefulness in predicting activities and mathematical–statistical methods are used to do that. However, no structure generation method is used for this work [14].

In this chapter, we have described in detail a graph theory-based method, developed recently by our research group [15], for combinatorial generation of chemical structures from activity-related substructural topological information. This approach [15] has been found to be useful in generating structures of active antitubercular compounds from activity-related vertices of the molecular graphs representing different other active antitubercular compounds. For developing the present method [15], we have leveraged primarily a non-isomorphic rooted tree generation algorithm [16] and a cycle enumeration method [17] to design novel bioactive compounds in the form of reconstructed molecular graph as outlined earlier [18, 19]. In the proposed integrated method, activity-related vertices are first identified by using the rule-based method [18, 19] where topological distance-based vertex indices are used as local molecular descriptors in data sets having the biological activities of interest. Once the activity-related vertices are identified, a suitable vertex is taken for structure generation using the distance distribution associated with the vertex which gives the topological distances of all the vertices in molecular graph from that vertex (say, the root vertex). A large number of rooted trees are thus generated de novo [15]. Subsequently, 2D molecular structures containing cycles of different size are created by joining vertices of the tree graphs. In this way, all the generated structures contain this activity-related substructure, and therefore, there is a possibility that some of generated structures may be classified as active. Furthermore, to get complete 2D structures of the compounds, user-defined parameters are used to add multiplicity of bonds (e.g. double and triple bonds) between pairs of vertices and add chemical nature of the atoms (nitrogen, oxygen, etc.) represented by the vertices. Canonicalization is used to identify unique structures which are further used for screening of potential active compounds.

It may be noted that scaffold hopping [7] is embedded in the method since the generated structures are different from the starting compound and are expected to have diverse topological architecture. Also, since both compound generation and activity prediction are done using the same vertex index (substructural/local descriptor), the method may also be regarded as an attempt to address the inverse quantitative structure–activity relationship (iQSAR) problem [13] in its integrated framework. Furthermore, in order to relax the condition for structure generation from distance distribution as outlined earlier [18, 19] and to make it more flexible, we have developed an algorithm for generating sub-trees by adding or deleting vertices from the tree structures generated on the basis of a given distance distribution associated with an activity-related vertex. To our knowledge, this is the first time that a method [15] has been developed and used for drug discovery through database searching using rooted tree and sub-tree matching algorithms.

The method has already been used to investigate its usefulness for a series of 41 acid alkyl ester (AAE) derivatives and three known antitubercular drugs [15]. In this chapter, we have furnished new results obtained for a series of 19 convulsant and anticonvulsant barbiturates [18], 20 nucleoside analogues (NA) for their activities against HIV [20, 21], and a data set of 3779 compounds (named GTB data set) for which minimum inhibitory concentration (MIC) values have been measured against H37Rv strain of *Mycobacterium tuberculosis* (*Mtb*) [22]. The GTB data set may be obtained from the link [23] given in the reference section. The results described here will therefore substantiate the findings obtained earlier [15]. Regarding activity prediction, results have been reported for NA and barbiturate data sets. For barbiturates data set, we have considered the same training set and test set as used in an earlier study [18]. However, for the NA data set, we have identified a reasonably well-performing training set–test set split and have reported the results for individual compounds present in that split. For prioritization of the generated active compounds that help screen potential active compounds, Molecular Priority Score (MPS) [15] has also been used and the results obtained for NA and barbiturate series of compounds have been given in the tables alongside their activity prediction results. We have carried out combinatorial generation of structures using topological distance-based substructural information associated with identified activity-related vertices (atoms) in some compounds of the data set. We have been able to reconstruct the structures of active NA and barbiturate compounds from the substructural information associated with activity-related vertices of other active NA and barbiturate compounds. Regarding substructure searching exercise, we have reported identified potential active compounds from GTB data set [22, 23] considering activity-related atoms (vertices) in the structures of Isoniazid and Streptomycin, both of which are known antitubercular drugs in use.

It appears from the outcome of the results that the integrated method would find a place as a useful drug discovery tool for designing and discovering novel bioactive compounds. In particular, the method is believed to be of much help in situations where novel drug candidates having very different structural characteristics/scaffolds are sought for particularly to overcome the drug resistance problem.

2 Methods

In this section, we have described in detail different mathematical approaches/tools which have been used to develop the present integrated drug discovery method and the related computer programs. Examples with tables and figures have been used to illustrate underlying concepts of the methods used. While we have leveraged few existing mathematical aspects for the present purpose, we have introduced some new algorithms as well.

2.1 Computation of Vertex Index

Let G be the carbon skeleton of n-butane and $D(G)$, the corresponding distance matrix is shown in Fig. 1. Computation of D^{-4} indices for the vertices of $D(G)$ has been illustrated below.

Fig. 1 Graph G representing vertex labelled carbon skeleton of n-butane and the corresponding topological distance matrix $D(G)$

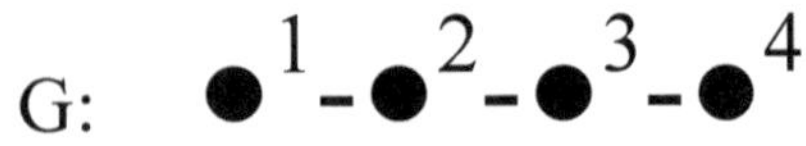

D(G):

	1	2	3	4
1	0	1	2	3
2	1	0	1	2
3	2	1	0	1
4	3	2	1	0

Therefore, D^{-4} index for the four vertices v_i, $i = 1, 2, \ldots, 4$ of G may be computed as:

$$D^{-4}(v_1) = 1^{-4} + 2^{-4} + 3^{-4} = 1.0749$$
$$D^{-4}(v_2) = 1^{-4} + 1^{-4} + 2^{-4} = 2.0625$$
$$D^{-4}(v_3) = 1^{-4} + 1^{-4} + 2^{-4} = 2.0625$$
$$D^{-4}(v_4) = 1^{-4} + 2^{-4} + 3^{-4} = 1.0749$$

One can, therefore, compute the values of D^{-4} index for all the atoms (vertices) of all the compounds (molecular graphs) in a data set considering the molecular graphs (hydrogen-suppressed or hydrogen-filled) of the compounds. Hydrogen-suppressed graphs may be considered for generating structures from the distance distribution associated with a vertex since structure generation using information about the vertices of hydrogen-filled graphs may pose computational bottlenecks during the process because of a large number of structures that are usually generated in this way. Moreover, if chemical information of the vertices is provided, one can always create the hydrogen-filled graphs from the corresponding hydrogen-suppressed graphs.

2.2 Rule-Based Activity Prediction

In order to carry out activity prediction studies using the present method, a data set containing both active and inactive compounds for a biological endpoint of interest is gathered. The data set is then divided suitably into a training set and a test set. The biological activities of the compounds are then predicted for both the training set and the test set using a rule-based system [18, 19]. In order to make the activity prediction, ranges of vertex index values coming from active and inactive compounds are first found out using some rules [18, 19] and the activity is predicted on the basis of the number of vertex index values falling in these ranges as defined in the rule-based system [18, 19]. For the present purpose, the values of vertex index D^{-4} are computed for the vertices of the training set compounds (molecular graphs). Once the indices are computed, they are arranged in an ascending order and ranges of values coming from both active and inactive compounds are found in the ordering and are tagged as "Active" and "Inactive" ranges by applying certain rules [18, 19] given below:

1. Three or more consecutive vertex index values coming exclusively from active compounds and exclusively from inactive compounds are said to form an "active range" and an "inactive range", respectively. However, at least three index values in a range have to be distinct if they come from the same compound and at least two index values in a range have to be distinct if they come from different compounds.
2. Some single vertex index value coming from both active and inactive compounds is not considered to form an "active range" or "an inactive range" by itself or along with other vertex index values unless two-thirds of that single vertex index comes from active compounds or inactive compounds, respectively.

It has been discussed earlier [24] in connection with identifying ranges that the vertices which correspond to the vertex index values forming active ranges may be regarded as topological features responsible for making the compounds active. In other words, they may be regarded as a set of features forming "Topological Biophore " which are responsible for exhibiting a given biological activity of the compound under consideration. From this point of view, it may be said that if the index values of some (or, all) of the vertices of a compound fall in active ranges, then those vertices may be regarded as forming certain topological biophore which make the compound active. Presumably, some of the vertex index values of a compound may fall in inactive ranges as well. Thus, in order to predict activity from the occurrences of the vertex index values in active and inactive ranges, another set of rules [18, 19], given below, are applied:

A compound is predicted "ACTIVE" if all or some of its vertices fall:

1. Only in active ranges or
2. In both active and inactive ranges, the number of index values falling in active ranges is greater than those falling in inactive ranges.

Otherwise, the compound is predicted "inactive".

In order to use this rule-based system for activity prediction, a set of bioactive compounds with known activities (e.g. experimentally determined activities) have to be collected (from the literature or an experimental laboratory). A training set is then formed by picking up compounds from the data set suitably to train the system to learn the structural requirement for a compound to be active. A fewer number of compounds are also kept for testing purposes (test set). Once the training is done, activity predictions for both training set compounds (retrofit studies) and test set compounds are carried out. For predicting the activities of the test set compounds, the D^{-4} index values for the test set compounds are computed. If the system is found to produce high (acceptable) percentage of correct activity predictions for both the training set and the test set compounds along with none or very few (acceptable) wrong activity predictions, it may be regarded as standardized for the prediction of activity of chemical compounds for the biological endpoint for which the system is standardized.

2.3 Training Set–Test Set Split

It is always important that a suitable training set be obtained from a data set of bioactive compounds such that the structural characteristics of the compounds, present in the data set, is reflected in the training set, and the learning of the (expert) system/prediction tool is as adequate as possible for getting useful activity predictions by the method used in this purpose. In general, researchers look for the diversity present in the structures in creating a training set from a given data set. Presumably, some intuition or expertise of the drug designer/medicinal chemist may be required to do that or some mathematical diversity analysis may be carried out in obtaining a suitable training set. However, it appears that generating a large number (e.g. 1000) of training set–test set splits (combinations) and reporting the successful predictions of all or some (e.g. top 20, 25) of the best-predicting splits for a given data set of bioactive compounds would be a very straightforward and useful approach for identifying a suitable training set. Having obtained various top performing splits, one can select a suitable split that gives high percentage of successful predictions for both training set and test set and obtains activity prediction for the compounds present in both the sets. Although such splits have been used [24, 25] for evaluating the performance of vertex indices and a rule-based method for activity prediction [18, 19] considering small and large data sets, no algorithm is available to report the activity predictions for different splits. We have incorporated this algorithm in the program for reporting the outcome of activity predictions for different splits so that one can consider a suitable split for further work such as structure generation. This can be done for both quantitative data and qualitative data (active–inactive type). It may also be noted that the computer program can be used for the identification of training set–test set splits and activity predictions by considering both hydrogen-filled (H-filled) and hydrogen-suppressed (H-suppressed) molecular graphs of the compounds under consideration.

2.4 Compound Prioritization

The present method [15] also contains a section that can be used for prioritization of potentially active compounds. This may be particularly useful for screening few highly active compounds from a big database, e.g. from a set of combinatorially generated compounds (described in the next section). This method is based on some of the characteristics of active and inactive ranges found in the ordering of vertex index values. Therefore, one has to look into some details of such ranges. In doing that, two factors may be given special attention—(1) the number of vertex index values in an active range (active range length: *ARL*); (2) the number of compounds contributing to form the range (active range weight: *ARW*). By applying one's intuition too, it becomes apparent that a joint effect of these two factors may help prioritize predicted active compounds. Therefore, we first propose a measure, active range value (*ARV*), as the algebraic sum of *ARL* and *ARW* values given by:

$$ARV = (ARL + ARW) \tag{1}$$

Clearly, a range larger in length and contributed by more number of compounds in forming the range would have higher *ARV* value. We define such a range of higher *ARV* value a "STRONGER" range compared to those which have lower *ARV* values. Now, let us assume that *M* out of *N* vertices of a molecular graph *G* (representing a chemical compound) have fallen in different active ranges. If the vertices are denoted by $v_1, v_2, \ldots, v_M$, one would get *M* number of *ARV* measures as $ARV(v_1), ARV(v_2), \ldots, ARV(v_M)$. In order to get a measure of the contribution of the vertices falling in different active ranges (i.e. contribution of activity-related vertices), we further propose a molecular activity index (*MAI*) as:

$$MAI(G) = \sum_{i=1}^{M} ARV(v_i) \tag{2}$$

It may also be noted that while considering the length of an active range and the number of compounds contributing to form the range, some single values that come from both active and inactive compounds are taken into account since they are part of the active range according to the second rule of range selection mentioned earlier.

At the same time, there is a possibility that some of the vertex indices of molecular graph *G* may fall in inactive ranges too (the second rule for activity prediction) and that may be considered to pose a negative effect on the activity of the compound. For the prediction purpose, therefore, vertices falling in inactive ranges have to be considered. For doing that, let us assume that M' vertices of *G*, viz. $u_1, u_2, \ldots, u_{M'}$ fall in inactive ranges. We, thus, propose a measure, molecular de-activity index (*MDI*) for *G* and it may be defined as:

$$MDI(G) = \sum_{j=1}^{M'} IRV(u_j) \tag{3}$$

In Eq. 3, *IRV* stands for inactive range value and is the sum of *IRL* (inactive range length) and *IRW* (inactive range weight) which is in line with the definitions used for such measures of active ranges. Computation of *IRV* can be done using Eq. (4) given below:

$$IRV = (IRL + IRW) \tag{4}$$

Therefore, by considering a combined effect of *MAI* and *MDI*, one can prioritize the newly generated active compounds and curate some high-ranking compounds for further studies. Thus, in order to get a measure of combined effect of the vertices falling in active ranges and inactive ranges (if any) and prioritizing (ranking) the compounds according to their activities, we propose a measure, Molecular Priority Score (*MPS*), for *G* and it may be computed using Eq. (5):

$$MPS(G) = MAI(G) - MDI(G) \tag{5}$$

Considering MPS value as a measure for prioritization of active compounds, a compound with higher *MPS* value will occupy a higher position in the ranking. Therefore, a compound may be regarded as more active if it gets higher *MPS* value. This will then help screen some top-ranking compounds. However, ranking of active compounds using *MPS* is not mandatory. One may always wish to consider all the predicted active compounds for further studies particularly if the number of highly ranked compounds (in terms of MPS value) is very small. At the same time, there is no need to prioritize those compounds which are predicted inactive since the idea is to screen potentially highly active compounds for a given biological endpoint.

2.5 Combinatorial Structure Generation from Root Vertex

In developing the structure generation method, we have used an algorithm for generating rooted trees [16] which have been extended to the generation of cyclic compounds and finally a complete 2D structure of chemical compounds. The structure generation exercise starts off as generating all possible canonical trees for any given number of vertices. Subsequently, topological distance restriction on the generated tree structures is used to filter and keep only those trees having a desired distance distribution. Further, for the application of relaxed distance criteria for compound structures having increased or decreased number of vertices (non-hydrogen atoms), the matching criteria of distance distribution have been suitably changed to accommodate the addition, deletion and migration of the vertices over the tree structures with exact distance restriction. The theories and implementation details are described in the following subsections.

2.5.1 Structure for a Given Distance Distribution

A molecular graph represents topological connections between the atoms of the molecules. A spanning tree of the graph can provide the basic skeleton over which additional edges can be inserted to introduce cycles and thereby produce the entire molecular structure. The multiplicity of bonds can be considered as edge weights and can be dealt by assigning weights 1, 2 and 3 for single, double and triple bonds, respectively. Similarly, heterogeneous atoms, with their valency information, can also be introduced as nodes, which are by default considered to be carbon atoms in our discussions.

It is clear from above that the starting point of structure generation for a given number of vertices (atoms) is the generation of rooted trees since the structure generation will be carried out with respect to a particular atom in a molecule in our current approach based on topological distances from a particular vertex. Moreover, to prevent duplicate structures, only non-isomorphic trees should be generated.

For the purpose of illustration, consider the chemical structure and the corresponding graphical and tree representation as shown in Fig. 2.

The numbering of vertices has no structural significance apart from that it is done to obtain the rightmost tree having node 1 as the root and pre-order numbering for the other vertices and is merely for array representation of the tree structure. The tree can be represented by the following parent and level array representations:

$$\text{parent} = [0, 1, 2, 3, 1, 5, 5] \quad \text{level} = [1, 2, 3, 4, 2, 3, 3]$$

where for a given vertex i, parent$[i] = j$ means vertex j is the parent of vertex i except for root vertex 1 having no parent vertex and is represented by 0 as its parent. Similarly, for a vertex i, level$[i] = j$ means vertex i is at level j, where root vertex 1 has a level 1 and other vertices have level one greater than the level of its parent vertex. The root vertex can sometimes be considered to have level 0 and the levels of the subsequent vertices follow.

With the illustrated example and the terms introduced in consideration, the different steps in structure generation are explained in the following points:

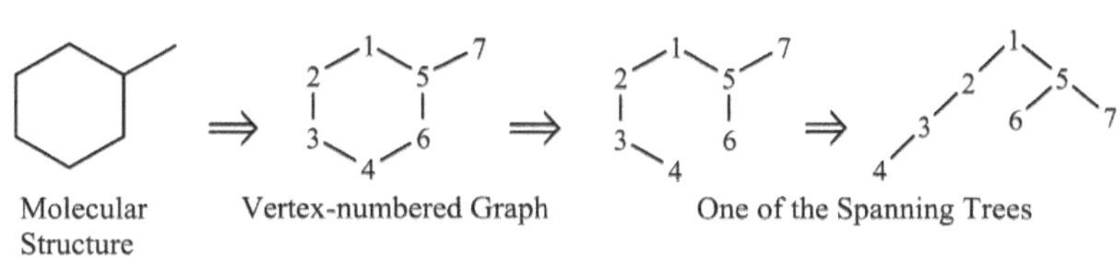

Fig. 2 Graph and tree illustration

(a) **Non-isomorphic canonical tree generation:**

Beyer and Hedetniemi [16] have proposed an iterative algorithm to reverse lexicographically generate non-isomorphic canonical trees for a given number of vertices. The algorithm achieves this transformation through a successor function defined below.

Let $L(T) = [l_1 l_2 \ldots l_n]$ be a level sequence containing an element greater than 2. Let p be the rightmost position of such an element, i.e. $p = \max\{i : l_i > 2\}$. Let q be defined as the rightmost position preceding p such that $l_q = l_p - 1$, i.e. $q = \max\{i : i < p,\ l_i = l_p - 1\}$. Hence, the vertex corresponding to position q is the parent of vertex corresponding to position p. Then the successor of $L(T)$, i.e. $\mathrm{succ}(L(T)) = [s_1 s_2 \ldots s_n]$ is defined such that:

(i) $s_i = l_i$ for $1 \leq i < p$
(ii) $s_i = s_{i-(p-q)}$ for $p \leq i \leq n$.

The algorithm can be used successively generating all the non-isomorphic canonical level representation of trees from a provided starting level sequence to the last possible reverse lexicographic sequence, i.e. $\left[1, \underbrace{2, 2 \ldots 2}_{n-1 \text{ times}}\right]$. If no starting level sequence can be provided, the algorithm can start with the lexicographically largest sequence $[1, 2, 3 \ldots n]$.

The trees generated by the aforementioned algorithm can in general have any number of children for any parent vertex. In context of chemical structures of carbon atoms, only those trees are being filtered and kept where the root has at most four children and the rest of the vertices have at most three children. This restriction can later be further refined for hetero-atoms in accordance with their valency.

(b) **Cycle introduction by addition of edges:**

The generated rooted trees are graphical models of acyclic compound structures. Cycles can be introduced by adding edges between any two vertices, say i and j, such that:

$$\mathrm{parent}[i] \neq j \quad \text{and} \quad \mathrm{parent}[j] \neq i$$

The size or the number of sides in the cycle so introduced can be obtained by the following relation:

$$\begin{aligned} \text{num\_cycle\_sides} = {} & \mathrm{level}[i] + \mathrm{level}\,[j] \\ & - 2 * \mathrm{level}[\text{lowest\_common\_anscester}(i,j)] + 1 \end{aligned}$$

In general, cycles of size 3 onwards will be possible. For more than one cycles to be introduced, a combination of these identified edge introductions can be simultaneously carried out.

However, introduction of multiple edges may lead to fused or bridged cycles and the size of cycle may become different than intended. Consider the case of starting structure generation from the tree in Fig. 2. If it is required to have two cycles which can have size 5, or 6, it can be seen (Fig. 3) that the edge introductions between vertices 3 and 6 and vertices 4 and 6 individually satisfy the size criteria, but in combination, they inadvertently lead to having a 3-sided cycle.

On the other hand, edge introductions between vertices 3 and 7 and vertices 4 and 6 satisfy the size criteria individually as well as in combination (Fig. 4).

Thus, in order to detect and remove cases similar to the first multiple introductions discussed before, it will be required to check the cycle size validity criteria considering all the elementary cycles, e.g. in the case being considered of multiple edge introductions, the elementary cycles present are C_1 (1–2–3–6–5–1), C_2 (1–2–3–4–6–5–1) and C_3 (3–4–6–3), having sizes 5, 6 and 3, respectively, even though the intended cycles were only C_1 and C_2. In graph theoretical terms, C_1 and C_2 are the fundamental sets of cycles while C_3 is a derived cycle. The term elementary cycles here has the standard graph theoretical definition, and from now on, the term cycle is considered to be an elementary cycle unless stated otherwise.

It will thus suffice to identify the fundamental set of cycles corresponding to the smallest sizes. The starting fundamental set of cycles corresponds to the cycles directly resulting from edge introductions. Any cycle enumeration algorithm can then be used to enumerate all the cycles present. We have considered the algorithm by Gibbs [17] which is a cycle vector space method in which the cycles of the fundamental set form the basis of the cycle vector space. With this vector space

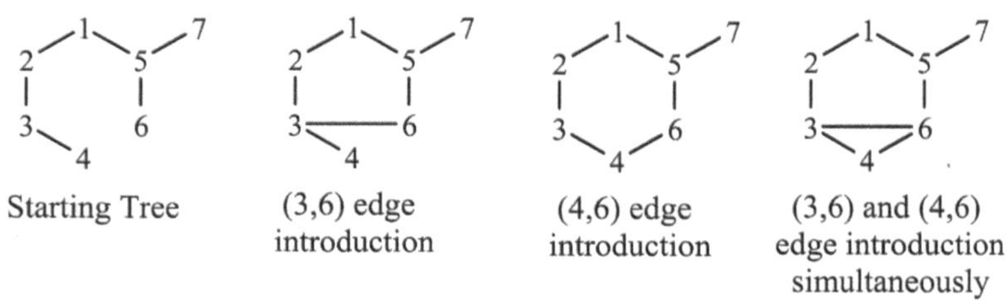

Fig. 3 Multiple cycle introduction example (1)

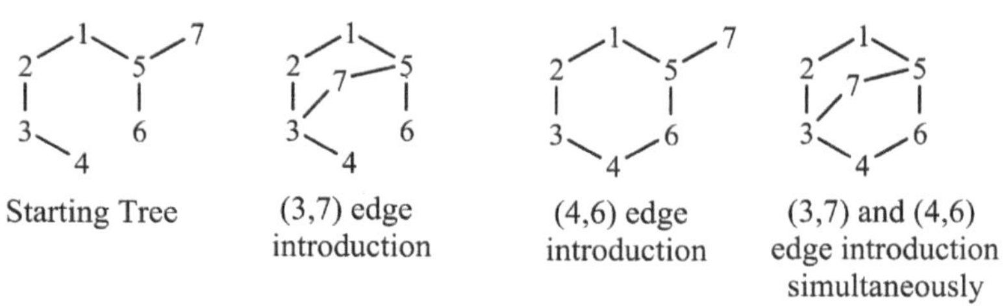

Fig. 4 Multiple cycle introduction example (2)

construct, one cycle, say C_3, can be obtained from two other cycles, say C_1 and C_2 from the previous example by a symmetric cycle-plus operation $\oplus$ defined below:

Let an edge between vertices i and j be denoted by e_{ij}. Let a cycle be denoted by the set of all such edges present in the cycle. Then for any two cycles A and B, the result of cycle-plus operation is:

$$A \oplus B = \left\{ e_{ij} \middle| e_{ij} \in A \cup B,\ e_{ij} \notin A \cap B \right\} = (A \cup B) \backslash (A \cap B)$$

The same operation can be performed computationally faster when all the edges present in the graph are assigned a unique number and a given cycle is represented by a bit string where bit positions from right are set "on" corresponding to the unique numbered edges in the cycle. The cycle-plus operation is then exactly analogous to the bit-wise XOR (^) operation, i.e. $A \oplus B \Leftrightarrow A^\wedge B$.

At this point, it is worthwhile to note that the following property, henceforth called *Property* 1, of the cycle-plus operator holds, which is proved using XOR operation on bit string representation of cycles A and B:

$$\begin{aligned} A \oplus (A \oplus B) &\Leftrightarrow A^\wedge(A^\wedge B) \\ &\Leftrightarrow (A^\wedge A)^\wedge B \quad \text{By associative property} \\ &\Leftrightarrow 0^\wedge B \Leftrightarrow B \end{aligned}$$

Hence, $A \oplus (A \oplus B) = B \quad \text{Property}(1)$

In terms of cycles, the result of the cycle-plus operation can either be another cycle or a union of cycles having no common edges. Thus, all the cycles present in the graph can be obtained by linear combination of cycles taken two at a time in the fundamental set, supplemented successively by the increasing number of cycles and union of cycles obtained through cycle-plus operation. In the end, the entries that supplemented the fundamental set should only be cycles and the edge disjoint union of cycles should be removed. The final set so obtained will be the set of all cycles, say in the considered example the final set will be $\{C_1, C_2, C_3\}$ starting from the fundamental set $\{C_1, C_2\}$.

It is easy to comprehend and evident from the previous example that the final set may contain cycles smaller in size than those in the starting fundamental set of cycles. Moreover, as the cycles are generated by linear combination over two cycles at any given time using cycle-plus operator and as Property (1) holds, any resultant cycle in combination with a fundamental cycle will yield the other fundamental cycle from which it was produced. This is to say, in previous case, C_2 can be obtained from C_1 and C_3.

Thus, the entire fundamental set can be changed to another fundamental set which contains only the cycles of non-decreasing number of sides starting from the smallest sized cycle, so that all the cycles in the final set can still be generated. Henceforth, the term fundamental set will correspond to this newly constructed set. It can be noted, though, that the cardinality of the fundamental set does not get altered. In the examples considered so far, this will lead to a change of fundamental set from $\{C_1, C_2\}$ to $\{C_1, C_3\}$ while the set of all cycles will still remain $\{C_1, C_2, C_3\}$. This, arguably, is just an instance of change of basis in the cycle vector space.

It will now suffice to check the sizes of the cycles in the fundamental set against the required sizes and keep or discard the generated structure accordingly. This decision made, considering the fundamental set only, is in accordance with the IUPAC convention of the number of rings in polycyclic systems [26] where the number of rings is equal to the minimum number of scissions required to convert the system into an open chain compound or structure. Following this convention of ring count, the example corresponding to Fig. 4 will be a valid structure against the cycle size restriction either being 5 or 6.

(c) **Removal of duplicate cyclic structures using graph canonicalization:**

Although the trees generated by the algorithm given by Beyer et al. [16] are non-isomorphic (hence distinct structures), it is easy to comprehend that introduction of edges may lead to generating more than one chemical structure of same topology. As the entire process starts with tree structure, consider the case of the rightmost tree representation shown in Fig. 2, and two different edge introductions for a given cycle size constraint of 6 and cycle count constraint of 1 as shown in Fig. 5.

Although the presented example is basic in nature, the problem aggravates when the number of nodes is fairly large and such node pairs lie in different branches, sometimes far apart. For example, the molecules with 30 or more non-hydrogen atoms are fairly common in organic compounds developed as pharmaceutical entities. Moreover, even when the graph topology is uniquely fixed, the combinatorial imposition of node colours for imparting heterogeneity by introducing different atoms and the imposition of multiplicity of bonds can again lead to duplicate structures. Hence, any duplicate elimination strategy should consider the complete graph along with heterogeneity and bond multiplicity.

In the above context, molecular graph canonicalization algorithms can be used to identify the duplicate structures and eliminate them during generation. As we intend to store the molecules in SMILES notation format, it has been decided to use the algorithm proposed for generation of unique SMILES by Weininger et al. [27], which tackles the molecular graph canonicalization by extended connectivity through an unambiguous function using product of primes.

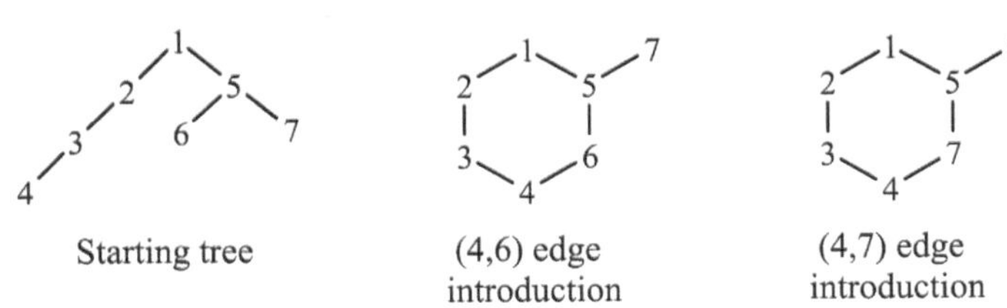

Fig. 5 Duplicate cyclic structures

The algorithmic steps leading to unique SMILES generation is discussed below:

(I) ***Initializing Rank of the Graph Vertices***—The rank initialization of the vertices is achieved using combined invariants which in turns are combinations of several individual atomic invariants. A total of 6 such atomic (node) invariants in the order of their priority are produced below:

(i) Number of connections
(ii) Number of non-hydrogen bonds
(iii) Atomic Number
(iv) Sign of Charge
(v) Absolute Charge
(vi) Number of attached hydrogen atoms.

It may be noted that the number of invariants can be varied based on the desired distinguishing properties [27]. The combined invariant will be the number obtained by successively concatenating the individual invariants such that higher priority invariants are to the left of lower priority invariants in the decimal system. For example, a methyl carbon (CH_3) in a molecule will have the individual invariants 1, 01, 06, 0, 0, 3 listed in the order of their priority while the combined invariant will be 10106003. The distinct combined invariants in the molecule are then sorted and mapped to their position in increasing order, hereafter referred to as consecutive ranks. The mapped position becomes the initial ranks of the atoms. For example, in case of n-Pentane, i.e. (C_1–C_2–C_3–C_4–C_5), where the subscripts denote the vertex labels, the combined invariants are 10106003–20206002–20206002–20206002–10106003 while the initial rank is 1–2–2–2–1.

(II) ***Extended Connectivity through an Unambiguous Function using Product of Primes***—The initial rank will not be able to identify the vertex symmetries. In the case of n-Pentane, vertices 2 and 4 are equivalent in terms of vertex symmetry while vertex 3 is not equivalent to them but is still initially ranked the same. To resolve this, rank of an atom is replaced by the result of an operation of a given function over its neighbours. This result is a representation of extended connectivity. A simple and elegant function is the product of primes corresponding to the rank of the neighbouring atoms. For example, in the n-Pentane case discussed so far, the updated rank of vertex 2 will now be prime number corresponding to rank of vertex 1 multiplied by prime number corresponding to the rank of vertex 3, i.e. $1\text{st}\,\text{prime} \times 2\text{nd}\,\text{prime} = 2 \times 3 = 6$, as ranks of vertices 1 and 3 are 1 and 2, respectively. Similarly, the rank of vertex 3 will be updated to $2\text{nd}\,\text{prime} \times 2\text{nd}\,\text{prime} = 9$. Subsequently, the revised rank will become 3–6–9–6–3 which can be remapped to consecutive ranks 1–2–3–2–1. This procedure of rank update is repeated and is stopped when the updated rank for each atom of the molecule remains same as the previous rank. It may be noted that in the end, the connectivity symmetrical vertices will be ranked the same.

(III) ***Tie Breaking***—The product of corresponding primes will yield same rank for connectivity symmetrical vertices. In such cases, the ties can be broken by arbitrarily choosing a node corresponding to the smallest repeating rank, doubling all the ranks and then reducing only the rank of the chosen vertex by one. The non-consecutive ranks so obtained are then remapped to form consecutive ranks, and the extended connectivity procedure using product of primes is performed to update ranks as described in the previous step. This step of breaking ties followed by rank updates is repeated until all the ties are broken and highest rank becomes equal to the number of vertices in the graph. The completion of this step also marks the completion of canonicalization of the graph.

(IV) ***Initial Vertex Selection and Branching Decisions for Traversal***—With the completion of graph canonicalization, the only steps required for unique SMILES generation is depth-first traversal sequence and identification of ring closures and their order in traversal. To start with, the lowest ranked atom is chosen for traversal. At a branching vertex, the branches are followed in the increasing order of the ranks of the neighbouring vertices; i.e. the branch corresponding to the lowest ranked neighbour is traversed first, then the second lowest ranked neighbour is followed and so on. It may be noted that Weininger et al. [27] also suggest giving branching preference towards the double or triple bonds in a ring even though the rank corresponding to such a vertex may be greater than other neighbouring vertices. However, this further complicates the final traversal sequence in the case of polycyclic compounds while the omission of this preference will save some computation time but will still generate unique SMILES.

V) ***Two-pass Approach***—Although, initially, the ring closures for the compounds are the edges that were introduced by joining vertices in the canonical trees, those edges will not be the ring closures under the depth-first traversal approach of the canonicalized graph and the traversal rule as given in the previous step. Additionally, the rings are to be numbered in the opening order in which they are encountered during traversal. In order to meet these requirements, the graph is traversed two times. During the first pass, the ring closures and their ordering are identified for the canonicalized graph and are stored as auxiliary data. The edges corresponding to these new ring closures will now be treated as if they were the edges introduced to complete the cyclic structure, while the tree obtained by removal of such edges is treated now as the spanning tree. Subsequently, the second pass is undertaken for SMILES string generation using the previously obtained auxiliary data.

2.5.2 Structure for a Relaxed Distance Distribution

The approach taken so far suffers from the drawback that only those compound structures will be generated that have the same number of non-hydrogen atoms as the starting molecule from which the distance distribution was obtained. This subsection tries to tackle this drawback by slightly relaxing the distance distribution matching criteria for the trees with number of vertices deviating from the source or starting distribution. This deviation can either lead to increased or decreased number of vertices.

(a) **Non-Isomorphic Canonical Tree Generation with Relaxed Distance Distribution**

The first step involves specifying the number of vertices (after factoring in the deviation) and then generating the trees. Positive deviation means required number of vertices is greater than that in the current tree while negative deviation means the required number of vertices is lesser. However, since exact distance distribution matching is not possible in this case, two variants of relaxed distribution matching are considered as explained below:

Strong matching—This situation arises when the distance distribution of the generated tree can be obtained from the starting/source distance distribution by either adding or deleting vertices at any level (named node deviation) although simultaneous insertion or deletion of vertices is not allowed for a given deviation. In essence, the obtained distance distribution corresponds to a pruned tree of the source distance distribution if the node deviation is negative and vice versa if the node deviation is positive.

Thus, to put it mathematically, if trees are to be generated by decreasing or increasing n number of vertices, then only n deletions or insertions are allowed so that:

$$\left|\sum_{i=1}^{e}\left(c_i^s - c_i^p\right)\right| = n$$

where c_i^s is the count of vertices at level i in the source distance distribution; c_i^p is the count of vertices at level i in the present distance distribution under consideration; and e is the maximum of the eccentricity of the source and present distance distribution.

Weak matching—In this case, the distance distribution matching criteria is further relaxed in that one can add and delete vertices simultaneously at any level. This, in effect, executes migration of vertices from one level to another (named node migration). If this is allowed without a cap on the number of node migrations, then all the possible structure generation will be considered a match which will include the linear chain too. Presumably, in order to match the source distance distribution closely using weak matching criterion, number of allowed node migrations should be provided preferably of low value.

For this exercise, if the trees are to be obtained by decreasing or increasing n number of vertices, then n deletions or insertions along with m migrations are allowed that satisfies the following criteria:

$$\left|\sum_{i=1}^{e}\left(c_i^s - c_i^p\right)\right| = n$$

and

$$\min\left(m_p, m_n\right) = m$$

where

$$m_p = \sum_{i=1}^{e} max\left(\left(c_i^s - c_i^p\right), 0\right)$$

$$m_n = \left|\sum_{i=1}^{e} \min\left(\left(c_i^s - c_i^p\right), 0\right)\right|$$

Here c_i^s, c_i^p and e have the same meaning as defined in the case of strong matching while m_p is the sum of vertex surplus and m_n is the sum of vertex deficit in the source distance distribution over the present distance distribution.

The procedure of cycle introduction, canonicalization and unique SMILES notation generation is the same as done before.

Now, once the structures are generated using the methods described above, one can use some user-defined parameters incorporated in the computer program to restrict the number and size of the cycles to be created in the 2D structures. Few other user-defined parameters, available in the program, may also be used to add multiplicity of bonds (double and triple bonds) between pairs of vertices and other hetero-atoms (e.g. nitrogen, oxygen, halogens) in order to get complete 2D structures of the compounds. The output of the generated structures may be saved in SMILES notations and can be viewed using a molecular modelling software that is capable of getting molecular structures from SMILES notation. Subsequently, the activities of the generated structures may be predicted using the rule-based method [18, 19] standardized for a biological endpoint of interest and can be prioritized and screened from their MPS values. In this way, one may be able to screen some potential bioactive compounds from the bigger set of combinatorially generated molecular structures using topological distance information associated with activity-related vertices present in the active compounds of a data set under consideration. It may be worth noting at this point that this newly developed method [15] is essentially a molecular topology-based approach and activity prediction is done using molecular graphs of the compounds where bond multiplicity and atom types are not required. However, since bond multiplicity and atom types can be introduced in the combinatorially generated topological structures using the options available in the program and those structures can be saved in SMILES format, one can always use these generated structures for any 2D and 3D drug design/discovery applications.

3 Results and Discussion

We furnish in this section the results obtained using the method, described in the previous section, that can generate chemical structures combinatorially using activity-related substructural topological information, predict activity for the biological endpoints under consideration, prioritize compounds and screen them to help discover novel therapeutic candidates. The results given here are for a series of 19 convulsant–anticonvulsant barbiturates [18], a series of 20 nucleoside analogues (NA) having anti-HIV activities [20, 21] and a data set of 3779 compounds [22, 23] for which minimum inhibitory concentration (MIC) values have been measured against H37Rv strain of *Mycobacterium tuberculosis* (Mtb).

3.1 Activity Prediction–Compound Prioritization–Molecular Design

We describe in this section the results obtained for combinatorial structure generation from the substructural information of activity-related vertices (atoms), activity prediction using a rule-based system [18, 19] and prioritization and screening of potential drug candidates using a newly defined Molecular Priority Score (MPS) [15]. The application of different algorithms incorporated in the computer program developed using the method, and the results obtained therefrom are given here and discussed accordingly. In particular, the method has been used for activity prediction, compound prioritization using MPS and structure generation considering barbiturates and the NA series of compounds. On the other hand, structure matching algorithm based on distance distribution has been used for searching potential antitubercular compounds from the data set of 3779 compounds mentioned above.

3.1.1 Studies with Barbiturates

The activity prediction for the series of barbiturates [18] considered for the present study is reported here using the rule-based method [18, 19] considering hydrogen-filled (H-filled) graphs of the compounds. Along with activity prediction considering H-suppressed graphs, the method also supports activity prediction using H-filled graphs and that option available in the computer program has been used for the activity prediction studies with the barbiturates. The R-groups of the barbiturates considered here and built on the core structure shown in Fig. 6 are given in Table 1.

Activity prediction for this series of compounds has already been reported [18] by considering information theoretical vertex indices V^d (vertex distance complexity) and V_n^d(normalized V^d), which are also available in this software for use. Although V_n^d has produced very high percentage of correct predictions [18], we

Table 1 A series of 19 barbiturates[a] considered for the present study

	R-group		R-group
1.	$-(CH_2)_3CH_3$	11.	$-(CH_2)_3C_6H_{11}$
2.	$-CH(CH_3)(CH_2)_2CH_3$	12.	$-(CH_2)_2CH{=}C_6H_{10}$
3.	$-(CH_2)_2CH(CH_3)_2$	13.	$-(CH_2)_2CH{=}C_5H_8$
4.	$-CH(CH_3)CH_2CH(CH_3)_2$	14.	$-CH_2C_6H_5$
5.	$-CH{=}CHCH_2CH_3$	15.	$-CH_2CH(CH_3)C_6H_5$
6.	$-C(CH_3){=}CHCH_2CH_3$	16.	$-CH{=}(CH)_2(CH_3)_2$
7.	$-CH_2CH{=}CHCH_3$	17.	$-C(CH_3){=}(CH)_2(CH_3)_2$
8.	$-CH(CH_3)CH{=}CHCH_3$	18.	$-(CH_2)_3C_6H_5$
9.	$-CH_2CH{=}C(CH_3)_2$	19.	$-(CH_2)_2C_6H_5$
10.	$-CH(CH_3)CH{=}C(CH_3)_2$		

[a]The data have been taken from Klopman and Raychaudhury [18]

Table 2 Assigned and predicted activities using D^{-4} index and Molecular Priority Score (MPS) of 19 barbiturates divided into 15 training set and 4 test set compounds

Sr. no.	Compound no.	Activity[a]		MPS[b]
		Assgn.	Pred.	Value
Training set				
1	1	+	+	93
2	9	+	+	10
3	10	+	+	56
4	12	+	+	178
5	13	+	+	168
6	15	+	+	34
7	2	−	−	−132
8	3	−	−	−102
9	4	−	−	−132
10	5	−	−	−113
11	6	−	−	−100
12	7	−	−	−120
13	8	−	−	−74
14	11	−	−	−149
15	14	−	−	−64
Test set				
1	17	+	+	6
2	19	+	+	53
3	16	−	−	−97
4	18	−	−	10

[a](+) means active and (−) means inactive
[b]Computation of MPS value is described in methods section

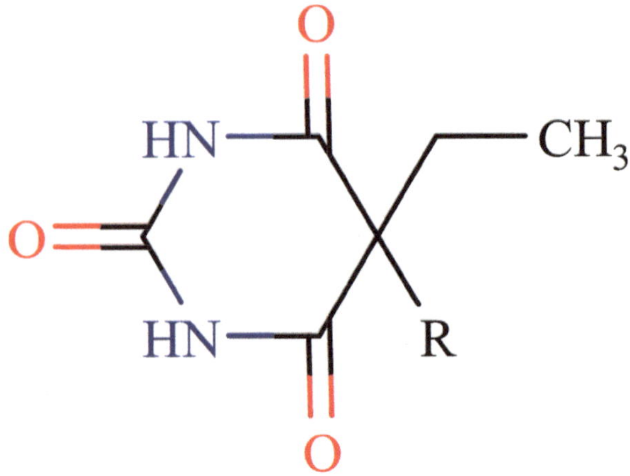

Fig. 6 Barbiturate core structure with R-group (Table 1) attachment point (R)

present here the results obtained using distance exponent index (D^{-4}) to see how this index performs for this series of compounds. The activity prediction results along with MPS values using D^{-4} index, computed for the hydrogen-filled graphs of the compounds, are shown in Table 2. It may, however, be noted that the indices of only non-hydrogen atoms have been considered for ordering of index values, range selection and activity prediction purposes. Thus, the indices computed for the hydrogen atoms in the H-filled graphs have not been used for this purpose.

Activity Prediction and Compound Prioritization for Barbiturates
For the prediction of activity and prioritizing the compounds on the basis of MPS values, we have considered the same set of compounds as well as the same training set and test set for the present study as used earlier [18]. In may be noted that, in this data set, the convulsant barbiturates are tagged active and the anticonvulsant barbiturates as inactive.

It can be observed that accuracy of activity prediction using D^{-4} index in the barbiturate data set is 100% for both training set and test set which equals the prediction obtained using V_n^d index reported earlier [18]. This further substantiates earlier findings [15] using this vertex index, rule-based method and MPS value about the usefulness of the method for activity prediction and compound prioritization. This is believed to help scientists work on the crucial issues related to convulsion and help drug designers find novel therapeutic agents in the area of anticonvulsant drug discovery.

Structure Generation for Barbiturates
The structure generation exercise has been carried out for the barbiturate data set with the same training set and test set split as considered earlier [18]. The index computation for the non-hydrogen atoms (vertices) has been performed considering hydrogen-filled graphs. As described in the method section, the D^{-4} index values computed for the training set compounds are arranged in an ascending order to find active and inactive ranges in order to get a "*strong*" range to identify an

Table 3 Details of the range in which vertices 17 and 18, in the molecular graph of compound no. 13, lie in

Serial no.	D^{-4} index value	Compound no. (Atom no.)	Activity
1	4.40994	13(16)	+
2	4.40994	13(19)	+
3	4.430099	12(16)	+
4	4.430099	12(20)	+
5	4.430937	13(17)	+
6	4.430937	13(18)	+
7	4.440002	1(14)	+
8	4.441781	13(13)	+
9	4.444924	12(13)	+
10	4.449867	12(18)	+
11	4.451095	12(17)	+
12	4.451095	12(19)	+

(+) means active, (−) means inactive

activity-related vertex to start structure generation considering that vertex as the root vertex. It has been observed that the vertices 17 and 18 (the numbers correspond to those in the respective SMI file used to work with the compounds considered) in the molecular graph representing compound no. 13 (Table 1), an active compound, fall in a strong range. Interestingly, when these two vertices are chosen

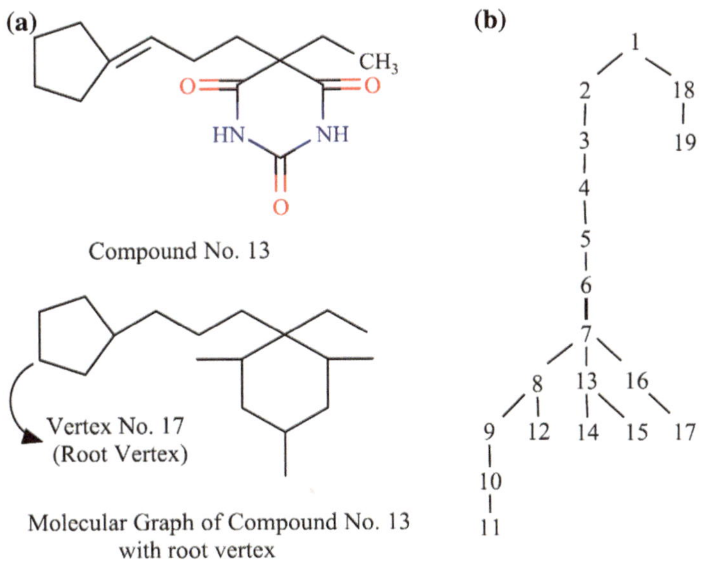

Fig. 7 **a** Compound no. 13 (Table 1), its molecular graph and the root vertex (vertex no. 17). **b** Sample rooted tree structure generated. In the tree, the root vertex is labelled as vertex 1

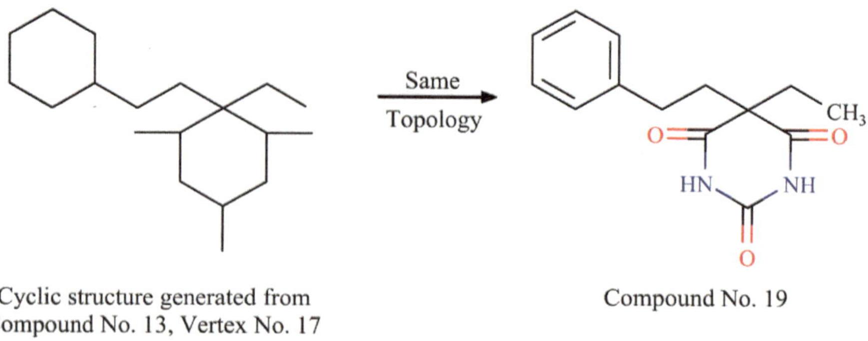

Fig. 8 One of the structures generated, from compound no. 13, which resembles the topology of compound no. 19 (Table 1)

for structure generation, both of them lead to the generation of a topological structure of another active compound. The details of the strong active range are given in Table 3 and the structure generation details in Fig. 8.

The compound no. 13 along with its molecular graph and the chosen structure generation vertex (root vertex) is given in Fig. 7a. The distance distribution associated with this vertex (Vertex No. 17) starting with distance 0 is (1, 2, 2, 1, 1, 1, 1, 3, 5, 1, 1). A sample rooted tree is shown in Fig. 7b with the corresponding distance distribution.

Considering any rooted tree, cycles can be introduced (described in the methods section) to generate the topology of the structural formula of variety of chemical compound while still maintaining the distance distribution. In the present study, we have chosen to generate structures containing two cycles, having number of sides 5 or 6, to investigate whether we are able to generate any other active compound present in the studied data set. A number of structures are generated in the process, and it has been found that the structures generated from the root vertex of compound no. 13 contain one such structure that matches with that of compound no. 19 (Fig. 8). It is interesting to note that compound no. 19 is an active compound from the test set (Table 2) which shows that the method can generate a structure that it has not seen in the training set. Therefore, one can expect to design novel structures using this method.

3.1.2 Studies with Nucleoside Analogues

For the nucleoside analogues (NA), we have carried out activity prediction and structure generation studies. It may be noted that for this series of compounds, we have investigated the performance of the training set–test set identification tool using the corresponding algorithm incorporated in the computer program. As mentioned earlier, in this way we are able to obtain a suitable training set for the system's learning and predict activities of the compounds on the basis of this

Table 4 A series of 20 nucleoside analogues [a] considered for the present study

	Compound Name		Compound Name
1.	3′-deoxyadenosine	11.	2′-deoxyinosine
2.	2′-deoxycytidine	12.	2′,3′-dideoxythymidine
3.	2′-deoxyadenosine	13.	2′,3′-dideoxyuridine
4.	2′,3′-dideoxyadenosine	14.	2′,3′,5′-trideoxyadenosine
5.	2′,3′-dideoxycytidine	15.	3′-amino-2′,3′-dideoxycytidine
6.	3′-fluoro-2′,3′-dideoxythymidine	16.	3′-amino-2′,3′-dideoxyadenosine
7.	3′-azido-2′,3′-dideoxythymidine	17.	2′-deoxyguanosine
8.	2′,3′-dideoxyinosine	18.	3′-azido-2′,3′-dideoxyadenosine
9.	2′,3′-dideoxyguanosine	19.	3′-azido-2′,3′-dideoxycytidine
10.	5′-iodo-2′-deoxycytidine	20.	3′-azido-3′-deoxyadenosine

[a]Data were taken from Raychaudhury et al. [20, 21]

training. This section, therefore, contains the results of the performance of training set identification and activity prediction. We have also reported here the results of structure generation for some of the NA series compounds in the same way as it has been done for the barbiturate series. For identifying a suitable training set–test set combination for the purpose of identifying a suitable training set that can produce high percentage of successful activity predictions, the program generates 1000 such combinations. The program has the option of getting the output on the basis of best test set predictions (starting from no misprediction) and best training set predictions. It has been observed that there are combinations where no mispredictions are found for the training set although there are 2 or more mispredictions for the test sets. On the other hand, there are combinations where there is one misprediction each for both the training set and the test set and it seems quite reasonable to consider such a balanced combination for activity prediction of newly generated compounds. We have reported here the activity predictions and MPS values of such a balanced outcome in Table 5 for the nucleoside analogues (NA) considered for the present study given in Table 4. The structural information of the compounds has been taken from the corresponding MOL files.

Activity Prediction for Nucleoside Analogues

For carrying out activity prediction and prioritization studies for NA series of compounds, we have used training set–test set split algorithm and the prediction results for split that has given one misprediction each for the training set and the test set are reported here.

It can be seen that for this NA series, activities of 92.86% (13 out of 14) of the training set compounds and 83.33% (5 out of 6) of the test set compounds have been predicted correctly, compound no. 10 of the training set and compound no. 13 of the test set being the lone mispredictions in each case. It is interesting to note that in both the cases the inactive compounds have been predicted to be active which may be regarded as an important factor in situations where a drug designer

Table 5 Assigned and predicted activities using D^{-4} index and Molecular Priority Score (MPS) of 20 nucleoside analogues divided into 14 training set and 6 test set compounds

Sr. no.	Compound no. #	Activity[a]		MPS[b]
		Assigned	Predicted	Value
Training set				
1	4	+	+	65
2	5	+	+	83
3	6	+	+	8
4	7	+	+	103
5	9	+	+	55
6	18	+	+	97
7	19	+	+	98
8	1	−	−	−56
9	2	−	−	−36
10	3	−	−	−48
11	10	−	+	8
12	14	−	−	−13
13	15	−	−	−36
14	16	−	−	−48
Test set				
1	8	+	+	65
2	12	+	+	65
3	20	+	+	50
4	11	−	−	−48
5	13	−	+	83
6	17	−	−	−6

[a](+) means active, (−) means inactive and (#) means incorrect prediction
[b]The details for the computation of MPS value are described in methods section
#Compound numbers are correspond to those in Table 4

does not want to lose any potential active compound/drug candidate particularly the one like the mispredicted compound of the test set (compound no. 13) which has a high MPS value (MPS = 83). Clearly, a number of active compounds have got high MPS values including compound no. 8 which represents a potent anti-HIV drug—Didanosine—and is a test set compound (Table 5). The method has also produced high MPS values for a number of training set active compounds too like compound nos. 5, 7, 18, 19 (Table 5). Therefore, picking at least a couple of top scoring (from MPS values) compounds out of them from prioritization point of view may help screen useful drug candidates using the present method. This finding therefore indicates that this method can be used for creating suitable splits in getting a reasonably useful training set from an available data set and help screen putative active compounds for drug discovery.

Structure Generation for Nucleoside Analogous
As done for the barbiturates, structure generation from various starting points, i.e. compound no., atom no., was carried out for the NA series of compounds too. In doing that, activity-related vertices have been picked up from the strong ranges in the ordering of D^{-4} index values for the vertices (atoms) of the H-suppressed graphs of these compounds. It has been found that a few carbon skeletons resembling the structure of other active compounds than the ones from where the activity-related vertices and the corresponding distance distribution values are taken have been generated.

For the purpose of illustration, the structure of the compound no. 6 and the generated structure which corresponds to compound no. 8 are shown in Fig. 9. It can be seen that in this case too, the algorithm is able to generate a structure with significantly different scaffold than the starting compound and has a higher MPS value (MPS = 65) too compared to that (MPS = 8) of the starting structure indicating that this generated structure has the potential of being highly active and therefore may be picked/prioritized for further studies. In fact, compound no. 8 is a potent anti-HIV drug—Didanosine. Therefore, the method may be regarded as a useful tool for generating, prioritizing and discovering potent anti-HIV compounds. Moreover, the generated compound belongs to the test set indicating that the structure of a compound that has not been used for training the system can also be

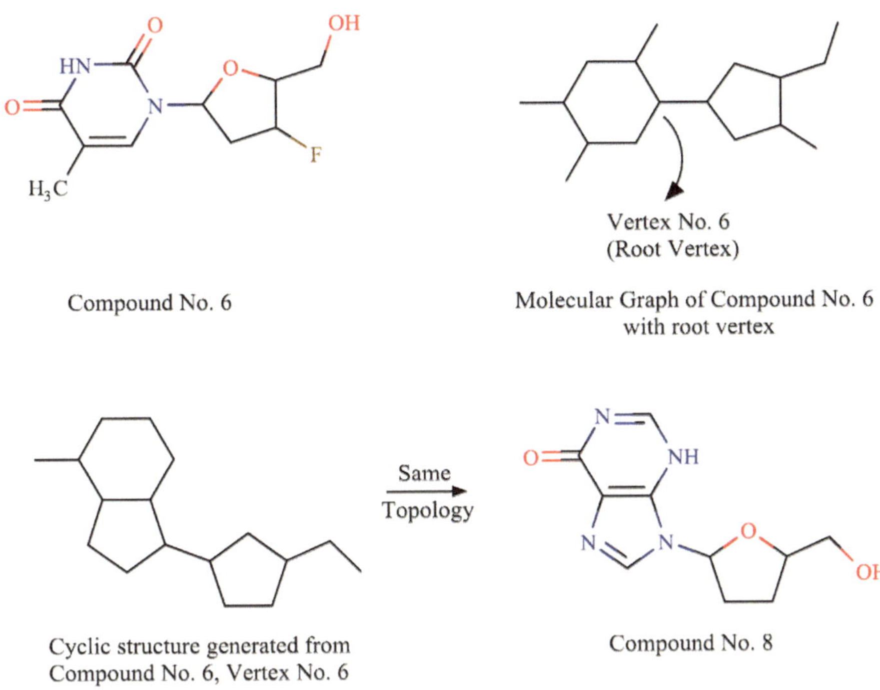

Fig. 9 Compound no. 6, its molecular graph with root vertex and one of the structures generated from compound no. 6 that resembles the topology of compound no. 8

designed by this method which may be believed to carry higher importance for discovering novel therapeutic candidates.

3.2 Rooted Substructure Searching for Drug Discovery

In the previous section, we showed how the exact matching algorithm can help find structures of active compounds which could be obtained from the trees generated from the topological distance distribution information of activity-related vertices obtained from other active compounds. In this section, we describe the use of two other matching algorithms—strong matching and weak matching—along with exact matching algorithm for searching active compounds in a data set in the form of tree and sub-tree matching. As given in the method section, these sub-trees are obtained by means of applying node deviation and node migration in the actual tree obtained from the distance distribution associated with an activity-related vertex. The presence of such trees and sub-trees are then searched for in the compounds present in a data set to identify potential drug candidates. In doing that, we have considered two known TB drugs—Isoniazid and Streptomycin—to describe the usefulness of the present method in finding potential antitubercular compounds from a data set (named GTB data set) of 3779 compounds [22, 23] for which *MIC* values against H37Rv strain of *Mtb* have been measured. The authors have made *MIC* = 5.0 as the cut-off point and the *MIC* value of any compound which is higher than 5.0 give an inactive compound in the data set. It therefore seems reasonable to consider the same cut-off value for the present purpose. We will first furnish the results obtained for Isoniazid which will be followed by those obtained for Streptomycin. It may be noted that the activity-related vertices for both Isoniazid and Streptomycin have been taken from the literature information and not by using rule-based method in the ordering of vertex indices which has been done for the barbiturate and NA series of compounds. In fact, it shows that the method can be used successfully in identifying potential drug candidates by picking activity-related vertices by other means than by the rule-based method.

3.2.1 Studies with Isoniazid

Isoniazid is a known first line drug for the treatment of tuberculosis. However, it may become resistant in situations, and therefore, this leads researchers look for novel drug candidates to overcome drug resistance problem for the treatment of tuberculosis . We have described in this subsection how structures generated from activity-related vertex information of Isoniazid using the present method can help search for potential TB drugs from a data set of 3779 compounds [22, 23]. It is known that the chemical/biochemical reaction takes place at the point of the first nitrogen (N) atom (underlined) of the fragment (–NH–NH2) in isoniazid molecule to convert this pro-drug into its metabolite that works as the effector molecule. Therefore, this vertex (N atom) may be regarded as an activity-related vertex for Isoniazid. Accordingly, the distance distribution associated with the vertex representing this nitrogen (N) atom has been considered for generating structures. In order to screen out potential antitubercular compounds having high activities, the exact, strong and weak matching algorithms (method section) have been applied on the GTB data set of 3779 compounds considered for the present study. A number of highly active compounds have been obtained in the process and the information for some of them obtained applying different node deviation and node migration on the tree obtained from the distance distribution associated with the root vertex are shown in Table 6 along with the structures of Isoniazid (with root vertex specified) and the screened compounds. As said earlier, in their studies [22], the researchers have considered a compound having *MIC* value less than 5.0 to be active. In this way, data set is composed of almost equal number of active and inactive compounds implying no bias for active or inactive compounds in forming the data set. Accordingly, compound nos. 1–1890 are active compounds and the other compounds are inactive. Considering the same cut-off value, one can see that only compound no. 3296 has *MIC* value higher than 5.0 and the rest of the compounds may be screened out as potential active compounds. In particular, compound no. 180 which is obtained by two types of node deviation and node migration in generating structures from the root vertex has quite low *MIC* value which identifies it as a highly active compound. Therefore, the result clearly shows that the method may be used to successfully screen potentially highly active antitubercular compounds from this data set starting from Isoniazid.

3.2.2 Studies with Streptomycin

Streptomycin is another antitubercular drug in use, an antibiotic. For this compound, the removal of even one of the two guanidino groups present in the structure reduces the activity of the compound. Considering that, we have taken the vertex representing the nitrogen (N) atom in one of the guanidino groups as the root vertex to start generating/designing novel structures. Out of a number of structures designed using the present method, i.e., using exact matching as well as strong matching and weak matching algorithms in relation to node deviation and node migration on the trees obtained from the distance distribution associated with the root vertex, information about some of these compounds are given in Table 7 along with the structures of Streptomycin having root vertex indicated and the matched/searched compounds from GTB data set. It is found from this table that all the compounds shown here are active according to the adopted criterion (*MIC* $\leq$ 5.0 is active) with compound no. 183 being the most active among them. Therefore, it appears from this finding that the method may be used successfully to screen potentially highly active antitubercular compounds from the data set of 3779 compounds starting from Streptomycin.

Table 6 Screened compounds obtained from the matching of trees/sub-trees obtained from the generated structure from the root vertex (indicated) of Isoniazid molecular graph

Source Compound	
	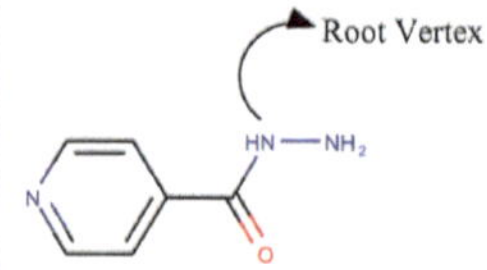 ***Isoniazid***

Compounds (in the Global TB data set) whose structures topologically matched with the source compound with the node deviation and node migration mentioned alongside

S. no.	Node deviation	Node migration	Matched compound
1	0	0	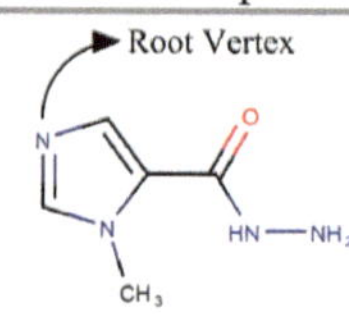 ***Compound No 1387***
2	0	0	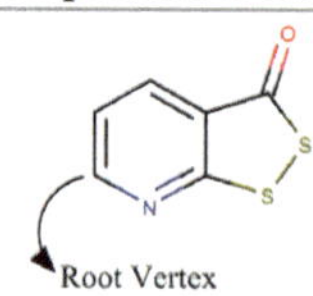 ***Compound No. 3296***
3	1	0	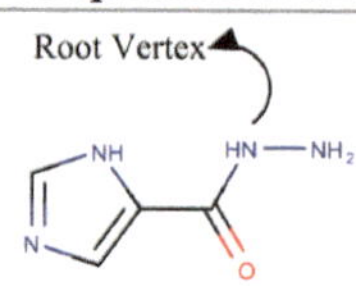 ***Compound No. 180***
4	1	0	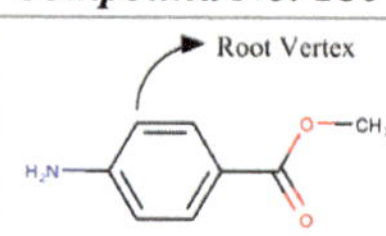 ***Compound No. 1174***
5	1	1	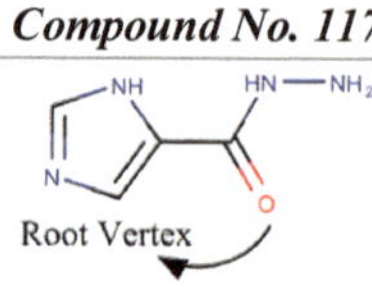 ***Compound No. 180***
6	1	1	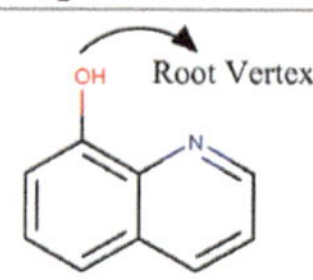 ***Compound No. 1192***

(continued)

Table 6 (continued)

Source Compound			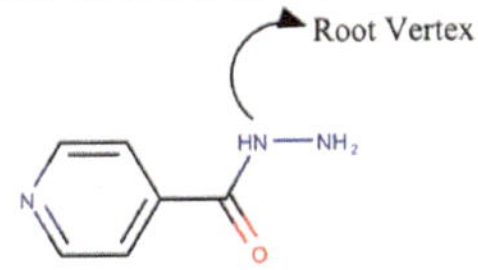 *Isoniazid*
Compounds (in the Global TB data set) whose structures topologically matched with the source compound with the node deviation and node migration mentioned alongside			
S. no.	Node deviation	Node migration	Matched compound
7	2	0	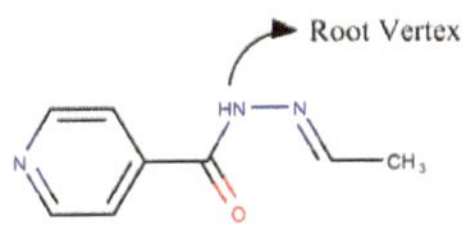 *Compound No. 524*
8	2	0	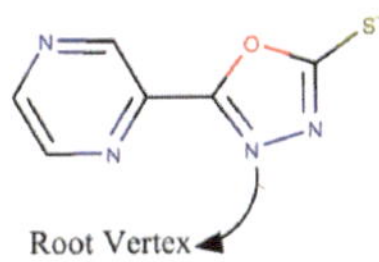 *Compound No. 928*
9	2	2	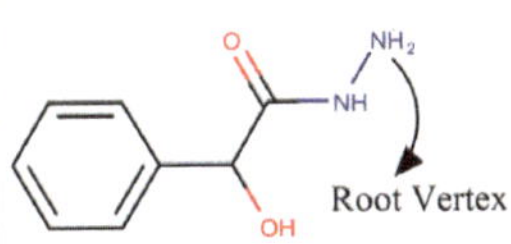 *Compound No. 661*
10	2	2	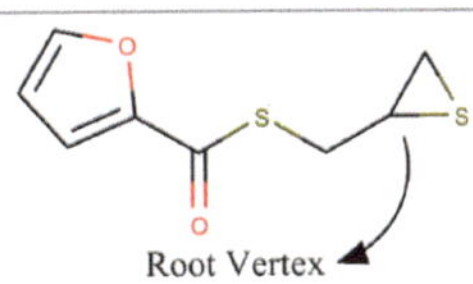 *Compound No. 1333*

Table 7 Screened compounds obtained from the matching of trees/sub-trees obtained from the generated structure using the root vertex in Streptomycin molecular graph

Source compound	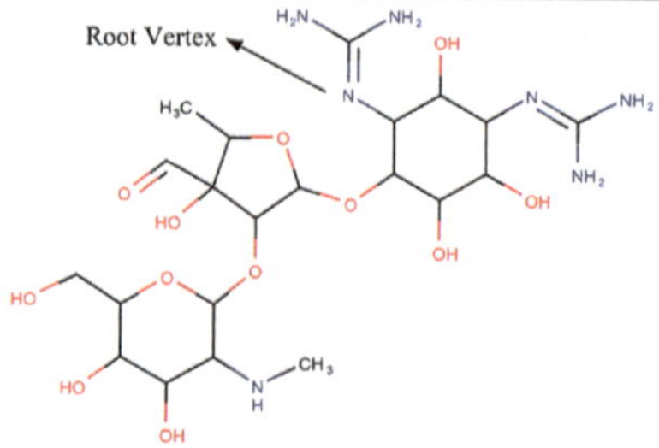 ***Streptomycin***

Compounds (in the Global TB data set) whose structures topologically matched with the source compound with the node deviation and node migration mentioned alongside

S. no.	Node deviation	Node migration	Matched compound
1	0	0	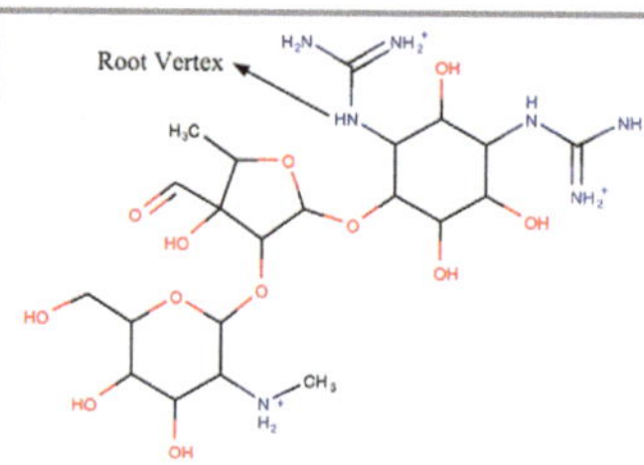 ***Compound No 183***
2	2	0	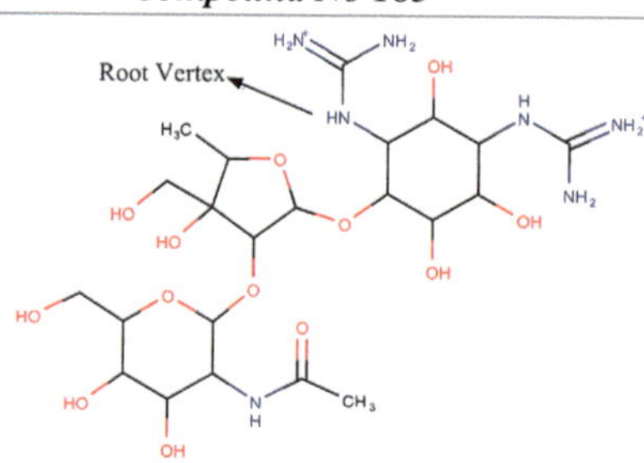 ***Compound No. 1483***
3	2	1	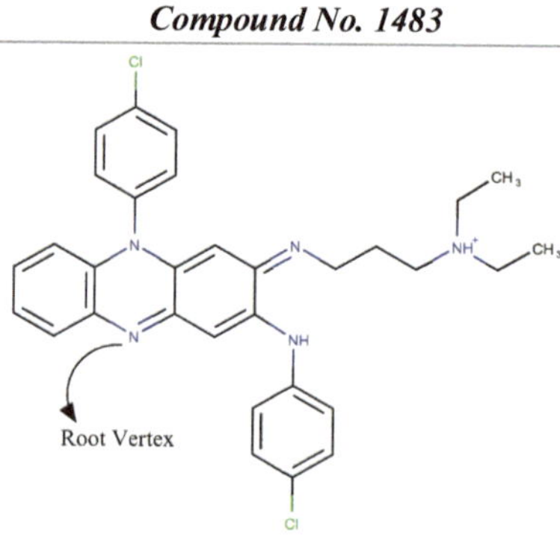 ***Compound No. 1059***

(continued)

Table 7 (continued)

Source compound	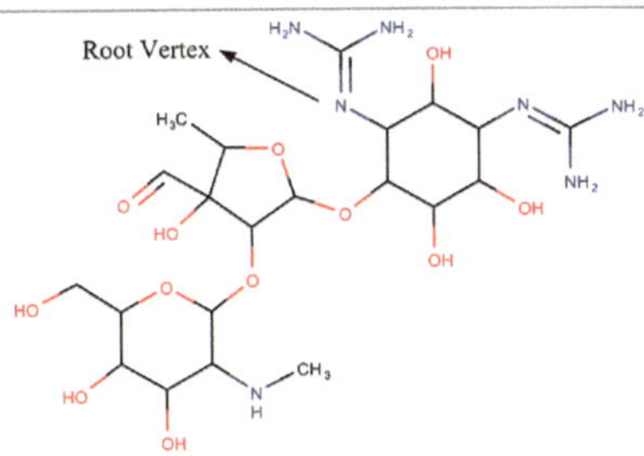 ***Streptomycin***

Compounds (in the Global TB data set) whose structures topologically matched with the source compound with the node deviation and node migration mentioned alongside

S. no.	Node deviation	Node migration	Matched compound
4	2	2	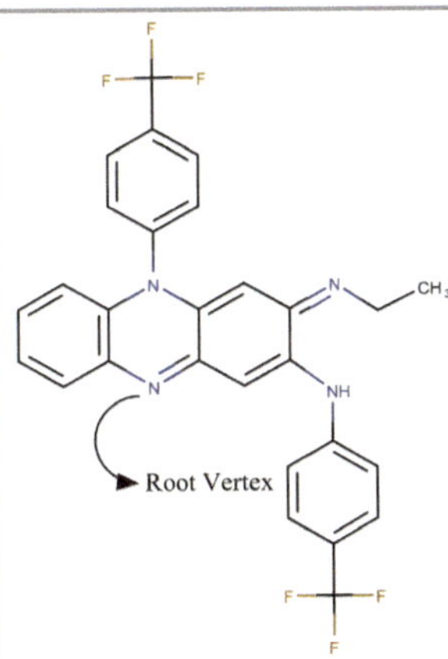 ***Compound No. 468***
5	3	1	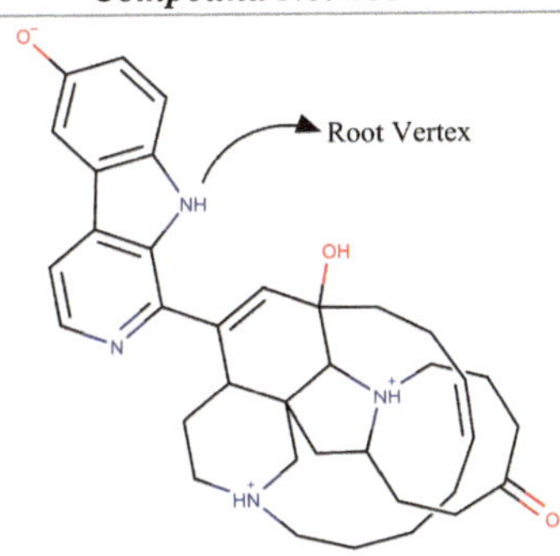 ***Compound No. 1006***

(continued)

Table 7 (continued)

Source compound	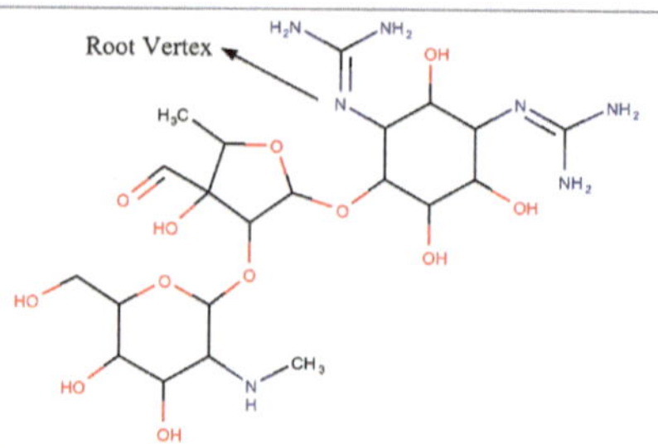 ***Streptomycin***

Compounds (in the Global TB data set) whose structures topologically matched with the source compound with the node deviation and node migration mentioned alongside

S. no.	Node deviation	Node migration	Matched compound
6	3	2	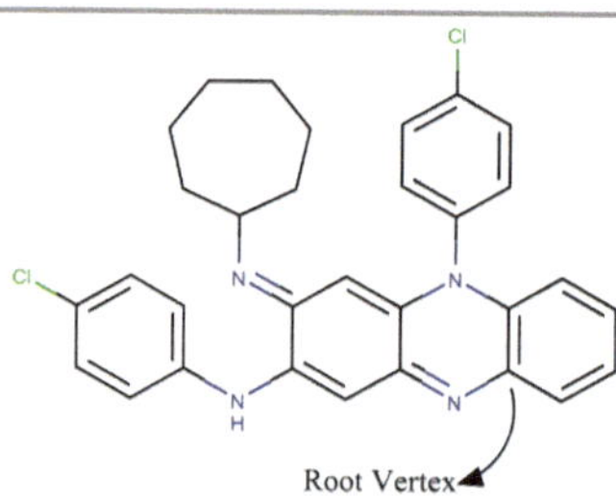 ***Compound No. 671***
7	4	1	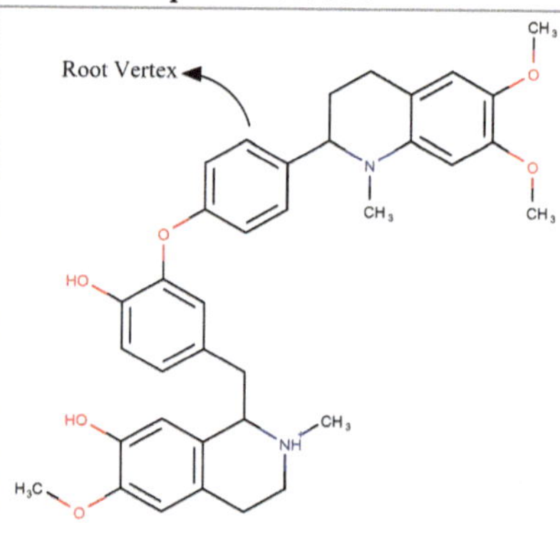 ***Compound No. 1287***

(continued)

Table 7 (continued)

Source compound	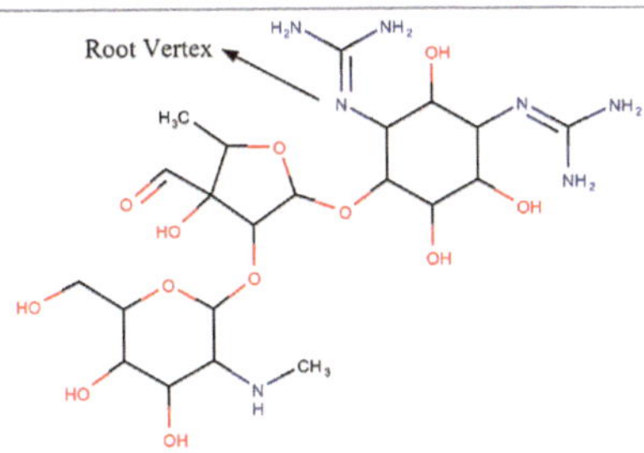 ***Streptomycin***

Compounds (in the Global TB data set) whose structures topologically matched with the source compound with the node deviation and node migration mentioned alongside

S. no.	Node deviation	Node migration	Matched compound
8	4	2	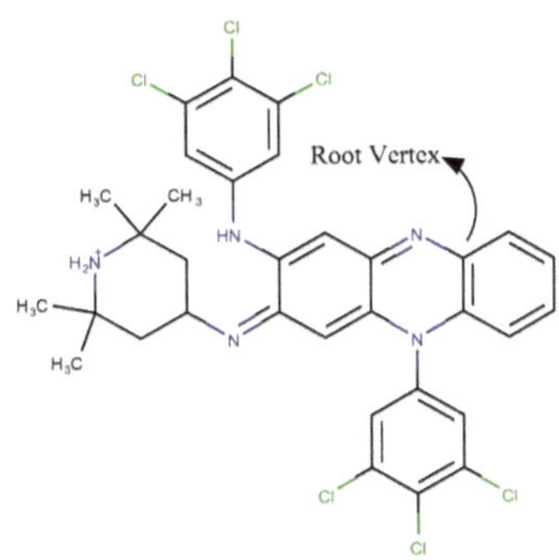 ***Compound No. 211***
9	5	0	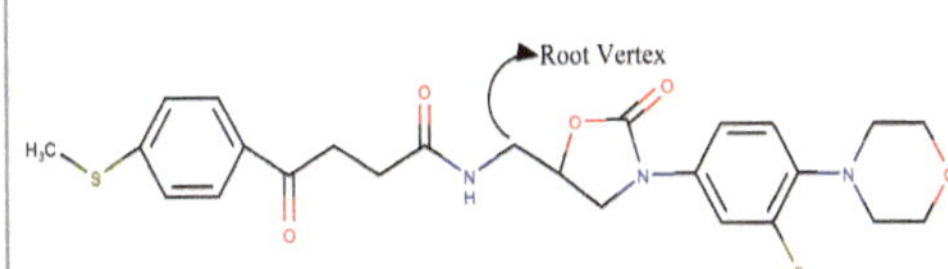 ***Compound No. 1086***
10	5	1	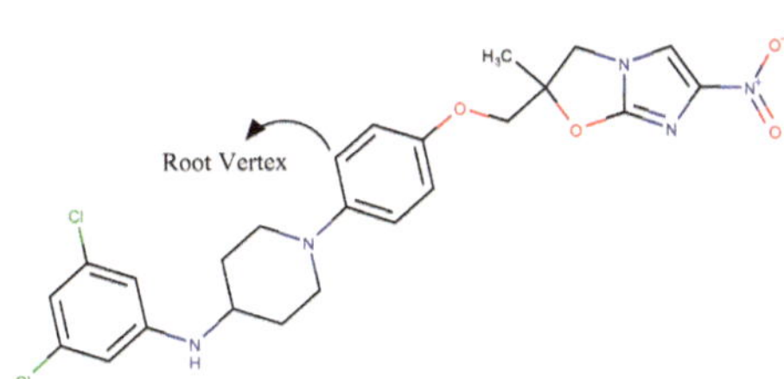 ***Compound No. 335***

(continued)

Table 7 (continued)

<table>
<tr><td>Source compound</td><td colspan="3">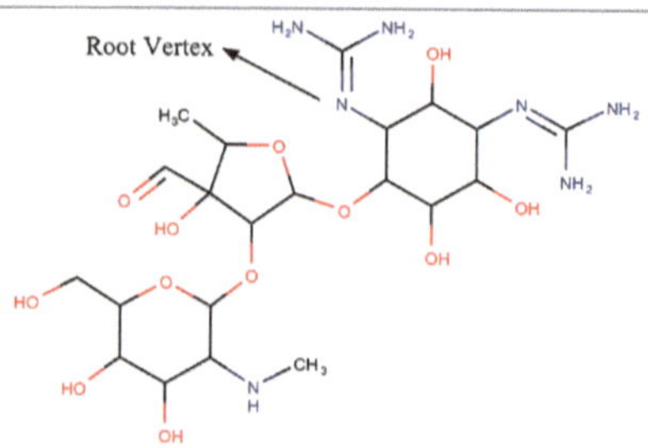

Streptomycin</td></tr>
<tr><td colspan="4">Compounds (in the Global TB data set) whose structures topologically matched with the source compound with the node deviation and node migration mentioned alongside</td></tr>
<tr><td>S. no.</td><td>Node deviation</td><td>Node migration</td><td>Matched compound</td></tr>
<tr><td>11</td><td>5</td><td>2</td><td>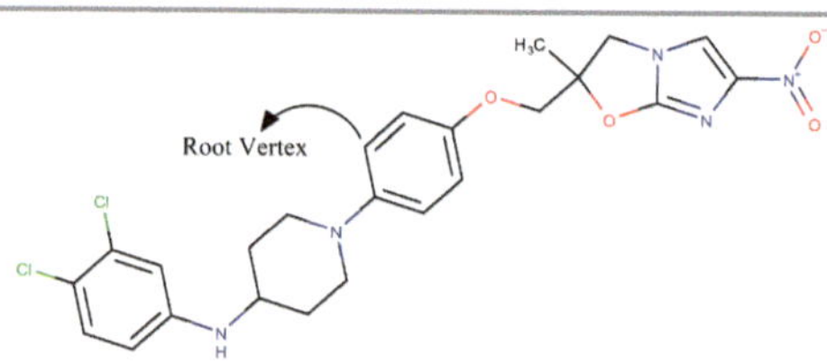

Compound No. 232</td></tr>
</table>

4 Conclusions and Future Prospect

The results obtained for different series of compounds using recently developed graph theory-based drug design/drug discovery method by our group [15] for combinatorial drug design from substructural topological information have been described in this chapter. Its application and usefulness for different series of antitubercular compounds have already been reported [15]. In this chapter, we have presented some new results for designing active compounds for barbiturates [18, 19] and nucleoside analogues [20, 21]. We have also reported some new results obtained for discovering novel active compounds from a data set using rooted tree/sub-tree searching/matching algorithms. In doing that, a data set (GTB) of 3779 potential antitubercular compounds [22, 23] has been taken for this study and the method has helped search a number of potentially highly active antitubercular compounds from this data set. Thus, to our knowledge, we have introduced here a method that can be used for searching databases to discover novel drug molecules using rooted tree and sub-tree matching algorithms. Furthermore, the usefulness of newly proposed Molecular Priority Score (MPS) for prioritizing and screening highly active compounds has also been described for the studies with a series of convulsant–anticonvulsant barbiturates and a series on nucleoside analogues for their activities against HIV. It is also found that the proposed method is capable of generating structures of known active compound that has scaffold different from that of the starting one. Furthermore, the structure generation starts from a vertex which plays a role in predicting biological activity. These observations seem to address the relationship of the present method [15] with two important aspects of modern-day drug discovery research—scaffold hopping and inverse QSAR (iQSAR) problem. Therefore, it appears that this newly developed method [15] may find useful applications in designing novel therapeutic candidates and may be helpful for working with drug resistance problems where compounds of very different molecular architecture may be sought for.

Our work presents an interesting alternative to "3D" drug discovery, where actual molecular coordinates in Cartesian space is used. Combinatorial design and generation in three-dimensional space would be far more expensive compared to our approach. Interestingly, one can always follow up on "3D" drug discovery based on molecule predictions from our method. This would allow a far tractable approach to drug discovery compared to a seemingly infinite exploration of molecules in actual "3D" Cartesian space.

Regarding future work, it may be worth exploring whether application of any quantitative measure for activity prediction can help screen potential bioactive compounds more effectively. Also, incorporation of new rooted tree-based compound generation and searching algorithms in the existing computer program would be another important aspect to work on. Finally, it would be of special interest to see how incorporation of ADME/Tox and drug-able property filters in the computer program can help discover drug molecules having desired pharmacological and undesired toxicological activities using the present method.

References

1. Ruddigkeit L, Van deursen R, Blum LC, Reymond JL (2012) Enumeration of 166 billion organic small molecules in the chemical universe database GDB-17. J Chem Inf Model 52:2864–2875
2. Hansch C, Sammes PG, Taylor JB, Ramsden C (1990) Comprehensive medicinal chemistry: quantitative drug design, vol 4. Pergamon Press
3. Kier LB, Hall LH (1986) Molecular connectivity in structure-activity analysis. Research Studies Press
4. Stuper AJ, Brügger WE, Jurs PC (1979) Computer assisted studies of chemical structure and biological function. Wiley
5. Kitchen DB, Decornez H, Furr JR, Bajorath J (2004) Docking and scoring in virtual screening for drug discovery: methods and applications. Nat Rev Drug Discov 3:935–949
6. Cramer RD (2003) Topomer CoMFA: a design methodology for rapid lead optimization. J Med Chem 46:374–389
7. Sun H, Tawa G, Wallqvist A (2012) Classification of scaffold-hopping approaches. Drug Discovery Today 17:310–324
8. Tanwar J, Das S, Fatima Z, Hameed S (2014) Multidrug resistance: an emerging crisis. Interdiscip Perspect Infect Dis 2014

9. Gálvez J, García-Domenech R (2010) On the contribution of molecular topology to drug design and discovery. Curr Comput Aided Drug Des 6:252–268
10. Gugisch R, Kerber A, Kohnert A, Laue R, Meringer M, Rücker C, Wassermann A (2014) MOLGEN 5.0, a molecular structure generator. In: Advances in mathematical chemistry and applications, vol 1. Bentham Publishers, pp 113–138
11. Harary F (1972) Graph theory. Addison-Wesley
12. Faulon JL, Bender A (2010) Handbook of chemoinformatics algorithms. CRC press
13. Wong WW, Burkowski FJ (2009) A constructive approach for discovering new drug leads: using a kernel methodology for the inverse-QSAR problem. J Cheminform 1:4
14. Klopman G (1994) Artificial intelligence approach to structure-activity studies: computer automated structure evaluation of biological activity of organic molecules. J Am Chem Soc 106:7315–7321
15. Raychaudhury C, Rizvi MIH, Pal D (2018) Combinatorial design of molecule using activity-linked substructural topological information as applied to antitubercular compounds. Curr Comput Aided Drug Des https://doi.org/10.2174/1573409914666180509152711
16. Beyer T, Hedetniemi SM (1980) Constant time generation of rooted trees. SIAM J Comput 9:706–712
17. Gibbs NE (1969) A cycle generation algorithm for finite undirected linear graphs. J ACM 16:564–568
18. Klopman G, Raychaudhury C (1990) Vertex indexes of molecular graphs in structure-activity relationships: a study of the convulsant-anticonvulsant activity of barbiturates and the carcinogenicity of unsubstituted polycyclic aromatic hydrocarbons. J Chem Inf Comput Sci 30:12–19
19. Raychaudhury C, Pal D (2012) Use of vertex index in structure-activity analysis and design of molecules. Curr Comput Aided Drug Des 8:128–134
20. Raychaudhury C, Klopman G (1990) New vertex indices and their applications in evaluating antileukemic activity of 9-anilinoacridines and the activity of 2′, 3′-dideoxy-nuclosides against HIV. Bull Soc Chim Belg 99:255–264
21. Raychaudhury C, Dey I, Bag P, Biswas G, Das B, Roy P, Banerjee A(1993) Use of a rule based graph-theoretical system in evaluating the activity of a class of nucleoside analogues against human immunodeficiency virus. Arzneim Forsch Drug Res 43:1122–1125
22. Prathipati P, Ma NL, Keller TH (2008) Global bayesian models for the prioritization of antitubercular agents. J Chem Inf Model 48:2362–2370
23. GTB data set. http://pallab.cds.iisc.ac.in/gtb_data.mol
24. Kandel DD, Raychaudhury C, Pal D (2014) Two new atom centered fragment descriptors and scoring function enhance classification of antibacterial activity. J Mol Model 20:2164
25. Raychaudhury C, Kandel DD, Pal D (2014) Role of vertex index in substructure identification and activity prediction: a study on antitubercular activity of a series of acid alkyl ester derivatives. Croat Chem Acta 87:39–47
26. Moss G (1999) Extension and revision of the von Baeyer system for naming polycyclic compounds (including bicyclic compounds). Pure Appl Chem 71:513–529
27. Weininger D, Weininger A, Weininger JL (1989) SMILES. 2. Algorithm for generation of unique SMILES notation. J Chem Inf Comput Sci 29:97–101

In Silico Structure-Based Prediction of Receptor–Ligand Binding Affinity: Current Progress and Challenges

Shailesh Kumar Panday and Indira Ghosh

Abstract Structure-based in silico studies aiming to predict affinity of a set of ligands to their cognate receptor have been enjoying keen interest and attention of researchers in drug design around the globe since many decades, and made significant progress to increase its predictive power, even it has emerged as a complementary field to in vivo and in vitro studies in recent years. Structure-based drug discovery (SBDD) process whose success heavily relies on a careful selection of structure of receptor and ligands and its accuracy, completeness, and rigor of chosen model, imitation of the physiological condition in such in silico models, e.g., pH and solvation. Appropriateness of selected mechanism of binding concept and the realization in mathematical terms used in scoring methods have a strong influence on the accuracy too. However, constant identification of new targets using systems approach like genomics, proteomics, metabolomics, and network biology has led a paradigm shift from single or a couple of targets toward the appreciation of emerging role of a network of targets. The application of such strategies in study of complex diseases is gaining attention. Identification of binding sites of receptor and their characterization is important to be able to portray its interacting features. It involves the search of ligands which are able to possess the features, present them complementary to the binding site, so by docking the set of ligands to the binding pocket of the receptor, activity can be evaluated. In silico receptor–ligand binding affinity prediction from docking has witnessed rigid-receptor rigid-ligand to flexible-ligand rigid-receptor treatment, and nowadays docking studies, through sampling side chain rotations of the binding site residues, also account for the flexibility of binding pocket of the receptor in indirect way. Literature survey has shown progress in ranking ligands in order of affinity using reliable scoring functions to find potent scaffolds which can be further optimized to gain more affinity. Many methods include effect of solvation in binding processes, like considering conserved water positions in active sites (water maps), explicit water simulation in presence of ligand with receptor, free energy perturbation, and thermodynamic

S. K. Panday · I. Ghosh (✉)
School of Computational and Integrative Sciences (SCIS),
Jawaharlal Nehru University, New Delhi 110067, India
e-mail: indira0654@gmail.com

C. G. Mohan (ed.), *Structural Bioinformatics: Applications in Preclinical Drug Discovery Process*, Challenges and Advances in Computational Chemistry and Physics 27, https://doi.org/10.1007/978-3-030-05282-9_5

integration. Availability of many conformers of receptors and ligands in solution suggests the importance of entropy in estimation of binding affinity, but entropy component of binding free energy directly is not included in such studies. In spite of unprecedented advancement of computational modeling, faster simulation techniques, accurate solvation models and current best practices, the dependence of binding affinity on pH, estimation of entropy along with enthalpy in binding affinity, inclusion of conformational entropy of ligand and receptor, and modulation of flexibilities during complex formation are important challenges lying ahead. Therefore, an account of prowess and challenges in structure-based prediction of binding affinity addressed in present review will provide directions for its appropriate application, understanding its limitations and getting important feedbacks for its betterment.

Keywords Structure-based drug design · X-ray crystal structure
Scoring function · Docking · Simulation · Structure validation
MM-PBSA · Entropy · Free energy

1 Introduction

The advancement of molecular understanding of the disease processes and their manifestations, along with computational advancement like in silico studies, aiming to predict high-affinity molecules/scaffolds binding to the target, grew as a promising complementary field of study mainly because of its cost-effectiveness and speed. It facilitated virtual high-throughput screening (vHTS) to narrow down the search space for further experimental work by making predictions about the ligand–receptor affinity [1]. Advancements in systems biology along with network biology helped identifying targets for diseases [2], and crystallography [3] and nuclear magnetic resonance (NMR) [4] techniques enabled solving structural models of the target molecules with higher resolution setting foundation of structure-based drug designing (SBDD). Docking is one of such computational studies, which aims to search high-affinity molecules from a library of chemicals and predict relative orientation (pose) of the molecule to the target. It also tries to rank the set of molecules/poses in a sorted affinity order [5]. Knowledge about the structure of receptors made binding site identification easier and enabled to screen the small-molecule libraries against the target seeking complementarity with the ligand.

Docking and scoring methods due to its promising applicability prospect has been extensively developed, critically evaluated, and constantly refined with the time, it has now shaped into a field of research; several software tools have been developed and are available for academic and industry research [5–11]. Recently, Taylor et al. [12] have reviewed the broad spectrum of major techniques amenable to the field of non-covalent docking studies, classifying them into molecular dynamics, Monte Carlo methods, genetic algorithms, fragment-based methods, point complementarity methods, distance geometry methods, tabu searches, and systematic searches. They briefly presented algorithms and validations of models and techniques using test cases as examples. The study has concluded that hybrids of various types of algorithms employing novel search for appropriate poses and consensus scoring are better for large-scale docking [12]. It has been observed that rigid receptor and flexible ligand models achieved success rates of 70–80%. It can be influenced by the fact that programs implementing these algorithms were well established at that time [12]. However, they pointed out that possible reason for failure is underestimation of conformational sampling of receptor flexibility [12]. In spite of great success of docking methods in discriminating ligands as good and bad, predicting the binding on the basis of their affinity towards cognate receptor is poor. Moreover, in certain cases, docking shows inability to reproduce experimental binding pose and it is a great concern in the technical aspects of the docking methodology and its current progress, so need to review time to time. In 2010, Huang et al. [13] have discussed currently practiced docking techniques, delineating the ways for ligand sampling, accounting protein flexibility and specific scoring functions.

During a docking study, one has to do many sequences of tasks/steps which influence the final outcome of the study and its success [14]. First and the foremost thing is to search for the potential binding sites on the receptor and characterize them; however, sometimes when binding site is not known blind docking can be done. Several cavity detection algorithms and software were built to help this. In parallel, right selection of the receptor structure is crucial [14]; thus, the quality of the structure and experimental conditions used for resolving the structure has to be taken care of, and structure resolved with experimental conditions closest to the actual functioning condition should be preferred if available [15]. Most often, hydrogen atoms are missing in the structure; thus, protonation states of the titratable receptor residues have to be fixed, and usually, it is borrowed from predictions made using different protonation state prediction tools [16, 17]. Apart from the protonation states of titratable residues of the receptor, ionization states of ligands to be docked have influence on correct model of binding [16, 18]. Scoring functions also greatly influence the final outcome of the docking studies, and there are many scoring functions available; some may be suitable to study the specific type of protein active site but less effective in other cases [19]. Inherent demand of fast evaluation of poses during docking enforces the scoring functions to adopt approximations and parameterization, which compromises predictivity [19]. Thus, it is tough to guess which scoring will be suitable for which kind of active site. However, chemical intuition and consensus scoring protocols can be adopted to get better results.

Although the correctness of ranking and order of predicted affinity more often fail to provide significant correlation with experimental ranking and observed pose [20], such limitation of the in silico high-throughput screening can be partially attributed to the multifaceted problems in current practices, e.g., selection of appropriate binding theory, selection of appropriate modeling data, and limited knowledge about the reaction mechanism. Many such challenges are discussed in the present article.

1.1 Targets Are Diverse

To be able to comprehend the challenges lying ahead on the way to drug design/ drug discovery, it is important to understand the diversity of the drug targets that have been exploited so far as well as the trend in new drug targets in recent history of drug discovery [21]. Mathias Rask-Andersen et al. performed a study on all drugs approved by FDA during 1983–2010. They took all 1542 drug entries as on May 2009 and filtered out 225 drugs with unknown targets, 192 with no human targets, and 609 non-therapeutic targets to yield a dataset of 435 therapeutic effect-mediating targets for humans and to account for the time lag between drug approval and their entry in DrugBank; drugs approved during 2007–2010 were taken from FDA data and included for analysis. Drug–target association was annotated by manual curation from literature data, and targets were kept in four classes (receptors, enzymes, transporters, and others) with receptor class has highest 193 targets, followed by enzymes with 124, transporters with 67, and others with 51 targets [21]. Analyzing curated drug–target association dataset, they found that every year 17.9 drugs targeting human proteins are approved by FDA, while 4.3 of them act on novel targets. The trend in FDA approval of drugs targeting new human proteins (novel target drugs: NTDs) does not decrease overall. Moreover, they noticed three peaks corresponding to durations 1990–1993, 1994–2000, and 2001–2008 when NTDs were plotted against years from 1983 to 2010; they called them first-, second-, and third-target "innovation peaks," respectively [21].

During the first innovation peak, it was observed that proportions of approved drugs for all major target groups—GPCRs, hydrolases, transferases, and isomerases—were similar to other two peaks. During second innovation peak, first time integrins appeared as drug target, while during the third innovation peak, asthma drug omalizumab-targeted Fc-receptors and imatinib appeared as kinase inhibitor [21].

Analysis of novel targets for drugs with time by Mathias Rask-Andersen et al. highlights the fact that with the passing time new drugs apart from targets belonging to earlier exploited classes, novel classes of targets are also being identified for new drugs. Thus, diversity in the classes of target molecules is expanding, and SBDD practices have to be optimized to improve success rates in such studies. Present review will attempt to enlighten and discuss the solutions for such relevant topics including the challenges upcoming ahead.

1.2 Targets Are More Diverse than Earlier

Genomic-wide association studies over a set of druggable genome, utilizing bioactivity data including approved drugs or clinical compounds and gene association data against these targets, can be used to come up with set of further druggable genes and gene combinations as target [22]. Recently in 2017, Finan et al. have performed a similar study and estimated that 4479 genes can be drugged or are druggable out of total 20,300 annotated protein-coding genes as per Ensembl version 73 (https://www.ensembl.org/) covering $\sim$22% of total. They reported that there could be 2282 genes more than earlier reports of the druggable human genome [22].

Systems biology approaches have been used for decades for predicting target genes in case of infectious diseases [2], studying systems approaches, e.g., metabolic control analysis (MCA) and flux balance analysis (FBA). Systems genetics approaches have also been used for identification of novel disease genes in rat and human [23]. Molecular networks information can be used for improving drug discovery projects at several stages from target identification utilizing information of existing data about drug–target association [24]. Metabolic and signaling pathway [25] and genome-wide association are studied in detail for identification of new target proteins and their interactions [26]. Genome-led methods provide a new pathway or a class of protein(s) as target.

Pharmacophore designed from ligands of a target protein can be looked for assessing binding site similarity for the proteins of same family as well as it can be used to compare binding site similarity for proteins from different families of proteins for selectivity. In recent times, several highly selective inhibitors of such protein(s) have been found to assess the multitarget activity. For example, c-Abl inhibitor imatinib [27] was approved as drug for chronic myeloid leukemia, but its clinical utility is widened after finding that it has shown significant activity against several other important targets, e.g., tyrosine-protein kinase kit (c-KIT or CD117). Similarly, sorafenib affects tumor proliferation and tumor angiogenesis pathways due to its multikinase inhibitory activity [28]. Sunitinib is also approved for being a multiprotein kinase inhibitor with similar effects as sorafenib [28].

1.3 Starting of Structure-Based Drug Design

One of the successful stories of the structure-based drug design started in the early eighties with purine nucleoside phosphorylase (PNP), targeted as a salvage enzyme important to inhibit, so that T-cell-mediated activation of immune system is suppressed. PNP is an important enzyme involved in purine salvage and catabolism [29]. Inactivity of PNP has been found to show adverse effect on T-cell proliferation [30]. Human PNP, a homotrimer with each subunit of molecular weight 97 kD, shows substrate specificity for guanine, inosine, and other 6-oxypurines analogs, while bacterial PNP shows specificity for adenine [30] also. PNP active site consists of three binding subsites: purine-binding site (Fig. 1, shown in cyan), hydrophobic site (or ribose-binding site, Fig. 1, shown in blue), and phosphate-binding site (Fig. 1, colored purple) [31]. In attempt to design potent PNP inhibitors, considering the features of three subsites of PNP binding site and three-dimensional structure of PNP as starting point, an iterative process of modeling inhibitor-bound structure, conformational search using Monte Carlo method followed by energy

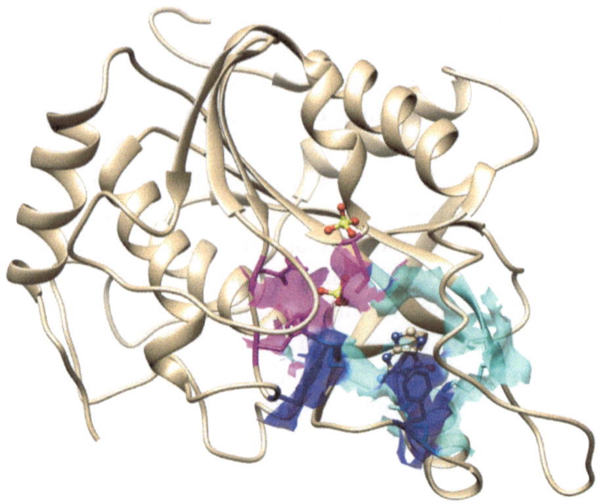

Fig. 1 Human purine nucleoside phosphorylase (PNP) monomer (PDB: 1ULB) in complex with guanine and sulfate ions. Guanine and sulfate ions are shown in ball and stick. Three subsites of PNP binding site: First subsite is called purine-binding site (shown in cyan surface, residues Ala116, Phe200, Glu201, Val217, Met219, Thr242, Asn243, Lys244), second subsite, i.e., hydrophobic site (or ribose-binding site consists of residues His86, Tyr88, Phe159 (from adjacent subunit of PNP trimer), Phe200, Met219) where Tyr88 and Phe200 are shown in blue surface. The third subsite termed phosphate-binding site (shown in purple surface residues Ser33, Arg84, His86, Ser220)

minimization and finally experimental determination of binding affinity and crystallization of complex structure was used. This iterative process yielded a series of potent and membrane-permeable 9-(arylmethyl)-9-deazapurines (2-amino-7-(arylmethyl)-4H-pyrrolo[3,2-d]-pyrimidin-4-ones) inhibitors of PNP [29]. Later, (S)-9-[l-(3-chlorophenyl)-2-carboxyethyl]-9-deazaguanin showed highest potency among all previously designed analogs [32]; however, the (R)-isomer was 30-fold less potent. This study exemplifies how structural information can be carefully used toward designing of potent inhibitors of the receptor of interest.

The enthalpy and entropy components of binding free energy together decide affinity of interaction between receptor and ligand. Therefore, affinity can be modulated favorably adopting following possible strategies: (i) decreasing the unfavorable entropy maintaining favorable enthalpy, (ii) increasing favorable enthalpy without introducing unfavorable entropy, and (iii) altering one or both of enthalpy and entropy favorably without losing proportionally on other component [33].

An example where first strategy has been used for optimizing affinity is inhibitors of PNP. Optimized picomolar-binding PNP inhibitors have also been reported [34]. The attention has been paid on reducing the entropic penalty, without sacrificing the enthalpy of binding to gain affinity. Hypoxanthine has K_i 4.3 μM, with enthalpy −30.5 kcal/mol, but 23.1 kcal/mol entropy penalty to result a −7.4 kcal/mol binding free energy [35], but optimized molecule SerMe-ImmH shows 5.2 pM K_i, with −20.2 kcal/mol enthalpy, but merely 4.7 kcal/mol entropy to result −15.5 kcal/mol binding free energy [34].

The second strategy has been utilized for optimizing HIV-1 protease inhibitors. After the FDA approval of Indinavir in 1995, which binds only because of −14.2 kcal/mol entropy despite 1.8 kcal/mol unfavorable enthalpy with binding free energy −12.4 kcal/mol, the process of affinity optimization started. The constant optimization of inhibitors for efficacy leads to Darunavir which binds with only −2.3 kcal/mol favorable entropy; however, −12.7 kcal/mol favorable enthalpy yielded binding free energy −15.0 kcal/mol. The free energy gain of −2.6 kcal/mol was reported where every −1.4 kcal/mol results ten times better binder [36, 37]. Another such example involves cholesterol-lowering drug statins to HMG-CoA reductase, and Fluvastatin binds only due to −9.0 kcal/mol favorable entropy despite zero contribution from enthalpy. However, newer drug Rosuvastatin binding has only −3.0 kcal/mol entropy contributions, but additional −9.3 kcal/mol enthalpy gain results −12.3 kcal/mol binding free energy, −3.3 kcal/mol better than Fluvastatin [38].

The third strategy is more tedious and challenging mainly because of enthalpy entropy compensation, more often enthalpy can be increased by introducing new hydrogen bonding groups as a strong hydrogen bond which provides ~4–5 kcal/mol enthalpy; however, introduction of hydrogen bond decreases favorable solvation and entropy by structuring regions involved in hydrogen bonding. Alternatively, in theory, introducing multiple hydrogen bonds targeting same structural regions of receptor has been suggested to mitigate the extent of enthalpy entropy compensation [33].

1.4 Flexibility and Adaptability of Target

Initially, the protein–ligand docking was modeled as a lock-and-key, where protein was treated as "lock" containing a binding site as "key-hole" which can host a complementary ligand or "key." However, later it was realized that lock-and-key model is not sufficient to characterize all binding events; thus, advanced models were proposed which can be put broadly in three groups: (i) lock-and-key (ii) induced fit (IF), and (iii) conformational selection (CS) [39]. The IF and CS models introduced to account for the receptor flexibility during the binding with ligands will be discussed in detail later. Although these models represent receptor–ligand binding in better way, still estimate only enthalpy of the interaction and the entropy component of the binding free energy remains to be estimated. It has been reported in the literature that entropic component of binding can be important in many interactions. A recent experimental and computational study of a human heat-shock protein 90 (HSP90) highlighted important alterations in binding properties of target on complex formation with small-molecule inhibitors [40]. Surprisingly, they found that compounds binding to helical conformation have increased target flexibility and gained entropy preference over compounds binding to loop conformation which was less flexible on complex formation [40].

1.5 *Knowledge of Target Structure Is Essential but not Sufficient*

In spite of success in structure-based drug discovery process [29] at several occasions, knowledge about the structure of the target involved in the disease does not necessarily lead to a drug for cure; β-Thalassemia is one such example. It is an inherited hematologic disease caused by less β-globin, largely reported in Mediterranean region, identified with the mutant β-globin [41]. The present treatment is continuous blood transfusions with chelation therapy [42] and less frequently, bone-marrow transplantation [43], because there is no drug treatment for cure. However, the first crystal structure of hemoglobin was known in 1968, and since then, more than 250 human hemoglobin structures are known [44]. Hence, druggability and understanding of disease is a field of research in itself, emerging as translational bioinformatics.

2 Challenges in Structure-Based Designing

As discussed in many review articles earlier, major steps to find in silico chemicals and design them for better inhibition of target macromolecule are identification of target protein or macromolecule of importance and associated functionally with the disease, characterization of its 3D structure and active site, mapping of interactions possible with chemical functional groups, docking, scoring, and finally ranking the possible chemicals to test experimentally. Each of these steps has many challenges which will be discussed here.

2.1 *Accuracy of Structures*

Before starting a docking study to screen, some library of compounds to come up with a set of molecules showing high binding affinity with the target receptor requires to have known 3D structure. The appropriate selection of the receptor structure can influence the success or failure of any screening study [14]. Therefore, a researcher needs a good structure to start with which could have been resolved mostly using X-ray or NMR. Sometimes, the structure of the desired receptor is not known. In such cases, a homology model of the structure can be used if a suitable template for the receptor can be found [14]. A template may be the same protein having similar function, showing high sequence similarity from different organism or even some other protein having same fold. If the structure of the receptor is known in advance, then there may be multiple structures resolved in different conditions, with varying resolution, varying model completeness, etc. In such a case, the most suitable structure has to be chosen [14]. In selecting receptor structure, one has to keep in mind that how well the structure resolution condition matches with the actual functioning condition of the receptor and resolution of the structure [14]. Apart from this, many questions may arise like whether the structure is ligand bound? Whether active conformation of the structure is solved? Whether the structure is solved at pH similar to the functioning pH? These can also be of importance to consider during docking. The receptor crystal structure selection has to be done with care considering the quality of the structure model. Some of the most important parameters for crystal structure assessment have been outlined in the literature [45] and listed in Table 1. Crystal structure resolution which is a measure of quality of electron density data collected is one of such parameters; structures resolved at less than 1 Å are considered high-quality one being able to resolve electron densities at atomic level while structures greater than 3 Å have smeared electron densities and atomic positions are not clearly identifiable. Hence, crystal structure with resolution in range: 1 Å < resolution < 3 Å can be

Table 1 List of important parameters for assessing quality of X-ray crystal structure

Parameter	Description	Preferred	Comment
Electron density and solved model quality			
σ-cutoff	σ-cutoff applied to the data	None	
Lower resolution	A minimum spacing (d) of crystal lattice planes that still provide measurable diffraction of X-rays.	20–50 Å	
Higher resolution	A minimum spacing (d) of crystal lattice planes that still provide measurable diffraction of X-rays and also $\langle I/\sigma(I)\rangle$ greater than 2 in high-resolution shell.	<3 Å	Higher is better
Completeness	The number of observed reflections divided by the theoretical maximum	~100%	Higher is better
$\langle I/\sigma(I)\rangle$	The average ratio of reflection intensity to its estimated error. Signal-to-noise ratio	>2	
R-factor	A measure of the global reliability factor or goodness-of-fit between the experimentally obtained structure factor amplitudes, F_{obs}, and the calculated structure factor amplitudes, F_{calc}, obtained from the model.	<25%	Smaller is better
R_{free}–R-factor	R_{free} is R-factor for random ~5% reflections, not used for model refinement. $R_{free} - R$-factor < 2, may be indication of overfitting while $R_{free} - R$-factor > 7 may be due to poor refinement of model	2–7%	Smaller is better
R_{O2A}	Observation to atom ratio		Higher is better
Geometric parameters of model quality			
RMSD (bonds)	Root mean square deviation of bond lengths from ideal values	0.15–0.25 Å	
RMSD (angles)	Root mean square deviation of bond angles from ideal values	1°–3°	
Ramachandran violations	Number of ϕ–ψ torsion pairs falling in disallowed regions of Ramachandran plot	0	Smaller is better

considered reasonable quality structures [46]. Apart from resolution, R_{value}, R_{free} and real-space R-value and real-value correlations are among the important parameters to assess the quality of crystal structure as discussed by Brown et al., in 2007 [45].

Geometric parameters and quality of structure: Apart from diffraction quality and structure refinement parameters, geometric and chemical parameters are equally important to consider while assessing its quality [15]. Atomic positions in model, planarity of peptide plane, stereoisomer of peptide bond, bond length, bond angle, and torsions angles should be checked for an unnatural occurrence [15]. Since all combinations of backbone torsions ϕ–ψ cannot occur in proteins, only those pairs which conform to the Ramachandran plot, thus number of ϕ–ψ pairs in disallowed regions of the Ramachandran plot which ideally should be zero, generally lesser violation considered better structure, are used as a critical parameter for the quality of the crystal/model structure as best practices.

Atomic occupancy and B-factor are among other important parameters to be considered while assessing the quality of structure. Occupancy of an atom is the fraction of molecules which occupy modeled position among all molecules in crystal. An occupancy 0.0 means modeled positions not observed in crystal, and 1.0 means modeled position is present in all molecules in crystal [47]. If some residues in crystal structure show more than one conformations in crystal structure, then conformation with highest occupancy should be preferred. In case of ligands, the occupancy is dependent on K_d value, e.g., for a ligand with K_d in range 10–100 mM, maximum achievable occupancy ranges 70–90% or 0.70–0.90 considering working ligand concentration <500 mM [48]. B-factor in theory represents the amplitude of oscillation of the atom around equilibrium position. It quantifies the dynamics of the atom; often, isotropic B-factors are reported in crystal structures; however, anisotropic B-factors may be reported in high-resolution structures. For high-resolution structures, anisotropic atomic displacement parameter (B-factor) can be substantiated only when resolution is higher than $\sim$1.4 Å [46]. Structural regions in crystal structure having B-factor higher than a threshold *B_max* should be carefully inspected because of their implications to high disorder in the region [49].

At times, in crystal structure water molecules play important role in binding and have to be considered for characterizing the binding site for its water interaction profiles [50]. However, identification of structurally important waters involved in receptor–ligand interaction is another challenge [51, 52].

Proteins are usually flexible molecules, and inherent dynamics characterizes its interaction. Moreover, a crystal structure is usually a time and space average of the conformers present in the crystal lattice [15]. Therefore, quite often it may not be the conformation presenting the best possible affinity for the given ligand due to the rigid treatment of the receptor. Thus, protein should be allowed to flex in such way that it could show best possible affinity with the ligand.

2.2 Comparative Homology Modeling and Role of Template

Very often the target protein crystallization is not possible, and no other way but homology or comparative modeling of structure becomes imperative. Many standard tools and directions are reviewed, and appropriate protocols are included [53, 54]. Many such tools to evaluate the modeled structures are also discussed in the literature [55, 56]. Here we shall cite a specific example showing importance of choice of template using homology modeling applied for *Mtb* isocitrate dehydrogenase (ICD).

Mycobacterium tuberculosis is known to use the glyoxylate shunt during the persistent stage [57]. Experiments have been performed to understand the glyoxylate shunt by considering the close analogy with *Escherichia coli* system [58]. For *E. coli*, glyoxylate shunt pathway is well studied and is initiated by phosphorylation of specific serine-105 residue of isocitrate dehydrogenase (ICD) [59]. *Mycobacterium tuberculosis* being a prokaryotic organism, same type of functionality was also expected for the glyoxylate bypass pathway [58, 60].

Phylogenetic analysis of the ICD sequences shows that *Mtb* has NADP-dependent ICD which belongs to subfamily II of ICD. Subfamily II has predominantly eukaryotic members, while *E. coli* ICD is classified in subfamily I [61]. Across the family, ICDs are found to be functional either monomers or dimers. *E. coli*, *Mtb*, and human all have functional homodimeric forms. Dimeric ICDs contain active sites which are contributed by the residues of both domains. Though *Mtb* ICD is regulated by phosphorylation process, it is more equivalent to eukaryotic ICDs. Eukaryotic ICDs are not found to be regulated by the phosphorylation, and also mammalian system does not possess glyoxylate shunt [62]. So overall evidence suggest that *Mtb* ICD has close similarity with eukaryotic system; however, the presence of glyoxylate shunt pathway makes this system closer to prokaryotic intracellular pathogenic survivor.

Understanding of shunt pathway shown that regulation of the *Mtb's* ICD depends upon the phosphorylation/de-phosphorylation state which is expected to be regulated by some of available 11 serine/threonine phosphatase/kinases [63]. In 2009, Vinekar et al. had performed molecular dynamics simulation-based analysis to understand the effect of selective phosphorylation of serine residues [62]. However, crystal structure of *Mtb* ICD was not available at that time (Table 2), so homology modeling had been done using different crystal structures as templates to select appropriate functional model.

The ultimate goal of the homology-based structure modeling is to model the structure from its sequence with an accuracy that is comparable to the best results achieved experimentally. As the crystal structure of *Mtb* ICD was unavailable, homology-based structure modeling was the preferred way to understand the structural features of the ICD. For ICD modeling, target sequence (UniProt ID: P9WKL1) was found to align with many sequences of already crystallized structures from both prokaryote and eukaryote. Based on the homology rules of %-identity, functionality, quality of the structure, and association with same taxonomy, three ICDs [64] were

Table 2 Comparison of the crystal structure of *Mtb*[a] with selected (template) prokaryote and eukaryote crystal structures

ICDH	*Mtb*	*Sus scrofa*	*E. coli*
Sequence length	409	413	416
Template PDB ID	4HCX [66]	1LWD [65]	3ICD [64]
Year of publication	2013	2002	1989
Template structure resolution (Å)	2.18	1.85	2.5
R_{free}	0.262	1.85 Å	NA
R_{work}	0.205	0.210	0.180
Ramachandran outliers (%)	1.8	0.2	0.5
Sequence **Identity** with respect to *Mtb* ICDH (UniProt ID: P9WKL1) (%)	100	65.2	23.6
Sequence **Similarity** with respect to *Mtb* ICDH (UniProt ID: P9WKL1) (%)	100	79.2	35.7

[a]Earlier modeled because structure was not available till 2013

selected as template structure for modeling. However, in cross-taxonomy (with eukaryote) 1LWD [65], same target sequence had higher sequence identity (Table 2) than *E. coli*. Both crystal structures (3ICD and 1LWD) have same Rossmann fold and a common dinucleotide-binding domain [64, 65].

In such case, where target structure from the same taxonomy is available and fulfills the most homology modeling criteria, it is not always true that model structure will also provide functional explanation. Model developed using *E. coli* is shown in Fig. 2a (dark gray color) with *E. coli* crystal structure (green color). Both structures are superimposed well with RMSD 4.68 Å. However, model structure (white color) developed using *Sus scrofa* (orange color) as template superimposes with crystal structure with RMSD of 0.57 Å (Fig. 2c). Both models are validated using PROCHECK [55], and more than 85% residues are found under Ramachandran region. So, both models follow homology criterion and passed by the structure validation tools.

In 2013, Quartararo et al. published the crystal structure of *Mtb* ICD dimer complex with NADPH. This structure is then used to understand the closeness of modeled structure of *Mtb* ICD with both *E. coli* and *Sus scrofa*. Superimposition of *Mtb* with *E. coli* and *Sus scrofa* is shown in Fig. 2b, d, respectively. Although all three have same folds, *Sus scrofa* is more close toward *Mtb* than *E. coli*. *E. coli* structure has 6.4 Å RMSD with *Mtb*, and major differences occurred in the beta-hairpin loop region where *E. coli* structure has helical element than beta-structure element. This region of dissimilarity known as clasp region between inter-subunit interface [64] plays important functional role during phosphorylation [61].

So, from this case study, it is very clear that one template cannot guarantee about the functional state of the homology model, so different templates may be used to develop appropriate functional model, as mentioned in comparative modeling review [53]. Key to the selection of the model is always to be associated with the

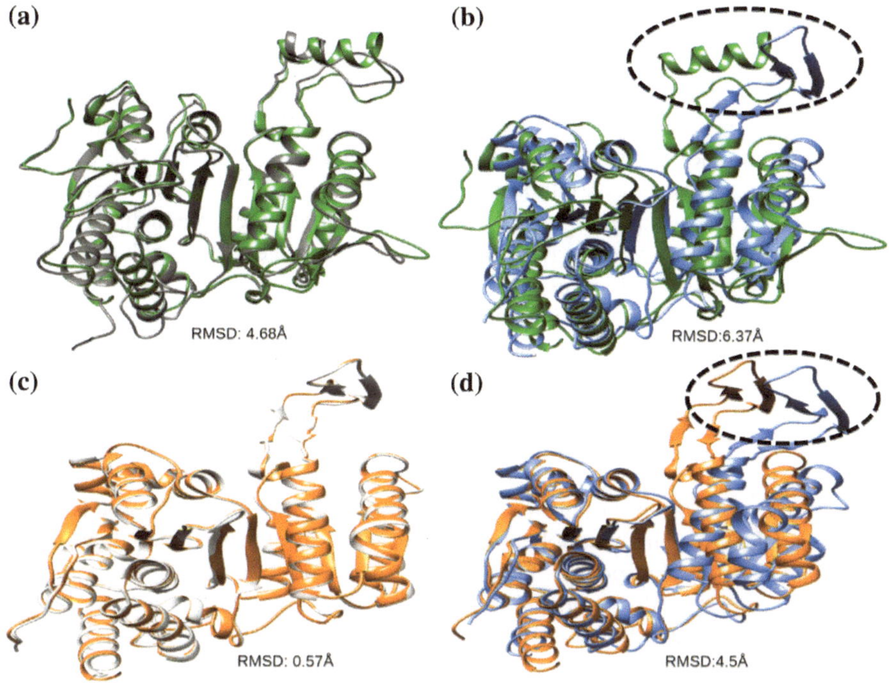

Fig. 2 Two homology-based models have been developed for *Mtb* ICD using two different crystal structures (one from *E. coli* and one from *Sus scrofa*). **a** Shows the modeled structure (dark gray color) superimposed with *E. coli* crystal structure (green color). Model fit well with 4.6 Å RMSD value. **c** Second model is developed using *Sus scrofa* structure (1LWD) and superimposed model structure (white color) is shown with 1LWD (orange color). When *Mtb* structure published in 2013, it is found that mammalian ICD is much closure to *Mtb* as shown in panel (**d**) than *E. coli* (panel **b**). Fold is well conserved in both models, but major differences are highlighted in clasp region (shown in black circle)

experimentally known functional features. It is also established that structure validation tools like PROCHECK [55] and WHAT IF [56] can only suggest the quality of the models not the functionality of the modeled protein. Other methods popularly known as ab initio designing of protein, alternate to template-based modeling, have been discussed in other reviews [67, 68]. A comparison of efficiency of modeling protein structure called CASP (critical assessment of methods of protein structure prediction) provides evaluation of such programs [69]. Recently, designing of protein structures has been successfully applied to model protein from genome sequence using an integrated pipeline by Jayaram and co-workers [70]. However, ensembles of model structure may provide a better docking success which has been cited in 2010 by Novoa et al. [71].

2.3 Ligand Flexibility

Apart from the traditional approach to look for potential inhibitor as small molecules for proteins, small peptides can also be strategically designed to complement interaction hot spots presented by receptor molecules, using knowledge about the structure of receptor and its interacting partner molecules. In a recent article published in Science, Kadam et al. [72] have exemplified the approach. The study focuses on influenza type 1 virus and their surface protein hemagglutinin (HA), which is associated with virus invasion of host cells. HA is composed of two domains HA1 and HA2, and functional unit is a homotrimer of HA. The interface of HA1 and HA2 forms a hydrophobic pocket. This HA-binding site, which is near the stem region of the HA membrane, is targeted by the broadly neutralizing antibodies (bnAbs) of the host and blocks large conformational rearrangement associated with membrane fusion and thus neutralize virus [72]. Structurally, analyzing the epitopes, at HA1/HA2 interface, a highly conserved site was found. This structural information allowed researches to synthesize novel proteins, e.g., HB80 and HB36, which could mimic bnAb paratope CR6261 and bind in the conserved hydrophobic pocket, by placing amino acid side chains in appropriate configuration and conformation. These proteins did show binding affinity comparable to CR6261 and inhibited low pH-induced conformational change in HA. Further, optimizations lead to improved analogues of HB36, which were also effective in protecting mice against lethal H1N1 infections [72].

Success of de novo designed protein inspired researcher to look for even smaller peptide like inhibitors seeking better drug-like properties, e.g., availability in blood stream with higher lifetime. Starting from the available structural and functional information about bnAbs, e.g., CR9114, CR6261, F10, A06, FI6v3, HCDR3 was selected which possesses major interactions as the starting point for design of smaller HA inhibitory peptides. After creating a pool of potential HA inhibitory peptides mimicking different structural features of the HCDR3 loop [72] and characterization of each peptide in terms of its thermodynamic (K_d) and kinetic parameters (k_{off} and $t_{1/2}$), a combination of all distinct structural features of these peptides into an 11-mercyclic peptide containing five non-proteinogenic residues was synthesized. This peptide showed better affinity and longer residence time for binding to HA. This study exemplified a novel approach, where compendium of available structure is utilized with chemical intuition of structure and function to yield a small cyclic peptide with better therapeutic prospect over existing inhibitory proteins, e.g., HB36 and its variants [72].

Alternatively, another novel idea has been floated by Young et al. of stapling small peptides to protect them from proteolytic cleavage and further designed a series of stapled peptides among which mimic of α-helical peptide ATSP-7041 was reported to be a potent and selective dual inhibitor of MDMX and MDM2 [73]. However, MDM2 and MDMX are suppressor of p53, thereby activates p53 pathway in tumors [74]. In a recent in silico study, where Garima et al. tried to study the mechanistic aspect of recognition of small stapled α-helical peptide ATSP-7041 with human serum albumin

(HSA) and compared it with mouse serum albumin (MSA) [75], starting from the crystal structures of HSA and ATSP-7041 in complex with MDMX. They used 50 ns molecular dynamics simulations to sample conformational states of HSA; simulation trajectories were clustered to give five clusters, and in these six (five cluster representatives and one crystal structure) HSA conformations were used for further docking studies. ATSP-7041 were fully blindly docked to above six HSA conformations using protein–peptide docking tool pepATTRACT [76] and generated ensemble ($\sim$24,000 poses) of possible docking poses for each; then these ensemble of poses was clustered using *k*-means algorithm to result 40 clusters for each of six HSA conformations. Further, they refined each of the 40 clusters representative poses for each of six HSA conformations and then performed MD simulation for 5 ns to assess the stability of the pose. Their study resulted four binding sites R1, R2, R3, and R4 which were most occupied and considered for further study. Moreover, representative poses of ATSP-7041 and HSA complex one for each site was simulated using explicit solvent, and binding affinity was estimated using MM-GBSA method. However, for MSA, no crystal structure was available, so they modeled it using swiss model choosing HSA as template. ATSP-7041 was kept in MSA at sites R1, R2, R3, and R4, and three replicates of 100 ns MD simulation in explicit solvent were performed. Their analysis of these results suggested that sites R2 and R3 were not stable for mouse in contrast to human which they attributed to sequence dissimilarity at the region in human and mouse serum albumins. Moreover, they also found that sites R1 and R4 have lesser affinity in case of mouse for ATSP-7041 serum albumin binding than HSA. They also predicted a list of residues in the binding pocket contribution to the difference in binding energy. The binding site R1 is canonical binding site overlaps with already known site called Sudlow's site II, but R4 appears to be a novel binding site. Such in silico studies try to provide computational protocols which can be carefully utilized to gain mechanistic detail into protein–ligand interaction processes. Flexible ligands, e.g., peptides, can show better complementarity by conformational adaptation to attain several weak interactions with the receptor [77]. Potential to gain affinity through modulation of flexibility of ligands has been sensed, and nowadays, smaller peptides are also being evaluated by researchers across the globe for their therapeutic usage.

2.4 Protein Flexibility During Binding

Proteins are generally flexible molecules. Therefore, flexibility of the receptor has to be accounted in in silico binding affinity prediction studies to better represent the physicochemical conditions. The enormous conformational space available to proteins is very challenging to exhaust in docking studies because of unrealistic sampling requirements. However, non-exhaustive but simplistic and computationally less demanding methods have been developed over the years as proxy for accounting the flexibility of the protein during the binding which can broadly be put in four classes: soft docking, side chain rotation, molecular relaxation, and docking to multiple structures.

Soft Docking: This technique allows small conformational relaxations by treating van der Waals which overlaps through a softened potential and is efficient in terms of computational cost, but it can only account for smaller relaxation in receptor structure during binding to ligand [78]. Ferrari et al. [78] applied this method using two cavities of T4 lysozyme and drug-target aldose reductase which undergo large conformation change during binding. Available Chemicals Directory (ACD) [78] was screened against chosen targets for evaluating the method. They reported, with single receptor conformation, soft potential was better in identifying known ligands, while with multiple receptor conformations, it was poor in identifying leads than hard function; this trend was similar for both receptor and more pronounced for aldose reductase. Soft docking gives better score for ligands and decoys thereby better scoring, but it misses true ligands [78]. Qualitatively, similar results were reported by soft-docking studies of protein–protein [79] and antigen–antibody [80] interaction studies.

Side chain rotation: Allowing side chains rotation of the binding site residues of the receptor is computationally costlier than soft docking but offers better ways to account flexibility of receptor through sampling side chain rotations of binding site residues and overcome the limitations of soft docking, avoiding unphysical van der Walls clashes in predicted poses [81]. Preliminary idea of incorporating side chain flexibility into docking through usage of rotamer states of the binding site residues with rigid ligand conformation by Leach et al. [82] has been carried forward and adapted in several studies. For example, approach of rigid anchor and flexible complementary growth of ligand in receptor-binding site is implemented in SLIDE by Schnecke et al. [83] and used it to screen for potential ligands of progesterone receptor, dihydrofolate reductase, and a DNA-repair enzyme from a dataset of 175,000 organic compounds. Another approach introduced by Dean and co-workers [84] is applied to successfully reproduce experimental pose of ligand in binding site by docking synthetic inhibitor RS-104966 to the S1' pocket of the human collagenase matrix metalloproteinase 1 (MMP-1) [84]. In this approach, an ensemble of binding site conformations was generated using side chain rotamer states of the binding site residues followed by identification of representative conformations combining principal component analysis and fuzzy clustering [84]. Frimurer et al. performed a study attempting to assess the extent of impact of flexible side chain conformations of binding site residues on predicted binding poses and affinity [85]. They chose protein, phosphatase tyrosine 1B co-crystalized with non-peptide inhibitors, and docked ligands to parent receptor structure, resulting correct poses to correlate with low predicted binding energy[85]. In the process, an ensemble of structures was generated using rotameric states of subset of binding site residues (Asp48, Lys120, and Phe182), and ligands were docked to each structure; correlation of binding affinity with predicted scores improved for correct poses [85]. The importance of considering side chain flexibility in docking is also highlighted in study of Gaudreault et al. They created a curated non-redundant dataset of 188 proteins where unbound- and bound-both structures were already crystallized. In their study, they found that 90% binding sites and side chain rotation were accounting the flexibility in it, and 30% of them were essential side chain rotation and only 10% binding sites are rigid [86].

Molecular relaxation: This concept takes one step further toward accounting protein flexibility from side chain rotation. In this approach, ligand is docked in the binding site of the receptor allowing potential atomic overlaps to certain extent followed by relaxation stage where docked pose of the ligand is energy minimized and complex is relaxed allowing backbone relaxation along with side chain using molecular dynamics or Monte Carlo simulation. Apostolakis et al. performed a study in which they tried to incorporate receptor flexibility to model induced fit in ligand and binding site over three challenging docking cases: (i) anti-steroid antibody DB3 with two ligands, a rigid-ligand progesterone (no rotatable bonds) and (ii) a flexible-ligand 5β–androstane-3,17-dione (having rotatable bonds), and (iii) N^{α}-(2-naphthyl-sulfonyl-glycyl)-D-*para*-amidino-phenyl-alanyl-piperidine (NAPAP) binding to human α-thrombin [87]. Progesterone and 5β-androstane-3,17-dione show two different binding modes, thus make a perfect test case. In this method, ligand was seeded to the center of binding pocket in random pose followed by a combination of minimization with shifted non-bonded interaction and Monte Carlo minimization; authors were able to successfully reproduce the crystalized pose for test cases with native structure of protein and without prior knowledge of structure of NAPAP in α-thrombin case [87]. This study highlighted the importance of considering receptor flexibility under the influence of ligands interaction field in docking. Davis and Baker [88] implemented a method in ROSETTALIGAND to account for the receptor backbone flexibility along with full-ligand flexibility and showed that on a challenging cross-docking test case of Meiler and Baker [89] (10 co-crystallized receptor–ligand pairs, with large flexible ligands and multiple side chains with changing rotamer), their new method reproduces binding poses better (lower RMSD for best-scoring docked poses) in comparison to their rigid-backbone docking.

Multiple structure docking: McCammon and co-workers [90] used relaxed complex method to dock fully flexible version of prospective drug molecules JE-2147 wild-type and V82F/I84V drug-resistant mutants of HIV-1 protease ensemble of conformations. In both cases, wild-type and mutant HIV-1 protease, an ensemble of 2200 conformation from 22 ns all atom explicit solvent MD simulation of closed conformers of apo structures of receptor and coordinates were saved every 10 ps; in both cases, crystal structure poses were successfully reproduced. Later, JE-2147 was docked to each 2200 conformation for both wild-type and mutant cases and optimized the protocol. To synthesize test inhibitors, same protocol was applied to dock 23 newly designed potential inhibitor (called JE.D.I. series molecules) to 700 conformations of the HIV-1 protease mutant. Based on high binding free energy of four compounds of the JE.D.I., which were significantly different from their parent compound JE-2147 as well rest members of the series; four new compounds with potentially better pharmacological properties were suggested for test [90].

Similar concept but using MD simulation to dock and identify the interactions between domain motions to influence the inhibitor/ligand binding has been attempted in case of Fe-artemisinin adduct binding to PfATP6, a Ca^{++} transporter well-known target in *Plasmodium falciparum* [91].

Sarco-endoplasmic reticulum membrane calcium ATPase (SERCA) is Ca^{++} transporting ATPase; it is found in the mammalian systems and regulate the Ca^{++} flow between cytoplasm and membrane-bound stores [92]. SERCA-type transporter is also found in *P. falciparum* and is known as PfATP6. PfATP6 is large multidomain Ca^{++} channel receptor and only orthologous receptor to mammalian SERCA [92]. Importance of this channel receptor highlighted in 2003 when it was found that artemisinin (one of the most effective antimalarial drug) targets this receptor [93]. To understand the plausible mechanism of artemisinin action on PfATP6, extensive molecular dynamics simulation-based study has been performed [91]. This computational study shows that activated artemisinin (Fe-Artemisinin adduct) enforced large conformational changes in the extracellular domains (Fig. 3). Artemisinin adduct binds in the membrane-bound helical region and makes a hydrogen bond network which connects it with extracellular nucleotide

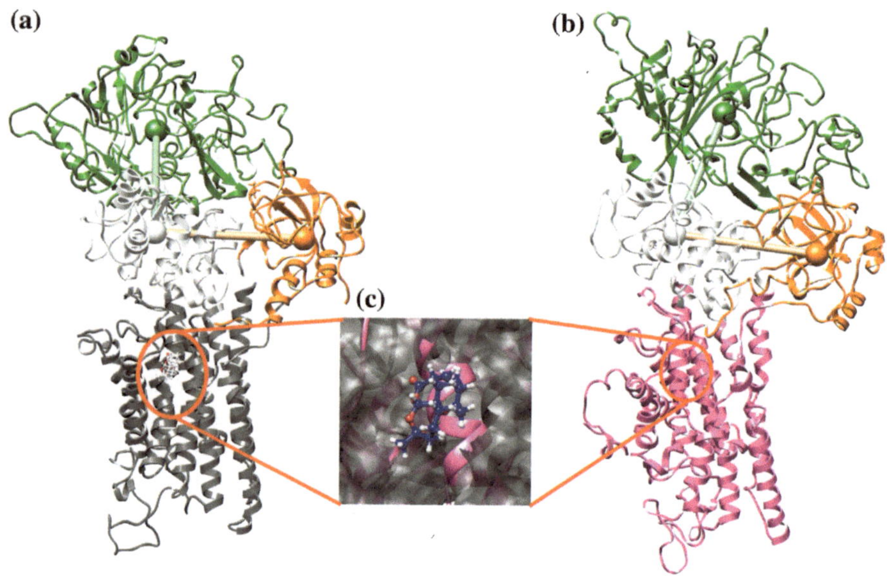

Fig. 3 Importance of receptor flexibility as observed in case of Fe-artemisinin adduct binding to *Plasmodium falciparum* ATP6 (PfATP6). Region spanning residues 364–799 shown in green contains nucleotide domain (N), region of residues 1–45 and 130–253 shown in orange contains actuator domain (A), region of residues 800 to 959 shown in white contains phosphorylation domain (P), and transmembrane region is shown in dark gray and pink colors in panel A and B, respectively. Ca^{++} and ligand binding sites are in the transmembrane region. Centroids of domain N, P, and A domains are shown with green, white, and orange spheres, respectively. The angle between centroid of domains N-P-A comes down to 78.5° (panel B) from 89.6° in open form (panel A), and distance N-A in closed conformation comes down to 44.9Å from open conformation distance 53.7 Å (see panels B and A, respectively). **a** Open-form receptor is shown in ribbon, Fe-artemisinin adduct in ball and stick with carbons in white and rest atoms colored by atom types. **b** Shows closed form or receptor; **c** dark gray surface shows ligand-binding site in open form, and pink ribbon shows closed ligand-binding site due to movement in domains shown in green and orange colors. Ligand is shown in ball and stick representation in blue color

(N) and actuator domain (A) [91]. This case study shows the selectivity gain by bound inhibitor, utilizing the domain flexibility of receptor [94].

2.5 *Effect of pH on Binding Affinities*

Protonation states of the titratable groups participating in the binding can have significant effect on the binding affinity of the interaction [16]. Waelbroeck [95] presented a model with assumptions that correct ionization state of all active groups is the requisite for binding, and ionization state of non-binding residues does not affect binding to study quantitative effect of pH change on binding affinity of the receptor–ligands interaction. They chose pH dependency of insulin and insulin analogs binding to their cellular receptor to study their model [95].

$$\log(K) = \log(K_{\text{real}}) + \log(R^*/R) + \log(L^*/L) \quad (1)$$

where log(K) is pH-dependent affinity, $\log(K_{\text{real}})$ is reference affinity, R^*/R is proportions of active and total receptor concentrations, and L^*/L is proportions of active and total ligand concentrations. Their model under given assumptions allowed them to attribute binding affinity change only due to change in proportions of active receptor and hormone with changing pH, and express pH dependence as function of number and ionization constants of active groups. Performing binding affinity measurement experiments at varying pH for different insulin analogs binding to their receptors, and analyzing data with modeled relationship [95]. Waelbroeck [95] detected two active groups responsible for marked pH dependence in the normal pH range and suggested that these groups could either belong to the receptor or common residues among porcine insulin, casiragua insulin, hagfish insulin, and desalanine–desasparagine insulin analogs [95]. This study opens up a field in medically relevant design of insulin.

A pH-dependent catalytic activity through hydrolyzing cleavage of type-1 transmembrane protein amyloid precursor protein (APP) of the β-secretase BACE-1 result amyloidogenesis in Alzheimer's disease has been reported by McCammon and co-workers. Enzymatic activity of the BACE-1 is highly dependent to the pH, with peak activity at pH 4.5, while significantly active in pH ranges 4–5 only [96]. The in silico study using constant pH replica exchange molecular dynamics simulation [97] (CpHMD) showed pH dependence of binding affinity of BACE-1 with its inhibitors [98]. The experimental binding affinity measured at pH 4.5 was taken as reference for in silico binding affinity predictions in pH range 1–12, for different inhibitor-bound BACE-1 complexes. CpHMD simulations enabled authors to study influence of conformational dynamics on the protonation equilibria and thereby pH dependence on binding affinity. The microscopic pK_a values of the aspartyl dyad residues Asp32 and Asp228 in apo- and holo-BACE-1 can be estimated from CpHMD simulation data, and protonation changes were observed in apo- and holo-forms suggesting their thermodynamic linkage. They also studied effect of protonation equilibria on conformational dynamics for the apo BACE-1 with fixed protonation states for titratable residues using conventional molecular dynamics (cMD) in acidic (pH range 1–3) and basic (pH range 9–11) conditions and observed that in acidic condition, two major conformations open and closed were populated while in basic condition, only widely open flap conformation was significantly populated. In another similar in silico study, again using CpHMD replica exchange simulation Ellis and Shen [96] reported that BACE-1 majorly occupies three conformations (so called Tyr-inhibited, binding-competent, and Gln-inhibited) and conformational population shift with varying pH causes the pH dependence of the inhibitors binding affinity to BACE-1 [96]. They showed that Gln-inhibited and binding-competent conformational states are separated by small (<1 kcal/mol) free energy barrier, and Gln-inhibited state has consistently low population (<25%) for entire pH range; thus, they focused on only remaining two of the conformational states, suggesting that substrate BACE-1 binding follows a conformational selection model [96].

2.6 *Effect of Solvation*

Almost all biological functions occur in cytosol in cell, but some of them are membrane-associated phenomena, water solubility of inhibitors showing significant binding affinity toward its cognate receptor poses another challenge in SBDD [99], since low solubility causes low bioavailability of the inhibitor to target. Similar problem surfaced with the potent non-peptide cyclic urea analogs of HIV-1 protease inhibitor, e.g., DMP-323, the carbonyl oxygen of cyclic urea of DMP-323 mimics a structural water in the binding site by providing similar hydrogen-binding features and therefore gains affinity by displacing the water. The low-molecular-weight compound was expected to have high bioavailability [100], but unexpectedly low bioavailability was observed later on, and poor solubility of DMP-323 in water and lipid milieu was suggested the reason for it [99]. Therefore, to increase water solubility, benzylic-substituted cyclic urea with strong acid or basic groups were designed, but highly basic group analogs were unsuccessful as inhibitory effect of such compounds is lowered by 1000-fold [99]. However, a neutral form binding, weak-basic derivative bis-meta-aminobenzyl, i.e., DMP-450 showed enhanced affinity. DMP-450 has enhanced water solubility and also found to show better oral bioavailability in animal species, rat and human [99].

2.7 *Covalent Inhibitors*

Non-covalent inhibitors bind to the target reversibly in concentration dependent manner. However, ~30% of FDA approved drugs are covalent binders, which make covalent bond with the target [101]. Aspirin induces irreversible acetylation of a serine residue (Ser516) in the cyclooxygenase site of the human prostaglandin endoperoxide H synthase-2 (hPGHS-2) [102], β-lactam antibacterials forms covalent bond with the active site serine of penicillin-binding proteins which inhibits cell wall synthesis of bacteria and causes its death, and tetrahydrolipstatin a fat absorption inhibitors acts by inhibiting activity of pancreatic lipase [103]; these are among the blockbuster drugs and examples of covalent inhibitors. Although non-covalent docking is more common, recently resurgence of covalent docking has been observed [101]. The covalent docking is more complicated mainly because their action between receptor and ligand has to be taken care of. Selectivity of the inhibitor toward target is important to avoid cross-reactivity. However, selective targeting via ligands equipped with different warheads makes covalent inhibition important [104]. In covalent inhibition, an electrophilic ligand binds to a nucleophilic target receptor via forming a covalent bond. Theory and application aspects of covalent docking have been reviewed elsewhere [101]. A comparative study of recent methods and tools, e.g., CovDock [105], AutoDock4 [106], FITTED [107], MOE [108], ICP-Pro [109], and GOLD [110] for covalent docking has also been recently published [104].

2.8 Functionally Relevant Structure

Biologically important molecules are involved in very diverse functions and possess the structural, modular, and interactional diversity to carry their functions in the cell. Numerous enzymes are monomer, while several of them are functional only as homo-/hetero-multimeric forms, e.g., PNP is a homotrimer [29], HIV-1 protease is a homodimer but has slight difference in structural features of the two monomers [90]. A large number of macromolecules catalyze enzymatic reactions, e.g., BACE-1 is responsible for catalyzing hydrolytic cleavage of amyloid precursor protein (APP) [111], some of them modulate their functions, e.g., MDMX/MDM2 complex suppresses activity of p53 and activate p53 pathway in tumor cells [74], some of them regulate, and some of them are not related to enzymatic activities at all, like ion channels and signaling related proteins. When we are designing structure-based drug, we are to face challenges posed by structural, functional, and reactional mechanistic diversity of target molecules as well.

The purine nucleoside phosphorylase (PNP) is a homotrimer and hosts three active sites each near the interface between two monomers, with monomer consisting an α/β-fold formed from a β-sheet of four strands, a β-sheet of six strands forming a distorted barrel, and eight α-helices [34]. The interaction between monomers will influence the binding of ligands.

HIV-1 protease is a homodimer consisting of 198 residues. McCammon and co-workers proposed a terminology to describe the topology as follows: flap (43–58), ear (35–42), cheek (cheek turn = 11–22 and cheek sheet = 59–75), eye (23–30), and nose (6–10) [112]. The active site of HIV-1 protease is covered by β-hairpin flaps of the two monomers and is involved in controlling polypeptides' access to the active site before binding and closing the active site during the cleavage and then release of the cleaved substrates. The flexibility of the flap plays a crucial role in the catalytic activity of the enzyme [113].

Isocitrate dehydrogenases (ICDs) are another group of interesting enzymes with two isoforms—one $NADP^+$-dependent homodimer and another NAD^+-dependent heterotetrametric isoform consisting of two α-subunits one β-subunit and one γ-subunit. As observed in understanding the mechanism of action during phosphorylation, the structural motions facilitate the flap to cover or open the active site, thus providing two different structures of dimmers; hence, the designing needs to take care of such two state structures of receptor [61].

3 Mapping Interaction at Binding Site

The primary focus of structural biology has been to study the relationship between structure and function of macromolecules. The evolution of protein structure to confer specificity and affinity is still not completely understood. Analysis of related structures has potential to yield local structural regions which are conserved and those which diverge. Such knowledge can potentially be translated into understanding proteins evolution to attain specificity or protein acquiring completely new function by matching curvature along the protein backbone to find structurally active site regions [114].

3.1 Identification of Active Site or Binding Site

The binding sites of most proteins are extremely specific and can determine even very small structural differences among putative binding patterns [114]. Folding of a protein can be considered to be a process which generates specific binding site or cavity from an unstructured polymer, driven and stabilized by thermodynamic forces [115]. Knowledge of protein cavities provide clue about the structure and shape of binding molecule [116]. Ligand-binding sites of protein provide insights to its biological function and reaction mechanism. Identification and application of druggable active sites of target proteins are pivotal in in silico drug design [117]. A very diverse active site of a protein is particularly useful for target-based drug discovery as it serves as a prerequisite for protein–ligand docking, which is integral part of structure-based drug design. Accurately predicting the binding modes of inhibitors in the active sites of protein is still observed as a challenge in drug discovery [10].

All the methods which identify the active site of receptor use the concept of accessible surface area as defined by Lee and Richards [118]. The accessible surface (ASA), also known as solvent-accessible surface area (SASA) if water is used as the probe, of a protein is stated as the locus of the center of the solvent molecule as it rolls along the protein, making the maximum permitted van der Waals contacts

Fig. 4 Surface area definition (courtesy: Wikipedia)

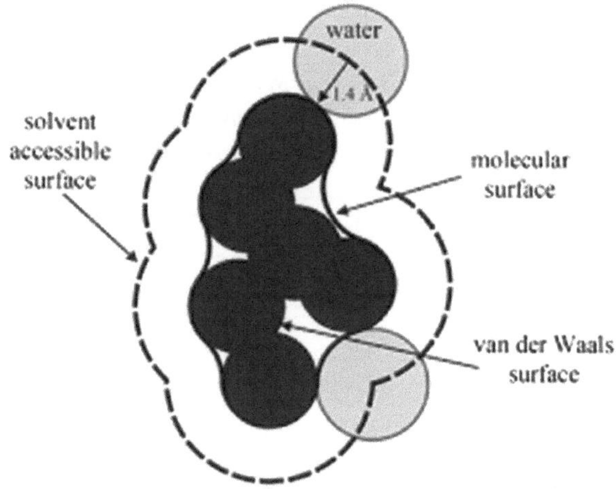

without penetrating any other atom. The ASA is closely related to the concept of the solvent-excluded surface also known as the molecular surface or Connolly surface. The cavity identified in protein molecules, effectively the inverse of the solvent-accessible surface, is the binding site as to be used by ligand to satisfy the available physical and chemical interactions. This is pictorially shown in Fig. 4.

Major methods to find the shape of active site using the 3D coordinate of protein or receptors can be classified as approximate and exact method depending on their numerical depth and accuracy in calculation involving the coordinates exclusively. Most of the approximate methods rely on numerical integration where some of them are analytical [119]. Connolly in 1983 [120] introduced the exact analytical methods for computing the accessible surface area. The computational efficiency and robustness has been improved in recent years, but the reduction in overlapping surfaces remains computationally expensive. The difference between approximate and exact computation is applied to existing methods evident from the detail calculation of the derivatives of the surface area with respect to atomic coordinates. All well-known methods used for computing the active site mapping by surface area suffer from the reproducibility problems. A method called Alpha shape [121] uses Delaunay triangulations and computes the surface area and volume of proteins as well as detects and measures cavities in proteins, as described by Edelsbrunner [122], to reduce the overlap. The Alpha shapes method employs a precision geometric method called triangulation to evade numerical problems by systematically resolving all singularities without explicitly perturbing positions of centers of spheres [123]. To provide fast calculation, an extension of the Alpha shapes method that includes the efficient, robust, and exact analytical computation of the derivatives of surface area terms has also been worked out [124].

Based on shape and ASA, many Web-based and stand-alone software are available as listed in Table 3 to find cavity and identify active site of known protein structures.

Table 3 A list of some popular Web servers and stand-alone tools based on shape and ASA formalisms

SN	Programs	Based	Web site links
1	CASTp [125]	Web	http://sts.bioe.uic.edu/castp/index.html?2cpk
2	CCCPP [126]	Desktop	http://petitjeanmichel.free.fr/itoweb.petitjean.freeware.html#CCCPP
3	LIGSITE^{csc} [127]	Web	http://projects.biotec.tu-dresden.de/pocket/
4	KVFinder [128]	Desktop	http://lnbio.cnpem.br/facilities/bioinformatics/software-2/
5	PASS [129]	Web	http://www.ccl.net/cca/software/UNIX/pass/overview.shtml
6	PrinCCes [130]	Desktop	http://scholar.semmelweis.hu/czirjakgabor/s/princces-download/#t1
7	POCASA [131]	Web	http://altair.sci.hokudai.ac.jp/g6/Research/POCASA_e.html
8	RosettaHoles [132]	Desktop	https://www.rosettacommons.org/
9	SURFNET [133]	Desktop	http://www.cgl.ucsf.edu/chimera/current/docs/ContributedSoftware/surfnet/surfnet.html
10	VOIDOO [134]	Desktop	http://xray.bmc.uu.se/usf/voidoo.html

3.2 Characterization of Active Site

Identification of active sites in large binding pockets in protein or macromolecules does not assure the correct or native poses of ligand binding because many subsites interaction influence the binding of ligands, which has been exploited favorably in case of designing combinatorial ligands of monoamine G-protein coupled receptors (GPCRs) [135]. To design a ligand which effectively come out to be a functional inhibitor requires prior knowledge of interacting subsites and their role to k_{on}/k_{off} kinetics of binding, which until recently [136, 137] hardly have been explored. Our study using kinases, from *P. falciparum* and from human, shows the selectivity of subsite also residing in active site [138]. Using ser/thr kinase sequences of human and plasmodial species those having PDB structure, a phylogenic tree was constructed. Human kinase proteins (22 of them having structural superimpossibility <2 Å RMSD of main chain atoms) shown in Table 4 are listed by sequence as

Table 4 Binding site clustering using sequence of human and plasmodial ser/thr kinase

Plasmodial kinases	Neighboring human kinases
Pfpk5	h_CDK4, h_CDK5, h_CDK3, h_CDK2, h_CDC2
Pfpk6 Pfmrk	h_CDKL1, h_CDKL4, h_CCRK, h_p38a, h_p38b, h_ERK1, h_ERK2, h_CDK10, h_p38d, h_CDK6, h_CDK7, h_p38g, h_CDK9
Pfpk7	h_SmMLCK, h_NEK1, h_LATS1, h_LATS2

Table 5 Selective binding site clustering using structure of human and plasmodial ser/thr kinase, uncommon one shown in bold face font and underlined

Plasmodial	Human		
	Kinase domain	ATP-binding site	Substrate-binding site
Pfpk5 Pfpk6 Pfmrk	CDK5, CDC2, CDK3, CDK9, ERK2, ERK1, p38-γ, p38-β, GSK3-β, DYRK1A, MAPK8	p38-δ, CDK5, p38-γ, ***CDK7***, ***MAPK6***, CDK3, ERK2, ***GSK3-β***, ***MAPK8***, CDK2, ERK1, ***CDK9***, p38-β, CDC2, ***DYRK1A***	CDK5, CDK3, ERK2, CDK2, ERK1, p38-γ, p38-β, p38-α, p38-δ, CDC2, ***CDK6***, ***PAK1***
Pfpk7	MAPK6, PAK1, PAK4, PAK7, PKC iota	***PAK1***, PAK4, PAK7, PKC iota	PAK4, PAK7, ***CDK9***, ***CDK4***, PKC iota, ***CDK7***
Non-plasmodial cluster	CDK2, CDK4, CDK6, p38-α, p38-δ	CDK6, Cdk4, p38-α	DYRK1A, MAPK8, MAPK6, GSK3-β

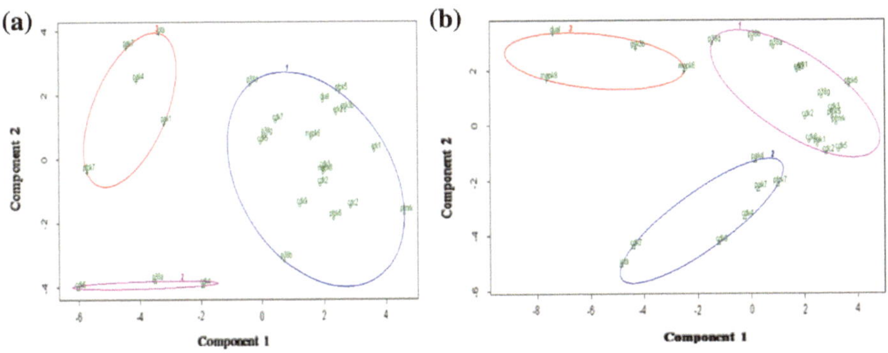

Fig. 5 Structure-based clustering of human kinases associated with *plasmodium* using **a** ATP-binding site and **b** using substrate-binding site. It clearly depicts different combinations of selectivity (listed in Table 5)

nearest neighbors of specific plasmodial kinases; their 3D structures are used for finding selectivity profile at the active sites. Separately, the ATP-binding and substrate-binding site domains of these kinases are extracted on the basis of Hunk and Hunter classification [139], and their structures are superimposed for clustering on the basis of RMSD matrix and are shown in Table 5 and Fig. 5.

It is interesting to note that three of the *plasmodium* kinases occur in the largest cluster containing most of human kinase, like MapK and CDKs; but PfPK7 occurs in different cluster in both ATP & substrate specific clustering, it signifies the selective functioning of this kinase. Hence, to achieve selectivity in favor of malarial ligand requires subsite exploitation and using appropriate designing strategy for docking compounds in search of both specific and selective ligand. In a recent review [39], such small active site differences are discussed under the context of how the entropy and enthalpy balances are carried out in free energy estimation in case of HIV proteases binding with ligands that differ by single functional group, by Freire et al. [140]. It may also happen that all available features in the active sites are not satisfied or they may be satisfied by different orientations or conformation of complementary features in ligands. Hence, it is imperative to have prior knowledge of biological function of active site of receptor and detail mapping/association of the subsites with different functional groups in ligand, before starting the docking of large number of ligands to evaluate the binding competency.

Using cliques of favorable interaction points at active site, emerging from probes of different chemical features among a class of protein, specificity pharmacophore has been generated [141, 142].

This novel method provides a complementary map of a class of active sites for designing new chemical entities which specific as well as selective for the receptors. Figure 6 provides an expanded series of such pharmacophores designed from four plasmepsins, acid proteases of *plasmodium*. Using such tools, designing of ligands is possible which can satisfy all the complementary features available in active sites; this can be used to design compounds with better binding capacity. This method can be applied to design the pharmacophores in search of novel inhibitor

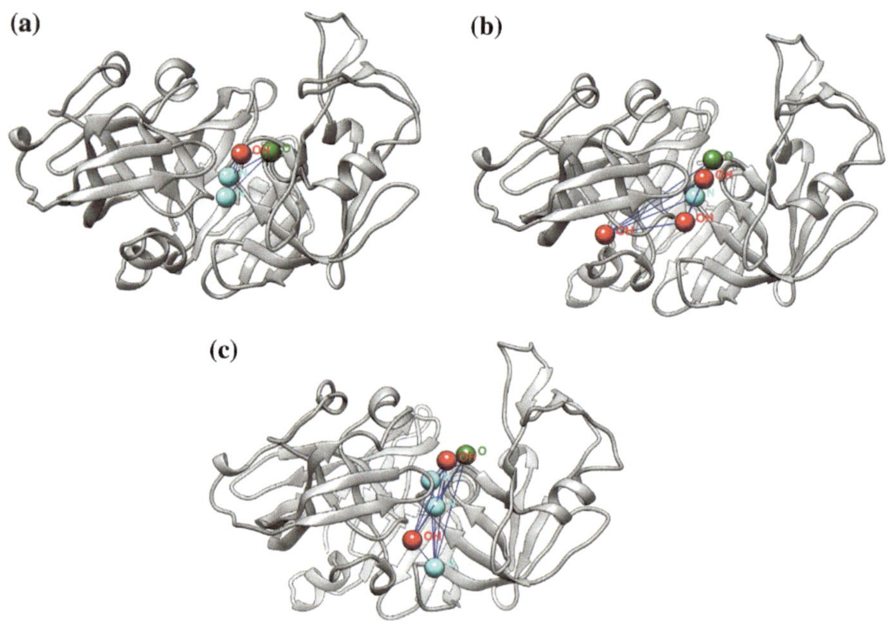

Fig. 6 Utilization of binding site information of class of aspartic protease (human cathepsin, pepsin proteases, and four *plasmodium* plasmepsins) for development of de novo pharmacophore features using in-house program CliquePharm. **a** Four-point, **b** five-point, **c** six-point pharmacophore features, all are shown in cavity of *plasmodium* plasmepsin II (PDB: 1SME), respectively. Nodes are shown as spheres with amide probe in cyan, hydroxyl probe in red, carbonyl probe in green, respectively, and edges are connected

for the multitarget structure-based designing like bacterial multidrug efflux pump and AcrAB-TolC pump [143].

3.3 *Why Different Poses?*

While docking of different chemical ligand at the known active site, one can generate different orientation for the same ligand, which is defined as "pose" due to the fact that many features available in the active sites may or may not be satisfied by the complementary features available in docked ligand. Such variations in interaction between protein and ligand may also occur due to the flexibility of active site residues [14].

Lock-and-key: The lock-and-key model of enzyme substrate interaction proposed by Emil Fisher in 1894. It assumes enzyme-binding site as a cavity with specific set of shape and physicochemical interaction features analogous to the key-hole of a lock, while ligands are potential molecules which possess shape and interaction feature of key, i.e., complementarity [39]. Generally, receptor–ligand interactions are considered to imitate this model during binding. This model was the early motivation for development of docking and scoring studies. However, many interactions associated with the flexibility of ligand upon binding to receptors and vice versa; hence, other models are proposed [36].

Induced fit: The idea of induced fit model (Fig. 7) of binding occurred as many cases the binding site of the protein undergoes subtle arrangements of key residues side chain orientations or conformational changes sensing the presence of ligand in the vicinity under the influence of its interaction fields [144]. For example, drug-target aldose reductase undergoes large conformation change during binding of ligand [79]. Several other cases of this model of ligand–receptor binding are discussed in the section Protein Flexibility.

Conformational selection: This model proposes that the receptor maintains an ensemble of conformations in equilibrium, rather than being into some particular conformational state before binding (as in lock-and-key) or changing conformation sensing the ligand (as in induced fit), whereas ligand binds to the conformation presenting best complementarity at the binding site [39]. For example, BACE-1 binding to and showing significant activity only at narrow pH range 4–5 is actually in equilibrium of at least three Tyr-inhibited, binding-competent and Gln-inhibited significant conformations [96]. However, only binding-competent conformation being conformationally compatible for binding has the highest population at the specified activity pH range 4–5, but the population of these conformation at $pH < 4$ or $pH > 5$ is decreased and hence the activity [96]. Another model known as conformational isomerism is found in the literature [14] and has been a special case of the conformational selection, where one or more conformational isomers of the receptor exist in equilibrium and ligand binds to only conformationally compatible isomeric form of the receptor, and binding shifts the conformational equilibrium in

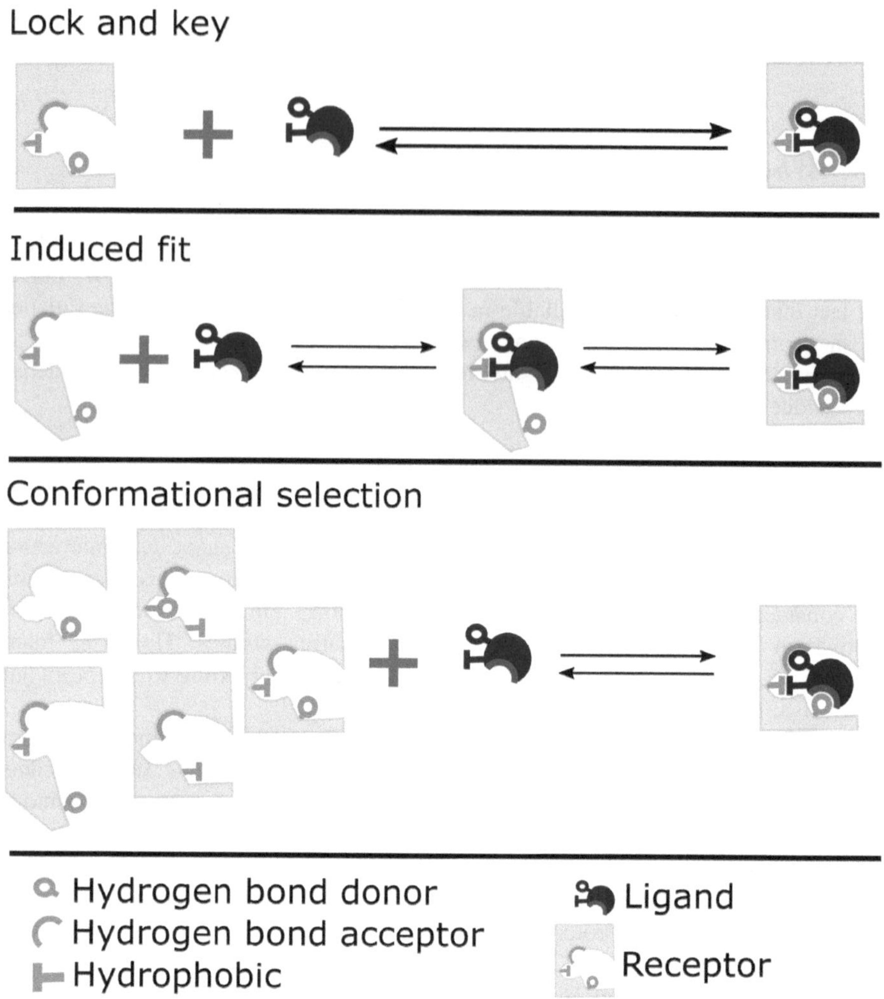

Fig. 7 Schematic representation of enzyme substrate-binding models. Ligand is shown in black color and receptor in gray color. Different binding site features/ligand features are described at the bottom

the direction to establish the equilibrium among conformational isomers. Earlier reported that binding to Fab antibody and catalysis of substrate is restricted to one of the conformation and not to others [145]. In recent paper [146], enzyme catalysis has been prescribed due to conformational dynamics of enzyme active site.

Prediction of poses of ligand with receptor from docking study may differ due to several reasons. For example, model of enzyme action (lock-and-key/induced fit/ conformational selection) assumed for the study may not be appropriate to capture the underlying binding mechanism, e.g., assuming lock-and-key for an actual induced fit or conformational selection case [14]. Existence of possibly alternate interaction features in binding site could provide complementarity for even structurally very similar ligands but provide different poses; several such cases have been reviewed by Teague et al. [147]. Another case could be enthalpy–entropy compensation due to receptor–ligand flexibility for different poses of ligand [147]. Although docking and scoring lack capability to account entropy, considering receptor–ligand flexibility in docking can be a poor proxy for entropy to certain extents.

3.4 *Flexibility of Ligand Provides Complementarity*

Generally, small molecules can adopt a number of conformations within few kcal/mol energy gap from the global minimum conformation. Thus, a number of conformations of ligands are generated and docked into the receptor to seek optimal complementarity between receptor-binding site and the ligand conformation to yield most probable pose. Therefore, several conformation generation schemes which can be broadly put in two groups, (a) systematic search and (b) random search, have been suggested and are routinely employed in docking studies [148]. Systematic search tries to generate all the conformation corresponding to the rotational states for the rotatable bonds of the molecule, but exponential increase of the number of conformations of the molecule with number of rotatable bonds turns out to be limiting for most of the practical uses. Random search tries to generate different ligand conformations using randomized schemes like genetic algorithm [14, 149].

Small-molecule ligands often interact with binding site presenting complementary features [150]. However, small size of such ligands at times has limited possibilities to interact with neutral binding pockets, because neutral binding site has weak electrostatics interactions and hydrogen bonding capabilities [151]. Neutral and wide open hydrophobic pockets can not present interactions strong enough to portray desired high affinity for small-molecule ligands. On the other hand, peptide ligands due to their flexibility can adopt a wide range of conformations to gain higher affinity in such cases by making more hydrogen bond interactions and through many weak hydrophobic interaction from several hot spots in the pocket [151, 152].

3.5 *Is Estimate of Binding Affinity Sufficient?*

In case of receptor binding processes, the stability of the binding is accounted by difference of Gibbs free energy between bound and unbound states. The equilibrium dissociation constant K_d which is ratio of unbinding process k_{off} and binding process k_{on} is associated with thermodynamic properties of the reactants/product, whereas the activation energy for the process influenced by kinetic properties [153].

Thus, in silico calculated affinity of receptor–ligand binding contains information about thermodynamic parameters and does not include kinetic parameters. All the methods aiming to measure/predict binding affinity would miss the kinetic aspect of the reaction. The kinetic aspect of the process is related to diffusion of the solute molecules under influence of the entropy of the system. Collision of the receptor molecule with the ligand is the requisite for the process to happen. Bigger solute molecules collide with small water molecules and undergo random Brownian motion, and their encounter allows reaction to happen [153].

The dissociation constant K_d represents the ligand concentration in which half of the protein binding pockets are occupied and relate to Gibbs free energy [154] by $\Delta G = RT \ln(K_\mathrm{d})$. Gibbs free energy is a state function and does not depend on the thermodynamic path followed during reaction; it only depends on the initial/final chemical potential of the reactants/products [154]. Association and dissociation rates k_on and k_off depend on transition states encountered on the pathway during the chemical reaction. Specifically, they depend on highest free energy barrier for the transition state that separates bound and unbound states [154].

Even if the reasonable accuracy in predicting affinity is achieved, it is not sufficient to characterize the protein–ligand-binding process completely [154]. Kinetic aspect of the process can be modeled by mimicking the protein–ligand diffusional encounter in the solvent under thermal fluctuation, which will be discussed later [155].

4 Estimation of Interactions

Scoring functions aim to predict the interaction energy between the receptor and the ligand in a given conformational pose, by summation of weighted interaction features. Scoring functions required to rank chemicals implemented in various docking tools use different assumptions to evaluate modeled complexes [8]. Simplification is achieved at the cost of neglecting full domain flexibility, entropic effects, and solvation effect [8].

4.1 Different Types of Scoring Functions

In the literature, wide choice of scoring function is available which can be classified as force-field-based scoring functions, empirical scoring functions, knowledge-based scoring functions, and descriptor-based scoring functions [156]. Force-field-driven scoring functions are based on the molecular mechanics and utilize atomic properties like atomic charge and vdW forces which are already parameterized such as AMBER [157] or CHARMM [158]. Dock6 [159], AutoDock [160], G-score [161], and GOLD [110] are a few popular ones in this class. In scoring functions, only intermolecular interactions are modeled, vdW interactions are expressed using Lennard-Jones potential function, and electrostatics interactions are calculated using Coulombic formulation. Empirical scoring function [162] on other hand is based on the available physicochemical properties which corresponding to hydrogen bonding interactions, hydrophobic interactions, entropic changes, and interactions with metal ions [162]. Binding free energy is estimated using the sum of various uncorrelated (sometime parameterized) terms derived from the regression analysis using experimentally determined binding energies from the already known crystallized complex structures [163]. ChemScore [163], LUDI [164], Glide score [165], X-score [166], etc., are major tools implemented with such empirical scoring function. Knowledge-based scoring functions [167] are derived from the crystallized protein–ligand complexes using statistical regression principles. The binding free energy of the complex is assumed to be the sum of free energies (potentials of mean force) of interatomic contacts calculated from the frequencies of these interatomic distances in a database of experimental structures from statistical methods [168]. As compared to empirical scoring function, knowledge-based potential function does not require known binding affinity and so are free to explore large and diverse structural complex information to derive the more accurate and less biased scoring function parameters. These functions are expected readily transferred to systems that have not been used in the development of the scoring function. Examples of knowledge-based scoring functions include PMF [169] and DrugScore [170].

4.2 Nonlinear Relation Between IC_{50} and Score Values

A standard scoring function is given in kJ/mol by Eq. 2.

$$\Delta G = 5.4\,\Delta G_0 - 4.7\,\Delta G_{\mathrm{HB}} - 8.3\,\Delta G_{\mathrm{ionic}} - 0.17\,\Delta G_{\mathrm{lipo}} + 1.4\,\Delta G_{\mathrm{flex/rot}} \quad (2)$$

Assumed to be linear, where coefficients present the weightage of each contribution as mentioned by suffix, in a case study out of 45 known ligand receptors from PDB, the standard deviation having +7.9 kJ/mol or 1.4 log unit error in binding constant. But this is not reflecting reality, which has been observed while comparison of actual and predicted values of binding across the range of activity. Correlation between the binding energies predicted by the docking programs like AutoDock, GOLD and FlexX [171–173] with the experimentally determined binding free energies is analyzed among a set of known ligands in the literature [110, 174]. Prediction of affinity using scoring function has been used for ranking compounds, while high-throughput screening but compared with known experimental data it has been observed that high-affinity compounds ($\sim$nM) are predicted with lower errors than weak binders (μM to mM). Generally, the weak binders are overpredicted, whereas tight binders (pM) are underpredicted [171, 175]. It may require implementing functions to address negative co-operativity so that present scoring functions are trained to penalize weak binding. Tight binders required to be associated with positive co-operativity. However, a measurement of applicability is done using reproduction of geometry from complexed crystal structure, comparing close relation with binding affinity (experimental) and scoring and ranking and two other important parameters known as enrichment factors (EF) and receiving operators characteristic (ROC) [176–178].

4.3 Does Scoring Function Reflect Binding Activity?

Scoring functions can only predict the binding affinity of a receptor with its ligands in isolation [156], but the cellular environment is significantly different, where it may be interacting with other molecules which may alter its affinity toward its ligand, e.g., activation of tumor suppressor protein p53 activation is regulated by MDM2/MDMX [74]. Inhibition measure of a ligand for its receptor is the end result of several pharmacokinetic factors as well other than affinity, e.g., bioavailability [73, 99]. Therefore, docking score of a ligand for its receptor may not be the actual measure of its inhibitory potential always. The similar kind of evidence emerged, when it was noticed that urea analog DMP-323 had shown good affinity and predicted inhibitory potential for HIV-1 protease [100], but it could not succeed because of its very low bioavailability due to its poor solubility [99]. In the follow-up study, a new analog DMP-450 with higher water solubility was designed and found to show better inhibition of HIV-1 protease [99]. As detailed in Sect. 2.3, in the similar way to save from proteolytic cleavage, α-helical clipped peptide was designed from human serum protein HSP's variants, as inhibitor of the MDM2 and MDMX complex [73]. The proteolytic cleavage was hampering its bioavailability; thus, clipped α-helical peptide achieved improved pharmacokinetics, thus ensured better efficacy in human and rat models [73].

5 Limitations of Methods

5.1 Appropriate Structure of Receptor to Select

While selecting a receptor structure for initiating docking study, parameters listed in Table 1 can be used to prioritize structures if more than one structure is available, and to choose appropriate structure. In present case, we have summarized some of the structure validation results for two different structures of HIV-1 protease (PDB ids: 1FQX and 4ZIP) in Fig. 8 and crystal structure details shown in Table 6. Analysis of structures is available from RSCB PDB [179] (https://files.rcsb.org/pub/pdb/validation_reports/fq/1fqx/1fqx_full_validation.pdf and https://files.rcsb.org/pub/pdb/validation_reports/zi/4zip/4zip_full_validation.pdf).

In general, structure for which different parameter values are in blue zone in horizontal bars for it is preferable. These horizontal bars represent statistical likelihood of reported structure to be in acceptable/unacceptable range. The range of

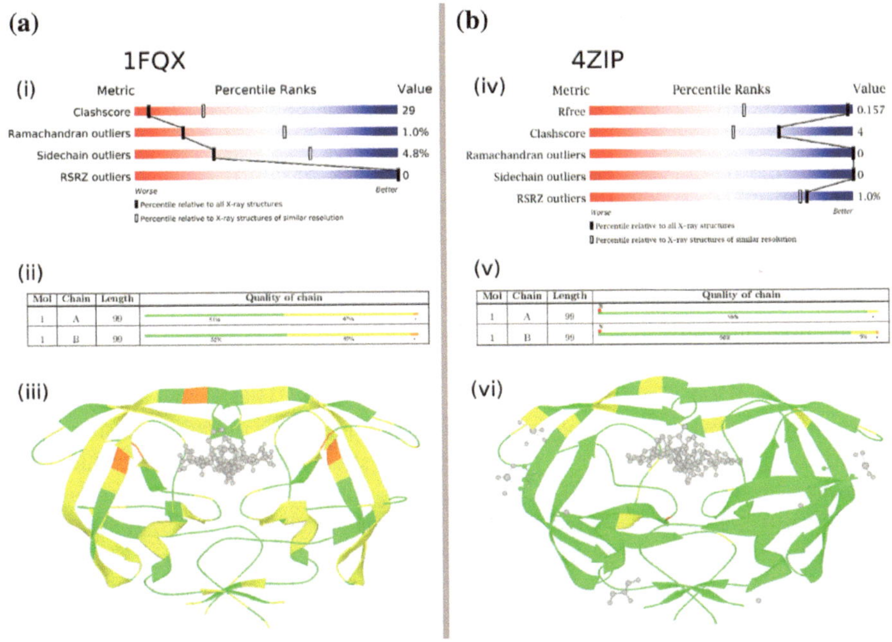

Fig. 8 Two crystal structures of HIV-1 protease shown **a** 1FQX.pdb and **b** 4ZIP. (i) and (iv) show structure quality summary obtained from RCSB Protein Data Bank (PDB). (ii) and (v) show conformance to geometric quality criterion of model residues: 0, 1, 2, and ≥ 3 geometric quality criterion outliers are shown in green, yellow, orange, and red colors, respectively. (iii) and (vi) show mapping of model validation results with electron density over 3D structure for PDBs 1FQX.pdb and 4ZIP.pdb, respectively

Table 6 Crystal structure parameters for HIV-1 protease structures with RCSB PDB (www.pdb.com) codes 1FQX and 4ZIP

Parameter	1FQX	4ZIP
Resolution range low[a]	26.00	50
Resolution high[b]	3.1	1.11
Completeness	Not available	91.7%
R_{work}	0.180	0.130
R_{free}	Not available	0.154
RMSD (bond lengths)	0.080	0.015

[a]A minimum spacing (d) of crystal lattice planes that still provide measurable diffraction of X-ray

[b]Additionally, $\langle I/\sigma(I) \rangle$ greater than 2 in high-resolution shell

parameter value is determined from all the structures already deposited in PDB of similar resolution range. As we see from the report that clash score, Ramachandran outliers and side chain outliers' values are higher than acceptable and are in red zones (statistically unfavorable) of their respective bars [180] for 1FQX. While in the case of 4ZIP, all the parameter values are in the blue zone (statistically favorable). Again, when looking at the geometric quality criterion for two structures, 1FQX only (chain A: 51% and chain B: 52% residues) does not have any outlier, while rest (chain A: 47% and chain B: 40%) have at least one outlier. The geometric quality for 4ZIP seems better as in this case 96 and 90% residues (chains A and B, respectively) do not have any geometric outlier. Further considering fit quality of the model to electron density, 1FQX has certain residues which has at least two outliers and a significant percentage of residues with at least one outlier, while in case of 4ZIP, there are no residues which have two outliers and only a small fraction of residues with only single outlier. Considering all above points among 1FQX and 4ZIP, 4ZIP should be preferable over 1FQX as receptor structure for any docking study.

In Fig. 9, the docking using Dock6 of ligand GRL-0648A to two different receptor structures of HIV-1 protease (4ZIP: high resolution and 1FQX: low resolution) is performed to assess the effect of receptor structure quality on outcome. Results show that when ligand was docked to native receptor structure (4ZIP), it reproduces the crystallized pose (RMSD: 0.40 Å, see Fig. 9a), with dock score of approximately −125. When we docked ligand to poor receptor structure (1FQX), it docked in different poses where core group adopts similar pose but the 5-atom ring (1 nitrogen, one oxygen) containing methyl adopts different poses and leans over Gly48 on chain B, score is significantly low (−14) and RMSD: 2.71 Å (Fig. 9b). This observation suggests that high-quality receptor structures are more likely to present better interaction complementarity, saving from predicting high-affinity binders mistakenly as poor-affinity ligand.

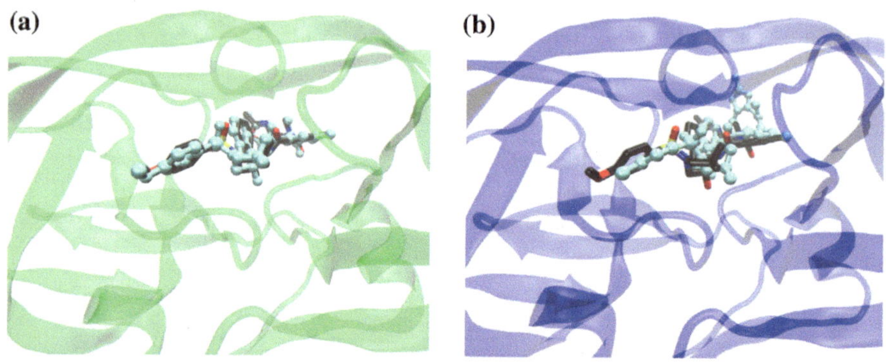

Fig. 9 HIV-1 protease-binding site structures shown **a** HIV-1 protease structure (PDB: 4ZIP) in complex with GRL-0648A (isophthalamide-derived P2-ligand), receptor-binding site is shown in green ribbon and crystallized pose of GRL-0648A in black stick. GRL-0648A is docked to the receptor using Dock6 and docked pose is shown with ball and stick representation and carbons colored in cyan, RMSD of docked pose with reference to crystallized pose is 0.40 Å over 49 non-hydrogen atoms. **b** HIV-1 protease structure (PDB: 1FQX) with GRL-0648A crystallized pose (taken from 4ZIP after superimposing receptor structures) shown in black stick, docked pose of GRL-0648A shown in ball and stick representation with carbons in cyan color, docked pose RMSD 2.71 Å over 49 non-hydrogen atoms

5.2 Analysis of Docking Tools

As discussed above, it is fruitful to analyze the ligands binding efficiency using many methods like AutoDock, GOLD, Glide, LibDock, and HADDOCK; all these tools are different in the method of docking as well as scoring.

There are several open-source commercial but free for academic use, and complete commercial docking programs available from different software vendors. In particular, fifty-one stand-alone and nineteen Web servers for docking employing diverse set of novel features are listed at http://www.click2drug.org/index.html#Docking (accessed on Dec 2017). To select suitable program(s) for docking studies for receptor(s) of interest requires insight and expertise [117] in the method. However, we shall discuss only a few selectively chosen methods based on popularity and diversity of strategies implemented in them as shown in Table 7.

Here we are discussing the in-house case study (unpublished work) of four docking programs used to dock already experimentally known inhibitors of *P. falciparum* protein kinase 5 (PfPK5) with IC_{50} values ranging from 130 to 15000 nM. PfPK5 is a ser/thr kinase and homolog of human CDK2 [185]. Chosen inhibitors are olomoucine (OLM), indirubin-5-sulfonate (INR), staurosporine (STA), and purvalanol B (PVB), respectively. Crystal structures of two of the inhibitors (INR and PVB) in complex with PfPK5 are available [185]. We have chosen LibDock v2.3, Gold v5.2, Dock v6.7, and Glide v7.0 for the comparison study. Different docking programs use different scoring schemes, e.g., Glide score and Dock score assign high negative score to high-affinity ligands, while LibDock and Gold assign high positive score to high-affinity ligands. Pose reproduction and also scoring/ranking of docked poses of these inhibitors is a good case to assess comparative performance of each of the selected docking program and also with experimental values.

The best-scoring poses predicted by each of the programs were compared with the crystallized poses for selected available complex of PfPK5 with PVB as in PDB (1V0P). Predicted poses for PVB obtained from LibDock, Gold, Dock, and Glide showed 0. 60, 1.01, 0.88, and 1.87 Å RMSDs with crystallized pose, respectively. In present case, all the selected programs were able to reproduce observed binding mode within RMSD of 2 Å.

Docking and scoring results obtained from the chosen programs show that none of these could predict the correct ranking against the experimentally known activity of chosen inhibitors (see Table 8). The best binder (PVB) among four inhibitors is predicted to be best binder as rank 1, by Gold and Dock6, while LibDock and Glide have ranked 2. LibDock is unable to discriminate between the OLM and INR and predicts them as rank 3 and rank 4, while experimentally found ranks would be 4 and 3, respectively. Again, LibDock does not discriminate between STA and PVB and predicted ranks are opposite to the experimental ranks. Gold predicts correct ranks for best and worst inhibitors, while is unable to discriminate between mid-ranged inhibitors INR and STA. Dock6 predicts correct ranks for better binders STA and PVB, while does not discriminate between weak binders OLM and INR. Predicted ranks from Glide did not match with experimental rank for any of the four

Table 7 Summary of docking approach used, techniques for ligand and/or receptor flexibility, and major features available in some chosen popular docking programs

Programs	Ligand flexibility	Receptor flexibility	Major features in brief
AutoDock [181]	Genetic algorithm	modeling flexible residues	Force field-based scoring function, uses averaged interaction energy grid to account for receptor conformations and simulated annealing for ligand conformations
DOCK [159]	Incremental build	Yes (through AMBER score)	Force field- and contact score-based scoring functions; docks either small molecules or fragments, include solvent effects
Glide [165]	Exhaustive search	No	Empirical score. Although, receptor flexibility can be used in Induced fit, Docking (IFD workflow) with Glide and side chain rotations through PRIME
GOLD [110]	Genetic algorithm	Side chain flexibility and ensemble docking	Empirical score, highly configurable allowing to utilize chemical intuition and domain expertise to improve pose prediction and virtual screening
HADDOCK [182]	Yes	Semi-flexible torsion angle refinement	Uses biochemical and/or biophysical interaction data such as chemical shift perturbation data resulting from NMR titration experiments, mutagenesis data, or bioinformatic predictions
LibDock [183]	Rigid docking can use programs in suit to generate conformation	No	Docks a pre-generated set of conformations for the ligand followed by a final flexible gradient-based optimization of the ligand in the protein binding site
LigandFit [184]	Monte Carlo	No	Empirical score, ligand conformation docked into an active site based on shape, followed by further CHARMm minimization

Table 8 Summary of docking scoring/ranking results of chosen four inhibitors with known IC_{50} values to PfPK5

Inhibitor	IC_{50} (in nM)	RT $\ln(IC_{50})$ (kcal/mol)	Docking score			
			LibDock[a]	Gold[a]	Dock6[b]	Glide[b]
OLM	15,000	−6.622	107.08(3)	55.56(4)	−56.46(3)	−5.75(3)
INR	5,500	−7.220	106.80(4)	64.56(2)	−55.04(4)	−8.65(1)
STA	1,000	−8.236	132.82(1)	60.98(3)	−64.55(2)	−4.99(4)
PVB	130	−9.453	130.25(2)	78.40(1)	−71.41(1)	−7.98(2)

Docking score from programs is given in cells of table, while rank is given in pair of parentheses. Four docking programs, LibDock v2.3, Dock v6.7, Glide v7.0, and Gold v5.2, were used to dock inhibitors in the PVB bound structure of PfPK5, after removing PVB. Inhibitors are tabulated from top to bottom in increasing affinity order

[a]Higher positive score represents higher affinity

[b]Higher negative score represents higher affinity

inhibitors. A limited study like this brings out the uncertainty in pose and rank prediction by popular tools.

5.3 Selection of Appropriate Database

Chemical databases are selected from the ensemble of the small organic and synthetic molecules, used for ligand docking, constituents of such chemical libraries influence the final outcome in the drug designing process. In general, chemical library databases are created to aid the drug discovery process by providing innovation in new lead structures selection. After the establishment of the in silico drug designing protocol, chemical databases are screened to identify the probable inhibitors which can be tested by experimental methods. Success rate in finding true inhibitor by in silico means depends upon both screening protocol and chemical databases used. So, before the selection of the chemical libraries, basic biological target specific chemical features should be marked. For the virtual screening purposes, the compound database may be selected in such a manner so that maximum structurally diverse chemicals can be utilized against the studied biological target(s). Chemo-informatics tools are mainly used not only for diversity analysis [186, 187] but also for converting them into focused chemical libraries [188].

Various chemical compounds databases are available which include databases of general organic compounds intended for screening, drugs, commercial databases, and databases with known biological activity, crystal structure information, and various physicochemical properties information [189, 190]. Table 9 shows some of the commonly used chemical databases which are categorized based on the different features like associated bioactivity information, known drug information, and having target specific information. Most of these databases provide chemical information using 1D representatives such as SMILE and InChI Key, or 2D structural coordinate information stored in SD file format. These databases are also provided online interface to access the whole chunk of chemical compounds for similarity-based screening. These functionalities intended to search close analogues of known bioactive compounds and thereby advances the lead optimization process.

Though different chemical databases are available for virtual high-throughput screening (vHTS), it is recommended to convert any chemical library to "target or focus" chemical library to avoid the false hits selection as novel inhibitor [191]. In the literature, several characteristic properties of small molecules have been discussed that are followed by the "lead-or drug-like" molecules and are considered to be important for a drug to be successful [192]. Currently, list of open-source Chemoinformatics tools is available which can be utilized for drug-like properties calculation and chemical databases filtration [193]. Well-known physicochemical properties which are used as empirical rules are Lipinski's "Rule of Five" [194], "Rule of Three" [195], and Pfizer's "Rule of 3/75" [196] (Table 10). Apart from filtering for lead-like properties, it is also important to exclude known toxicophores or metabolically liable moieties which can interfere with the assay and detection protocol.

Table 9 Some commonly used chemical databases

Databases	Web link
Bioactivity data	
Binding activity database	https://www.bindingdb.org/
ChEMBL	https://www.ebi.ac.uk/chembldb/
NCI	https://cactus.nci.nih.gov/download/nci/
PDB bind database	http://sw16.im.med.umich.edu/databases/pdbbind/index.jsp
PubChem	https://pubchem.ncbi.nlm.nih.gov/
Patents	
IBM	www-935.ibm.com/services/us/gbs/bao/siip/
SureChEMBL	www.surechembl.org
Drugs	
DrugBank	www.drugbank.ca
FDA	http://fdasis.nlm.nih.gov/srs/srs.jsp
Available for vHTS	
ZINC	http://zinc.docking.org
ChemSpider	http://www.chemspider.com
eMolecules	www.emolecules.com
MDL drug data report (MDDR)	http://accelrys.com/products/collaborative-science/databases/bioactivity-databases/mddr.html
BioPrint	http://www.cerep.fr/cerep/users/pages/ProductsServices/bioprintservices.asp
Target specific	
Pfaldb	http://pfaldb.jnu.ac.in/Malaria/homeHit.action
Mycobacterium DB	http://tbnetindia.in/
Therapeutic target database	http://bidd.nus.edu.sg/group/cjttd/TTD_HOME.asp
KLIFS	http://klifs.vu-compmedchem.nl/
Kinase profiling inhibitor database	http://www.kinase-screen.mrc.ac.uk/kinase-inhibitors
Structural databases	
Cambridge crystallographic data center	https://www.ccdc.cam.ac.uk/
Crystallography open database	http://www.crystallography.net/cod/

There is a well-recognized need of creating standard datasets for which experimental bioactivity of the ligands is already known for receptors coming from various functional classes [197] in the research community. Availability of standard dataset for benchmarking docking would potentially aid to spot limitations and non-optimal parameter sets used for docking and scoring with the concerned docking program and thereby allowing tracing and possibly fixing of issues in earlier phases of the study. Development of benchmarking datasets for docking and scoring has been reviewed recently [197, 198]. Primary attempts toward docking was made by Bissantz et al., a dataset contained estrogen alpha receptor (ERα) and thymidine kinase (TK) with one PDB structure, ten active compounds, and 990 randomly selected decoys from pre-curated Advanced Chemical Directory (ACD) which was considered for each of receptors to evaluate DOCK, FlexX, and GOLD programs and seven scoring functions (Dock, FlexX, GOLD, PMF, ChemScore, Fresno, and Score) [197].

5.4 *Consensus Evaluation of Docking*

Docking studies performed using different programs which do not necessarily agree with each other as discussed earlier, mostly because each program carries different subtasks of docking with potentially different approach [199]. Thus, when results disagree among themselves, then selection of the final compounds to test becomes indecisive. Matthew and co-workers [199] suggested selection of results based on consensus followed by rationalization through physicochemical intuition. As discussed later, such strategies should be projected as standard to increase confidence in docking results and decrease failure rate of docking studies.

Benchmarking of docking studies is very important for unbiased evaluation of various docking methodologies and their implementations in docking programs. To address this issue, Huang et al. [176] conducted a study along with creating a directory of useful decoys (DUD) [176]. They choose total 40 different targets with eight nuclear hormone receptors, nine kinases, three serine proteases, four metalloenzymes, two folate enzymes, and ten other enzymes. The crystal structures of all targets except one kinase (PDGFrb) were available in PDB. They used 2950 ligands, creating 36 physically similar but topologically different decoys for each ligand. Docking was done using DOCK 3.5.54, with flexible ligand and a force-field-based scoring function accounting van der Waals and electrostatics interaction energies corrected for ligand desolvation. Authors reported that for most of the targets, with MDDR (Elsevier MDL, San Leandro CA) databases, enrichment were almost half log better than DUD, which supported their conclusion that generally databases have bias.

Another protocol is known as checking with cross-docking which aims to summarize the overall success of docking study [200], it captures ligands specificity for its cognate receptor at diagonal of the matrix, and off-diagonal entries represent enrichments against off-diagonal targets. The off-diagonal enrichments could also be indicative of promiscuity of the ligand, or the similarity of the off-diagonal targets [201]. The cross-docking performed in the process highlighted striking results that ligands having very good enrichment for their cognate receptor had good enrichments against a few other receptor sets, while ligands with poor enrichment for their receptor had poor enrichment against others [202–204].

Overall, it has been found that the interaction-based classification and estimation of accuracy of poses during docking are in better agreement with the experimental results [205].

5.5 Selection of Suitable Scoring Function

Whether to select just a scoring function or a consensus scoring function? A suitable scoring function has important role to play to extract correct poses while docking. Poses should be evaluated by the docking score or the ranks are better for evaluation of docking; these are critical aspects influencing the final outcome of the docking results. None of the available scoring functions appears to be fit in all cases [206]. James B. Matthew and co-workers performed a study to evaluate performance of four individual scoring functions DOCK, GOLD, PMF, and FlexX and several forms of consensus scores (CScore) derived from them, over a dataset of twelve HIV protease and nine thermolysin complexes with known crystal structure and experimental binding affinity [199]. Since DOCK and GOLD scoring functions were not available in FlexX, they implemented these scoring functions according to their open descriptions in the literature and will be referred by D-SCORE and G-Score. They found that none of the considered scoring functions was consistently good for all active sites [206], but the CScore (consensus score) was better than all individual scoring function [199]. Secondly, they studied these scoring functions for scoring candidate ligand configurations over a set of five known receptor ligand complexes (2-MQPA or NAPAP into thrombin (1ETR and 1DWD), 1–3-phenyllactic acid into carboxypeptidase A (2CTC), 1-deoxynojirimycin into glucoamylase (1DOG), and DANA into neuraminidase (1NSD) each of the ligand was docked to cognate receptor, and top thirty configurations with most favorable FlexX scores were chosen for further study, each of these configurations were scored using D-SCORE, G-SCORE, PMF, rank-score, deprecated rank-sum (rank-sum after leaving out worst rank), worst-best and CScore methods. They found that average scores from several methods are better than individual score [199]. Apart from this, their study highlighted that there could be alternate poses for NAPAP binding in thrombin and DANA in neuraminidase as predicted by FlexX along with crystal structure poses reproduced in Fig. 10a, b respectively.

Table 10 Typical physicochemical properties which are used to filter the chemical databases

Properties	Lead-likeness
Molecular weight (MW)	200–500
Lipophilicity (cLogP)	−4/4.2
H-bond donor	≤ 5
H-bond acceptor	≤ 10
Polar surface area (PSA)	≤ 170 Å^2
Number of rotatable bonds	≤ 10
CACO-2 membrane permeability	≥ 100
Solubility in water (log S)	−5/0.5
Others	Absence of both toxic and reactive fragments

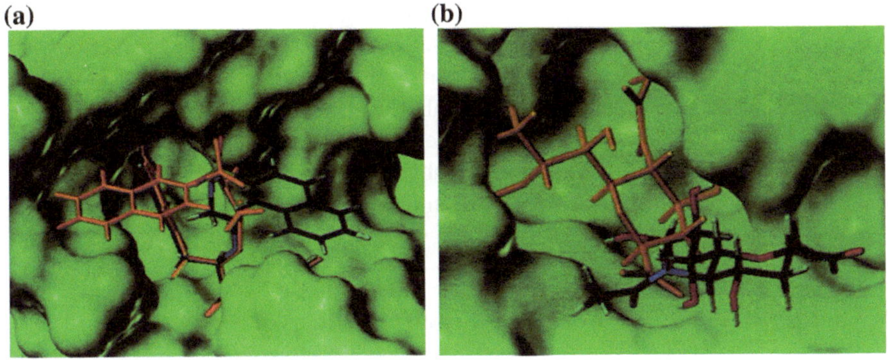

Fig. 10 Alternative docking mode for identified by FlexX and CScore. The alternative configuration is colored by atom type, whereas the binding mode found in the crystal structure in colored orange. **a** NAPAP in thrombin (1DWD) and **b** DANA in neuraminidase (1NSD) [200]. Reproduced with permission

5.6 Consensus Scoring

Despite availability of variety of scoring functions, none of them is universally good for assessment of all receptor ligand binding using docking. Therefore, several attempts [174, 199, 207] have been made by researchers to investigate several scoring functions and their combinations using different consensus schemes. In particular, Oda et al. used two force field-based (Dock score and GOLD score), two knowledge-based (DrugScore and PMF score), and five empirical (FlexX score, ChemScore, PLP, Screen Score, and X-Score) scoring functions and systematically assessed performances of all 511 (2^9 −1) consensus scores over a test set where structures were available in PDB for all chosen 220 protein–ligand complexes. For the sake of comparison, either all the candidate poses scored by a scoring function were ranked assigning best-scoring pose a rank 1 or the scores were scaled to span range 0–1, with best-scoring pose assigned 0 and worst assigned 1. These schemes were consistently used for all the scoring functions, except for X-Score, since it assigns a higher value to better pose in contrast to rest of others. Therefore, S-Score was multiplied by −1 before scaling or ranking [207]. Oda et al. [207] used six different averaging schemes for consensus score with three different ways of model selection (selecting models with consensus score $\leq x_{\text{threshold}}$, top $y_{\text{threshold}}$ models from sorted list of consensus scores in increasing order, and top $z_{\text{threshold}}$ % models from sorted list of consensus score in increasing order) combined with two ways (by rank and by scaled score) of mapping score to common scale. Prefixes number-by-, rank-by-, and percent-by- were used to denote way of model selection, and suffix rank and number were used to denote ways of mapping scores. Apart from these six, three more double thresholds (one for model selection from $x_{\text{threshold}}$, $y_{\text{threshold}}$, and $z_{\text{threshold}}$ and other number of minimum votes for electing the model)-based vote-by-consensus scores were also evaluated [207]. Considering the accuracy and efficiency balance in selecting poses rank-by-number and percent-by-number are more useful, while for accuracy number-by-number and vote-by-number approaches are more pertinent to pose selection [207]. GOLD score and Dock score were poor individually but were useful in consensus scoring [207]. Consensus score involving all nine scores or five CScore functions were useful without any optimization and suitable for practical usage [207]. However, Free energy and empirical scoring has been used together in the recent paper [174].

5.7 Inclusion of Flexibility of Ligand and Receptor

In computer-assisted drug discovery process such as structure-based drug design and ligand-based drug design, ligand flexibility plays key role for pharmacophore features extraction and model generation [208], 3D-QSAR analysis [209], molecular docking-based studies [210], shape similarity [211], and so on. In these cases, the outcome results largely depend upon the ability to achieve those conformers that represent the bound state. Hence, it is important to achieve bioactive conformational space of each compounds under study [212]. The term "bioactive conformation generation" specifies the generation of pool of all possible molecular structures that are found in the bound state of the complex macromolecules. Various studies suggest that during the interaction with the receptor, small molecules generally adopt low-energy conformation [213].

The literature suggests two major classes of methods that are utilized to explore the conformational landscape of the small molecules [214]. These approaches include stochastic sampling, systematic or deterministic sampling. Deterministic approaches attempt to generate full range of minimum energy conformations by adopting systematic exhaustively space search approach. This type of space search methods largely dependent upon the number of rotational bonds a small molecule has. Due to combinatorial explosion in torsion angle combinations, this approach is feasible only for very small molecules [214]. Stochastic sampling tries to explore various energy landscapes by incorporating randomness during the search process. Monte Carlo-type (MC) simulations and genetic algorithms (GAs) are the major techniques of this type of sampling methods [214]. A detailed review of these approaches can be found in the following papers [215].

Using above-mentioned approaches, various conformation generation programs have been developed and utilized in drug discovery process cited in Table 11. These programs generally adopt heuristics to overcome combinatorial explosion in case of systematic search and random perturbations and selection in stochastic search.

Ligand being usually smaller in size with lesser number of rotatable bonds exhaustive sampling of available conformational space is achievable with current computational capabilities; but proteins being large macromolecules, available conformational space is vast due to large number of degrees-of-freedom (DOFs) and its exhaustive sampling is almost infeasible. Therefore, techniques seeking to

Table 11 A brief summary of major programs for small-molecule conformation generation

Program	Type	Algorithm	Cost/license	References
Balloon_GA	Stochastic	Genetic algorithm	Free/ proprietary	[216]
CAESAR	Systematic	Incremental search of torsion angles combined with distance geometry	Commercial	[217]
Confgen	Stochastic	Random walk on energy surface	Commercial	[218]
Confab	Systematic	Torsion driving approach	Open source	[219]
Corina	Systematic	Knowledge-based rules derived from CSD	Commercial	[220]
ETKDG	Stochastic	Distance geometry and knowledge base	Open source	[221]
Frog2	Stochastic	Monte Carlo	Open source	[222]
MS-Dock	Systematic	Brute force, anchor, and grow	Open source	[223]
MOE	Stochastic	Random perturbations of rotatable bonds in increments biased around 30°	Commercial	[108]
OMEGA	Systematic	Knowledge-based, complete enumeration	Commercial	[212]
RDKit	Stochastic	Distance geometry	Open source	[224]

incorporate protein flexibility during binding has been attempted, but they incorporate receptor flexibility only to a limited extent, focusing on sampling only most plausible/relevant portion of the conformational space, e.g., through side chain flexibility, conformational relaxation, and multiple structure docking, as already discussed in protein flexibility section. However, newer techniques, e.g., supervised molecular dynamics (SuMD) can be useful to incorporate receptor flexibility, because they allow receptor to experience thermal fluctuation and supervision of ligand toward binding site from unbound state might allow receptor to adopt induced conformational changes sensing the ligand in vicinity of binding site under influence of its interaction field [225].

6 Binding Ability and Free Energy Calculation

The binding free energy of ligand to receptor is the thermodynamic signature of the interaction affinity. Therefore, accurate prediction of binding free energy has been attempted from long times. The free energy calculation methods can be grouped into relative binding free energy calculation methods and absolute binding free energy methods [226]. Relative binding free energy methods aim to calculate binding free energy of one ligand (reference ligand) relative to another ligand (target ligand) both binding to same receptor, by summing up the work carried to convert one ligand to another in bound and free states in solution [226]. This method can be significantly efficient when reference ligand is very similar to target ligand, but if they are dissimilar then defining and sampling along the conversion path may pose severe computational demand [226]. Since reference ligand to target state conversion path is artificial, these methods are also called alchemical methods, and excellent review on popular methods of this class already exists [227]. Absolute binding free energy methods estimate standard binding free energy of interaction by computing reversible work done in process of transferring it from binding site into solution [226]. Absolute binding free energy methods have been reviewed by Shirts et al. [228]. Practical aspects of free energy calculation have also been recently reviewed [229, 230]. The accuracy of the binding free energy calculations is influenced by adequacy of sampling (theoretically, accurate results require infinite sampling), force field used for sampling, and correctness of the molecular model used, e.g., usually simulation is performed using fixed protonation states of titratable residues, while protonation states might change in experimental conditions [226].

6.1 Calculation of Enthalpy by MM-PBSA

The end-state free energy methods explained here are most common approaches to calculate binding free energy. Linear response approximation (LRA), linear interaction energy (LIE), and molecular mechanics Poisson–Boltzmann surface area (MM-PBSA), molecular mechanics generalized Born surface area (MM-GBSA) [231] are such methods available in the literature. End-state free energy methods are computationally less demanding, but the speed gain in CPU comes at cost of compromised accuracy of the results [231]. These methods are required to be plugged with estimation of configurational entropy which usually is obtained by rigid-rotor approximation and normal mode analysis or quasi-harmonic analysis to yield binding free energy [232]. However, these methods can be good for evaluating binding enthalpy for ligand–receptor interaction. In MM-PBSA/MM-GBSA approaches (schematically shown in Fig. 11), the binding energy is calculated by taking energy difference of free-form of protein (P), and ligand (L) from protein–ligand complex form (PL) [232].

The free energy of each of the molecular species (say X) can be expressed as sum of their molecular mechanics energy in gas phase $E_{\mathrm{MM}}(X)$, solvation free energy $G_{\mathrm{solv}}(X)$, and entropic part—TS(X). The $E_{\mathrm{MM}}(X)$ contribution can be expressed as sum of bonded, electrostatics, and van der Waals energies, i.e., $E_{\mathrm{MM}}(X) = E_{\mathrm{bond}}(X) + E_{\mathrm{elec}}(X) + E_{\mathrm{vdW}}(X)$ [231]. Similarly, $G_{\mathrm{solv}}(X)$ can be expressed as sum of polar and non-polar contributions $G_{\mathrm{polar}}(X)$ and $G_{\mathrm{non\text{-}polar}}(X)$, where $G_{\mathrm{polar}}(X)$ can be accounted using Poison–Boltzmann or its simplified version

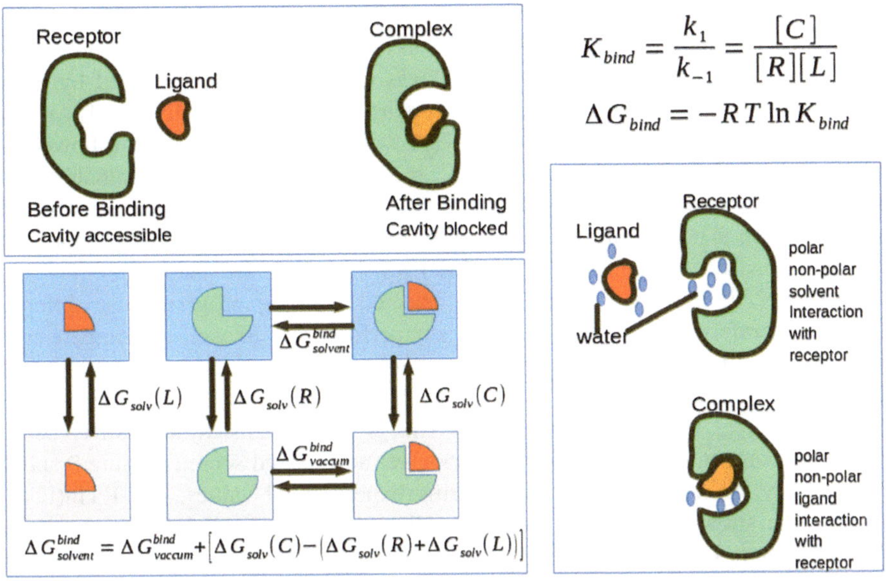

Fig. 11 Schematic representation of the end-state free energy using molecular dynamics Poisson–Boltzmann surface area method for estimating binding energy for receptor ligand binding

generalized Born method as $G_{\text{PB}}(X)$ or $G_{\text{GB}}(X)$, while non-polar is taken to be proportional to accessible surface area change $G_{\text{SASA}}(X)$ [231].

$$\Delta G_{\text{bind}} = G(PL) - G(P) - G(L) \tag{3}$$

The dynamics of the PfPK5 kinase structure complexed with the inhibitor(s) described earlier in docking section is used here as case study using MD simulations. Starting structures of PVB-PfPK5 [185] and INR-PfPK5 [185] complex were taken from crystal structures 1V0P and 1V0O, while OLM-PfPK5 and STA-PfPK5 were taken as consensus pose obtained from docking study using Gold, Glide, and Dock6 as mentioned above. All the systems were prepared using AmberTools14 [233] for MD simulation, and AM1-BCC charges for ligands and GAFF [234] force field parameters with ff14SB [235] parameters for protein. Equilibration was performed using standard protocol [236]. For each case, 12 independent (starting from different starting velocities) MD simulations in NPT ensemble each with length 254 ns were done, initial 4 ns run were discarded to allow for equilibration, bond lengths involving hydrogen were constrained using SHAKE [237] to allow use of 2 fs time step, temperature was controlled using Langevin thermostats with collision frequency 1 ps, and pressure was regulated with Berendsen scheme at target pressure 1 atmosphere using cuda version of program pmemd available in Amber14 [238] MD simulation package. Coordinates were saved every 1 ps. These trajectories were concatenated to yield 3 μs MD simulation for each case containing

3000,000 frames. Every 100th frame was taken for MM-PBSA analysis using MMPBSA.py [239] program in AmberTools14 [233].

The gas phase binding energy ΔE_{MM} was highest for INR followed by STA, PVB, and OLM, but the solvation penalty was also highest for INR and least for STA. In terms of enthalpy of binding, STA was predicted to be best, followed by PVB, INR, and OLM, respectively. The inconsistency of binding enthalpy with IC_{50} indicates possible role of entropy in this case. There may be role of solvation as well which is not rigorously captured in solvation terms considered proportional to buried surface area on binding in MM-PBSA method; see Table 12.

However, it may be criticized that selected docking programs use different scoring, therefore to be able to assess their performance as well as compare with experiment values is not possible. So, another attempt was done by normalizing all the scores, by converting all of them to positive scores (normalized using $(\text{score} - \text{min}_{\text{score}})/(\text{max}_{\text{score}} - \text{min}_{\text{score}})$). This yields a consistent normalized score, where weakest and strongest binder ligands get normalized scores ranging 0 and 1, respectively. Same is used for normalizing experimental values, i.e., $RT\ln(IC_{50})$. Results are shown in Fig. 12, Dock6 predicted scores for all ligands are within 1-sigma range, Gold and LibDock each predicted one outlier, and Glide predicted two outlier scores. In present case, Dock6, Gold, and LibDock appear to perform better than Glide. These results may not be sufficient to capture docking/scoring capabilities of chosen programs, as only four ligands are studied and they bind to only one target. A more diverse target set and a large ligand set could better comprehend the features and/or limitation of individual programs; this will be discussed later also.

The binding enthalpy predicted using MM-PBSA method consists of two outliers from $1-\sigma$ range (computed as discussed earlier), and it does not agree fully with docking scores obtained from any of the four chosen programs, as expected. However, strong and weak binders predicted using MM-PBSA is same as predicted by LibDock, and second strong binder predicted using these two is similar in affinity. While MM-PBSA results agree with Gold results for two weak binders and not for strong binders. Glide agrees on experimentally found strong and weak binders with MM-PBSA. Score using Dock6 agrees better than MM-PBSA (Fig. 12). As observed in the present case, the scoring by Docking methods as well

Table 12 Enthalpy component of binding free of selected inhibitors of PfPK5, calculated using MM-PBSA method

Inhibitor	IC_{50} (in nM)	RT ln(IC_{50}) (kcal/mol)	Predicted (kcal/mol)		
			ΔE_{MM}	ΔG_{Solv}	Total: ΔH_{PBSA}
OLM	15,000	−6.622	−58.5 ± 7.9	25.4 ± 5.9	−33.1 ± 4.1
INR	5500	−7.220	−102.7 ± 8.7	62.4 ± 5.9	−40.3 ± 4.2
STA	1000	−8.236	−69.0 ± 5.5	19.8 ± 4.0	−49.2 ± 4.7
PVB	130	−9.453	−65.7 ± 8.9	22.6 ± 6.0	−43.1 ± 4.7

These values are computed for 3000 snapshots extracted from 3-μs-long MD simulations for each inhibitors in complex with PfPK5, internal dielectric constant was taken 2, and ionic strength zero

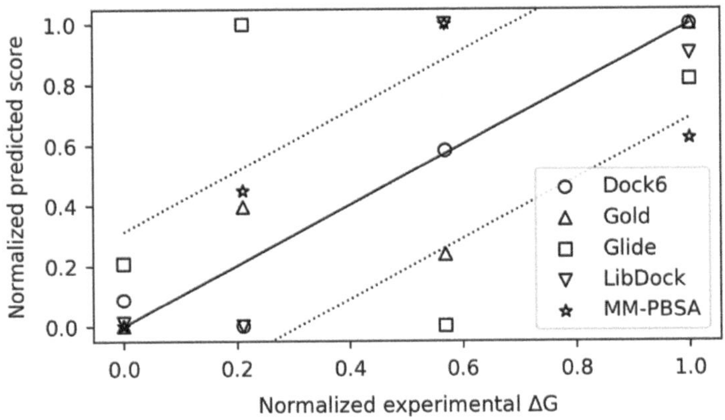

Fig. 12 All the scores have been normalized as discussed in text, to compare the predicted affinities for chosen four inhibitors of PfPK5 obtained using docking with Dock6, Gold, Glide and LibDock and MM-PBSA against experimental binding affinity. Solid line shows perfect correlation of scores with experimental results, and dotted lines above and below show one-σ range of error for predicted affinity

as the end-state Free Energy methods show discrepancies with experimental results, which emphasizes the effect of entropic contribution in case of flexible Kinase binding to ligands.

6.2 *Effect of Entropy to Ligand Binding*

Gibbs free energy (ΔG) of binding has two components enthalpy (ΔH) and entropy ($-T\Delta S$) as given by Eq. 4:

$$\Delta G = \Delta H - T\Delta S \tag{4}$$

Enthalpy of the protein ligand interaction is assumed to be the major determinant of the binding free energy assuming entropic contributions for smaller ligands binding to the same receptor would have similar entropic profile. However, this assumption can be seen as an attempt to simplify the scenario, as entropy estimation of binding process still lacks direct and reliable experimental/computational methods [240]. Experimental methods seek to estimate this quantity from the conformation flexibility as proxy for it and relate NMR relaxation parameter to calibrate it with conformation part of the biding entropy; conformation entropy is again assumed to be linearly correlated with the total binding entropy [241]. While, computation methods also try to estimate configurational entropy on similar line-of-thought, using molecular fluctuation data generated from molecular mechanics as a proxy for the entropy and thereby try to estimate configurational entropy from it [242–245]. Normal mode analysis (NMA) tries to infer conformational entropy as function of the vibration modes where DOFs are modeled as a set of simple harmonic oscillators, vibrating independently [246], but with the growing understanding of the nature of vibrational modes of biomolecules, it was realized that NMA is not the most suitable theory [247] for understanding entropy. Thus, methods utilizing internal coordinates for molecular description in conjunction with approximations representing full dimensional probability density function as a series of marginal PDFs of fluctuation of DOFs got attention of research community. This theory has been successfully applied to estimate entropy for small molecules [248], peptides [249, 250], to protein–peptide binding study with at least qualitative insight, while quantitative aspect still remains to be debatable [251, 252]. In some case, even for the set of ligands binding to the same receptor, entropic components are surprisingly quite different and play a crucial role in deciding the rank/affinity order of ligands.

As mentioned above, we found out that for a set of experimentally known ligands binding to the *P. falciparum* protein kinase PfPK5, docking scores yielded very poor correlation with experimental affinity, even inclusion of end-state free energy using MM-PBSA [253] method using 3 μs simulation data for each of the ligands, no significant improvement in computed affinity was observed. However, when configurational entropy for the ligands was included with the MM-PBSA estimates, a significant improvement in the bonding affinity was observed (manuscript in preparation).

As shown in (Fig. 13), achieving convergence to reduce error in estimation of entropy takes longer trajectories i.e., covering larger configuration space. Using a distance cutoff-based adaptation of Maximum Information Spanning Tree (MIST) called Neighbor Approximated Maximum Information Spanning Tree (A-MIST)

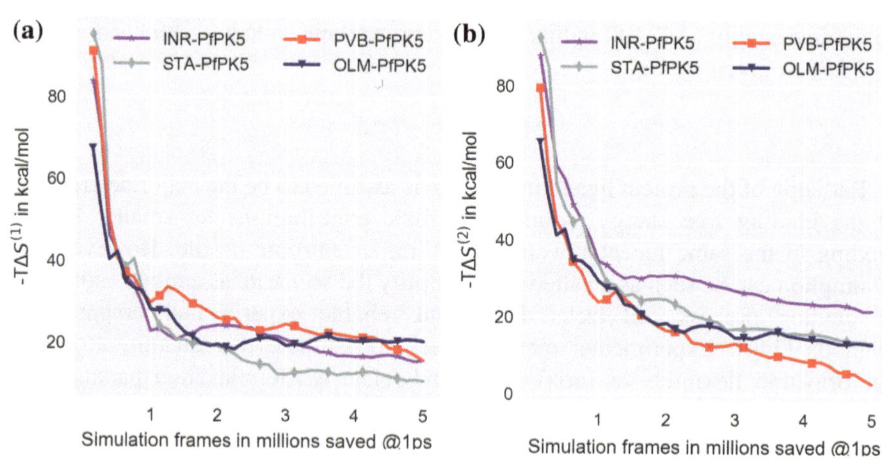

Fig. 13 Binding configurational entropy estimated using A-MIST methods with a distance cutoff of 14 Å and convergence of estimate with simulation time is shown. **a** Convergence of first order (assuming DOFs are uncorrelated) is shown. **b** Convergence of second order (accounting pair-wise correlations DOFs) is shown

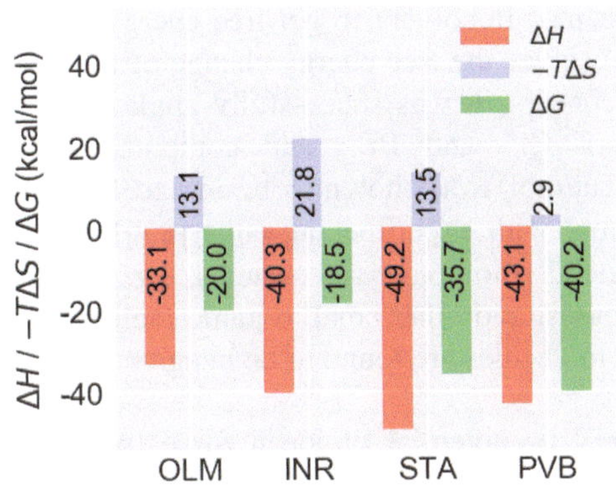

Fig. 14 Enthalpy, configurational entropy and free energy of binding of chosen inhibitors is shown in kcal/mol. Inhibitors are arranged in increasing experimental affinity (RTln(IC50)) order from left to right. Enthalpy is calculated using MM-PBSA method as discussed earlier. Here, temperature is taken to be 300 K

[254], the configurational entropy was estimated using MD dataset of ~5 μs, adding enthalpy, Free energy was calculated. It indicates that largely omitted entropic contributions can play important role and even deciding factor in case of small ligands binding to the flexible proteins (Fig. 14).

As shown in Fig. 14, combining enthalpy (ΔH) and configurational entropy $\left(-T\Delta S_{\text{config}}\right)$ of binding for chosen inhibitors, the binding free energy (ΔG) for best binder PVB is highest. However, binding free energy does not discriminate between OLM and INR, where experimentally OLM is weakest binder. Lower ΔG for INR (−18.5 kcal/mol) in comparison to OLM (−20.0 kcal/mol) may be attributed to the role of solvation free energy which is not accounted rigorously in MM-PBSA methods. Variations in configurational entropy of binding from 21.8 kcal/mol to mere 2.9 kcal/mol suggest that different ligands modulate and influence receptor flexibility in their own different way while forming complex, highlighting importance of receptor flexibility in binding affinity prediction studies; recently more attentions are attracted in this field.

6.3 Thermodynamic Methods

Relative binding free energy for a ligand formed by a chemical group substitution relative to parent compound can be computed using free energy perturbation molecular dynamics simulation [255]. This technique requires constructing a path from parent ligand L_1 to analog ligand L_2, which binds to a common receptor R, in two steps as follows. First, by carrying out a sequence of simulations in solvent and mutating L_1 to L_2 through several intermediate points and adding up the free energy changes along hypothetical intermediate points to yield free energy (say A_s) of mutating L_1 to L_2 in solvent, then similarily, mutating the ligand L_1 to L_2 in the binding pocket of receptor in solvent to get free energy change (say A_p). Finally, subtracting A_p from A_s gives the free energy change of the binding [255]. As early as 1985, to test the concept, it was successfully applied to calculate relative solvation free energy of Cl^- and Br^-, and computed Helmholtz free energy $\Delta\Delta A$ (3.35 ± 0.15 kcal/mol) was shown to be in excellent agreement $\Delta\Delta A_{\mathrm{hydr}} \approx \Delta\Delta G_{\mathrm{hydr}} = 3.3\,\mathrm{kcal/mol}$ with experimental value [256]. Further, the applicability of the method was extended to non-trivial systems, e.g., amino acids and their side chains, nucleic acid bases, and other small organic molecules; computed solvation free energies of these molecules are found to be in agreement with experiment [257, 258].

Relative free energy or potential of mean force (pmf, $w(r_c)$)-based methods relate it to the distribution of a chosen reaction coordinate (r_c), the direct sampling along r_c, and constructing its distribution function $g(r_c)$. The distribution function of reaction coordinate $g(r_c)$ can be related to pmf $(w(r_c))$ as

$$w(r_{\mathrm{c}}) = -k_{\mathrm{B}}T\ln g(r_{\mathrm{c}}) + \mathrm{constant} \tag{5}$$

However, barrier on the $w(r_{\mathrm{c}})$ can limit the sampling thereby the estimated pmf. Therefore, techniques like Umbrella sampling and Importance Sampling were introduced. But, choosing the right biasing function and ability to verify the adequacy of sampling for simulation widow is still challenging. A brief review of these methods is presented by Jorgensen et al. [259]. Statistical perturbation theory (SPT)-based methods which estimate free energy difference between systems i and j are related to the average of function of energy difference between systems i and j where sampling is based on system i [259]. Authors summarized several applications of SPT-based methods, e.g., for relative solvation free energy, relative pK_a values, study of solvent effect on conformational equilibria, study of binding and molecular recognition, and study of reactions in solvent [259]. The computational cost of carrying out SPT-based calculations inspired cost-effective semi-empirical methods using MD simulation samples for binding free energy calculation [260]. Aqvist et al. divided the binding free energy in two independent components electrostatic and non-polar, where electrostatic component $\Delta G^{\mathrm{el}}_{\mathrm{solv}}$ was taken to be half of the solvent–ion interaction energy [260]. For non-polar component, linearity between solvent size sigma and non-polar van der Waals energy and corresponding solvation energy, empirical parameter α was derived to relate vdW component of solvation free energy $\Delta G^{\mathrm{vdW}}_{\mathrm{solv}}$ with average of vdW component of interaction potential for transferring ligand from binding site (i) to solvent (s) given by $\Delta G^{\mathrm{vdW}}_{\mathrm{solv}} = \alpha\langle\Delta \mathrm{V}^{\mathrm{vdW}}_{\mathrm{i}\to\mathrm{s}}\rangle$ to yield expression for binding free energy [260] as: $\Delta G_{\mathrm{bind}} = 1/2.\langle \mathrm{V}^{\mathrm{el}}_{\mathrm{i}\to\mathrm{s}}\rangle + \alpha\langle \mathrm{V}^{\mathrm{vdW}}_{\mathrm{i}\to\mathrm{s}}\rangle$. This new semi-empirical method was tested on aspartic protease endothiapepsin and five small-molecule inhibitors with one as reference for which binding data and also crystal structure were available. It was reported that predicted relative binding free energy has mean unsigned error of 0.39 kcal/mol with highest for one of five inhibitors being 0.53 kcal/mol with parameter $\alpha = 0.161$ [260]. Application of such methods in details was discussed by Warshel and co-workers who have systematically examined performance of protein dipoles Langevin dipoles (PDLD) and other techniques using phosphorylcholine analogs binding to murine myeloma protein (McPC603) [261].

7 Molecular Recognition and Brownian Dynamics

As earlier discussed diffusional encounter of reacting substrates is the prerequisite for the binding interaction to happen [153]. Diffusional encounter is basically controlled by the long-range electrostatic interaction between participating chemical species [262]. Generally, the timescale of such encounter is from micro- to millisecond, which is tough to achieve with existing hardware technologies using molecular dynamics even for small- to moderate-sized biomolecules [263]. Therefore, simplified coarse-grained models of biomolecules can be simulated using Langevin dynamics and Brownian dynamics [262]. Brownian dynamics has been successfully applied to study ion permeations through ion channels [264] and enzymatic reactions [265]. However, to gain kinetic insight into receptor–ligand recognition, BD can be utilized [266–269], but BD being computationally very expensive is practically challenging [263]. This has called for alternate methods with simplistic approaches to study recognition process.

Supervised molecular dynamics (SuMD), a tabu-like search algorithm, aims to predict the pose of the ligand in the binding site of its cognate receptor, monitoring ligand-binding site distance along a series of short MD simulation has been proposed [225]. SuMD has been successfully applied to study a variety of molecular recognition processes [270–272]. In particular, Moro and co-workers applied to study molecular recognition process of four globular receptor–ligand systems and two transmembrane receptor ligand systems; in all these cases, experimental crystal structures and binding affinity values (IC_{50}, K_i or K_d) were already known [271]. In the study, it is observed that using SuMD, binding from unbound state (where ligand is placed at >30–50 Å away from binding site) of above ligands to their cognate receptor can be simulated; moreover, various interaction hot spots (metastable states) during recognition are possible to explore, which may be important in providing insight into kinetics of the recognition process, hence better designing of ligand [271]. In another study, the effect of allosteric modulator LUF6000 on adenosine binding with A_3 adenosine receptor (A_3AR) was reported. In this study, recognition of allosteric modulator LUF6000 to A_3AR and adenosine to A_3AR in presence and absence of LUF6000 was studied using SuMD. It is observed that adenosine visited a metastable site between helices EL3 and EL2, participating in hydrogen bonds with Val259 and Gln261, and it triggers an orientation change in adenosine mediated through hydrophobic interactions before occupying the binding site [270]. In future, such techniques along with Free Energy perturbation method will provide more accurate estimation of free energy binding of ligands to receptors which will include the flexibility of both partners.

8 Ligand Becomes Drug!

Drug research encompasses by various pipelines to achieve common goal, i.e., new therapeutic molecules. After the successful identification of the novel ligand or lead molecules by either computational or medicinal chemistry approach, each molecule must be characterized for absorption, distribution, metabolism, excretion, and toxicity (ADME-Tox) properties along with pharmacokinetic/pharmacodynamic (PK/PD) activity that decides the success rate of the drug [273, 274]. Evaluation of these properties belongs to the pre-clinical stage, and result of this stage decides the advancement of novel chemical entity (NCE) to clinical stage. Failure of the drug is dependent on the targeted therapeutic area; comparatively drug targeted to cardiovascular has maximum chance of success than CNS targeted [261]. So, successful candidates have to fulfill the essential criteria of potency, selectivity, oral bioavailability, therapeutic efficacy, along with an acceptable side effect profile [275]. Testing of thousands of leads molecules, found to be active against any disease, requires huge amount of money and time, and also it is not always easy to perform every test [276]. Understanding from the already prescribed drugs and knowledge from the failure rate during the different clinical stages has provided directions and specified various properties of chemicals which can be utilized to assess the lead molecules before performing costly and complex clinical tests [277].

Detailed information about ADME-Tox and its role in successful drug design is reviewed and available in many recent literatures [273, 278, 279]; however, major application of these properties is related to reduction in clinical drug failures from 40 to 10% [280]. This reduction has been seen with the advancement in the chemoinformatics and computational application in drug development process. As mentioned in the ligand design libraries, various physiochemical properties based on rules have been set to develop the lead-like and drug-like libraries to screen [281–284]. Along with these filters, for further libraries optimization filters like Pan Assay Interference Compounds (PAINS) and ALARM-NMR have been developed to remove known toxicophores or metabolically liable moieties which can interfere with the assay protocol [285, 286].

9 Summary

In this review, we have summarized many methods related to structure of receptor, characterization of active sites and subsites, binding affinity calculations, docking with specific poses, ranking chemicals and elucidated existing challenges in these methods. In spite of many mathematically and computationally elegant tools to understand and perform efficiently docking and scoring for large number of compounds, the success of identifying novel inhibitor of infectious disease and challenges thereof is still significantly high. Some of the solutions are already evident but many are yet to find. Still to ponder, how to estimate efficiently the effect of ions, pH dependency, and Brownian dynamics, which are playing significant role in Free energy of binding to receptor. Many relevant receptors are not crystallized yet, it is clearly evident that, errors occurring in in silico model structure and plurality of interactions with the binding site play a dominant role in correctly identify any novel inhibitor. Prior knowledge of physico-chemical interactions at active site and the functional importance of interacting residues influence the pose of binding of inhibitors to flexible receptors. A prior knowledge about the mechanism of binding provides lead towards the accuracy and effective binding of docked ligand. Flexible peptides derived structures provide higher affinity and in future, emerging field of study will be designing of such restrained chemicals driven by highly active peptides. Free energy estimation, rather than scoring (however accurate it may be), provides better designing capability. Knowledge of mechanism of inhibition is mandatory for innovation of novel chemical structure to lead the drug design, even in dominant era of artificial intelligence.

In conclusion, we have attempted to highlight the existing challenges in estimating the ligand receptor binding and critically inspect the methods applied day in and day out in the field of structure-based drug design. Summarization of tools and case studies are not the scope of the review. Most important aspect is that this field evolved largely using efficient algorithm and computational tools, however, effective use requires more indulgence of chemistry and biology, in future to progress successfully.

Acknowledgements We sincerely thank the facilities under Center of Excellence and Builder project, Department of Biotechnology (DBT), for supporting the computational work and DST purse for software support. SKP is supported by DBT-BINC fellowship. The authors thank all group members and Pawan Kumar for valuable input on improving the manuscript.

References

1. Seifert MHJ, Wolf K, Vitt D (2003) Virtual high-throughput in silico screening. Biosilico 1:143–149. https://doi.org/10.1016/s1478-5382(03)02359-x
2. Engin H, Gursoy A, Nussinov R, Keskin O (2014) Network-based strategies can help mono- and poly-pharmacology drug discovery: a systems biology view. Curr Pharm Des 20:1201–1207. https://doi.org/10.2174/13816128113199990066
3. Jones PG (1984) Crystal structure determination: a critical view. Chem Soc Rev 13:157–172. https://doi.org/10.1039/cs9841300157
4. Billeter M, Wagner G, Wüthrich K (2008) Solution NMR structure determination of proteins revisited. J Biomol NMR 42:155–158. https://doi.org/10.1007/s10858-008-9277-8
5. Halperin I, Ma B, Wolfson H, Nussinov R (2002) Principles of docking: an overview of search algorithms and a guide to scoring functions. Proteins Struct Funct Genet 47:409–443. https://doi.org/10.1002/prot.10115
6. Marrone TJ, Briggs JM, McCammon JA (1997) Structure-based drug design: computational advances. Annu Rev Pharmacol Toxicol 37:71–90. https://doi.org/10.1146/annurev.pharmtox.37.1.71
7. Perola E, Walters WP, Charifson PS (2004) A detailed comparison of current docking and scoring methods on systems of pharmaceutical relevance. Proteins Struct Funct Genet 56:235–249. https://doi.org/10.1002/prot.20088

8. Kitchen DB, Decornez H, Furr JR, Bajorath J (2004) Docking and scoring in virtual screening for drug discovery: methods and applications. Nat Rev Drug Discov 3:935–949. https://doi.org/10.1038/nrd1549
9. Yuriev E, Ramsland PA (2013) Latest developments in molecular docking: 2010–2011 in review. J Mol Recognit 26:215–239. https://doi.org/10.1002/jmr.2266
10. Yuriev E, Agostino M, Ramsland PA (2011) Challenges and advances in computational docking: 2009 in review. J Mol Recognit 24:149–164. https://doi.org/10.1002/jmr.1077
11. Yuriev E, Holien J, Ramsland PA (2015) Improvements, trends, and new ideas in molecular docking: 2012–2013 in review. J Mol Recognit 28:581–604. https://doi.org/10.1002/jmr.2471
12. Taylor RD, Jewsbury PJ, Essex JW (2002) A review of protein-small molecule docking methods. J Comput Aided Mol Des 16:151–166. https://doi.org/10.1023/a:1020155510718
13. Huang SY, Zou X (2010) Advances and challenges in protein-ligand docking. Int J Mol Sci 11:3016–3034. https://doi.org/10.3390/ijms11083016
14. Chen Y-C (2015) Beware of docking! Trends Pharmacol Sci 36:78–95. https://doi.org/10.1016/j.tips.2014.12.001
15. Acharya KR, Lloyd MD (2005) The advantages and limitations of protein crystal structures. Trends Pharmacol Sci 26:10–14. https://doi.org/10.1016/j.tips.2004.10.011
16. Petukh M, Stefl S, Alexov E (2013) The role of protonation states in ligand-receptor recognition and binding. Curr Pharm Des 19:4182–4190. https://doi.org/10.2174/1381612811319230004
17. Onufriev AV, Alexov E (2013) Protonation and pK changes in protein–ligand binding. Q Rev Biophys 46:181–209. https://doi.org/10.1017/s0033583513000024
18. Srivastava J, Barreiro G, Groscurth S et al (2008) Structural model and functional significance of pH-dependent talin-actin binding for focal adhesion remodeling. Proc Natl Acad Sci U S A 105:14436–14441. https://doi.org/10.1073/pnas.0805163105
19. Ferrara P, Gohlke H, Price DJ et al (2004) Assessing scoring functions for protein-ligand interactions. J Med Chem 47:3032–3047. https://doi.org/10.1021/jm030489h
20. Warren GL, Andrews CW, Capelli AM et al (2006) A critical assessment of docking programs and scoring functions. J Med Chem 49:5912–5931. https://doi.org/10.1021/jm050362n
21. Rask-Andersen M, Almén MS, Schiöth HB (2011) Trends in the exploitation of novel drug targets. Nat Rev Drug Discov 10:579–590. https://doi.org/10.1038/nrd3478
22. Finan C, Gaulton A, Kruger FA et al (2017) The druggable genome and support for target identification and validation in drug development. Sci Transl Med 9:1–16. https://doi.org/10.1126/scitranslmed.aag1166
23. Civelek M, Lusis AJ (2014) Systems genetics approaches to understand complex traits. Nat Rev Genet 15:34–48. https://doi.org/10.1038/nrg3575
24. Csermely P, Korcsmáros T, Kiss HJM et al (2013) Structure and dynamics of molecular networks: a novel paradigm of drug discovery: a comprehensive review. Pharmacol Ther 138:333–408. https://doi.org/10.1016/j.pharmthera.2013.01.016
25. Zhu P, Aliabadi HM, Uludağ H, Han J (2016) Identification of potential drug targets in cancer signaling pathways using stochastic logical models. Sci Rep 6:23078. https://doi.org/10.1038/srep23078
26. Torkamani A, Topol EJ, Schork NJ (2008) Pathway analysis of seven common diseases assessed by genome-wide association. Genomics 92:265–272. https://doi.org/10.1016/j.ygeno.2008.07.011
27. Druker BJ, Tamura S, Buchdunger E et al (1996) Effects of a selective inhibitor of the Abl tyrosine kinase on the growth of Bcr-Abl positive cells. Nat Med 2:561–566. https://doi.org/10.1038/nm0596-561
28. Fabian MA, Biggs WH, Treiber DK et al (2005) A small molecule-kinase interaction map for clinical kinase inhibitors. Nat Biotechnol 23:329–336. https://doi.org/10.1038/nbt1068

29. Montgomery JA, Niwas S, Rose JD et al (1993) Structure-based design of inhibitors of purine nucleoside phosphorylase. 1. 9-(arylmethyl) derivatives of 9-deazaguanine. J Med Chem 36:55–69. https://doi.org/10.1021/jm00053a008
30. Ealick SE, Babu YS, Bugg CE et al (1993) Application of X-ray crystallographic methods in the design of purine nucleoside phosphorylase inhibitors. Ann N Y Acad Sci 685:237–247
31. Ealick SE, Babu YS, Bugg CE et al (1991) Application of crystallographic and modeling methods in the design of purine nucleoside phosphorylase inhibitors. Proc Natl Acad Sci U S A 88:11540–11544. https://doi.org/10.1073/pnas.89.20.9974c
32. Erion MD, Niwas S, Rose JD et al (1993) Structure-based design of inhibitors of purine nucleoside phosphorylase. 3. 9-arylmethyl derivatives of 9-deazaguanine substituted on the arylmethyl group. J Med Chem 36:3771–3783. https://doi.org/10.1021/jm00076a004
33. Freire E (2008) Do enthalpy and entropy distinguish first in class from best in class? Drug Discov Today 13:869–874. https://doi.org/10.1016/j.drudis.2008.07.005
34. Ho M-C, Shi W, Rinaldo-Matthis A et al (2010) Four generations of transition-state analogues for human purine nucleoside phosphorylase. Proc Natl Acad Sci 107:4805–4812. https://doi.org/10.1073/pnas.0913439107
35. Edwards AA, Mason JM, Clinch K et al (2009) Altered enthalpy-entropy compensation in picomolar transition state analogues of human purine nucleoside phosphorylase. Biochemistry 48:5226–5238. https://doi.org/10.1021/bi9005896
36. Ohtaka H, Freire E (2005) Adaptive inhibitors of the HIV-1 protease. Prog Biophys Mol Biol 88:193–208. https://doi.org/10.1016/j.pbiomolbio.2004.07.005
37. Muzammil S, Armstrong AA, Kang LW et al (2007) Unique thermodynamic response of tipranavir to human immunodeficiency virus type 1 protease drug resistance mutations. J Virol 81:5144–5154. https://doi.org/10.1128/jvi.02706-06
38. Carbonell T, Freire E (2005) Binding thermodynamics of statins to HMG-CoA reductase. Biochemistry 44:11741–11748. https://doi.org/10.1021/bi050905v
39. Du X, Li Y, Xia Y-L et al (2016) Insights into protein-ligand interactions: mechanisms, models, and methods. Int J Mol Sci 17:144. https://doi.org/10.3390/ijms17020144
40. Amaral M., Kokh DB, Bomke J, Wegener A, Buchstaller HP, Eggenweiler HM, Matias P, Sirrenberg C, Wade RC, Frech M (2017) Protein conformational flexibility modulates kinetics and thermodynamics of drug binding. Nat Commun. https://doi.org/10.1038/s41467-017-02258-w
41. De Sanctis V, Kattamis C, Canatan D et al (2017) β-thalassemia distribution in the old world: an ancient disease seen from a historical standpoint. Mediterr J Hematol Infect Dis 9:1–14. https://doi.org/10.4084/mjhid.2017.018
42. Goss C, Giardina P, Degtyaryova D et al (2014) Red blood cell transfusions for thalassemia: results of a survey assessing current practice and proposal of evidence-based guidelines. Transfusion 54:1773–1781. https://doi.org/10.1111/trf.12571
43. Michlitsch J, Walters M (2008) Recent advances in bone marrow transplantation in hemoglobinopathies. Curr Mol Med 8:675–689. https://doi.org/10.2174/156652408786241393
44. Human Hemoglobin Structures (PDB) (2018) http://www.rcsb.org/pdb/results/results.do?tabtoshow=Current&qrid=205B68A9. Accessed 15 Feb 2018
45. Brown EN, Ramaswamy S (2007) Quality of protein crystal structures. Acta Crystallogr D Biol Crystallogr 63:941–950. https://doi.org/10.1107/s0907444907033847
46. Wlodawer A, Minor W, Dauter Z, Jaskolski M (2008) Protein crystallography for non-crystallographers, or how to get the best (but not more) from published macromolecular structures. FEBS J 275:1–21. https://doi.org/10.1111/j.1742-4658.2007.06178.x
47. Muller P (2009) Practical suggestions for better crystal structures. Crystallogr Rev 15:57–83. https://doi.org/10.1080/08893110802547240
48. Deller MC, Rupp B (2015) Models of protein-ligand crystal structures: trust, but verify. J Comput Aided Mol Des 29:817–836. https://doi.org/10.1007/s10822-015-9833-8
49. Carugo O (2018) How large B-factors can be in protein crystal structures. BMC Bioinform 19:1–9. https://doi.org/10.1186/s12859-018-2083-8

50. Li Z, Lazaridis T (2005) The effect of water displacement on binding thermodynamics: Concanavalin A. J Phys Chem B 109:662–670. https://doi.org/10.1021/jp0477912
51. Young T, Abel R, Kim B et al (2007) Motifs for molecular recognition exploiting hydrophobic enclosure in protein–ligand binding. Proc Natl Acad Sci 104:808–813. https://doi.org/10.1073/pnas.0610202104
52. Snyder PW, Mecinovic J, Moustakas DT et al (2011) Mechanism of the hydrophobic effect in the biomolecular recognition of arylsulfonamides by carbonic anhydrase. Proc Natl Acad Sci 108:17889–17894. https://doi.org/10.1073/pnas.1114107108
53. Sánchez R, Sali A (1997) Advances in comparative protein-structure modelling. Curr Opin Struct Biol 7:206–214
54. Eswar N, John B, Mirkovic N et al (2003) Tools for comparative protein structure modeling and analysis. Nucl Acids Res 31:3375–3380
55. Laskowski RA, MacArthur MW, Moss DS, Thornton JM (1993) PROCHECK: a program to check the stereochemical quality of protein structures. J Appl Crystallogr 26:283–291. https://doi.org/10.1107/s0021889892009944
56. Vriend G (1990) WHAT IF: a molecular modeling and drug design program. J Mol Graph 8:52–6, 29
57. McKinney JD, Zu Bentrup K Höner, Muñoz-Elias EJ et al (2000) Persistence of *Mycobacterium tuberculosis* in macrophages and mice requires the glyoxylate shunt enzyme isocitrate lyase. Nature 406:735–738. https://doi.org/10.1038/35021074
58. Tanjore Soundarajan Balganesh; Santanu Datta; Indira Ghosh (2004) WO2004087943A1
59. Nelson K, Wang FS, Boyd EF, Selander RK (1997) Size and sequence polymorphism in the isocitrate dehydrogenase kinase/phosphatase gene (aceK) and flanking regions in *Salmonella enterica* and *Escherichia coli*. Genetics 147:1509–1520
60. Garnak M, Reeves H (1979) Phosphorylation of isocitrate dehydrogenase of *Escherichia coli*. Science 203(80):1111–1112. https://doi.org/10.1126/science.34215
61. Vinekar R, Verma C, Ghosh I (2012) Functional relevance of dynamic properties of dimeric NADP-dependent isocitrate dehydrogenases. BMC Bioinform 13:S2. https://doi.org/10.1186/1471-2105-13-s17-s2
62. Vinekar R, Ghosh I (2009) Determination of phosphorylation sites for NADP-specific isocitrate dehydrogenase from *Mycobacterium tuberculosis*. J Biomol Struct Dyn 26:741–754. https://doi.org/10.1080/07391102.2009.10507286
63. Av-Gay Y, Everett M (2000) The eukaryotic-like Ser/Thr protein kinases of *Mycobacterium tuberculosis*. Trends Microbiol 8:238–244
64. Hurley JH, Thorsness PE, Ramalingam V et al (1989) Structure of a bacterial enzyme regulated by phosphorylation, isocitrate dehydrogenase. Proc Natl Acad Sci U S A 86:8635–8639. https://doi.org/10.1073/pnas.86.22.8635
65. Ceccarelli C, Grodsky NB, Ariyaratne N et al (2002) Crystal structure of porcine mitochondrial $NADP^+$-dependent isocitrate dehydrogenase complexed with Mn^{2+} and isocitrate: Insights into the enzyme mechanism. J Biol Chem 277:43454–43462. https://doi.org/10.1074/jbc.m207306200
66. Quartararo CE, Hazra S, Hadi T, Blanchard JS (2013) Structural, kinetic and chemical mechanism of isocitrate dehydrogenase-1 from *Mycobacterium tuberculosis*. Biochemistry 52:1765–1775. https://doi.org/10.1021/bi400037w
67. Hardin C, Pogorelov TV, Luthey-Schulten Z (2002) Ab initio protein structure prediction. Curr Opin Struct Biol 12:176–181
68. Bonneau R, Baker D (2001) Ab initio protein structure prediction. Annu Rev Biophys Biomol Struct 30:173–189
69. Moult J, Fidelis K, Kryshtafovych A et al (2017) Critical assessment of methods of protein structure prediction (CASP)—round XII. Proteins Struct Funct Bioinform 82:1–6
70. Singh A, Kaushik R, Mishra A et al (2016) ProTSAV: a protein tertiary structure analysis and validation server. Biochim Biophys Acta Proteins Proteomics 1864:11–19. https://doi.org/10.1016/j.bbapap.2015.10.004

71. Novoa EM, de Pouplana LR, Barril X, Orozco M (2010) Ensemble docking from homology models. J Chem Theory Comput 6:2547–2557
72. Kadam RU, Juraszek J, Brandenburg B et al (2017) Potent peptidic fusion inhibitors of influenza virus. Science 358(80):496–502. https://doi.org/10.1126/science.aan051
73. Chang YS, Graves B, Guerlavais V et al (2013) Stapled α-helical peptide drug development: a potent dual inhibitor of MDM2 and MDMX for p53-dependent cancer therapy. Proc Natl Acad Sci 110:E3445–E3454. https://doi.org/10.1073/pnas.1303002110
74. Shadfan M, Lopez-Pajares V, Yuan Z-M (2012) MDM2 and MDMX: alone and together in regulation of p53. Transl Cancer Res 1:88–89. https://doi.org/10.3978/j.issn.2218-676x.2012.04.02
75. Tiwari G, Verma CS (2017) Toward understanding the molecular recognition of albumin by p53-activating stapled peptide ATSP-7041. J Phys Chem B 121:657–670. https://doi.org/10.1021/acs.jpcb.6b09900
76. Schindler CEM, De Vries SJ, Zacharias M (2015) Fully blind peptide-protein docking with pepATTRACT. Structure 23:1507–1515. https://doi.org/10.1016/j.str.2015.05.021
77. Leder L, Berger C, Bomhauser S et al (1995) Spectroscopic, calorimetric, and kinetic demonstration of conformational adaptation in peptide-antibody recognition. Biochemistry 34:16509–16518. https://doi.org/10.1021/bi00050a035
78. Ferrari AM, Wei BQ, Costantino L, Shoichet BK (2004) Soft docking and multiple receptor conformations in virtual screening. J Med Chem 47:5076–5084. https://doi.org/10.1021/jm049756p
79. Totrov M, Ferna J, Abagyan R (2002) Soft protein–protein docking in internal coordinates. Protein Sci 11:280–291. https://doi.org/10.1110/ps.19202.ical
80. Li CH, Ma XH, Chen WZ, Wang CX (2003) A soft docking algorithm for predicting the structure of antibody-antigen complexes. Proteins Struct Funct Genet 52:47–50. https://doi.org/10.1002/prot.10382
81. Alonso H, Bliznyuk AA, Gready JE (2006) Combining docking and molecular dynamic simulations in drug design. Med Res Rev 26:531–568. https://doi.org/10.1002/med.20067
82. Leach AR (1994) Ligand docking to proteins with discrete side-chain flexibility. J Mol Biol 235:345–356. https://doi.org/10.1016/s0022-2836(05)80038-5
83. Schnecke V, Kuhn LA (2000) Virtual screening with solvation and ligand-induced complementarity. Perspect Drug Discov Des 20:171–190. https://doi.org/10.1023/a:1008737207775
84. Källblad P, Dean PM (2003) Efficient conformational sampling of local side-chain flexibility. J Mol Biol 326:1651–1665. https://doi.org/10.1016/s0022-2836(03)00083-4
85. Frimurer TM, Peters GH, Iversen LF et al (2003) Ligand-induced conformational changes: Improved predictions of ligand binding conformations and affinities. Biophys J 84:2273–2281. https://doi.org/10.1016/s0006-3495(03)75033-4
86. Gaudreault F, Chartier M, Najmanovich R (2012) Side-chain rotamer changes upon ligand binding: common, crucial, correlate with entropy and rearrange hydrogen bonding. Bioinformatics 28:423–430. https://doi.org/10.1093/bioinformatics/bts395
87. Apostolakis J, Plückthun A, Caflisch A (1998) Docking small ligands in flexible binding sites. J Comput Chem 19:21–37. https://doi.org/10.1002/(sici)1096-987x(19980115)19:1%3c21::aid-jcc2%3e3.0.co;2-0
88. Davis IW, Baker D (2009) RosettaLigand docking with full ligand and receptor flexibility. J Mol Biol 385:381–392. https://doi.org/10.1016/j.jmb.2008.11.010
89. Meiler J, Baker D (2006) ROSETTALIGAND: protein-small molecule docking with full side-chain flexibility. Proteins Struct Funct Bioinform 65:538–548. https://doi.org/10.1002/prot.21086
90. Perryman AL, Lin JH, McCammon JA (2006) Optimization and computational evaluation of a series of potential active site inhibitors of the V82F/I84V drug-resistant mutant of HIV-1 protease: an application of the relaxed complex method of structure-based drug design. Chem Biol Drug Des 67:336–345. https://doi.org/10.1111/j.1747-0285.2006.00382.x

91. Shandilya A, Chacko S, Jayaram B, Ghosh I (2013) A plausible mechanism for the antimalarial activity of artemisinin: a computational approach. Sci Rep 3:2513
92. Li J, Zhou B (2010) Biological actions of artemisinin: insights from medicinal chemistry studies. Molecules 15:1378–1397
93. Eckstein-Ludwig U, Webb RJ, Van Goethem IDA et al (2003) Artemisinins target the SERCA of *Plasmodium falciparum*. Nature 424:957
94. Pawan K, Shandilya A, Jayaram B, Ghosh I (2016) Integrative method for finding antimalarials using in silico approach. In: Kholmurodov KT (ed) Computer design for new drugs and materials. Nova Science Publishers, New York, NY, pp 13–38
95. Waelbroeck M (1982) The pH dependence of insulin binding. A quantitative study. J Biol Chem 257:8284–8291
96. Ellis CR, Shen J (2015) pH-dependent population shift regulates BACE1 activity and inhibition. J Am Chem Soc 137:9543–9546. https://doi.org/10.1021/jacs.5b05891
97. Mongan J, Case DA, McCammon JA (2004) Constant pH molecular dynamics in generalized Born implicit solvent. J Comput Chem 25:2038–2048. https://doi.org/10.1002/jcc.20139
98. Kim MO, Blachly PG, McCammon JA (2015) Conformational dynamics and binding free energies of inhibitors of BACE-1: from the perspective of protonation equilibria. PLoS Comput Biol 11:1–28. https://doi.org/10.1371/journal.pcbi.1004341
99. Hodge CN, Aldrich PE, Bacheler LT et al (1996) Improved cyclic urea inhibitors of the HIV-1 protease: synthesis, potency, resistance profile, human pharmacokinetics and X-ray crystal structure of DMP 450. Chem Biol 3:301–314. https://doi.org/10.1016/s1074-5521(96)90110-6
100. Lam PY, Jadhav PK, Eyermann CJ et al (1994) Rational design of potent, bioavailable, nonpeptide cyclic ureas as HIV protease inhibitors. Science 263(80):380–384. https://doi.org/10.1126/science.8278812
101. Kumalo HM, Bhakat S, Soliman MES (2015) Theory and applications of covalent docking in drug discovery: merits and pitfalls. Molecules 20:1984–2000. https://doi.org/10.3390/molecules20021984
102. Lecomte M, Laneuville O, Ji C et al (1994) Acetylation of human prostaglandin endoperoxide synthase-2 (cyclooxygenase-2) by aspirin. J Biol Chem 269:13207–13215
103. Hadváry P, Lengsfeld H, Wolfer H (1988) Inhibition of pancreatic lipase in vitro by the covalent inhibitor tetrahydrolipstatin. Biochem J 256:357–361. https://doi.org/10.1042/bj2560357
104. Scarpino A, Ferenczy GG, Keserű GM (2018) Comparative evaluation of covalent docking tools. J Chem Inf Model acs.jcim.8b00228. https://doi.org/10.1021/acs.jcim.8b00228
105. Zhu K, Borrelli KW, Greenwood JR et al (2014) Docking covalent inhibitors: a parameter free approach to pose prediction and scoring. J Chem Inf Model 54:1932–1940. https://doi.org/10.1021/ci500118s
106. Bianco G, Forli S, Goodsell DS, Olson AJ (2016) Covalent docking using autodock: two-point attractor and flexible side chain methods. Protein Sci 25:295–301. https://doi.org/10.1002/pro.2733
107. Corbeil CR, Englebienne P, Moitessier N (2007) Docking ligands into flexible and solvated macromolecules. 1. Development and validation of FITTED 1.0. J Chem Inf Model 47:435–449. https://doi.org/10.1021/ci6002637
108. MOE: Molecular Operating Environment (2018) http://www.chemcomp.com/MOE-Molecular_Operating_Environment.htm. Accessed 7 Feb 2018
109. Katritch V, Byrd CM, Tseitin V et al (2007) Discovery of small molecule inhibitors of ubiquitin-like poxvirus proteinase I7L using homology modeling and covalent docking approaches. J Comput Aided Mol Des 21:549–558. https://doi.org/10.1007/s10822-007-9138-7
110. Verdonk ML, Cole JC, Hartshorn MJ et al (2003) Improved protein–ligand docking using GOLD. Proteins Struct Funct Bioinform 623:609–623. https://doi.org/10.1002/prot.10465

111. Domínguez JL, Christopeit T, Villaverde MC et al (2010) Effect of the protonation state of the titratable residues on the inhibitor affinity to BACE-1. Biochemistry 49:7255–7263. https://doi.org/10.1021/bi100637n
112. Perryman AL, Lin J (2004) HIV-1 protease molecular dynamics of a wild-type and of the V82F / I84V mutant: possible contributions to drug resistance and a potential new target site for drugs. Protein Sci 13:1108–1123. https://doi.org/10.1110/ps.03468904.importance
113. Trylska J, Tozzini V, Chang CEA, McCammon JA (2007) HIV-1 protease substrate binding and product release pathways explored with coarse-grained molecular dynamics. Biophys J 92:4179–4187. https://doi.org/10.1529/biophysj.106.100560
114. Liu Z-P, Wu L-Y, Wang Y et al (2008) Bridging protein local structures and protein functions. Amino Acids 35:627–650
115. Ratnaparkhi GS, Varadarajan R (2000) Thermodynamic and structural studies of cavity formation in proteins suggest that loss of packing interactions rather than the hydrophobic effect dominates the observed energetics. Biochemistry 39:12365–12374
116. DesJarlais RL, Sheridan RP, Seibel GL et al (1988) Using shape complementarity as an initial screen in designing ligands for a receptor binding site of known three-dimensional structure. J Med Chem 31:722–729
117. Anderson AC (2003) The process of structure-based drug design. Chem Biol 10:787–797. https://doi.org/10.1016/j.chembiol.2003.09.002
118. Lee B, Richards FM (1971) The interpretation of protein structures: estimation of static accessibility. J Mol Biol 55:379–400
119. Connolly ML (1983) Analytical molecular surface calculation. J Appl Crystallogr 16:548–558
120. Connolly ML (1983) Solvent-accessible surfaces of proteins and nucleic acids. Science 221:709–713
121. Liang J, Woodward C, Edelsbrunner H (1998) Anatomy of protein pockets and cavities: measurement of binding site geometry and implications for ligand design. Protein Sci 7:1884–1897
122. Edelsbrunner H (1995) Smooth surfaces for multi-scale shape representation. In: International conference on foundations of software technology and theoretical computer science, pp 391–412
123. Mücke EP (1998) A robust implementation for three-dimensional Delaunay triangulations. Int J Comput Geom Appl 8:255–276
124. Bryant R, Edelsbrunner H, Koehl P, Levitt M (2004) The area derivative of a space-filling diagram. Discrete Comput Geom 32:293–308
125. Dundas J, Ouyang Z, Tseng J et al (2006) CASTp: computed atlas of surface topography of proteins with structural and topographical mapping of functionally annotated residues. Nucleic Acids Res 34:W116–W118
126. Benkaidali L, André F, Maouche B et al (2013) Computing cavities, channels, pores and pockets in proteins from non-spherical ligands models. Bioinformatics 30:792–800
127. Huang B, Schroeder M (2006) LIGSITE csc: predicting ligand binding sites using the Connolly surface and degree of conservation. BMC Struct Biol 6:19
128. Oliveira SHP, Ferraz FAN, Honorato RV et al (2014) KVFinder: steered identification of protein cavities as a PyMOL plugin. BMC Bioinform 15:197
129. Brady GP, Stouten PFW (2000) Fast prediction and visualization of protein binding pockets with PASS. J Comput Aided Mol Des 14:383–401
130. Czirják G (2015) PrinCCes: continuity-based geometric decomposition and systematic visualization of the void repertoire of proteins. J Mol Graph Model 62:118–127
131. Yu J, Zhou Y, Tanaka I, Yao M (2009) Roll: a new algorithm for the detection of protein pockets and cavities with a rolling probe sphere. Bioinformatics 26:46–52
132. Sheffler W, Baker D (2009) RosettaHoles: rapid assessment of protein core packing for structure prediction, refinement, design, and validation. Protein Sci 18:229–239
133. Laskowski RA (1995) SURFNET: a program for visualizing molecular surfaces, cavities, and intermolecular interactions. J Mol Graph 13:323–330

134. Kleywegt GJ, Jones TA (1994) Detection, delineation, measurement and display of cavities in macromolecular structures. Acta Crystallogr Sect D: Biol Crystallogr 50:178–185
135. Jacoby E, Fauchère J-L, Raimbaud E et al (1999) A three binding site hypothesis for the interaction of ligands with monoamine g protein-coupled receptors: implications for combinatorial ligand design. Mol Inform 18:561–572
136. Deganutti G, Moro S (2017) Estimation of kinetic and thermodynamic ligand-binding parameters using computational strategies. Future Med Chem 9:507–523
137. Chiu SH, Xie L (2016) Toward high-throughput predictive modeling of protein binding/unbinding kinetics. J Chem Inf Model 56:1164–1174
138. Singh A (2014) Selectivity and specificity profiling of binding sites of SER/THR kinases: case study of homo sapiens and *Plasmodium falciparum*. Jawaharlal Nehru University
139. Hanks SK, Hunter T (1995) Protein kinases 6. The eukaryotic protein kinase superfamily: kinase (catalytic) domain structure and classification. FASEB J 9:576–596. https://doi.org/10.1096/fasebj.9.8.776834
140. Freire E (2015) The binding thermodynamics of drug candidates. In: thermodynamics and kinetics of drug binding. Wiley Online Library, pp 1–13
141. Cross S, Baroni M, Carosati E et al (2010) FLAP: GRID molecular interaction fields in virtual screening. Validation using the DUD data set. J Chem Inf Model 50:1442–1450
142. Kaalia R, Kumar A, Srinivasan A, Ghosh I (2015) An ab initio method for designing multi-target specific pharmacophores using complementary interaction field of aspartic proteases. Mol Inform 34:380–393
143. Nakashima R, Sakurai K, Yamasaki S et al (2013) Structural basis for the inhibition of bacterial multidrug exporters. Nature 500:102
144. Koshland DE (1958) Application of a theory of enzyme specificity to protein synthesis. Proc Natl Acad Sci 44:98–104. https://doi.org/10.1073/pnas.44.2.98
145. Debler EW, Müller R, Hilvert D, Wilson IA (2008) Conformational isomerism can limit antibody catalysis. J Biol Chem 283:16554–16560. https://doi.org/10.1074/jbc.m710256200
146. Doshi U, McGowan LC, Ladani ST, Hamelberg D (2012) Resolving the complex role of enzyme conformational dynamics in catalytic function. Proc Natl Acad Sci 109:5699–5704. https://doi.org/10.1073/pnas.1117060109
147. Teague SJ (2003) Implications of protein flexibility for drug discovery. Nat Rev Drug Discov 2:527–541. https://doi.org/10.1038/nrd1129
148. Hawkins PCD (2017) Conformation generation: the state of the art. J Chem Inf Model 57:1747–1756. https://doi.org/10.1021/acs.jcim.7b00221
149. Grinter SZ, Zou X (2014) Challenges, applications, and recent advances of protein-ligand docking in structure-based drug design. Molecules 19:10150–10176. https://doi.org/10.3390/molecules190710150
150. Mobley DL, Dill KA (2009) Binding of small-molecule ligands to proteins: "what you see" is not always "what you get." Structure 17:489–498
151. London N, Movshovitz-Attias D, Schueler-Furman O (2010) The structural basis of peptide-protein binding strategies. Structure 18:188–199. https://doi.org/10.1016/j.str.2009.11.012
152. London N, Raveh B, Schueler-Furman O (2013) Peptide docking and structure-based characterization of peptide binding: from knowledge to know-how. Curr Opin Struct Biol 23:894–902. https://doi.org/10.1016/j.sbi.2013.07.006
153. Bernetti M, Cavalli A, Mollica L (2017) Protein–ligand (un)binding kinetics as a new paradigm for drug discovery at the crossroad between experiments and modelling. Med Chem Commun 8:534–550. https://doi.org/10.1039/c6md00581k
154. Pan AC, Borhani DW, Dror RO, Shaw DE (2013) Molecular determinants of drug-receptor binding kinetics. Drug Discov Today 18:667–673. https://doi.org/10.1016/j.drudis.2013.02.007
155. Case DA (1988) Dynamical simulation of rate constants in protein-ligand interactions. Prog Biophys Mol Biol 52:39–70. https://doi.org/10.1016/0079-6107(88)90007-7

156. Liu J, Wang R (2015) Classification of current scoring functions. J Chem Inf Model 55:475–482. https://doi.org/10.1021/ci500731a
157. Cornell WD, Cieplak P, Bayly CI et al (1995) A second generation force field for the simulation of proteins, nucleic acids, and organic molecules. J Am Chem Soc 117:5179–5197
158. MacKerell AD Jr, Bashford D, Bellott M et al (1998) All-atom empirical potential for molecular modeling and dynamics studies of proteins. J Phys Chem B 102:3586–3616
159. Allen WJ, Balius TE, Mukherjee S et al (2015) DOCK 6: impact of new features and current docking performance. J Comput Chem 36:1132–1156. https://doi.org/10.1002/jcc.23905
160. Simonson T, Archontis G, Karplus M (2002) Free energy simulations come of age: protein-ligand recognition. Acc Chem Res 35:430–437
161. Kramer B, Rarey M, Lengauer T (1999) Evaluation of the FLEXX incremental construction algorithm for protein–ligand docking. Proteins Struct Funct Bioinform 37:228–241
162. Böhm H-J (1994) The development of a simple empirical scoring function to estimate the binding constant for a protein-ligand complex of known three-dimensional structure. J Comput Aided Mol Des 8:243–256
163. Eldridge MD, Murray CW, Auton TR et al (1997) Empirical scoring functions: I. The development of a fast empirical scoring function to estimate the binding affinity of ligands in receptor complexes. J Comput Aided Mol Des 11:425–445
164. Böhm H-J (1992) LUDI: rule-based automatic design of new substituents for enzyme inhibitor leads. J Comput Aided Mol Des 6:593–606
165. Friesner RA, Banks JL, Murphy RB et al (2004) Glide: a new approach for rapid, accurate docking and scoring. 1. Method and assessment of docking accuracy. J Med Chem 47:1739–1749. https://doi.org/10.1021/jm0306430
166. Wang R, Lai L, Wang S (2002) Further development and validation of empirical scoring functions for structure-based binding affinity prediction. J Comput Aided Mol Des 16:11–26
167. Sippl MJ (1995) Knowledge-based potentials for proteins. Curr Opin Struct Biol 5:229–235
168. Lyne PD (2002) Structure-based virtual screening: an overview. Drug Discov Today 7:1047–1055. https://doi.org/10.1016/s1359-6446(02)02483-2
169. Muegge I (2000) A knowledge-based scoring function for protein-ligand interactions: probing the reference state. Perspect Drug Discov Des 20:99–114
170. Gohlke H, Hendlich M, Klebe G (2000) Knowledge-based scoring function to predict protein-ligand interactions. J Mol Biol 295:337–356
171. Head RD, Smythe ML, Oprea TI et al (1996) VALIDATE: A new method for the receptor-based prediction of binding affinities of novel ligands. J Am Chem Soc 118:3959–3969
172. Rarey M, Kramer B, Lengauer T, Klebe G (1996) A fast flexible docking method using an incremental construction algorithm. J Mol Biol 261:470–489
173. Morris GM, Goodsell DS, Halliday RS et al (1998) Automated docking using a Lamarckian genetic algorithm and an empirical binding free energy function. J Comput Chem 19:1639–1662
174. Spyrakis F, Amadasi A, Fornabaio M et al (2007) The consequences of scoring docked ligand conformations using free energy correlations. Eur J Med Chem 42:921–933. https://doi.org/10.1016/j.ejmech.2006.12.037
175. Oprea TI, Marshall GR (1998) Receptor-based prediction of binding affinities. Perspect Drug Discov Des 9:35–61
176. Huang N, Shoichet BK, Irwin JJ (2006) Benchmarking sets for molecular docking. J Med Chem 49:6789–6801. https://doi.org/10.1021/jm0608356
177. Empereur-Mot C, Guillemain H, Latouche A et al (2015) Predictiveness curves in virtual screening. J Cheminform 7:52. https://doi.org/10.1186/s13321-015-0100-8
178. Bauer MR, Ibrahim TM, Vogel SM, Boeckler FM (2013) Evaluation and optimization of virtual screening workflows with DEKOIS 2.0—a public library of challenging docking benchmark sets. J Chem Inf Model 53:1447–1462

179. Berman HM, Westbrook J, Feng Z et al (2000) The protein data bank. Nucl Acids Res 28:235–242. https://doi.org/10.1093/nar/28.1.235
180. Gore S, Sanz García E, Hendrickx PMS et al (2017) Validation of structures in the protein data bank. Structure 25:1916–1927. https://doi.org/10.1016/j.str.2017.10.009
181. Morris GM, Huey R, Lindstrom W et al (2009) AutoDock4 and AutoDockTools4: automated docking with selective receptor flexibility. J Comput Chem 30:2785–2791. https://doi.org/10.1002/jcc.21256
182. Van Zundert GCP, Rodrigues JPGLM, Trellet M, Schmitz C (2016) The HADDOCK2. 2 web server: user-friendly integrative modeling of biomolecular complexes. J Mol Biol 428:720–725. https://doi.org/10.1016/j.jmb.2015.09.014
183. Diller DJ, Merz KM (2001) High throughput docking for library design and library prioritization. Proteins Struct Funct Genet 43:113–124. https://doi.org/10.1002/1097-0134(20010501)43:2%3c113:aid-prot1023%3e3.0.co;2-t
184. Venkatachalam CM, Jiang X, Oldfield T, Waldman M (2003) LigandFit: a novel method for the shape-directed rapid docking of ligands to protein active sites. J Mol Graph Model 21:289–307. https://doi.org/10.1016/s1093-3263(02)00164-x
185. Holton S, Merckx A, Burgess D et al (2003) Structures of P. falciparum PfPK5 test the CDK regulation paradigm and suggest mechanisms of small molecule inhibition. Structure 11:1329–1337. https://doi.org/10.1016/j.str.2003.09.020
186. Engels MFM, Gibbs AC, Jaeger EP et al (2006) A cluster-based strategy for assessing the overlap between large chemical libraries and its application to a recent acquisition. J Chem Inf Model 46:2651–2660. https://doi.org/10.1021/ci600219n
187. Krier M, Bret G, Rognan D (2006) Assessing the scaffold diversity of screening libraries. J Chem Inf Model 46:512–524. https://doi.org/10.1021/ci050352v
188. McGregor MJ, Muskal SM (1999) Pharmacophore fingerprinting. 1. Application to QSAR and focused library design. J Chem Inf Comput Sci 39:569–574. https://doi.org/10.1021/ci980159j
189. Reymond J-L, van Deursen R, Blum LC, Ruddigkeit L (2010) Chemical space as a source for new drugs. Medchemcomm 1:30. https://doi.org/10.1039/c0md00020e
190. Williams AJ (2008) A perspective of publicly accessible/open-access chemistry databases. Drug Discov Today 13:495–501. https://doi.org/10.1016/j.drudis.2008.03.017
191. Lavecchia A, Di Giovanni C (2013) Virtual screening strategies in drug discovery: a critical review. Curr Med Chem 20:2839–2860. https://doi.org/10.2174/09298673113209990001
192. Hann MM, Oprea TI (2004) Pursuing the leadlikeness concept in pharmaceutical research. Curr Opin Chem Biol 8:255–263. https://doi.org/10.1016/j.cbpa.2004.04.003
193. Villoutreix Bruno O, Renault Nicolas, Lagorce David et al (2007) Free resources to assist structure-based virtual ligand screening experiments. Curr Protein Pept Sci 8:381–411. https://doi.org/10.2174/138920307781369391
194. Lipinski CA, Lombardo F, Dominy BW, Feeney PJ (2012) Experimental and computational approaches to estimate solubility and permeability in drug discovery and development settings. Adv Drug Deliv Rev 64:4–17. https://doi.org/10.1016/j.addr.2012.09.019
195. Congreve M, Carr R, Murray C, Jhoti H (2003) A "rule of three" for fragment-based lead discovery? Drug Discov Today 8:876–877. https://doi.org/10.1016/s1359-6446(03)02831-9
196. Hughes JD, Blagg J, Price DA et al (2008) Physiochemical drug properties associated with in vivo toxicological outcomes. Bioorg Med Chem Lett 18:4872–4875. https://doi.org/10.1016/j.bmcl.2008.07.071
197. Lagarde N, Zagury J-F, Montes M (2015) Benchmarking data sets for the evaluation of virtual ligand screening methods: review and perspectives. J Chem Inf Model 55:1297–1307. https://doi.org/10.1021/acs.jcim.5b00090
198. Réau M, Langenfeld F, Zagury J-F et al (2018) Decoys selection in benchmarking datasets: overview and perspectives. Front Pharmacol 9 https://doi.org/10.3389/fphar.2018.00011
199. Clark RD, Strizhev A, Leonard JM et al (2002) Consensus scoring for ligand/protein interactions. J Mol Graph Model 20:281–295

200. Cox JAG, Mugumbate G, Del Peral LVG et al (2016) Novel inhibitors of *Mycobacterium tuberculosis* GuaB2 identified by a target based high-throughput phenotypic screen. Sci Rep 6:1–10. https://doi.org/10.1038/srep38986
201. McInnes C (2007) Virtual screening strategies in drug discovery. Curr Opin Chem Biol 11:494–502. https://doi.org/10.1016/j.cbpa.2007.08.033
202. Kumar A, Zhang KYJ (2018) A cross docking pipeline for improving pose prediction and virtual screening performance. J Comput Aided Mol Des 32:163–173. https://doi.org/10.1007/s10822-017-0048-z
203. Thilagavathi R, Mancera RL (2010) Ligand- protein cross-docking with water molecules. J Chem Inf Model 50:415–421
204. Kilambi KP, Gray JJ (2017) Structure-based cross-docking analysis of antibody–antigen interactions. Sci Rep 7:8145
205. Kroemer RT, Vulpetti A, McDonald JJ et al (2004) Assessment of docking poses: interactions-based accuracy classification (IBAC) versus crystal structure deviations. J Chem Inf Comput Sci 44:871–881
206. Smith RD, Dunbar JB, Ung PM et al (2011) CSAR benchmark exercise of 2010: combined evaluation across all submitted scoring functions. J Chem Inf Model 51:2115–2131. https://doi.org/10.1021/ci200269q
207. Oda A, Tsuchida K, Takakura T et al (2006) Comparison of consensus scoring strategies for evaluating computational models of protein-ligand complexes. J Chem Inf Model 46:380–391. https://doi.org/10.1021/ci050283k
208. Dixon SL, Smondyrev AM, Knoll EH et al (2006) PHASE: a new engine for pharmacophore perception, 3D QSAR model development, and 3D database screening: 1. Methodology and preliminary results. J Comput Aided Mol Des 20:647–671. https://doi.org/10.1007/s10822-006-9087-6
209. Verma J, Khedkar VM, Coutinho EC (2010) 3D-QSAR in drug design—a review. Curr Top Med Chem 10:95–115. https://doi.org/10.2174/156802610790232260
210. Halgren TA, Murphy RB, Friesner RA et al (2004) Glide: a new approach for rapid, accurate docking and scoring. 2. Enrichment factors in database screening. J Med Chem 47:1750–1759. https://doi.org/10.1021/jm030644s
211. Hawkins PCD, Skillman AG, Nicholls A (2007) Comparison of shape-matching and docking as virtual screening tools. J Med Chem 50:74–82. https://doi.org/10.1021/jm0603365
212. Hawkins PCD, Skillman AG, Warren GL et al (2010) Conformer generation with OMEGA: algorithm and validation using high quality structures from the protein databank and Cambridge structural database. J Chem Inf Model 50:572–584. https://doi.org/10.1021/ci100031x
213. Takagi T, Amano M, Tomimoto M (2009) Novel method for the evaluation of 3D conformation generators. J Chem Inf Model 49:1377–1388. https://doi.org/10.1021/ci800393w
214. Ebejer J-P, Morris GM, Deane CM (2012) Freely available conformer generation methods: how good are they? J Chem Inf Model 52:1146–1158. https://doi.org/10.1021/ci2004658
215. Loferer MJ, Kolossváry I, Aszódi A (2007) Analyzing the performance of conformational search programs on compound databases. J Mol Graph Model 25:700–710. https://doi.org/10.1016/j.jmgm.2006.05.008
216. Vainio MJ, Johnson MS (2007) Generating conformer ensembles using a multiobjective genetic algorithm. J Chem Inf Model 47:2462–2474. https://doi.org/10.1021/ci6005646
217. Li J, Ehlers T, Sutter J et al (2007) CAESAR: a new conformer generation algorithm based on recursive buildup and local rotational symmetry consideration. J Chem Inf Model 47:1923–1932. https://doi.org/10.1021/ci700136x
218. Watts KS, Dalal P, Murphy RB et al (2010) ConfGen: a conformational search method for efficient generation of bioactive conformers. J Chem Inf Model 50:534–546. https://doi.org/10.1021/ci100015j

219. O'Boyle N, Vandermeersch T, Hutchison G (2011) Confab—generation of diverse low energy conformers. J Cheminform 3:P32. https://doi.org/10.1186/1758-2946-3-s1-p32
220. Gasteiger J, Rudolph C, Sadowski J (1990) Automatic generation of 3D-atomic coordinates for organic molecules. Tetrahedron Comput Methodol 3:537–547. https://doi.org/10.1016/0898-5529(90)90156-3
221. Riniker S, Landrum GA (2015) Better informed distance geometry: using what we know to improve conformation generation. J Chem Inf Model 55:2562–2574. https://doi.org/10.1021/acs.jcim.5b00654
222. Miteva MA, Guyon F, Tuffery P (2010) Frog2: efficient 3D conformation ensemble generator for small compounds. Nucl Acids Res 38:W622–W627. https://doi.org/10.1093/nar/gkq325
223. Sauton N, Lagorce D, Villoutreix BO, Miteva MA (2008) MS-DOCK: accurate multiple conformation generator and rigid docking protocol for multi-step virtual ligand screening. BMC Bioinform 9:1–12. https://doi.org/10.1186/1471-2105-9-184
224. Greg L (2018) RDKit: open-source cheminformatics. http://www.rdkit.org/. Accessed 7 Feb 2018
225. Sabbadin D, Moro S (2014) Supervised molecular dynamics (SuMD) as a helpful tool to depict GPCR-ligand recognition pathway in a nanosecond time scale. J Chem Inf Model 54:372–376. https://doi.org/10.1021/ci400766b
226. Mobley DL, Gilson MK (2017) Predicting binding free energies: frontiers and benchmarks. Annu Rev Biophys 46:531–558. https://doi.org/10.1146/annurev-biophys-070816-033654
227. Chodera JD, Mobley DL, Shirts MR et al (2011) Alchemical free energy methods for drug discovery: progress and challenges. Curr Opin Struct Biol 21:150–160. https://doi.org/10.1016/j.sbi.2011.01.011
228. Shirts MR, Mobley DL, Brown SP (2010) Free-energy calculations in structure-based drug design. In: Merz KM, Ringe D, Reynolds CH (eds) Drug design: structure-and ligand-based approaches. Cambridge University Press, New York, Cambridge, pp 61–86
229. Hansen N, Van Gunsteren WF (2014) Practical aspects of free-energy calculations: a review. J Chem Theory Comput 10:2632–2647. https://doi.org/10.1021/ct500161f
230. Christ CD, Mark AE, van Gunsteren WF (2009) Basic ingredients of free energy calculations: a review. J Comput Chem 31:1569-1582. https://doi.org/10.1002/jcc.21450
231. Genheden S, Ryde U (2015) The MM/PBSA and MM/GBSA methods to estimate ligand-binding affinities. Expert Opin Drug Discov 10:449–461. https://doi.org/10.1517/17460441.2015.1032936
232. Homeyer N, Gohlke H (2012) Free energy calculations by the molecular mechanics Poisson-Boltzmann surface area method. Mol Inform 31:114–122. https://doi.org/10.1002/minf.201100135
233. Case DA, Babin V, Berryman JT, Betz RM, Cai Q, Cerutti DS, Cheatham TE III, Darden TA, Duke RE, Gohlke H, Goetz AW, Gusarov S, Homeyer N, Janowski P, Kaus J, Kolossváry I, Kovalenko A, Lee TS, LeGrand S, Luchko T, Luo R, Madej B, Merz KM, Paesani F, Roe DR, Roitberg A, Sagui C, Salomon-Ferrer R, Seabra G, Simmerling CL, Smith W, Swails J, Walker RC, Wang J, Wolf RM, Wolf X, Kollman PA (2014) Amber14 Reference manual
234. Sprenger KG, Jaeger VW, Pfaendtner J (2015) The general AMBER force field (GAFF) can accurately predict thermodynamic and transport properties of many ionic liquids. J Phys Chem B 119:5882–5895. https://doi.org/10.1021/acs.jpcb.5b00689
235. Maier JA, Martinez C, Kasavajhala K et al (2015) ff14SB: improving the accuracy of protein side chain and backbone parameters from ff99SB. J Chem Theory Comput 11:3696–3713. https://doi.org/10.1021/acs.jctc.5b00255
236. Nguyen CN, Kurtzman Young T, Gilson MK (2012) Grid inhomogeneous solvation theory: hydration structure and thermodynamics of the miniature receptor cucurbit[7]uril. J Chem Phys 137:044101. https://doi.org/10.1063/1.4733951
237. Kräutler V, Van Gunsteren WF, Hünenberger PH (2001) A fast SHAKE algorithm to solve distance constraint equations for small molecules in molecular dynamics simulations.

J Comput Chem 22:501–508. https://doi.org/10.1002/1096-987x(20010415)22:5%3c501:aid-jcc1021%3e3.0.co;2-v
238. Salomon-Ferrer R, Case DA, Walker RC (2013) An overview of the amber biomolecular simulation package. Wiley Interdiscip Rev Comput Mol Sci 3:198–210. https://doi.org/10.1002/wcms.1121
239. Miller BR, McGee TD, Swails JM et al (2012) MMPBSA.py: an efficient program for end-state free energy calculations. J Chem Theory Comput 8:3314–3321. https://doi.org/10.1021/ct300418h
240. Marlow MS, Dogan J, Frederick KK et al (2010) The role of conformational entropy in molecular recognition by calmodulin. Nat Chem Biol 6:352–358. https://doi.org/10.1038/nchembio.347
241. Kasinath V, Sharp KA, Wand AJ (2013) Microscopic insights into the NMR relaxation-based protein conformational entropy meter. J Am Chem Soc 135:15092–15100. https://doi.org/10.1021/ja405200u
242. Diehl C, Engström O, Delaine T et al (2010) Protein flexibility and conformational entropy in ligand design targeting the carbohydrate recognition domain of galectin-3. J Am Chem Soc 132:14577–14589. https://doi.org/10.1021/ja105852y
243. Fenley AT, Muddana HS, Gilson MK (2012) Entropy-enthalpy transduction caused by conformational shifts can obscure the forces driving protein-ligand binding. Proc Natl Acad Sci U S A 109:20006–20011. https://doi.org/10.1073/pnas.1213180109
244. Chodera JD, Mobley DL (2013) Entropy-enthalpy compensation: role and ramifications in biomolecular ligand recognition and design. Annu Rev Biophys 42:121–142. https://doi.org/10.1146/annurev-biophys-083012-130318
245. Olsson TSG, Ladbury JE, Pitt WR, Williams MA (2011) Extent of enthalpy-entropy compensation in protein-ligand interactions. Protein Sci 20:1607–1618. https://doi.org/10.1002/pro.692
246. López-Blanco JR, Miyashita O, Tama F, Chacón P (2014) Normal mode analysis techniques in structural biology. In: John Wiley & Sons Ltd (ed) eLS. John Wiley & Sons, Ltd, Chichester, UK, p 9
247. Numata J, Wan M, Knapp E-W (2007) Conformational entropy of biomolecules: beyond the quasi-harmonic approximation. Genome Inform 18:192–205
248. Killian BJ, Yundenfreund Kravitz J, Gilson MK (2007) Extraction of configurational entropy from molecular simulations via an expansion approximation. J Chem Phys 127:024107. https://doi.org/10.1063/1.2746329
249. Numata J, Knapp E-W (2012) Balanced and bias-corrected computation of conformational entropy differences for molecular trajectories. J Chem Theory Comput 8:1235–1245. https://doi.org/10.1021/ct200910z
250. Suárez E, Díaz N, Méndez J, Suárez D (2013) CENCALC: a computational tool for conformational entropy calculations from molecular simulations. J Comput Chem 34:2041–2054. https://doi.org/10.1002/jcc.23350
251. Killian BJ, Kravitz JY, Somani S et al (2009) Configurational entropy in protein-peptide binding: computational study of Tsg101 ubiquitin E2 variant domain with an HIV-derived PTAP nonapeptide. J Mol Biol 389:315–335. https://doi.org/10.1016/j.jmb.2009.04.003
252. Fenley AT, Killian BJ, Hnizdo V et al (2014) Correlation as a determinant of configurational entropy in supramolecular and protein systems. J Phys Chem B 118:6447–6455. https://doi.org/10.1021/jp411588b
253. Fogolari F, Brigo a, Molinari H (2002) The Poisson-Boltzmann equation for biomolecular electrostatics: a tool for structural biology. J Mol Recognit 15:377–392. https://doi.org/10.1002/jmr.577
254. King BM, Silver NW, Tidor B (2012) Efficient calculation of molecular configurational entropies using an information theoretic approximation. J Phys Chem B 116:2891–2904. https://doi.org/10.1021/jp2068123
255. Tembre BL, Mc Cammon JA (1984) Ligand-receptor interactions. Comput Chem 8:281–283. https://doi.org/10.1016/0097-8485(84)85020-2

256. Lybrand TP, Ghosh I, McCammon JA (1985) Hydration of chloride and bromide anions: determination of relative free energy by computer simulation. J Am Chem Soc 107:7793–7794. https://doi.org/10.1021/ja00311a112
257. Bash P, Singh U, Langridge R, Kollman P (1987) Free energy calculations by computer simulation. Science 236(80):564–568. https://doi.org/10.1126/science.3576184
258. Kollman P (1993) Free energy calculations: applications to chemical and biochemical phenomena. Chem Rev 93:2395–2417. https://doi.org/10.1021/cr00023a004
259. Jorgensen WL (1989) Free energy calculations: a breakthrough for modeling organic chemistry in solution. Acc Chem Res 22:184–189. https://doi.org/10.1021/ar00161a004
260. Aqvist J, Medina C, Samuelsson JE (1994) A new method for predicting binding affinity in computer-aided drug design. Protein Eng 7:385–391
261. Lee FS, Chu ZT, Bolger MB, Warshel A (1992) Calculations of antibody-antigen interactions: microscopic and semi-microscopic evaluation of the free energies of binding of phosphorylcholine analogs to McPC603. Protein Eng 5:215–228. https://doi.org/10.1093/protein/5.3.215
262. Ermak DL, McCammon JA (1978) Brownian dynamics with hydrodynamic interactions. J Chem Phys 69:1352. https://doi.org/10.1063/1.436761
263. Adcock SA, McCammon JA (2006) Molecular dynamics: survey of methods for simulating the activity of proteins. Chem Rev 106:1589–1615. https://doi.org/10.1021/cr040426m
264. Bek S, Jakobsson E (1994) Brownian dynamics study of a multiply-occupied cation channel: application to understanding permeation in potassium channels. Biophys J 66:1028–1038. https://doi.org/10.1016/s0006-3495(94)80884-7
265. Sines J, Allison S, McCammon JA (1990) Brownian dynamics simulation of the superoxide-superoxide dismutase reaction: iron and manganese enzymes. J Phys Chem 94:959–961
266. Northrup SH, Erickson HP (1992) Kinetics of protein–protein association explained by Brownian dynamics computer simulation. Proc Natl Acad Sci U S A 89:3338–3342
267. Kozack RE, Subramaniam S (1993) Brownian dynamics simulations of molecular recognition in an antibody-antigen system. Protein Sci 2:915–926. https://doi.org/10.1002/pro.5560020605
268. Gabdoulline RR, Wade RC (1997) Simulation of the diffusional association of barnase and barstar. Biophys J 72:1917–1929. https://doi.org/10.1016/s0006-3495(97)78838-6
269. Gabdoulline RR, Wade RC (1998) Brownian dynamics simulation of protein–protein diffusional encounter. Methods 14:329–341. https://doi.org/10.1006/meth.1998.0588
270. Deganutti G, Cuzzolin A, Ciancetta A, Moro S (2015) Understanding allosteric interactions in G protein-coupled receptors using supervised molecular dynamics: a prototype study analysing the human A3 adenosine receptor positive allosteric modulator LUF6000. Bioorg Med Chem 23:4065–4071. https://doi.org/10.1016/j.bmc.2015.03.039
271. Cuzzolin A, Sturlese M, Deganutti G et al (2016) Deciphering the complexity of ligand-protein recognition pathways using supervised molecular dynamics (SuMD) simulations. J Chem Inf Model 56:687–705. https://doi.org/10.1021/acs.jcim.5b00702
272. Paoletta S, Sabbadin D, von Kügelgen I et al (2015) Modeling ligand recognition at the P2Y12 receptor in light of X-ray structural information. J Comput Aided Mol Des 29:737–756. https://doi.org/10.1007/s10822-015-9858-z
273. Lin JH, Lu AYH (1997) Role of pharmacokinetics and metabolism in drug discovery and development. Pharmacol Rev 49:403–449
274. Gallo JM (2010) Pharmacokinetic/pharmacodynamic-driven drug development. Mt Sinai J Med A J Transl Pers Med 77:381–388. https://doi.org/10.1002/msj.20193
275. Alavijeh MS, Chishty M, Qaiser MZ, Palmer AM (2005) Drug metabolism and pharmacokinetics, the blood-brain barrier, and central nervous system drug discovery. NeuroRx 2:554–571. https://doi.org/10.1602/neurorx.2.4.554
276. Altshuler J, Flanagan A, Guy P et al (2001) A revolution in R&D: how genomics and genetics are transforming the biopharmaceutical industry. Boston Consulting Group, Boston

277. Davis AM, Riley RJ (2004) Predictive ADMET studies, the challenges and the opportunities. Curr Opin Chem Biol 8:378–386
278. White RE (1998) Short-and long-term projections about the use of drug metabolism in drug discovery and development. Drug Metab Dispos 26:1213–1216
279. Eddershaw PJ, Beresford AP, Bayliss MK (2000) ADME/PK as part of a rational approach to drug discovery. Drug Discov Today 5:409–414
280. Kola I, Landis J (2004) Can the pharmaceutical industry reduce attrition rates? Nat Rev Drug Discov 3:711
281. Harris CJ, Hill RD, Sheppard DW et al (2011) The design and application of target-focused compound libraries. Comb Chem High Throughput Screen 14:521–531
282. Paricharak S, Méndez-Lucio O, Chavan Ravindranath A et al (2016) Data-driven approaches used for compound library design, hit triage and bioactivity modeling in high-throughput screening. Brief Bioinform 19(2):277–285. https://doi.org/10.1093/bib/bbw105
283. Chuprina A, Lukin O, Demoiseaux R et al (2010) Drug-and lead-likeness, target class, and molecular diversity analysis of 7.9 million commercially available organic compounds provided by 29 suppliers. J Chem Inf Model 50:470–479
284. Oprea TI, Allu TK, Fara DC et al (2007) Lead-like, drug-like or "Pub-like": how different are they? J Comput Aided Mol Des 21:113–119
285. Baell JB, Holloway GA (2010) New substructure filters for removal of pan assay interference compounds (PAINS) from screening libraries and for their exclusion in bioassays. J Med Chem 53:2719–2740. https://doi.org/10.1021/jm901137j
286. Metz JT, Huth JR, Hajduk PJ (2007) Enhancement of chemical rules for predicting compound reactivity towards protein thiol groups. J Comput Aided Mol Des 21:139–144. https://doi.org/10.1007/s10822-007-9109-z

Structure-Based Drug Design of *Pf*DHODH Inhibitors as Antimalarial Agents

Shweta Bhagat, Anuj Gahlawat and Prasad V. Bharatam

Abstract Structure-based drug design (SBDD) is being efficiently used for the design of antimalarial agents. It is a very effective tool for challenges like drug selectivity and resistance. Over the past decade, a considerable number of druggable targets have been explored—these include Na^+ ATPase 4 ion channel, cytochrome bc1, mitochondrial electron transport chain, phosphatidylinositol 4-kinase (*Pf*PI4 K), dihydroorotate dehydrogenase, hemozoin formation, dihydrofolate reductase inhibitors, etc. Among these, *Plasmodium falciparum* dihydroorotate dehydrogenase (*Pf*DHODH) is a new and very promising target. *Pf*DHODH has shown considerable potential in arresting growth of the parasite at blood stage by inhibiting pyrimidine biosynthesis. This chapter provides a review of all the SBDD efforts for the development of inhibitors against *Pf*DHODH.

Keywords *Plasmodium falciparum* · Structure-based drug design
Molecular docking · Virtual screening · Dihydroorotate dehydrogenase
Selectivity

List of Abbreviations

ACT	Aspartate carbamoyltransferase
ADME	Absorption, distribution, metabolism, and excretion
CoMFA	Comparative molecular field analysis
CoMSIA	Comparative molecular similarity index analysis
CoQ	Coenzyme Q (Ubiquinone)
CTP	Cytidine triphosphate
DBP	Docking-based pharmacophore
DHOtase	Dihydroorotase

S. Bhagat · P. V. Bharatam (✉)
Department of Medicinal Chemistry, National Institute of Pharmaceutical Education and Research (NIPER), S.A.S. Nagar 160062, Punjab, India
e-mail: pvbharatam@niper.ac.in

A. Gahlawat
Department of Pharmacoinformatics, National Institute of Pharmaceutical Education and Research (NIPER), S.A.S. Nagar 160062, Punjab, India

C. G. Mohan (ed.), *Structural Bioinformatics: Applications in Preclinical Drug Discovery Process*, Challenges and Advances in Computational Chemistry and Physics 27, https://doi.org/10.1007/978-3-030-05282-9_6

DHO	Dihydroorotate
DHODH	Dihydroorotate dehydrogenase
dTMP	Deoxyribose thymidine monophosphate
E. coli.	*Escherichia coli*
FAD	Flavin adenine dinucleotide
FMN	Flavin mononucleotide
G/PLS	Genetic partial least squares
GAT/CPS	Glutamine amidotransferase/carbamoyl phosphate synthetase
m-	*meta-*
MM/GBSA	Molecular mechanics/Generalized Born surface area
MSA	Molecular shape analysis
MLR	Multilinear regression
NAD	Nicotinamide adenine dinucleotide
OMPDC	Orotidine 5′-monophosphate decarboxylase
OPRT	Orotate phosphoribosyltransferase
ORO	Orotate
o-	*ortho-*
p-	*para-*
Pb	*Plasmodium berghei*
PDB	Protein Data Bank
Pf	*Plasmodium falciparum*
PRPP	Phosphoribosylpyrophosphate
QSAR	Quantitative structure–activity relationship
RMS	Root mean square
RNA	Ribonucleic Acid
SBDD	Structure-based drug design
SVM	Support vector machine
UMP	Uridine monophosphate
UTP	Uridine triphosphate

1 Introduction

Malaria is one of the most challenging communicable diseases caused by plasmodium parasite. It affects half of the world's population with 91 countries under the direct risk of transmission. Five million more cases of malaria were reported globally in 2016 compared to 2015. Children under the age of five are most susceptible to the disease with high death rate. The disease has a tropical and subtropical localization with prevalence in poor countries [1]. Due to the above facts and figures, it becomes essential to look into measures to limit the spread of disease and provide better solutions for curing malaria.

It is observed that majority of the drugs used to cure malaria have developed resistance within a span of 20 years of their introduction [2]. Other limitations of the drugs in market are compliance, safety, and cost. The drugs should be fast acting, curative within 3 days, and safe in pregnancy/early infancy [3]. Vaccine RTS,S has also been introduced for malaria but due to low efficacy (26–50%) is not recommended for babies between 6 and 12 weeks age [4].

To overcome these limitations, an extensive research is underway to discover new lead molecules with the potential of being introduced as new drug. Structure-based drug design (SBDD) is one such approach to discover new leads [5]. With the explosion of new information regarding the structure of many new targets, it has become easier to study the detailed structural aspects of the target. A thorough analysis of mechanism of their enzymatic activity, important amino acids responsible for molecular recognition and selectivity, mutated amino acids and their role in resistance as well as changes in enzyme efficiency due to mutations are some of the questions which are observed and answered during this process. Choosing appropriate drug hit/lead is based on this information along with the synthetic feasibility of the designed molecules. This is followed by biological activity analysis and selecting lead compounds. These leads can be further modified to improve the activity along with bioavailability and several such cycles of drug discovery process help in identifying molecules with improved target binding and specificity/selectivity.

*Pf*DihydrofolateReductase (*Pf*DHFR) is a very widely studied target which depicts an ideal example for SBDD approach. It was identified as the target for drugs like cycloguanil and pyrimethamine. The enzyme soon showed resistance within a span of 20 years of introduction of its inhibitors. It was then observed that the resistance occurred due to the mutation of Ser108 in the active site of the enzyme to Asn108, which shows steric clash with the *p*-chloro substitution at the phenyl ring of cycloguanil and pyrimethamine. This caused the emergence of two forms of double mutants and finally the most resistant quadruple mutant [6]. To avoid this steric clash, a linker chain was proposed to provide flexibility to the molecule and avoid close interaction with mutated amino acid Asn108. This led to the identification of WR99210 as lead molecule for in vitro *Pf*DHFR inhibition both in wild type and mutant form of the enzyme but failed during in vivo studies due to low bioavailability and toxicity [7]. It was further observed that due to high *pKa* of triazine moiety, WR99210 showed bioavailability problems. Further SBDD approach led to the identification of P218 as the lead molecule which was successful in both in vitro and in vivo studies and is currently undergoing clinical trials [8]. Molecular modeling studies on this enzyme led to the identification of key structural features that are essential for its selective inhibition. These include (i) H-bond donor head group for molecular recognition site, (ii) hydrophobic tail, and (iii) linker chain between the head group and tail [9]. These parameters were applied during SBDD approach for identification of new chemical head groups for *Pf*DHFR inhibitor design [10]. S-substituted guanylthiourea were identified which showed similar interactions as that of the WR99210 during molecular docking studies and later molecular dynamics studies [11]. In this series, two compounds were identified to in vitro activity against *Pf*DHFR and one compound was found to be curative against *Plasmodium berghei* during in vivo studies [12].

In 2017 itself, 21 crystal structures of recently discovered targets of *Plasmodium falciparum* were reported in RCSB Protein Data Bank. Given the extensive development in availability of high-resolution crystal structures of various important targets and techniques available to analyze these targets, the opportunities to carry out SBDD are enormous. Also, the availability of crystal structures of mutated enzymes offers opportunities to understand the reasons for mutation, its effect on the parasite and modifications required to overcome these mutations. The purpose of this review is to understand the SBDD efforts involved in the identification of new leads for malaria taking example of *Pf*DHODH as a target.

2 New Targets for Antimalarial Agent Design Employing SBDD

2.1 P. falciparum *ATP-Dependent Heat Shock Protein 90*

Heat shock protein 90 (HSP90) is a highly conserved molecular chaperone involved in the protein folding, stabilization, and protein–protein interaction of a variety of different proteins such as E3 ligases, transcription factors, various kinases, and many other proteins [6a]. It thus plays a major role in signal transduction and cell-cycle regulation of various species including *P. falciparum* [13]. Various inhibitors including geldanamycin (anticancer drug) compete with the natural substrate (ATP) for occupying the ATP-binding domain present at the N-terminal of the protein. This inhibition results in the arrest of the parasite growth in intra-erythrocytic phase by blocking the transition from immature ring-form stage to maturetrophozoite stage [13].

It is a homodimer protein and has three functional domains: (1) ATP-binding domain present at N-terminal, (2) a middle domain which facilitates ATP turnover, and (3) C-terminal domain which helps in dimerization. The ATPase cycle begins with the binding of substrate protein (like transcription factors, transducers) on the hydrophobic interface between the N-terminal and middle domains. This is followed by ATP binding and its subsequent hydrolysis that resulting in the compression of the substrate protein [14]. Its inhibitors bind to the ATP-binding site and halt conformational changes which are necessary to convert protein into compact one. The crystal structure reveals that *Pf*Hsp90 enzyme (PDB ID: 3K60) comprises of seven α-helices on one side and nine antiparallel β-sheets on other side of enzyme. The ATP binds to the solvent-accessible surface between a β-sheet and several α-helices. Hsp90 in all the species is characterized by the presence of an ATP lid (with different length, tertiary structure, and conformation) which is formed by the loop connecting the β-sheets and α-helices [13b].

Experimental high-throughput screening analysis performed over 4000 natural compounds recognized three new molecules, that includes Harmine, as selective inhibitor of *Pf*HSP90 [15]. Molecular docking and molecular dynamics analysis of 7-azaindole class of compounds (IND311 and IND31119) (Fig. 1) bound to *Pf*HSP90 active site have reported that the enzyme hydrophobic cavity is occupied by the side chains present at first and second position of IND31119 [6b]. The side chain of Asn37 residue forms two hydrogen bonds with secondary amine present at position 5 and carbonyl of amide present at position 3 of IND31119. The 7th position of IND31119 forms water-mediated hydrogen bond with Asn92 residue and there was another Ile96 residue which also interacts with structural water molecule. All these structural water molecules are conserved in enzyme active site [13b]. An in vitro study demonstrated that Geldanamycin exhibits inhibitory action on parasitic Hsp90 (with IC_{50} of 20 nM) [16]. The 7-azaindole (i.e., IND3) and several other its derivatives (i.e., IND31119 and IND311) possess in vitro and in silico selective antimicrobial activity against *Pf*HSP90. In 2018, Posfai et al. reported in vitro inhibitory activity of indazol-4(5H)-one class of compounds i.e. SNX-2112 (with K_i 5.9 nM) and Harmine (with K_i 27,000 nM) on *Pf*HSP90 target [17]. The molecular dynamics simulations performed on *Pf*Hsp90, human Hsp90, and mutated *Pf*Hsp90 indicated that human Hsp90 and mutated Hsp90 have more flexibility as compared to *Pf*Hsp90 [6b].

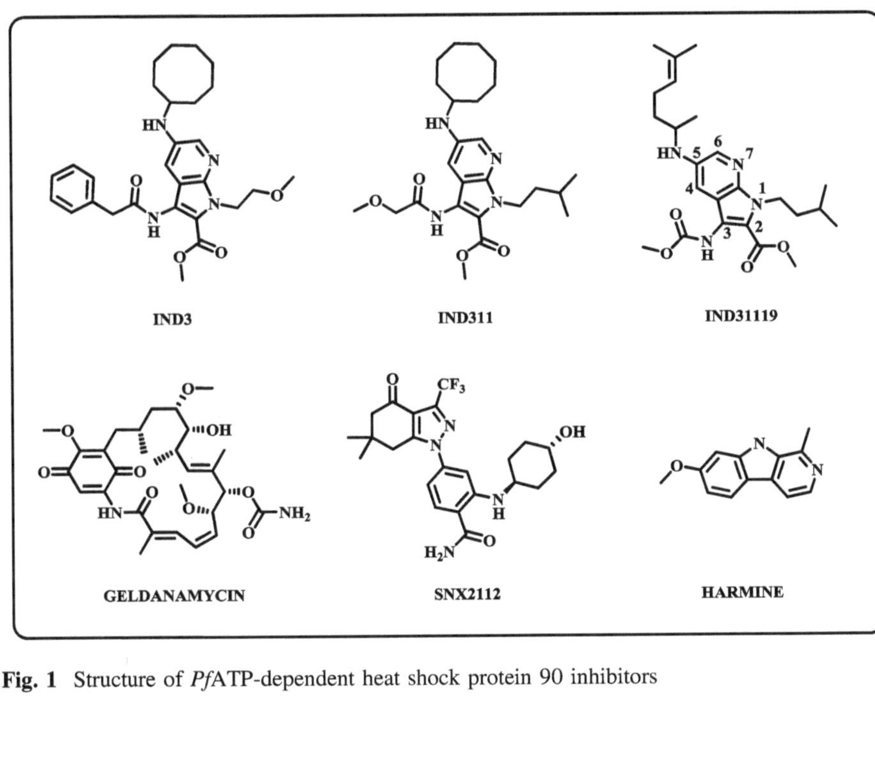

Fig. 1 Structure of *Pf*ATP-dependent heat shock protein 90 inhibitors

2.2 P. falciparum *Phosphatidylinositol 4-kinase (PfPI4 K)*

The phosphatidylinositol 4-kinase (PI4 K) enzyme catalyze the conversion of phosphatidylinositol into two essential phospholipids, i.e., phosphatidylinositol 4,5-bisphosphate and phosphatidylinositol 3,4,5-triphosphate by phosphorylation at one or more hydroxyl groups present in inositol moiety [18]. These phosphorylated products regulate numerous biological events, including intracellular signaling, vesicular transport, and cytoskeletal organization. Therefore, this biochemical reaction is essential for mammals and for the parasite. The parasite has only this enzyme to facilitate the phosphorylation of Phosphatidylinositol, but mammals have four enzymes for this biochemical reaction [19]. Inhibition of PfPI4 K in parasite cells lead to the deficiency of phospholipids in plasmodium leading to disruption of plasma membrane around developing merozoites and finally causing cell death. Also, lack of PI4 K in human erythrocytes ensure the unavailability of phospholipids in the vicinity of the parasite cells [20].

Rajkhowa et al. in 2017 developed an acceptable homology model of the catalytic domain of *Pf*PI(4) KIIIβ, consisting of 327 amino acids [19]. They selected X-ray crystal structure of *Hs*Phosphatidylinositol 4-kinase III β(PDB ID-4D0L) as a structural template with 44% sequence identity. Chain A was considered (with ligand PIK93 involved in antimalarial activity) for model development. The model was validated with the help of Ramachandran plot (favoured—90.5%, allowed—7.7%, and outlier—1.8%). The virtual screening analysis started with 178 compounds selected from PubChem database. After hERG and toxicity screening, ten compounds were selected for further molecular docking and molecular dynamics analysis (e.g., CHEMBL3355638 and CHEMBL2062798, Fig. 2). These ten compounds were docked into modeled *Pf*PI(4) KIIIb enzyme. The most active compound showed interaction with Lys66, Leu85, Tyr124, Val125, Thr128, Cys129, Ser130, Ser133, and Ile197 after docking. The molecular dynamics studies after 40 ns simulations using Gromacs package-4.6.6 showed important hydrophobic and polar interactions with Ile40, Leu44, Asn126, Asp198 residues [19].

An in vivo study on animal model suggested that imidazopyrazine class, KDU691 compound (Fig. 2) active against several drug-resistant strains (IC_{50} 27–70 nM) with

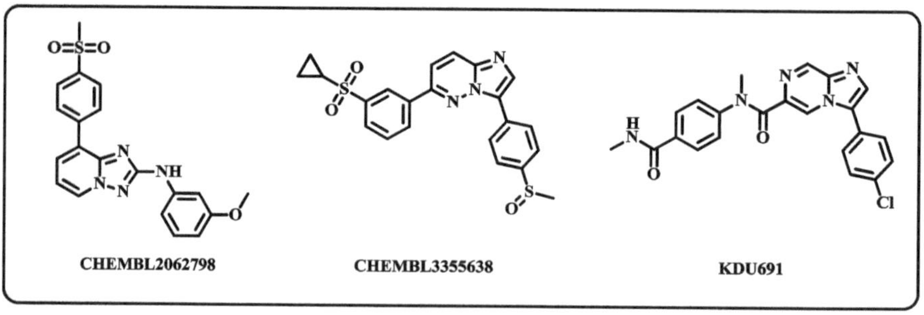

Fig. 2 Structure of compounds CHEMBL2062798, CHEMBL3355638, and KDU691

good potency [20] and exert their antimalarial effect through inhibitory interaction with the ATP-binding pocket of *Pf*PI4 K [21]. However, these compounds failed to provide prevention against parasite in human due to poor solubility and causes serious hERG liability (Arrhythmia). The reason for toxicity might be catalytic similarity between human and parasite enzyme as suggested by the homology model designed by Rajkhowa et al. [19].

2.3 **Pf***NDH2* **(P. falciparum** *NADH-Ubiquinone Oxidoreductase)*

*Pf*NADH-ubiquinone oxidoreductase is an enzyme of respiratory chain, present on the inner mitochondrial membrane. It is involved in the transfer of electrons from NADH to subsequent CoQ (ubiquinone) for CoQ–H_2 production and is coupled with translocation of proton or Na^+ across the membrane in mammals [22]. This enzyme is also present in *P. falciparum,* but it does not pump protons across the membrane; however, it still maintains the redox state of the cell. The human NDH2 is inhibited by rotenone, but *Pf*NDH2 enzyme was found be insensitive to it. A flavin reagent DPI (diphenyl iodonium chloride) inhibits the *Pf*NDH2 enzyme which leads to the depolarization of mitochondrial membrane potential and finally parasite cell death [23].

The *Pf*NDH2 enzyme exists in homodimer form comprising of four domains (1) C-terminal domain (CTD) helps in dimerization, conserved in plasmodium species; (2) two Rossmann fold domains, A and B, which bind to FAD and NADH cofactors, respectively; (3) domain C which shares no homology with other structures. Pidathala et al. in 2012 identified 2-bisaryl-3-methyl quinolone derivatives as *Pf*NDH2 inhibitors through in silico high-throughput screening studies [24]. In this series, 7-chloro-3-methyl-2-(4-(4-(trifluoromethoxy)benzyl)phenyl) quinolin-4(1H)-one (CK-2-68, Fig. 3) was identified to be the most potent compound with 16 nM IC_{50} value against *Pf*NDH2. However, due to poor solubility issues, this structure was modified with fluorine substitution on the quinolone ring to obtain 5-floro-3-methyl-2-(4-(4-(trifluoromethoxy)benzyl)phenyl)quinolin-4 (1H)-one (RYL-552, Fig. 3) [25]. It was found that RYL-552 binds as a

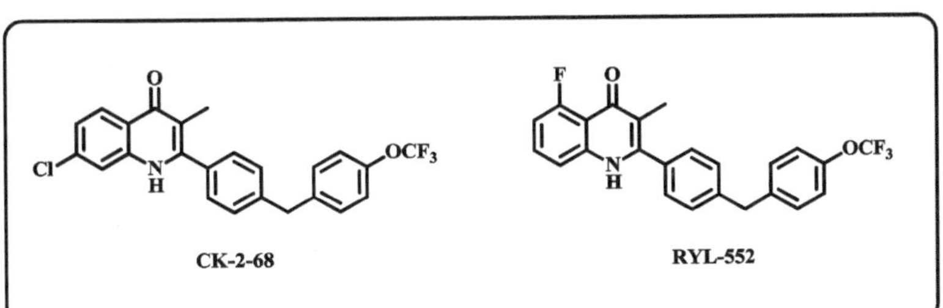

Fig. 3 Structure of *Pf*NDH2 inhibitors

non-competitive inhibitor at two different allosteric sites which were present close to C-terminal domain, causing conformational changes in NADH binding region that inhibits the binding of NADH to its pocket. These allosteric sites provide potential selectivity with minimal side effects.

Yang et al. in 2017 also reported that a total of four molecules of RYL-552 bind in the enzyme in homodimer state (PDB ID 5JWC). One molecule in each monomer and two molecules at dimer interface with difference in their binding poses. The first binding pose at the allosteric site between the two C-terminals of the homodimer (dimer interface) showed that two hydrogen bonds were formed by trifluoromethoxy group (with Tyr74 and Lys533), a hydrogen bond by carboxyl group (with Asn92), a water-mediated hydrogen bond by 4-oxo-5-fluoro group of quinolone with Gly87 and Lys523, and a hydrophobic interaction offered by bisaryl part (with Leu174, Val91, Ile170, and Ile532) of RYL-552. The other binding pose at the second allosteric pocket present in each monomer demonstrated that a water-mediated hydrogen bond is formed by quinolone ring nitrogen with Glu218 and Arg529 residues, two hydrogen bonds by 4-oxo-fluoro group (with Lys501), and two edge-to-face π-π stacking interaction by quinolone ring (with Trp500 and Tyr475 residue) [25]. Other compounds like HDQ (1-Hydroxy-2-dodecyl-4(1H)-quinolone), Aurachins A-D [26], RYL-552 [25], 7-chloro-3-methyl-2-(4-(4-(tri-fluoromethoxy) benzyl)phenyl) quinolin-4(1H)-one (CK-2-68) with IC_{50} 36 nM [24], and quinolone NQO2 (*p*-fluoro substituent has IC_{50} 9.6 nM) [27] were also reported as inhibitors of *Pf*NDH2.

2.4 **P. falciparum** ***Aspartate Carbamoyltransferase (PfACT)***

Aspartate carbamoyltransferase is an essential enzyme for the de novo pyrimidine biosynthesis in intra-erythrocytic stage, and it catalyzes the formation of N-carbamoyl-L-aspartate from carbamoyl phosphate and L-aspartate (Fig. 5). The malaria parasite cannot utilize salvage pathway for pyrimidine biosynthesis, as in the case of human cells, makes it a potential drug target for antimalarial drug design [28].

Banerjee et al. in 2011 developed a homology model of aspartate carbamoyl-transferase using amino acid sequence of the *P. falciparum* 3D7 (ID: XP_001350162.1 of NCBI database) as the target sequence and aspartate car-bamoyltransferase of *Pyrococcus abyssi* (PDB ID: 1ML4, Resolution: 1.8 Å) as template structure, with 38% overall sequence identity. The modeled structure was found to be stable during molecular dynamics studies performed using NAMD2.5 software. The validation of modeled structure was carried out using PROCHECK, WHATCHECK, WHATIF, VERIFY 3D, PROSA, ERRAT programs which suggested high quality of the model. The Q-SiteFinder software was used to find the possible active sites in the modeled protein. N-(phosphonacetyl)-L-aspartate (PALA) was known to inhibit *Escherichia coli* aspartate transcarbamylase. Therefore, PALA and its derivatives were selected for molecular docking in the predicted active site containing Ser107 using GOLD program. 3-(4-Hydroxy-phenyl)-2-(2-phosphono-acetylamino)-propionic acid was found to be the most suitable molecule docked in the predicted active site based on its good binding affinity toward the enzyme [28].

Lunev et al. first reported the crystal structure of *Pf*ACT (or *Pf*AspartateTranscarbamoylase, *Pf*ATC) (PDB ID 5ILQ) in 2016 without any bound ligand. In 2018, a comparative study was reported between the apo-form of the enzyme, *Pf*ACT complexed with citric acid at the active site (PDB ID 5ILN) and *Pf*ACT complexed with 2,3-naphthalenediol at the allosteric site (PDB ID 6FBA). The citrate-bound complex structure of the enzyme was found to be analogues to the liganded R-state of the enzyme. It was observed that the active site is highly conserved in this enzyme which might reduce the usefulness of this enzyme as a target using SBDD approach. The recently reported crystal structure with PDB ID 6FBA showed the presence of an allosteric site in which the enzyme is present in the T-state which is very similar to the apo-protein. Most of the amino acids in the allosteric site are non-conserved, thus making this site a potential target for SBDD [29].

2.5 P. falciparum *Thioredoxin Reductase* (Pf*TrxR)*

Thioredoxin reductase (TrxR) is a flavoenzyme (i.e., NADPH dependent) that maintains the enhanced oxidative stress in the erythrocytic stage of the parasite. It catalyzes the reduction of disulfide bridge of oxidized thioredoxin (Trx-S_2) into the thiol form, i.e., Trx-SH_2. This enzyme system is present in the cytosol, parasitophorous vacuole, and endoplasmic reticulum of the parasite. The *Pf*Thioredoxin (*Pf*Trx1), a biological substrate for *Pf*TrxR enzyme, is an important proton donor for the vital proteins like ribonucleotide reductase and other sets of peroxiredoxins [30]. *P. falciparum* lacks classical glutathione peroxidase and catalase enzymes which are present in eukaryotes to manage oxidative stress. Therefore, this enzyme plays an essential role for the survival of *P. falciparum* in the erythrocytic stages. The disruption of the redox state leads to antimalarial activity [31].

It is a homodimer containing three redox active centers to balance the redox state in *P. falciparum* (*Pf*TrxR) [31]. These are: (1) FAD-binding domain (2) N-terminal redox center near to the FAD-binding domain (Cys-88 and Cys-93), and (3) C-terminal redox center located on the flexible and accessible arm of other monomeric subunit (Cys-535 and Cys-540) which finally interacts with the thioredoxin substrate [30]. The homodimer is stabilized by Met105, Phe109, Ile108, Trp118, Phe120, and Leu123 (aromatic and hydrophobic) residues present at the interface. Met105 and Phe109, involved in the bending of interface helices, are conserved in the plasmodium species but not in the mammals. It is also reported that the buried interface of *Pf*TrxR enzyme is stabilized by more polar contacts than that of the human enzyme [32]. The *Pf*TrxR shows up to 40–42% sequence identity to the *Hs*TrxR and 77–80% sequence identity with other five species of the malaria parasite. The CVNVGC redox center at the N-terminal is conserved for all six species of plasmodium and human isoenzymes. However, the GCGGGKC region at the C-terminal is found to be preserved in all plasmodium isoforms but not in the *Hs*TrxR [33]. The parasite has an extended loop at C-terminal which provides flexibility and good interaction with *Pf*Trx1 substrate [30]. In the RCSB PDB, two crystal structures of *Pf*TrxR are available in complexation with Trx1 (PDB ID: 4J56, 4J57) and one crystal structure of *Pf*TrxR in the apo-form (PDB ID: 4B1B) [30, 32].

In the *Pf*TrxR, NADPH and FAD cofactors bind to their respective pockets in each monomer followed by hydride transfer from NADPHto FAD and then subsequently to N-terminal redox center (Cys88 and Cys93). This step is followed by the subsequent attack of Cys88 (present at the N-terminal of one monomeric subunit) at Cys540′ (C-terminal of the neighboring monomeric subunit). This in turn leads to nucleophilic attack of Cys540′ residue on the disulfide bond of thioredoxin leading to a mixed disulfide bond formation between Cys540′ of *Pf*TrxR and Cys30 of *Pf*Trx1. Finally, the mixed bond is broken with the help of Cys535′ (*Pf*TrxR) leading to the release of reduced substrate [30].

Boumis et al. in 2012 suggested the dimer interface cavity to be the site for non-competitive inhibitors binding in the *Pf*TrxR enzyme. In *Pf*TrxR, a narrow interface cavity is formed by Tyr101 and His104 residues which in case of *Hs*TrxR are much wider due to the presence of equivalent Gln72 and Leu75 residues [32]. It was also observed that smaller and slightly more amphipathic molecules could have selectivity toward parasite. The *Pf*TrxR enzyme cavity walls have less negative charge compared to the human isoforms which can be further exploited for selective inhibitor design [34]. Muniguntí et al. in 2013 studied the binding interactions of five known inhibitors of *Pf*TrxR enzyme (1,4-napthoquinone (1,4-NQ), bis-(2,4-dinitrophenyl)sulfide (2,4-DNPS), 4-nitrobenzothiadiazole (4-NBT), 3-dimethylaminopropiophenone (3-DAP), menadione (MD)) at dimer interface of *Pf*TrxR and *Hs*TrxR using molecular docking (AutoDock Vina software). It was observed that Tyr101 residue forms π-π stacking interaction with all the inhibitors while in *Hs*TrxR due to the presence of Gln72 at the equivalent position shows a different docking pose in order to avoid steric clashes. The other residues, i.e., Tyr116′ and Ile108, from both monomeric subunits (or only one subunit), form hydrophobic interaction with the docked inhibitors [34]. Munigunti et al. in 2014 reported similar residue interactions with curcuminoids at the dimer interface of *Pf*TrxR using Auto Dock software and suggested that the presence of methoxy group on curcumin structure reduces the interaction with Tyr101 residue [35]. The aculeatin-like analogues were also reported as inhibitors of *Pf*TrxR enzymes [36].

2.6 P. falciparum *Histone Deacetylase* (Pf*HDAC)*

Histone deacetylase (HDAC) posttranslationally modifies the histone proteins by removal of an acetyl group from the ε-nitrogen at the lysine side chain (present within the histone protein) and prevents the accessibility of DNA (wrapped around histones) for transcription factor [37]. Under normal circumstances in plasmodium parasite, only a few genes are expressed for transcription while the rest remain silent at a particular stage, and this phenomenon is known as stage-specific expression of genes. The inhibition of HDAC enzyme increases the acetylation that results in the loss of control over the gene expression which must be expressed in stage-specific manner in the parasite and, finally, implode the transcription cascade of parasite [37]. This enzyme is an attractive target for antimalarial drug therapy because, unlike mammalian cells, HDACs are more limited and potentially less redundant than plasmodium species [38].

The crystal structure of *Pf*HDAC-1 is yet to be elucidated. Therefore, a ligand-refined homology model of *Pf*HDAC-1 complexed with a hydroxamate-based inhibitor TrichostatinA (TSA) was generated, using human HDAC8 as template. It was observed that the modeled *Pf*HDAC-1 enzyme comprises of a single domain with open α/β class topology. The structure consists of eight β sheets surrounded by fourteen α helices and these secondary structures were linked by seven loops. *Pf*HDAC-1 comprises of hydrophobic upper region (lined with His24, Pro25, Thr96, Phe148, Phe203, Leu269, and Tyr301 residues) and a Zn^{2+} metal ion present in its catalytic site. This metal ion forms penta-coordinated geometry forming three bonds with the enzyme (side chain O(δ) of Asp174, Asp262, and the N(δ) of His176) and two bonds with ligand's hydroxamate group (i.e., carboxyl and hydroxyl oxygen) [39].

Mukherjee et al. in 2008 performed molecular docking and molecular dynamics studies on the homology model and found that hydroxamate group is essential for binding with *Pf*HDAC enzyme. The carbonyl oxygen of hydroxamate group forms hydrogen bond interactions with Tyr301, His138, and His139 residues. The TSA showed hydrophobic interaction with His24, Pro25, Phe148, Phe203, Leu269, and Tyr301 residues [39].

The development of compounds that are selective for parasitic HDACs over mammalian HDACs is still in relative infancy. SB939 (Fig. 4) was found to be a potent inhibitor of *Pf*HDAC enzyme (with IC_{50}—100 to 200 nM) in in vitro and in vivo studies, and its inhibitory effect was potentiated by aspartic protease inhibitor Lopinavir [38b]. Other compounds like 2-ASA-9, WR301801, MS-275, FR235222, LMK235 (and its derivatives) [40], Apicidin A [41], suberoylanilide-hydroxamic acid (SAHA, Vorinostat®), and a sulfonylpyrrolehydroxamate

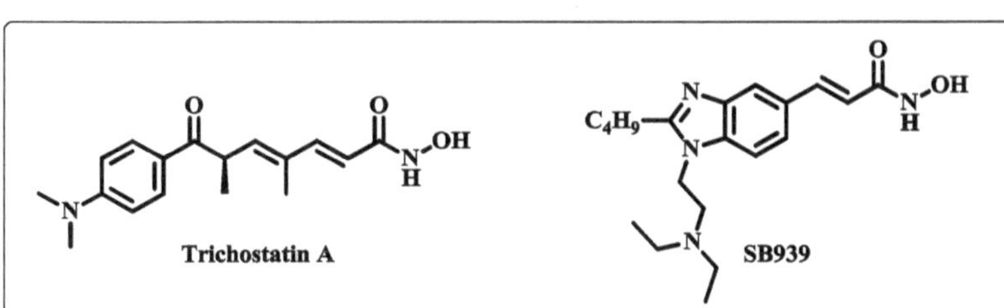

Fig. 4 Structures of *Pf*HDAC-1 inhibitors

(4SC-201, Resminostat) [39] were also found to be active against *Pf*HDAC1. These compounds were found to be less selective for the *Pf*HDAC1 and thus show off target activity.

2.7 P. falciparum *Glutathione S-Transferase (PfGST)*

Glutathione S-transferase (GST) is a detoxifying enzyme which catalyzes the conjugation of glutathione to electrophilic substrates to form conjugated products that are easily excreted out of body [42]. In plasmodium species, this enzyme relieves the oxidative stress during the intra-erythrocytic stage of the parasite [43]. Due to its abundance in the parasite cells and increased activity in chloroquine-resistant cells, it makes a potential target for antimalarial drug therapy [44].

Wolf et al. in 2003 found that *Pf*GST enzyme exists in dimer–tetramer transition state. Liebau et al. in 2005 observed that it favors tetrameric inactive state in the absence of reduce GST and other ligands. Liebau et al. in 2009 demonstrated that dimer–tetramer transition state is only present in case of *Pf*GST and absent for *Hs*GST. The active homodimer form is mainly assisted by hydrophobic interactions in which Phe56 residue of one subunit is buried inside the hydrophobic pocket of the other subunit formed by Trp131, Phe135, and Tyr134 residues. A hydrogen bond interaction between side chain of Arg77 and Asp97 residues, present at the two different neighboring monomeric subunits, also play important role. In typical μ-class human GSTs, the active (G- and H-) sites are present deep in the protein structure and are shielded by amino acids where as in case of *Pf*GST, these two sites (G- and H-sites) have more access to solvent. Furthermore, the two non-active dimers (i.e., inactive forms of the dimer) are interconnected by the loop 113–119 with the help of mainly hydrophobic and a few hydrophilic interactions, leading to the formation inactive tetramer state. These loop interactions block the active site of the enzyme and make it inactive [45]. Perbandt et al. later found that whole loop 113–119 is not important for the formation of inactive tetramer state, but Asn112 and Lys117 residues of neighboring subunits are most essential [45]. Other hydrogen bonds formed by Thr121 and Lys175 also aid to the tetramer formation. It was identified by Perbandt et al. that the non-substrate binding pocket was occupied by MES (2-(*N*-morpholino) ethanesulfonic acid) in its tetrameric form (PDB ID 4ZXG). The non-substrate binding pocket is outlined with Tyr25, Leu26, Leu196, Pro197, and Asn198 residues. These residues form a highly positively charged environment which attracts negatively charged ethane sulfonic moiety of MES by hydrogen bond formation with Asn198 and hydrophobic interaction with other residues of the cavity [45]. After mutation studies, Tyr9 was identified as an essential residue for selective inhibition of *Pf*GST [43]. The other inhibitors reported for the target were S-hexylglutathione [43], Protoporphyrin IX, cibacron blue, and menadione [46].

3 *P. falciparum* DHODH

Dihydroorotate dehydrogenase (DHODH) is one of the most validated and druggable targets. Miller et al. in 1968 first isolated L-dihydroorotate–ubiquinone reductase complex (from the beef liver) and was determined to be a mitochondrial enzyme [47]. The enzyme was recognized as dihydroorotate dehydrogenase (DHODH) and was further isolated from rat liver in 1976 by Chen et al. [48]. Its localization was determined to be the outer surface of the inner mitochondrial membrane which allows free diffusion of dihydroorotate (DHO) from cytosol into the mitochondria and orotate from mitochondria to the cytosol for further conversion to uridine monophosphate (UMP). Larsen et al. established in 1985 that the *E. coli* DHODH is a flavoprotein which catalyzes the conversion of dihydroorotate (DHODH) to orotate in the fourth and only redox reaction in de novo pyrimidine biosynthesis [49]. The DHODH enzymes can be classified into two different classes [50]:

Family I includes the cytosolic enzymes which utilize fumarate or NAD^+ as the terminal electron acceptor and deprotonation of alpha hydrogen occurred in the presence of cystein.
Family II includes membrane-bound enzymes that transfer electrons to ubiquinone (CoQ) and deprotonation occur in the presence of serine. Both human and plasmodium contain family II mitochondrial enzymes. In the host cells, pyrimidine biosynthesis occurs via salvage and de novo pathway, whereas in *P. falciparum* pyrimidines are synthesized only via *de novo* pathway. Thus, lack of salvage pathway in plasmodium makes it a vulnerable target [51].

McRobert and McConkey in 2002 reported the importance of DHODH enzyme in *P. falciparum* by performing RNA interference assay [52]. In 2002, Baldwin et al. conducted inhibitory studies of various known human DHODH inhibitors (Redoxal, dichloroallyllawsone (DCL), three analogs of A77-1726, and brequinar analogs) on malarial enzyme [53]. It was observed that the plasmodium enzyme showed 10^2–10^4 folds higher IC_{50} compared to the human enzyme. This study suggested that inhibition of DHODH enzyme is species specific and can be further explored to design *P. falciparum* selective DHODH inhibitors. Boa et al. in 2005 identified brequinar derivatives as non-selective and weakly selective *Pf*DHODH inhibitors proving the previous hypothesis and laying a base for further development of selective *Pf*DHODH inhibitors [54]. Baldwin et al. identified phenyl benzamide/naphthamides as selective *Pf*DHODH inhibitors in nanomolar range through high-throughput screening [55]. Heikkilä et al. used de novo design technology to identify six molecules as *Pf*DHODH inhibitors in micromolar range [56].

3.1 Functional Aspects of PfDHODH

Pyrimidines are essential metabolites that are precursors for DNA and RNA biosynthesis. Cells acquire pyrimidines either through de novo synthesis starting from ammonia (derived from L-Gln), bicarbonate, and L-asp, or by salvaging preformed pyrimidine base. *Plasmodium* species lack pyrimidine salvage enzymes and the de novo pathway provides the only source of pyrimidines for cell growth. In contrast, human cells are able to utilize both pathways. Inhibition of de novo pyrimidine synthesis in humans leads to immunosuppression and bone marrow depression. Immunosuppression is desirable in rheumatoid arthritis and organ transplant. However, immunosuppression and bone marrow depression during malaria may lead to life threatening situations which necessitate selective inhibition of parasite DHODH to be of utmost importance [57]. Pyrimidine biosynthesis requires six enzymes that are essential for the synthesis of UMP which is further utilized in generation of UTP, CTP, dTMP, and other metabolites of these nucleotides required by the cell. Enzymes involved are bifunctional glutamine amidotransferase/carbamoyl phosphate synthetase (GAT/CPS), aspartate carbamoyltransferase (ACT), dihydroorotase (DHOtase), DHODH, orotate phosphoribosyltransferase (OPRT) and orotidine 5′-monophosphate decarboxylase (OMPDC) (Fig. 5). The only redox step in the de novo synthesis of pyrimidines is the oxidation of DHO to ORO catalyzed by DHODH [58]. Reaction involves both a deprotonation and a hydride transfer converting DHO to ORO [59]. The reaction involves removal of acidic proton located at α position to the carbonyl group by an active base (Ser in family II enzymes) and the transfer of the hydrogen on C of DHO directly to N of the flavin as a hydride resulting in reduction of FMN to

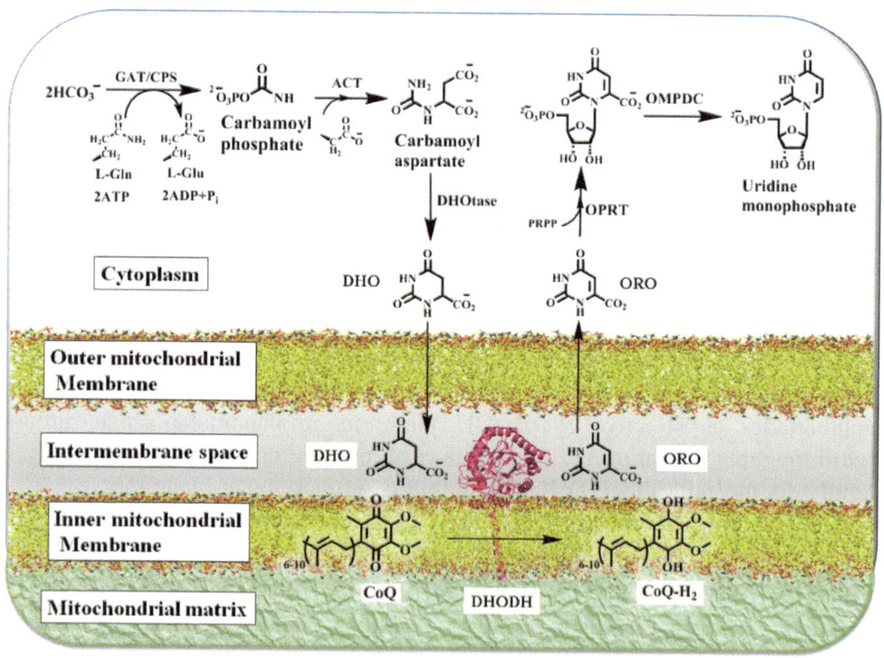

Fig. 5 Schematic representation of pyrimidine biosynthetic pathway

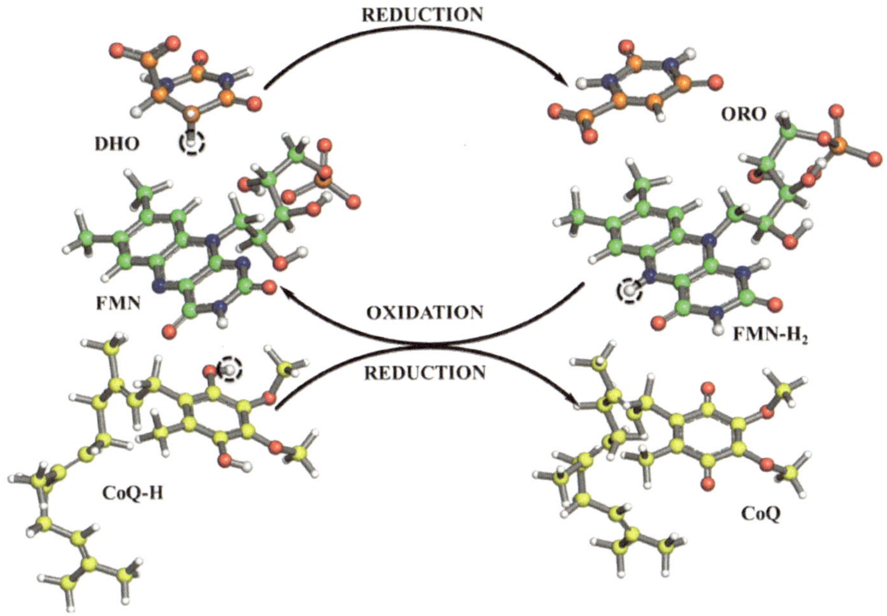

Fig. 6 Schematic representation of "hip-hop" redox mechanism involving transfer of hydride from C of DHO to N of FMN oxidizing dihydroorotate (DHO) to orotate (ORO) and reducing FMN to FMN-H_2. FMN-H_2 is reoxidised to FMN by co-substrate ubiquinone (CoQ) which itself gets reduced to ubiquinol (CoQ–H_2)

$FMNH_2$ (Fig. 6). $FMNH_2$ then gets reoxidised to FMN by ubiquinone (CoQ) which itself gets reduced to ubiquinol (CoQ–H_2). Inhibitors of DHODH affect the binding of this co-substrate ubiquinone with $FMNH_2$ [60].

A kinetic isotopic study on *E. coli* and human DHODH identified the mechanism of proton and hydride transfer with specific roles played by conserved amino acids. Two mechanisms were proposed, i.e., concerted and sequential in the absence of tunneling. It was observed that without tunneling, a concerted oxidation of DHO to orotate is not compatible. However, two stepwise mechanisms are still possible. If deprotonation precedes hydride transfer, then an enolate intermediate would form that could be stabilized by two conserved asparagine residues. If hydride transfer precedes deprotonation, ammonium intermediate would form that could hydrogen bond with another conserved asparagine residue [59, 61].

3.2 *Structural Details of* **PfDHODH**

*Pf*DHODH belongs to the DHODH family 2 located on the outer side of the inner mitochondrial membrane (mitochondrial intermembrane space) (Fig. 5) and is embedded in the membrane by a single α-transmembrane helix that holds the

Table 1 Crystal structures of *Pf*DHODH with various inhibitors, year of publication, resolution and enzyme inhibitory acitivity (IC_{50}) in μM

Sr. No	PDB ID	Hit structures	Year	Resolution (Å)	*Pf* IC_{50} (μM)	*Hs* IC_{50} (μM)	Refs.
1	1TV5	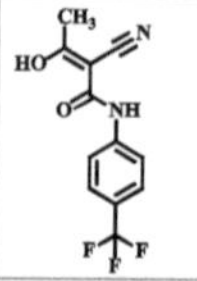	2005	2.40	190.1	0.26	[62]
2	3I6R	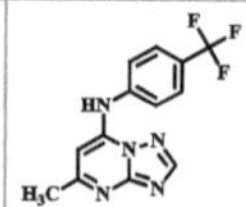	2009	2.50	0.28	>100	[64]
3	3I65	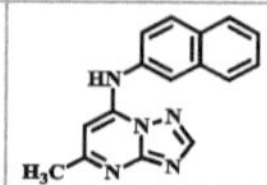	2009	2.00	0.047	>100	[64]
4	3I68	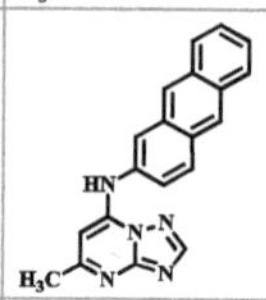	2009	2.40	0.056	>100	[64]
5	3O8A	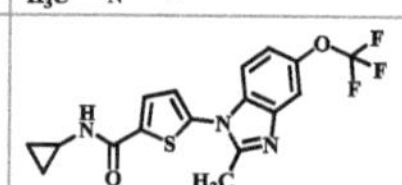	2010	2.30	0.022	>30	[65]
6	3SFK	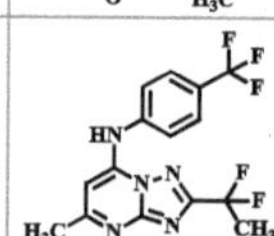	2011	2.90 Å	0.038	>100	[66]
7	4CQ8	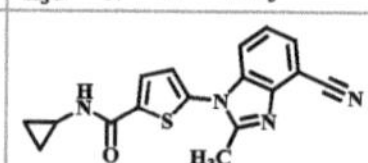	2014	1.98	0.08	>30	[67]
8	4CQ9	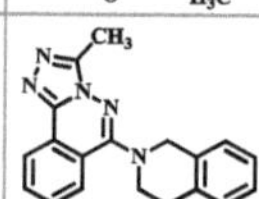	2014	2.72	3.5	3.8	[67]
9	4CQA	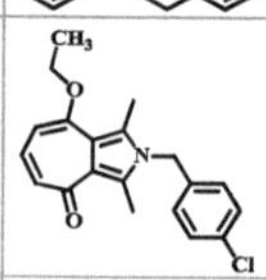	2014	2.82	13.5	>30	[67]
10	4ORM	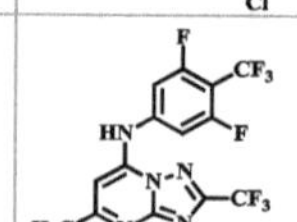	2014	2.07	0.022	1.6	[68]
11	4RX0	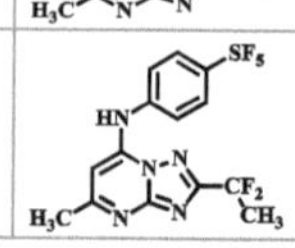	2015	2.25	0.033	>100	[69]

(continued)

Table 1 (continued)

Sr. No	PDB ID	Hit structures	Year	Resolution (Å)	*Pf* IC_{50} (μM)	*Hs* IC_{50} (μM)	Refs.
12	5BOO	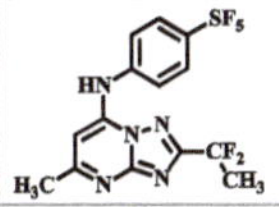	2015	2.80	0.033	>100	[69]
13	5DEL	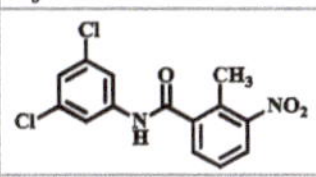	2015	2.20	0.016	>100	[70]
14	5FI8	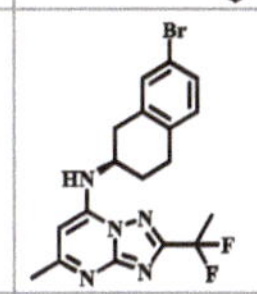	2016	2.32	0.0046	>100	[71]
15	5TBO	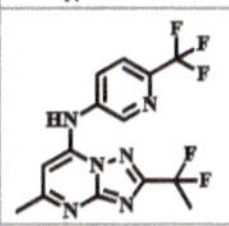	2016	2.15	0.053	>100	[72]

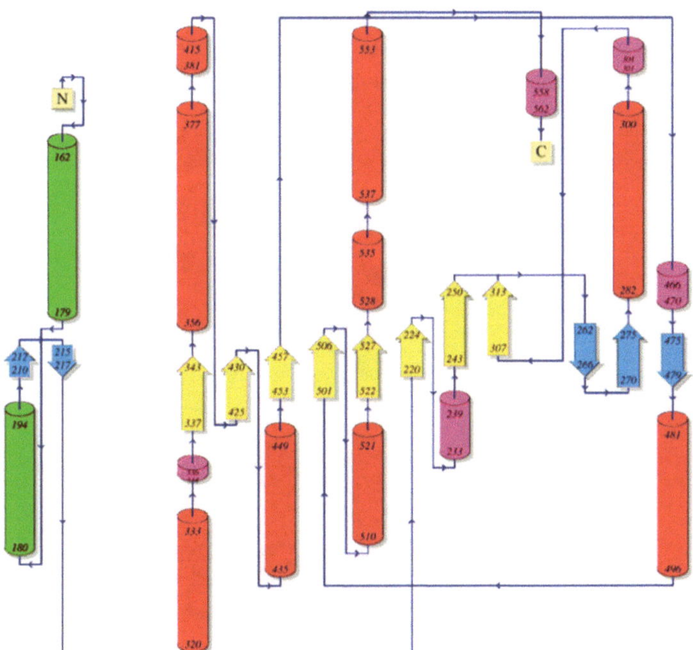

Fig. 7 Secondary structure of *Pf*DHODH. Cylinders represent α-helices and arrows represent β-sheets. Green α-helices form the N-terminal and red are part of C-terminal. Yellow color represents parallel β-sheets and cyan color represents antiparallel β-sheets. Light magenta color shows small helices with 3–5 amino acids

position of the enzyme in the membrane [62]. *Pf*DHODH has a total of 569 amino acids [50a] and till now 15 crystal structures are reported for this enzyme. Table 1 lists the reported crystal structure details from Protein Data Bank (PDB). All the crystal structures reported so far consist of full details of C-terminal domain but only the truncated details of N-terminal domain (amino acids 158–569). Amino acids 143–163 are part of the transmembrane helix and the remaining N-terminal part is located in the mitochondrial matrix for which no structural details are available (uniport ID Q08210). The secondary structure of the truncated enzyme (Fig. 7) (generated using Jpred 4 software using PDB ID 5FI8) consists of 13 β sheets and 16 α helices [63]. Details of secondary structure for amino acids 1–142 are not provided in the literature so far.

The most important structural feature of *Pf*DHODH is the presence of α/β-barrel core domain which is formed due to the almost parallel arrangement of eight β-sheets (Fig. 8). This β-barrel is surrounded by seven α-helices which provide protective layer to the core. The 3D structure is also characterized by the presence of a few short helices interspersed across the protein. The barrel is capped by a pair of antiparallel β-strands on one side and three β-strands on the other side [62].

The catalytic site is present near the cap with three β-strands. The cofactor FMN and substrate DHO bind in this region before undergoing redox catalytic reaction. There is a very unique tunnel in the 3D structure of DHODH. This is the tunnel through which a long-chain co-substrate with the quinone head group and six to ten repeating isoprene units (ubiquinone) travel through and reach the co-substrate

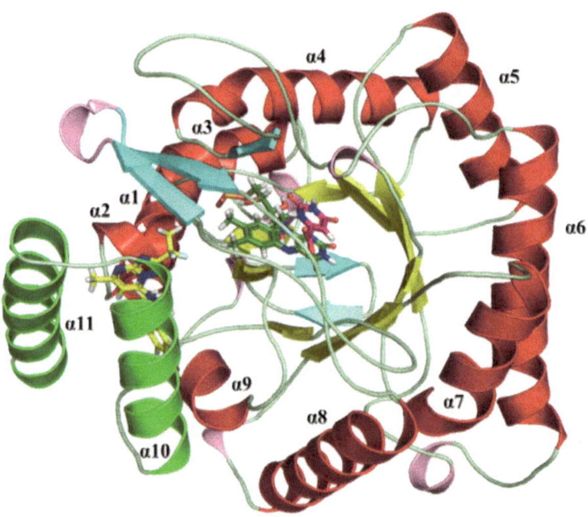

Fig. 8 3D structure of *Pf*DHODH showing central barrel formed by parallel β-sheets (in yellow) wrapped around with α-helices (in red, α1–α9). Both ends of the barrel are covered by anti-parallel β-sheets forming the lid (in cyan). The turns are represented in light magenta. The reaction site contains FMN as co-factor (in dark green) and dihydroorotate (in magenta) as substrate. This site is connected to the ubiquinone tunnel through two α-helices (in green, α10-α11) of the N-terminal and contains the inhibitor (DSM422) in yellow (PDB ID 5FI8) [71]

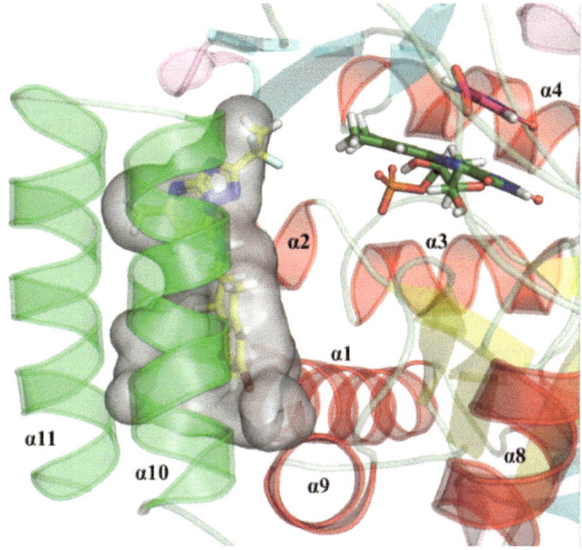

Fig. 9 Surface view of CoQ-binding tunnel (gray) with the inhibitor (PDB ID 5FI8)

binding site. The prosthetic group of FMN in the middle separates the dihydroorotate-binding site (substrate binding site) at the mouth of the barrel from the CoQ-binding tunnel (co-substrate binding site) at the outer surface of the barrel. The aromatic ring of DHO is almost parallel to FMN and is 3.2–3.8 Å from its *si*face. DHO's other face is completely covered by the Asn212–Gly226 loop [62]. The tunnel through which ubiquinone enters is formed by two α-helices (α10–α11) of the N-terminal domain (Fig. 8). These two α-helices of the N-terminal guide the entry of the CoQ co-substrate into the CoQ-binding tunnel to reoxidise $FMNH_2$ to FMN. CoQ is directed through the inner mitochondrial membrane with the help of transmembrane α-helix which is embedded into the membrane with a tilt of $8 \pm 7°$.

The X-ray crystal structure analysis of the *Pf*DHODH enzyme (PDB ID 1TV5) proposed the CoQ-binding tunnel to be the site of inhibitor binding and co-substrate binding (Fig. 9). In 2008, Malmquist et al. performed site-directed alanine mutagenesis studies of seven residues (His185, Phe188, Phe227, Arg265, Ile272, Tyr528, and Leu531) in the A77-1726 binding site. It was observed that the CoQ-binding site and species-selective inhibitor site do not overlap. It was suggested that the inhibitor acts by blocking the electron path between the FMN and CoQ or by stabilizing the enzyme conformation that excludes the ubiquinone-binding site [73].

3.3 Comparison with DHODH of Other Species

A total of 162 crystal structures of dihydroorotate dehydrogenase from different species are available. These crystal structures and sequence alignment studies clearly established that the enzyme forms two families (Table 2) [50b, 74].

Table 2 Classification and sub-classification of two families of DHODH enzyme and their respective properties [50b, 74]

PROPERTIES	FAMILY 1			FAMILY 2
Location	Cytosol			Outer Membrane of inner mitochondria
Final electron acceptor	Fumarate or NAD^+			Ubiquinone (CoQ)
Base	Cystein			Serine
Organisms	Mostly prokaryotes			Most eukaryotes
Differences	Core domain forms the entire protein			In addition to core domain, N-terminal domain (forms ubiquinone binding tunnel). Two α-helices of N-terminal are markers for family 2. This is preceded by a single transmembrane helix which anchors protein onto the inner mitochondrial membrane and a putative mitochondrial signaling sequence.
Properties	**1A**	**1B**	**1S**	
Structure	Homodimer	Heterodimer	Heteromeric	
Base	Cystein	NAD-dependent	Serine	
Electron receptor	Fumarate	An iron-sulfur cluster and FAD	CoQ and molecular oxygen	
Commonality	α/β-barrel core domain containing flavin prosthetic group that forms the active site			

Inhibition of *Pf*DHODH can be considered by either blocking the DHO (substrate)-binding site or CoQ (co-substrate)-binding site. In most of the species, the DHO-binding site is conserved which can lead to selectivity issues. High variability is observed in ubiquinone-binding site which renders it to be the preferential site for species-selective DHODH inhibition. Selectivity against *Hs*DHODH and *Pf*DHODH is the major requirement for designing antimalarial leads.

The CoQ-binding tunnel is divided into three regions, i.e., mouth, waist, and end of the tunnel. Structural and chemical composition differences in these regions lead to the species-selective inhibition. The mouth of the *Hs*DHODH is broader compared to *Pf*DHODH due to a slight kink provided by $Leu_{Hs}42$. This leads to an average r.m.s backbone displacement of 2.2 Å between the human and *Pf*DHODH protein domain. Also, substitution of $Phe_{Pf}171$ and $Met_{Pf}536$ for $Leu_{Hs}42$ and $Pro_{Hs}364$, respectively, brings the N-terminal of the first helix closer to the C-terminal of the second helix which is responsible for the narrow mouth of the tunnel in *Pf*DHODH. End of the tunnel in case of *Pf*DHODH is comparatively smaller than the *Hs*DHODH which is due to the replacement of $Val_{Hs}134$ and

Val_{Hs}143 with larger size hydrophobic residues, i.e., Ile_{Pf}263 and Ile_{Pf}272, respectively. This results in the ineffectiveness of the larger molecules such as brequinar and atovaquone to bind in *Pf*DHODH. Hurt et al. in 2005 reported the role of non-conserved residues in modifying the interactions of the inhibitor (A77-1726) with the conserved amino acids. The replacement of Met_{Hs}43 and Ala_{Hs}59 for Leu_{Pf}172 and Phe_{Pf}188, respectively, leads to alteration in the H-bond pattern of the inhibitor with the conserved residues (His_{Pf}185, Arg_{Pf}265, and Tyr_{Pf}528) at the end of the tunnel. Also, the replacement of Tyr_{Hs}147 for Cys_{Pf}276 leads to conformational changes in His_{Pf}185 resulting in changed interaction pattern of the inhibitor for the two enzymes [62]. In summary, the non-conserved residues present in the inhibitor binding site of the *Pf*DHODH structure are Phe_{Pf}171 (Leu_{Hs}42), Met_{Pf}536 (Pro_{Hs}364), Leu_{Pf}172 (Met_{Hs}43), Phe_{Pf}188 (Ala_{Hs}59), Leu_{Pf}176 (Gln_{Hs}47), Ile_{Pf}263 (Val_{Hs}134), Ile_{Pf}272 (Val_{Hs}134) [62, 75].

3.4 *Inhibition of DHODH*

L-DHO was observed to be the specific substrate of DHODH with a K_m value of 5.2 ± 0.6 μM. D-DHO is not a substrate but inhibits the enzyme competitively with K_i of 1.4 mM concentration [48]. For the oxidation from DHO to orotate, L-DHO diffuses passively from the cytosol to the intermembrane space of the mitochondria where it binds tightly to the enzyme due to low K_m value so that enzyme shows maximum efficiency even at low concentrations of DHO. It was observed that conversion of DHO to orotate is not the rate-limiting step, so substrate-competitive inhibition will not be effective. This leaves the researchers with two possibilities for inhibition of the enzyme, i.e., either by increasing the intracellular accumulation of orotate or a lack of oxygen/its equivalent (inhibition of electron receptor). However, intracellular accumulation of orotate inhibits dihydroorotase (enzyme catalyzing the formation of DHO from carbamylaspartate), thus controlling the intracellular concentration of DHODH. Thus, the main center for enzyme inhibition is obstructing the electron receptors [48].

Copeland et al. studied the role of N-terminal in enzyme inhibition in human DHODH. It was observed that the essential catalytic region and site of inhibition are located within 40 kDa area of truncated enzyme and the remaining 10 kDa of the truncated N-terminal portion of the protein does not significantly disturb the catalytic action or inhibitor binding ability of the enzyme [76]. However, it was later observed that the truncated enzyme only retains the activity under in vitro conditions and not under in vivo conditions [77]. This may be due to the removal of signaling peptide and transmembrane helix which are responsible for cellular localization and directing CoQ into the ubiquinone-binding tunnel.

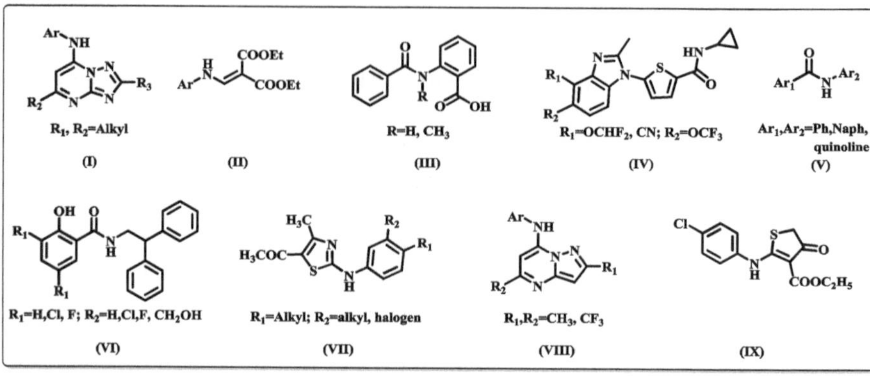

Fig. 10 Different classes of *Pf*DHODH inhibitors

3.5 Reported Classes of Compounds for PfDHODH *Inhibition*

There are nine chemical classes of *Pf*DHODH inhibitors known in the literature (Fig. 10). Different classes of *Pf*DHODH reported in the literature are triazolopyrimidine (**I**) [78], diethyl 2-((arylamino)methylene)malonate (**II**) [79], benzamide/naphthamide derivatives of anthranilic acid (**III**) [56], *N*-alkyl-5-benzimidazole thiophene-2-carboxamide (**IV**) [65, 80], benzamide (**V**) [55], *N*-substituted salicylamides (**VI**) [81], thiazole (**VII**) [82],7-arylaminopyrazolo[1,5-a]pyrimidines (**VIII**) [83], and dihydrothiophenone (**IX**) [84]. DSM265 [69] from triazolopyrimidine class is in clinical development phase, and two analogs of **Genz-667348** (*N*-alkyl-5-benzimidazole thiophene-2-carboxamide derivatives) [80b] are undergoing pilot toxicity testing to determine their suitability as clinical development candidates.

3.6 Structure-Based Drug Design of PfDHODH *inhibitors*

All the early efforts to the identified *Pf*DHODH inhibitors are based on the known *Hs*DHODH inhibitors. First effort was initiated by Boa et al. (in 2005), in which various analogues of Brequinar (Fig. 11 **XII**) (*Hs*DHODH inhibitor; immunosuppressive agent) were designed using analog-based methods [54]. These quinolone-4-carboxylic acid derivatives showed poor to moderate selectivity and activity in medium micromolar range. This study was followed by report of high-throughput screening studies on *Pf*DHODH inhibitors by Baldwin et al. (2005) [55]. The chemical classes included halogenated phenyl benzamide/naphthamides and napthyl or quinolinyl substituted urea-based compounds. Inhibitor binding site was confirmed by direct site mutagenesis (His185Ala and

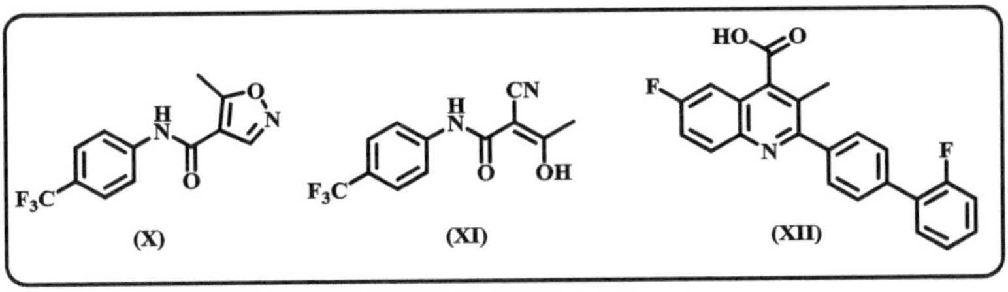

Fig. 11 Structures of *Hs*DHODH inhibitors; Leflunomide (**X**), Teriflunomide (**XI**), and Brequinar (**XII**)

Arg265Ala) which confirmed teriflunomide (Fig. 11**XI**) binding pocket to be the current binding site. Benzamide derivatives were found to be more potent in *Pf*DHODH compared to *Hs*DHODH (IC_{50} value range 50–520 nM) and selective (70–12, 500 fold selectivity toward *Pf*DHODH and against *Hs*DHODH). In benzamide derivatives, 2-nitro-3-methyl benzamide-based compounds showed high preference for the parasite enzyme. However, these molecules showed weak activity in cell-based assays indicating low absorption through the cells. In 2005, Hurt et al. reported the X-ray crystal structure of *Pf*DHODH with teriflunomide [62]. This provided opportunity for the SBDD approach for designing of various *Pf*DHODH inhibitors based on the detailed knowledge about the active site in which the inhibitors bind (the 3D structure of *Hs*DHODH is known since 2000) [60].

3.6.1 Benzamide/Naphthamide Derivatives of Anthranilic Acid

Anthranilic acid derivatives were designed using de novo molecular design program SPROUT [56]. De novo drug design is a part of structure-based drug design methods in which molecular fragments and atoms are made to interact with the binding pocket of the target enzyme and subsequently assembled in a stepwise manner based on the interactions on these fragments. This finally results in a template with novel chemotype and expanding the chemical library for A given target. The tractable synthetic route is also considered while preparing an in silico library of high-quality structures [85]. Sprout de novo design tool uses different modules to achieve these functions which particularly include (a) identification of the binding pocket; (b) recognizing the hydrophobic regions, probable polar regions and metal bonding possibilities; (c) docking of various functional groups, fragments, and atoms into the binding pocket; (d) joining all the fragments in the best possible way by satisfying the steric constrains; (e) finally, scoring the fragments and sorting out the templates based on their binding affinity, complexity, synthetic feasibility, and substructure search [86].

For this study, Heikkilä et al. (2006) [56] studied the reported X-ray crystal structures of teriflunomide (A77-1726) with *Hs*DHODH (PDB ID 1D3H) [60] and *Pf*DHODH (PDB ID 1TV5) [62]. It was observed that the inhibitor binding tunnel in the *Hs*DHODH is considerably flattened due to the methyl side chain protrusion

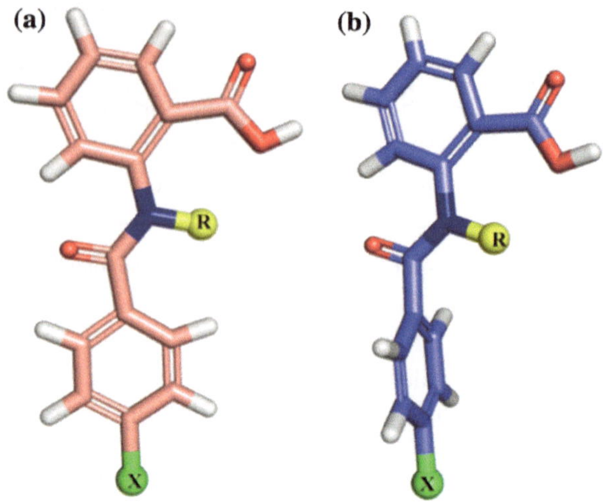

Fig. 12 Two observed conformers of the designed inhibitor template (Fig. 10, **III**). Methyl substitution at the amide nitrogen causes conformational restrictions

of the Ala59 at the position where the phenyl ring of the teriflunomide binds. Similar position in *Pf*DHODH cavity is comparatively less congested (Ala59 is replaced by Phe188) and seems to accommodate the inhibitors which might provide a cylindrical geometry. Considering this structural feature, six molecules which are amides of anthranilic acid were identified as potential inhibitors. The two major conformations of designed molecules were hypothesized to play an important role in selective inhibition of *Pf*DHODH (Fig. 12). The non-planar arrangement of the two phenyl rings (Fig. 12b) was considered to favor the *Pf*HDODH binding, whereas the planar phenyl rings (Fig. 12a) were more suitable for *Hs*DHODH binding. The conformer B (Fig. 12) was predicted to be more suitable for *Pf*DHODH inhibition and it was restricted by *N*-methyl substitution. Enzyme inhibitory assay showed that the *N*-methyl substituted biphenyl (*Pf*DHODHIC_{50} 42.6 μM) and bromonaphthylbenzamide derivatives (*Pf*DHODHIC_{50} 93.4 μM) were selective against *Pf*DHODH inhibition, whereas *N*-unsubstituted compounds were non-selective and more active against *Hs*DHODH inhibition.

3.6.2 Diethyl 2-((Arylamino)Methylene) Malonate

Heikkilä et al. (2007) proposed the design of multicyclic aromatic rings which are malonate and cyanoacrylate derivatives [79]. Ten compounds with mono-, bi, and

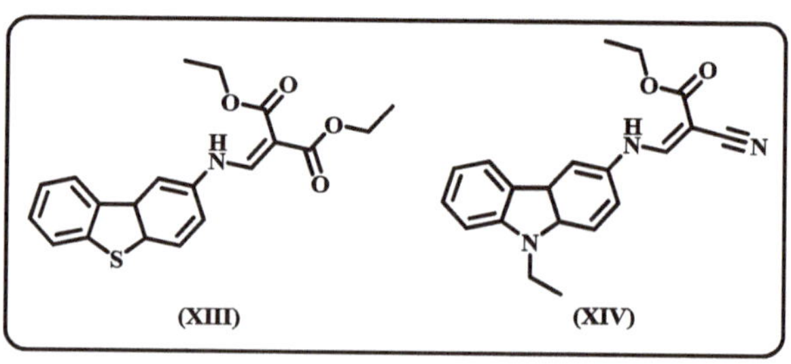

Fig. 13 Structure of compounds showing good inhibitory activity against *Pf*DHODH with selectivity against *Hs*DHODH

tricyclic heteroaromatic ring systems were synthesized and tested against *Pf*DHODH and *Hs*DHODH. It was observed that tricyclic derivatives, i.e., diethyl-2-((dibenzo[b,d]thiophen-2-ylamino)methylene)malonate(Fig. 13, **XIII**) and ethyl-2-cyano-3-((9-ethyl-9H-carbazol-3-yl)amino)acrylate (Fig. 13, **XIV**), show the most promising results with *Pf*DHODHIC$_{50}$ in lower micromolar range (0.16 and 0.44 μM, respectively) and relatively high selectivity toward *Pf*DHODH (182- and 1208-fold selectivity, respectively, against *Hs*DHODH). Heteroaromatic bicyclic compounds (indazole and benzimidazole derivatives), phenyl derivatives, and *m*-biphenyl derivatives have *Pf*DHODHIC$_{50}$ values in higher micromolar range and low selectivity (1–10 folds). The *p*-biphenyl derivatives were found to be inactive in both *Pf*DHODH and *Hs*DHODH. The molecular docking results of the tricyclic compounds (in the crystal structure with PDB ID 1TV5) demonstrated the importance of planar aromatic hydrophobic groups for π-stacking interaction with Phe188 (the selectivity is due to the presence of Ala59 in *Hs*DHODH in place of Phe188 in *Pf*DHODH). The non-planar biphenyl rings are not accommodated into the hydrophobic site, and hence, biphenyl derivatives are not suitable. The polar groups of the active compounds showed hydrogen bonding interactions with His185, Arg265, and Tyr528 amino acids at the end of the tunnel [79].

3.6.3 Triazolopyrimidine

Phillips et al. (2008) first reported triazolopyrimidine derivative obtained through high-throughput screening studies on *Pf*DHODH [78]. A total of 220,000 molecules were screened through colorometric enzyme assay from which **DSM1** (Fig. 14b) was identified (*Pf*DHODHIC$_{50}$ value of 0.047 ± 0.022 μM). This molecule showed an EC$_{50}$ value of 0.079 ± 0.048 μM and 0.14 ± 0.05 μM in whole-cell assay against non-resistant strain (3D7) and multidrug-resistant strain (Dd2), respectively. The hit also showed >5000-fold selectivity against *Hs*DHODH.

It was observed that primary amine is essential for the activity. Methyl group substitution is suitable for R and R$_1$ position (Fig. 14a). Naphthyl group at R$_3$

(a)

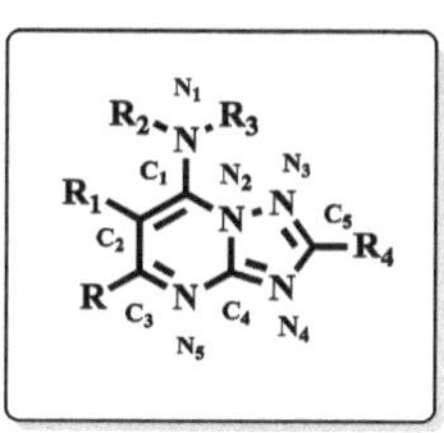

(b)

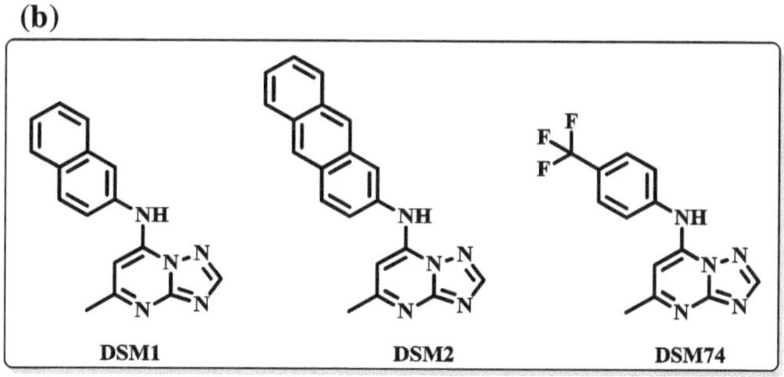

Fig. 14 **a** General structure of triazolopyrimidine class of compounds. **b** Structure of **DSM1**, **DSM2**, and **DSM74**

(Fig. 14a) position shows optimal activity and introduction of heteroatom in the naphthyl ring leads to decrease in enzymatic activity. A smaller aromatic group other than naphthyl and anthracene moiety leads to decrease in activity. Even though **DSM1** showed good in vitro and whole-cell activity, it was not active in vivo against *P. berghei*. It was further observed that the compound showed reduced plasma concentration on repeated exposure [78, 87]. In 2009, Gujjar et al. prepared a series of forty new compounds with different substituted phenyl moieties at R_3 position [87]. *Para* substitution was found to give active compound compared to the unsubstituted and *o-/m*-substituted analogues and large electron-withdrawing hydrophobic substituents were found to be preferred in the order of $CF_3 > Br > OCF_3 > CH_3 > NO_2 > F > Cl$. **DSM74** was found to be the best choice among the prepared series with *Pf*DHODH IC_{50} 0.28 ± 0.02 μM (*Pf*3D7 cells EC_{50} 0.34 ± 0.04 μM). The compound **DSM74** was equally potent in *P. falciparum* and *P. berghei* and showed good plasma exposure in mice *in vivo* studies. This hit was also more stable in human microsomes (in vitro). This study established the confidence that this class of compounds can be active in in vivo studies and there is a scope for further improvement of its metabolic profile. However, **DSM74** showed activity in mid-nanomolar range leaving a wide berth for further improvement [87].

In 2009, Deng et al. reported the crystal structures of lead *Pf*DHODH inhibitors (**DSM1**, **DSM2**, **DSM74**) in the inhibitor binding site of *Pf*DHODH (PDB ID 3I65, 3I68, 3I6R, entry 2–4, Table 1) [64]. The triazolopyrimidine ring in all three inhibitors binds to the polar region at the end of the inhibitor binding tunnel similar to the teriflunomide binding in crystal structure 1TV5. The amino acids involved are His185 forming hydrogen bond with N_1 (Fig. 14a) and Arg265 forming hydrogen bond with N_5. In case of **DSM2** and **DSM74**, the triazolopyrimidine ring tilts slightly inside the polar region with the slight reorientation of amino acid Leu176. Tyr528 show water-mediated hydrogen bond with N_3 of the inhibitor in all three crystal structures. The orientation of the inhibitor in the cavity is such that C_5 position lies closest to the FMN (6 Å distance) and there is a small channel which can be further exploited for structure-based drug design. The hydrophobic pocket in these three crystal structures is different from that of 1TV5. The amino acid residues of the hydrophobic pocket comprises of Ile237, Leu189, Leu197, Met536, Phe227, and Phe188. The large aromatic ring of **DSM2** is accommodated by small rotational changes of amino acid residues Leu197 and Met536, which results in the expansion of the hydrophobic cavity. The smaller phenyl group of **DSM74** does not fill the hydrophobic cavity completely and this might be the reason for its 10-fold less activity than that of **DSM1** and **DSM2**. There is an extended aromatic stacking network from FMN to Tyr528 carried forward toward the hydrophobic pocket through Phe227 which forms edge-to-face π-π interaction with the inhibitor naphthyl (**DSM1**)/phenyl (**DSM74**) ring followed by π-π interaction with Phe188. The mutagenesis studies of the inhibitor binding pocket suggested that His185 and Arg265 mutation to Ala raises the IC_{50} by 80–90-folds. Other mutations have less effect on the activity with a raise of 5- to 30-fold in IC_{50} value. Also, a very interesting observation was noted during the crystal structure of the inhibitor. It was observed that the N_1–C_1 (Fig. 14a) bond length of the inhibitor was between single and double bond (1.313 Å in **DSM1**) and partial positive charge was observed at N_1, indicated by the presence of chloride ion adjacent to the N_1 in the crystal structure. It was suggested that the triazolopyrimidine undergoes electron delocalization between N_1 and N_5, giving a low-range positive dipole at the N_1 center which enhances its interaction with His185 and N_5 acquires a slight negative dipole which allows it to form an ion pair with Arg265 [64]. This delocalization of charges might be the reason for inactivity of compounds with O and S as bridging atoms. Further modifications at the m- and p-positions led to the identification of **DSM161** (Fig. 14a; R = CH_3, $R_1/R_2/R_4$ = H, R_3 = 4-SF_5-Ph) and **DSM190** (Fig. 14a; R = CH_3, $R_1/R_2/R_4$ = H, R_3 = 3,5-diF-4-CF_3-Ph) with *Pf*DHODH IC_{50} to be 0.13 and 0.19 μM, respectively. These compounds showed better plasma exposure and improved efficacy in mouse model [88].

This activity was further improved by Coteron et al. (2011) with the design of **DSM265** (Fig. 14a; R = CH_3, R_1/R_2 = H, R_3 = 4-SF_5-Ph, R_4 = CF_2CH_3) which was found to be active against both sensitive and resistant strains of *P. falciparum* [66]. As discussed above [64], the crystal structure of *Pf*DHODH with triazolopyrimidines showed a narrow channel existing between the FMN and inhibitor. In order to improve the pharmacokinetics along with the activity, this information was utilized and modifications were done at the R_4 position. Small hydrophobic electron-withdrawing groups were found to fit in the narrow space, out of which CF_2CH_3 was found to be most suitable. It showed potency similar to chloroquine in humanized SCID mouse *Pf*model. The compound also showed excellent oral bioavailability, long half-life, and low clearance in humanized SCID mouse *Pf*model. **DSM265** was found to possess excellent in vivo efficacy with once a day dose in mice. Further extended studies gave very promising results in order to consider **DSM265** as a drug candidate [69]. The *Pf* and *Pb* IC_{50} were found to be 0.033 and 2.5 μM, respectively, with the *Pf* 3D7 cells EC_{50} to be 0.046 μM. It showed a high selectivity, >100 μM, against *Hs*DHODH. The compound was analyzed to act on both liver and blood stage of the parasite and active against isolated resistant strains. 200–400 mg dose for eight days is well tolerated in repeated dose with cardiovascular safety in mice and dogs. **DSM265** thus shows an excellent safety profile, blood–liver stage activity and a predicted long half-life in humans [69]. The crystal structure (PDB ID 4RX0, entry 11, Table 1) showed that the –CF_2CH_3 group shows van der Waals interactions with amino acid residues Ile263, Ile272, the hydrophobic potion of Arg265 side chain and Tyr528. Also, the electron-withdrawing effect of fluorine reduces the electron density on triazole ring nitrogens, which may be responsible for increased potency. Recently, Kokkonda et al. proposed tetrahydro-2-naphthyl and 2-indanyl substituted triazolopyrimidines with improved potency and selectivity over **DSM265** [71]. However, these compounds have high metabolic clearance and are proposed to be tolerated only in multi-dose regime.

In 2012, Bedingfield et al. proposed selectivity factors responsible which can be exploited to design *Hs*DHODH and *Pf*DHODH selective triazolopyrimidine class of inhibitors [75]. It was observed that $His_{Hs}56$ and $His_{Pf}185$ play important role in

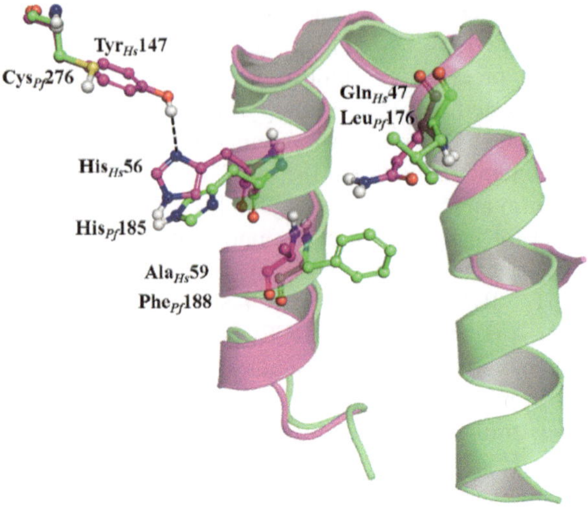

Fig. 15 Overlapped N-terminal of *Pf*DHODH (green) and *Hs*DHODH (magenta) showing important amino acids residues important for selectivity

selectivity. The amino acid residue Tyr$_{Hs}$147 form direct or indirect hydrogen bond with His$_{Hs}$56 which moves the δ nitrogen of His$_{Hs}$56 away from the inhibitor binding site (Fig. 15). In case of *Pf*DHODH, Tyr$_{Hs}$147 is replaced by Cis$_{Pf}$276 which cannot form hydrogen bond with His$_{Pf}$185, thus directing the δ nitrogen toward the inhibitor binding site forming a direct hydrogen bond with the inhibitor.

3.6.4 *N*-Alkyl-5-(1H-Benzimidazol-1-yl)Thiophene-2-Carboxamide

Compounds belonging to this class were designed based on the high-throughput screening studies from Genzyme library, by Patel et al. [80a]. In this study, 208,000 compounds were screened for *Pf*DHODH inhibitory activity. Thirty-eight compounds from this library were identified for showing *Pf*DHODH inhibition in

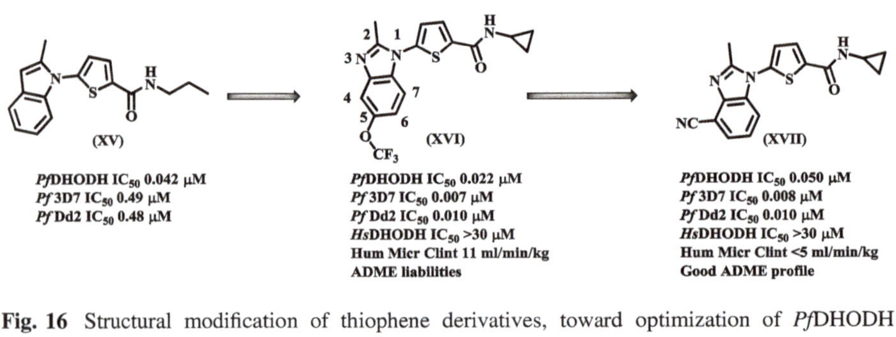

Fig. 16 Structural modification of thiophene derivatives, toward optimization of *Pf*DHODH activity using SBDD approach

sub-micromolar range out of which thirty-three were picked based on their selectivity against human DHODH. The selected molecules were further tested against 3D7 strain of *P. falciparum* from which five molecules showed activity in sub-micromolar range. The selected molecules were then tested against multidrug-resistant strains of *P. falciparum,* i.e., HB3 and Dd2. 5-(2-methyl-1H-indol-1-yl)-*N*-propylthiophene-2-carboxamide exhibited the most promising IC_{50} value (42 nM) against *Pf*DHODH (Fig. 16, **XV**).

Molecular docking results revealed that the inhibitor may bind to the same site as teriflunomide in X-ray crystal structure with PDB 1TV5. Additional modifications of the indole ring system with piperidine, piperazine, pyrazol, benzimidazole, and 2-methyl benzimidazole highlighted that multi-ring substitutions are important for activity. This might be due to the better interactions shown by multiple ring substitutions in the hydrophobic region of the inhibitor binding site. Replacement of indole moiety with benzimidazole moiety retained the *Pf*DHODH inhibitory activity in addition to which physiochemical properties were considerably improved [65]. Modifications in the amide region lead to the conclusion that secondary amide is essential for the activity, whereas primary and tertiary amides are inactive. Cyclopropyl group was found to be the best optimized group at this position as it occupies the hydrophobic pocket which was later found out to be different from that occupied by triazolopyrimidine class of compounds [80b]. Second position of the benzimidazole moiety was also optimized with various alkyl (ethyl, *n*-propyl) and polar (hydroxyl, dimethyl amino) substituents out of which simple methyl substitution was found to be the only suitable one and comparable to the unsubstituted derivative. Methyl substitution at this position also improves the in vitro hepatic metabolic stability in human microsomes. Replacement of the thiophene ring with various aromatic substituents (such as 2,5-substituted *N*-methyl pyrrole, pyrrole, thiazole, furan, oxazole, and *m*-/*p*-substituted phenyl) resulted in considerable decrease in the potency. Also, substitution at 2,5-position on the thiophene ring is essential and there is a loss of activity with substitution at 3,4-position (Fig. 16).

Further modifications were focused on the benzimidazole moiety at 4th to 7th position (Fig. 16, **XVI**). Simple methyl substitution at 4th, 5th, and 6th position resulted in a twofold increase in activity (*Pf* IC_{50} 39–56 nM) compared to the corresponding unsubstituted molecule (*Pf* IC_{50} 80 nM). Substitution at the 7th position causes an eightfold reduction in activity (*Pf* IC_{50} 609 nM). This can be interpreted to be due to the steric hindrance in the cavity of the enzyme caused by substitution at the 7th position (Fig. 16). Hydrophobic electron-withdrawing group (OCF_3 and CF_3) at the 5th position is more favorable, with 2–3-fold increase in activity (*Pf* IC_{50} 22–28 nM, respectively) compared to the 6th position substitution (*Pf* IC_{50} 52–98 nM, respectively). The 5-OCF_3-substituted derivative (*N*-cyclopropyl-5-(2-methyl-5-(trifluoromethoxy)-1H-benzimidazole-1-yl)thiophene-2-carboxamide, Genz-667348; Fig. 16, **XVI**) was further studied in the acute *P. berghei* (ANKA strain) efficacy model and was found to be curative when dosed orally. The X-ray crystal structure of *Pf*DHODH with Genz-667348 (PDB ID 3O8A, entry 5, Table 1) revealed an alternate hydrophobic binding pocket different from that of triazolopyrimidine analogues (Fig. 17) [65]. This may be due to more flexible

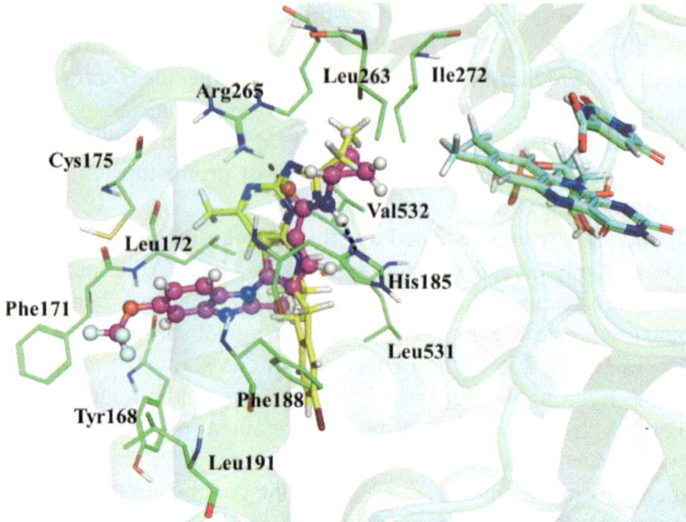

Fig. 17 Structural overlap of **Genz-669178** (magenta) and **DSM416** (yellow) in the X-ray crystal structures with PDB ID 3O8A and 5FI8, respectively, depicting the difference in hydrophobic cavity

nature of the inhibitor. The cyclopropyl ring interacts with the amino acids Val532, Ile272, and Ile263 present at the mouth of the inhibitor binding tunnel near to His185 which forms H-bond with the *N* of methylformamide. Arg265 also forms a hydrogen bond with the oxygen of methyl formamide. The alternate hydrophobic cavity consists of amino acid residues Tyr168, Cys175, Phe171, Leu172, Phe188, Leu191, and Leu531.

However, significant proof was obtained for **Genz-667348** to have ADME liabilities and most likely to cause drug–drug interactions. Thus, alterations were done at the 4th position with electron-withdrawing groups such as F, Cl, Br, and CN. This caused a fivefold increase in the activity (*Pf* IC_{50} 21–40 nM). 4-CN analogue was found to be the compound with best overall profile with suitable physiochemical properties, cardiovascular safety, and least drug–drug interaction possibility. This compound *N*-cyclopropyl-5-(2-methyl-4-(cyano)-1H-benzimidazol-1-yl)thiophene-2-carboxamide (**Genz-669178**; Fig. 16, **XVII**) (*Pf*IC_{50} 40 nM) was further studied in three species (mouse, rat, and dog) and three mouse models of malaria (*P. falciparum*, *P. berghei*, and *P. vivax*). The compound showed pan-parasitic activity with promising DHODH inhibitory activity in all three plasmodium species. It showed selectivity against human DHODH with $IC_{50} > 30$ μM. The cell-based IC_{50} values in *Pf* sensitive (3D7) and multidrug-resistant strain (Dd2) were observed to be 7 and 10 nM, respectively. The lead compound showed moderate oral bioavailability in rat and dog model (49 and 19%, respectively). The results were also promising in human microsomes and hepatocytes with low hepatic clearance. **Genz-669178** is further being tested for its toxicity profile and is studied for its clinical testing suitability [80b].

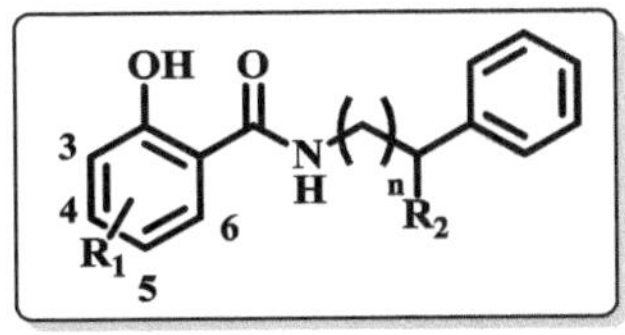

Fig. 18 General structure of *N*-substituted salicylamide analogues

3.6.5 *N*-Substituted Salicylamides

This class of compounds was reported by Fritzson et al. in 2011 [81]. It was observed that series with unsubstituted R_2 position (Fig. 18) showed human DHODH selective or non-selective property. Compound with phenyl-substituted R_2 and $n = 1$ was found to be *Pf*DHODH selective. This compound was developed into a complete series by varying the substituents on the phenyl ring of salicylamide. It was observed that substitutions at 3 and 6 positions are highly unfavorable, whereas substitutions at 4 and 5 positions are preferred (Fig. 18). Four compounds with substitutions 5-Cl (*Pf*DHODH IC_{50}9.1 μM), 3,5-dichloro (*Pf*DHODH IC_{50}7.0 μM), 3,4-difluoro (*Pf*DHODH IC_{50}9.9 μM), and 5-CH_2OH (*Pf*DHODH IC_{50}8.0 μM) were found to be most active. Interactions of unsubstituted analogue were observed after manual docking in crystal structure with PDB ID 3I65 [64] followed by energy minimization of the complex using MacroModel program (Schrödinger software). It was observed that the oxygen of the amide showed single H-bond with Arg265 and the biphenyl part of the molecule forms π-stacking interactions with Phe188 [81]. The four selected compounds were further tested for their cell-based activity against *Pf* 3D7 strain. 5-chloro-*N*-(2,2-diphenylethyl)-2-hydroxybenzamide was found to be the most promising compound for inhibiting parasite growth with the EC_{50} of 23 ± 4 μM.

3.6.6 Dihydrothiophenone Derivatives

Xu et al. in 2013 identified ethyl 2-((4-chlorophenyl)amino)-4-oxo-4,5-dihydrothiophene-3-carboxylate compound (lead molecule) to be showing *Pf*DHODH inhibitory activity (IC_{50} 1.11 μM) after virtual screening studies from SPECS database with ID AG-690/40639878 (Fig. 19) [84]. This virtual screening included glide-based molecular docking and prime MM-GBSA-based ΔE estimation both fall under SBDD methods. The molecular docking studies showed that the phenyl ring with *p*-Cl substitution exhibited hydrophobic and van der Waals interactions with amino acids Leu197, Ile237, Leu240, and Met536 along with π-π edge-to-face interactions with Phe227 at the entrance of the ubiquinone-binding tunnel. The heterocyclic ring and its substituents form polar interactions with His185 and Arg265.

The molecule was thus divided into two sections, namely the hydrophobic aromatic group and dihydrothiophenone section as hydrophilic group. Three approaches were considered for lead optimization, i.e., hydrophobic modification of

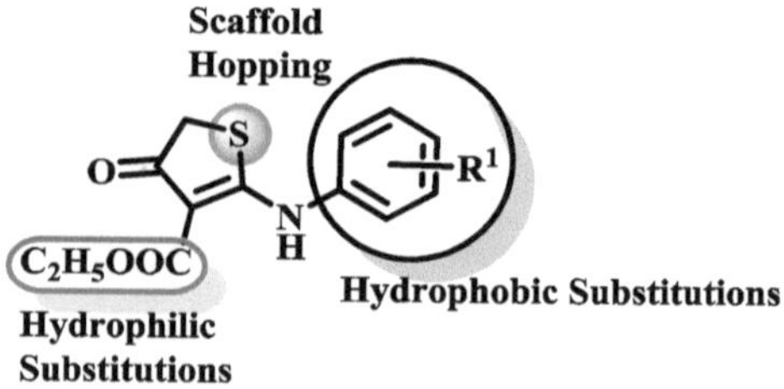

Fig. 19 Structural modification strategy of dihydrothiophenone class of compounds

the aromatic ring, hydrophilic modification of the substituents on dihydrothiophenone ring, and scaffold hopping of the dihydrothiophenone core. It was observed that *para* substitution is suitable at the phenyl ring of hydrophobic group. 4-*t*-butyl substitution provides best results in this group with *Pf*DHODH IC_{50} 0.23 μM. Single *m*-substitution causes loss of activity, whereas *m*- with *p*-substitution gives improved activity compared to the lead compound. Dual *meta*-substitution retains the activity comparable to the lead molecule. *Ortho* substitution causes loss in activity. Replacement of the phenyl ring with larger aromatic systems improves substantial activity with 2-nathyl group (ethyl 2-(naphthalen-2-ylamino)-4-oxo-4,5-dihydrothiophene-3-carboxylate) showing the best results with *Pf*IC_{50} of 0.02 μM activity (56-fold improvement in activity over the lead molecule). Introducing aromatic rings with heteroatom is an improvement over lead but less active than 2-naphthyl derivative [84].

Acid, amide, and ester substitutions were tried at the hydrophilic group. Removing the ethoxycarbonyl substitution leads to poor activity. Acid and amide substitutions lead to loss or poor activity. Esters larger than ethoxy also result in poor activity. It was observed that the ester moiety forms simultaneous hydrophobic interactions (with amino acid residues Ile263 and Ile272) and hydrogen bond (with Tyr528) in the inhibitor binding pocket of the enzyme emphasizing its importance. Scaffold hopping by replacement of sulfur with oxygen (ethyl 2-(naphthalen-2-ylamino)-4-oxo-4,5-dihydrofuran-3-carboxylate) improves the activity (IC_{50} 6 nM) by 3-fold compared to 2-naphthyl derivative and 185-fold compared to the lead. In conclusion, bicyclic ring systems are more promising in the hydrophobic region, ethoxycarbonyl ester substitution in the hydrophilic region is essential, and dihydrofuranone ring is more suitable compared to dihydrothiophenone ring. These results were also correlated in the *Pf*3D7 and *Pf*Dd3 cell-based assays.

3.6.7 Thiazole derivatives

Zhu et al. reported thiazole-based *Hs*DHODH inhibitors in 2015 [82]. However, the lead molecule used for optimization also showed *Pf*DHODH activity with IC_{50} value of 0.63 μM. It was observed that methyl substitution at R_1 position (Fig. 20) and *m*-/*p*-substitutions at the phenyl ring results in non-selective *Pf*DHODH inhibition in lower micromolar range. An overlap of crystal structure of *Hs*DHODH (PDB ID 4JGD) upon *Pf*DHODH (PDB ID 3I65) revealed a smaller tunnel of

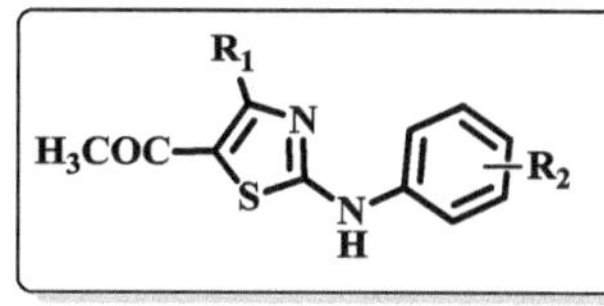

Fig. 20 General structure of thiazole class of *Pf*DHODH inhibitors

inhibitor binding site in *Pf*DHODH (Phe171, Met536, Ile263, and Ile272) compared to the *Hs*DHODH (Leu42, Pro364, Val134, and Val143) resulting in the inability of malarial enzyme to accommodate large substituents. Thus, phenyl or other bulky substituents at R_1 and R_2 position results in loss of inhibitory activity toward *Pf*DHODH.

3.6.8 7-Arylaminopyrazole Derivatives

Azeredo et al. proposed the bioisosteric replacement of trizolopyrimidine moiety with 7-aryl aminopyrazole analogues leading to the design and synthesis of a new class of *Pf*DHODH inhibitors [83]. Fifteen compounds were synthesized and tested for their *Pf*DHODH inhibitory activity. It was observed that the compounds of this series show activity in lower micromolar range with *Pf*DHODH IC_{50} 24–0.16 μM. It was observed that out of substituted phenyl and 2-naphthyl derivatives, naphthyl analogues were most promising. Alternate CF_3 and CH_3 substitutions were tried at R_1 and R_2 position. 2-naphthyl-substituted compound with CH_3 at R_1 and CF_3 at R_2 was observed to be the most active compound in this series (Fig. 21). Further docking studies in *Pf*DHODH (PDB ID 3I65) perceived that these series of compounds show similar binding interactions as that of the bound ligand. It was also indicated that series with R_1 CH_3 and R_2 CF_3 show similar hydrogen bonding interactions as that of the bound ligand in addition to the water-mediated hydrogen bond with Tyr528.

3.7 *Other in Silico Efforts*

Ojha et al. (2010) performed QSAR and molecular docking studies on triazolopyrimidine class of compounds [89]. A total of four models were prepared based on classical QSAR (two models), molecular shape analysis (MSA), and QSAR with combined set (2D and 3D) of descriptors using G/PLS spline technique.

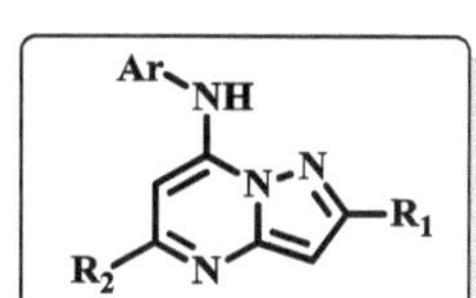

Fig. 21 General structure of 7-aryl aminopyrazole series of compounds as *Pf*DHODH inhibitors

This classical QSAR study mainly focused on physiochemical descriptors. The descriptors used by first classical QSAR model are in the following of $B1_p$ (Sterimol width parameter as the smallest width along *Z* axis; *p* indicates *para* position), L_o (Sterimol length parameter as the maximum length along the *X* axis), MR_p (molar refractivity impact in the *para* position), MR_m (molar refractivity impact in the *meta* position), π_m (lipophilicity substitution constant). The descriptors used by second classical QSAR model are $B1_p$, $B5_o$ (width parameter defined as maximum width from *X* axis), MR_p, $B1_m$, and π_p. The MSA involved use of molecular shape descriptors (DIFFV, COSV, NCOSV, Fo, Shape RMS) in addition to electronic (dipol-mag and Sr), spatial (radius of gyration, Jurs descriptors, area, PMI-mag, density, V_m), thermodynamic (ALogP, ALogP98, MolRef, MR, LogP) and structural (H-bond donor, H-bond acceptor, rotatable bonds) descriptors. Descriptors used by MSA are Fo (common overlap steric volume descriptor), MolRef (atom type molar refractivity), JursDPSA_3 (difference in atomic charge-weighted surface area), JursFPSA_1 (fractional charged partial positive surface area), LogP (partition coefficient) and JursPPSA_3 (atomic charge-weighted positive solvent-accessible surface area). Combined set descriptors (classical QSAR and MSA) include $B1_p$, π_p, $B1_o$ and V_m (molecular volume inside the contact surface). The entire data set consists of 29 compounds which are divided into training set ($n = 22$) and test set ($n = 7$) by using *k*-means clustering. All the models have predictive R^2 more than 0.5, thus passing the basic criteria. Classical QSAR with physiochemical descriptors was found to be the best model based on $r^2_{m(\text{overall})}$(0.733) and R^2_p(0.767). It was observed through this study that (1) unsubstituted ortho position is desirable; (2) moderate hydrophobicity and volume at *meta* position of the phenyl may enhance the *Pf*DHODH inhibitory activity; (3) *para* substitution is essential for inhibition with volume and hydrophobicity to be high but restricted.

Molecular docking was performed using LigandFit module under Discovery studio 2.1. Figure 22 shows the general 3D interactions of various polar and non-polar residues with the 5-methyl-*N*-phenyl- [1, 2, 4] triazolo[1,5-a]pyrimidin-7-amine. It was observed that the *p*-position of the phenyl substitution can accommodate hydrophobic groups with large volume. However, substitutions with phenyl ring at *p*-position were not able to attain an optimal position causing a bump with important amino acids such as His185 and Val532 [89]. Therefore, the substituents were restricted to non-aromatic groups such as CF_3, OCF_3, and CH_3 (Fig. 22).

Shah et al. reported 3D-QSAR on the same class of compounds. Thirty-five molecules of triazolopyrimidine class were selected for this study and molecular docking was performed using FlexX software in X-ray crystal structure with PDBID 3I68 [90]. The docking results highlighted two important structural features, a hydrophobic (aromatic) region which should have planar arrangement and a polar region. As reported by the other group, His185 and Arg265 played crucial role in polar interactions. Amino acid residues responsible for van der Waals interactions are Gly181, Cys184, His185, Phe188, Leu189, Phe227, Leu531, and Val532. Phe188 forms π-π interactions and Phe227 forms edge-to-face π

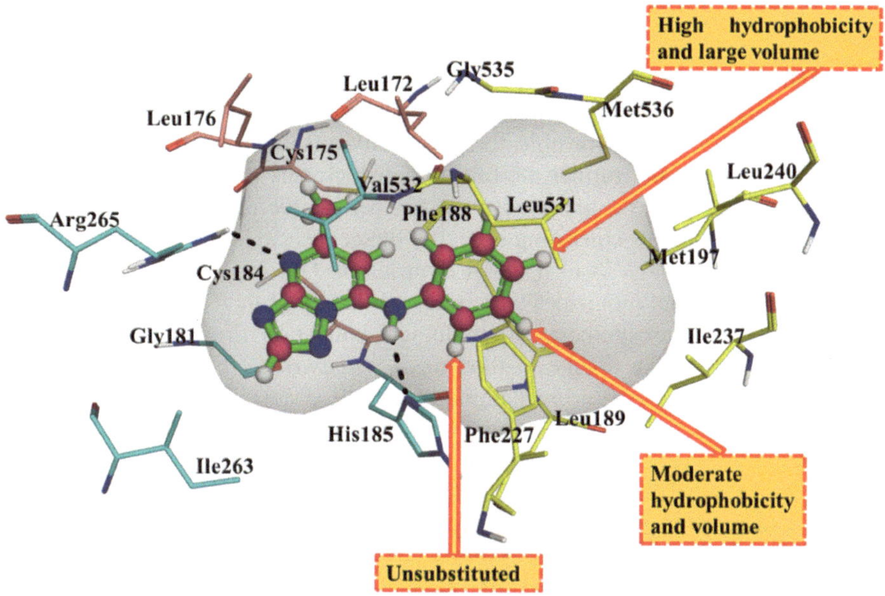

Fig. 22 Schematic 3D diagram of interactions of triazolopyrimidine derivatives in the active site of *Pf*DHODH. The *p*- and *m*-substituents of the phenyl ring interact with the residues in yellow. The methyl group of triazolopyrimidine ring interacts with the residues in peach color and residues in blue interact with the triazole ring portion. His185 and Arg256 form H-bonds with the primary amine linker and pyrimidine ring nitrogen, respectively

interactions. This was followed by generation of CoMFA and CoMSIA models (using genetic algorithm for optimization). The final two models were selected from each CoMFA and CoMSIA studies, which showed good predictive r^2 value (r^2_{pred} 0.99 and 0.94, respectively) after test set (eight molecules) run. The cross-validation coefficient was found to be q^2_{LOO} (leave-one-out) 0.841 for CoMFA (0.757 for CoMSIA) and q^2 (cross-validated) to be 0.818 for CoMFA (0.653 for CoMSIA). The total field contribution was determined to be 21.5% electrostatic field and 78.5% steric field (for CoMFA). It was observed that for *N*-2-naphthyl-substituted triazolopyrimidine derivative, favorable steric contours surrounded the naphthyl ring indicating good activity of the compounds with large steric bulk at this position. The compounds with good activity observed negative electrostatic contour at *meta* and *para* positions of the phenyl substituents indicating electron-rich substituents at these positions enhanced the activity. These results were also in correlation to Ojha et al. [89]. Molecular dynamics was also performed for most active compounds (Fig. 14a; **DSM125** (R = CH_3, $R_1/R_2/R_4$ = H, R_3 = 3-F-4-CF_3-Ph); **DSM1** (R = CH_3, $R_1/R_2/R_4$ = H, R_3 = naphthyl)) using GROMACS suite of programs with 2 ns of production run time. It was observed that the hydrophobic residues (Lue197, Ile237, Leu240, Leu531, and Met536 showed greater fluctuation in case of **DSM125** compared to **DSM1**. The *m*-fluoro substitution was found to form favorable interactions with Leu531. Also, His185 and Arg265 were observed to form stronger hydrogen bonding interactions with compound **DSM1** indicating that CF_3 substitution increased the polarity of the compound and thus favored electrostatic interactions. Both the above studies highlighted the factors responsible for good activity shown by triazolopyrimidine class of compounds and further modifications which can be possible in this class.

Desai et al. designed a library of *N*-phenylbenzamide (Fig. 10, **V**) derivatives with various substituents consisting of eighty molecules [91]. These were docked in *Pf*DHODH (PDBID 1TV5) using GOLD program. Out of these eighty designed molecules, fifteen final molecules were selected (based on their docking scores and interactions) for further synthesis and biological evaluation using cell-based *in vitro* assay (3H-hypoxanthine uptake method on *P. falciparum* NF54 (sensitive strain) and K1 (chloroquine and pyrimethamine resistance strain). This was followed by 3D-QSAR studies using CoMFA and CoMSIA models. One CoMFA and four CoMSIA models were generated from 89 models. The CoMFA model predicted that the *o*-position of the *N*-phenyl ring (Fig. 10, **V**(Ar_2)) of the *N*-phenylbenzamide and *m*-/*p*-positions of the carbonyl phenyl ring (Fig. 10, **V**(Ar_1)) can be substituted with electronegative atoms for improved activity. Similarly, the *o*- of carbonyl phenyl ring (Ar_1) and *p*-position of *N*-phenyl ring (Ar_2) show electropositive substitution. Bulky substituents are well tolerated at the *m*-/*o*-positions and not at the *p*-position of the carbonyl phenyl ring (Ar_1). The *m*-position of the *N*-phenyl ring (Ar_2) can only bear bulky substituents to a smaller extent and not at the *p*-position. The CoMSIA studies were depicted by hydrophobic contours based on model 3, i.e., SHE fields (steric, hydrophobic, and H-bond donor) and model 5, EDH fields (electrostatic, H-bond donor and hydrophobic) along with hydrogen bond donor and acceptor contours based on model 2, SDA fields (steric, H-bond donor, and H-bond acceptor). It was observed that the entire carbonyl phenyl ring (Ar_1) is surrounded by the hydrophobic contour except the *m*-position. In the *N*-phenyl ring (Ar_2), the hydrophobic contour is situated near the *m*- and *p*-position. H-bond acceptor contour is present at the *m*-position of the carbonyl phenyl ring (Ar_1), and H-bond donor contour is present at the *o*-region of the *N*-phenyl ring (Ar_2) of the *N*-phenylbenzamide. Out of the five models, r^2_{pred} for the CoMSIA model was more satisfactory than the CoMFA model. Two compounds, KMC-3 and KMC-15, were found to be active with IC_{50} value of 8.7 and 5.7 μM, respectively, against *P. falciparum*.

Vyas et al. in 2013 [92] performed 3D-QSAR on 5-(2-methylbenzimidazol-1-yl)-*N*-alkylthiophene-2-carboxamide derivatives reported by Brooker et al. [65]. A total of thirty-eight molecules were studied with thirty-five molecules in training set and five molecules in test set. The q^2 for the best CoMFA and CoMSIA models was determined to be 0.669 and 0.727, respectively. The prediction value (r^2_{pred}) obtained after validation by external test set was observed to be 0.799 and 0.815 for CoMFA and CoMSIA models, respectively. The CoMSIA model was observed to be better than the CoMFA model. In CoMSIA contour maps, a large favorable hydrophobic contour was observed near the C_2 position and nitrogen atom of the benzimidazole ring (Fig. 16). This was correlated with the presence of lipophilic group (methyl) at this position in active molecules. Hydrogen bond donor contour was observed near the nitrogen of the amide and other near the cyclopropyl group. A large hydrogen bond acceptor contour was observed near the oxygen of the amide.

Wadood et al. generated structure-based pharmacophore model using the X-ray crystal structure with PDBID 3O8A with 5-(2-methylbenzimidazol-1-yl)-*N*-cyclopropylthiophene-2-carboxamide as the co-crystallized ligand [93]. This pharmacophore model was used to identify molecules from ChemBridge database. Eighty-seven molecules were identified using this model system, and the hits were further screened using molecular docking and binding energy calculations using GOLD, and generalized Born interaction energies, and binding affinity using MOE docking software. Using these filters, twenty-five molecules with variable chemical classes were identified.

Tseng et al. worked on 3D-QSAR pharmacophore generation and docking-based pharmacophore development from a group of sixty-seven inhibitors of *Pf*DHODH belonging to different chemical classes [94]. The training set consisted of thirty-eight compounds and the test set includes twenty-five molecules. The pharmacophoric features used were hydrogen bond donor (HD), hydrogen bond acceptor (HA), hydrophobic group (H), and hydrophobic aromatic (HR) and were used during HypoGen during hypothesis generation. Hypo1 pharmacophoric model was considered to be the top model due to its high correlation coefficient (0.935), lowest RMS deviation (2.15) and successful prediction efficiency of training (89.4%) and test sets (72.4%). In docking-based pharmacophore generation, sixty-seven molecules (both training and test set) were docked using X-ray crystal structure of *Pf*DHODH with PDB ID 1TV5. The pharmacophore models generated were based on top scoring pose and genetic algorithm-based generation of 255 conformers. The scoring function was based on the interactions shown by the molecules with His185, Arg265, and Tyr528 [94, 95]. The docking-based pharmacophore model, DBP-All255 (docking-based pharmacophore (DBP) of all 255 conformers) was found to show comparable results as that of Hypo1. In Hypo1, the hydrophobic group feature was observed on the left side of the HA feature, and in DBP-All255, the hydrophobic feature appears on the right side of the HA feature. Both the models were able to predict the potential bioactive conformation of the inhibitors based on the structure activity relationship and binding mode of the inhibitors.

Hou et al. in 2016 performed QSAR on *Pf*DHODH inhibitors using multilinear regression (MLR) and support vector machine (SVM) [95]. A dataset of 255 molecules from ChEMBL database and literature, with *Pf*DHODH activity, was used. Most of the structures from the dataset contained triazolopyrimidine and benzimidazole as basic moiety. 161 molecules as training set, 94 molecules as test set, and 14 molecular descriptors (based on Pearson correlation) were selected. Four final computational models were generated showing good prediction quality with q^2 (leave-one-out) > 0.66, correlation coefficient (r) > 0.85 on both training and test sets. The mean square error (MSE) for training set is <0.32 and for test set is <0.37. It was observed through this study that the antimalarial activity of the inhibitors is mainly based on the hydrogen binding ability, atom polarizability, and ring complexity.

Pavadai et al. performed a systematic search of identification of species-selective *Pf*DHODH inhibitors by performing 3D-QSAR pharmacophore modeling followed by molecular docking-based virtual screening [96]. The final filtered compounds were tested for their *Pf*DHODH inhibitory activity and antimalarial activity. For 3D-QSAR model generation, a total of 38 compounds were selected from the work of Xu et al. on dihydrothiophenone class of compounds [84]. These are divided into training set (19 compounds) (activity range 6–39,450 nM) and test set (19 compounds). A total of 250 conformers were generated for these compounds. Hypogen algorithm was used to generate the pharmacophore hypotheses. Based on high correlation coefficient (r), low cost and low root mean square (RMS), a total of top ten hypotheses were selected. Based on the model, validation results based on test set prediction and cost values Hypo1 (test set correlation, $r = 0.933$) were selected for further virtual screening. 265,242 compounds from NCI database were used as query. The hit compounds were selected based on the features of the pharmacophore model and the hit compounds undergo molecular docking studies using *Pf*DHODH crystal structure (PDB ID 3I65) using Glide module of Schrödinger software. The molecules selected from standard precision docking based on the score and poses were filtered through extra precision docking. The top-ranked molecules obtained after this analysis were subjected to prime MM-GBSA calculations to calculate ΔG_{Bind}. The sixty-two compounds obtained after the thorough virtual screening process were subjected to biological testing in which three compounds were identified to exhibit *Pf*DHODH activity in the range of 0.38–20 μM IC_{50} value. The most active compound (NSC336047) showed selectivity against *Hs*DHODH ($IC_{50} > 100$ μM) and inhibition of parasite growth with 26 μM IC_{50} value [96].

4 Conclusions

SBDD proved to be a successful approach in the design of antimalarial agents. With the expanding knowledge into the new targets and enzymes essential for malarial parasite survival, it has become less straining in discovering or modifying the lead molecules based on the structural knowledge of the target. Some of these targets are *Pf*ATP-dependent heat shock protein 90, *Pf*phosphatidylinositol 4-kinase (PI4 K), *Pf*NADH dehydrogenase, *Pf*aspartate carbamoyltransferase, *Pf*thioredoxin reductase, *Pf*histone deacetylase, and *Pf* dihydroorotate dehydrogenase.

Taking an example of *Pf*DHODH, all the SBDD efforts related to this target were discussed. This enzyme was found to be important for the synthesis of pyrimidines in the parasite through de novo pathway. As this enzyme is also a part of the mitochondrial respiratory chain complex, its inhibition also affects the electron transfer in the inner mitochondrial membrane. The overall reaction involves the redox hip-hop mechanism in which DHO is oxidized by FMN to ORO.

The inhibitor binding site was predicted to be near to ubiquinone-binding site due to non-conserved nature of amino acids in this region and reoxidation of FMN was characterized to be the rate-limiting step.

The enzyme consists of a β-barrel core of eight parallel β-sheets surrounded by seven α-helices. The top and bottom of the barrel is covered by antiparallel β-strands, three on one side and two on side attached to N-terminal. The inhibitor binding site is located between the α10–α11 helices of the N-terminal. There are fifteen crystal structures reported for *Pf*DHODH till date. The first reported X-ray crystal structure with PDB ID 1TV5 containing teriflunomide as the co-crystallized inhibitor discusses about the important hydrogen bonding interactions shown by His185 and Arg265 at the end of the inhibitor binding tunnel. Later, two new hydrophobic binding pockets were reported in crystal structure 3I6R (**DSM74** as co-crystallized ligand) and 3O8A (**Genz-669178** as co-crystallized ligand).

A total of nine important chemical classes of *Pf*DHODH inhibitors are reported in the literature. Out of these triazolopyrimidine derivatives and *N*-alkyl-5-benzimidazole thiophene-2-carboxamide analogs were found to be most suitable with IC_{50} values in lower nanomolar range. The SBDD approach is facilitated in the designing of lead molecules with optimal in vitro and in vivo activity, good metabolic profile with minimum toxicity along with selectivity against *Hs*DHODH. **DSM265** has been is under clinical development and **Genz-669178** is understudy for clinical trials suitability. Various in silico studies are also reported which mainly include 3D-QSAR studies/molecular docking/molecular dynamics.

Acknowledgements The University Grants Commission is gratefully acknowledged for the financial support to Shweta Bhagat (UGC, Grant No. 43395). The authors thank Department of Science and Technology (DST), Government of India, New Delhi, India, for financial support.

References

1. World Malaria Report (2017) World Health Organization, Geneva. doi: ISBN 978-92-4-156552-3
2. [a] Gregson A, Plowe CV (2005) Mechanisms of resistance of malaria parasites to antifolates. Pharmacol Rev 57:117–145; [b] Harinasuta T, Suntharasamai P, Viravan C (1965) Chloroquine-resistant falciparum malaria in Thailand. The Lancet 2:657–660; [c] Sirawaraporn W, Prapunwattana P et al. (1993) The dihydrofolate reductase domain of *Plasmodium falciparum* thymidylate synthase-dihydrofolate reductase. Gene synthesis, expression, and anti-folate-resistant mutants. J Biol Chem 268:21637–21644
3. Wells TNC, van Huijsduijnen RH, Van Voorhis WC (2015) Malaria medicines: a glass half full? Nat Rev Drug Discov 14:424–442
4. Tinto H, D'Alessandro U et al (2015) Efficacy and safety of RTS, S/AS01 malaria vaccine with or without a booster dose in infants and children in Africa: final results of a phase 3, individually randomised, controlled trial. The Lancet 386:31–45
5. Anderson AC (2003) The process of structure-based drug design. Chem Biol 10:787–797

6. [a] Cowman AF, Morry MJ et al. (1988) Amino acid changes linked to pyrimethamine resistance in the dihydrofolate reductase-thymidylate synthase gene of *Plasmodium falciparum*. Proc Natl Acad Sci USA 85:9109–9113; [b] Foote SJ, Galatis D, Cowman AF (1990) Amino acids in the dihydrofolate reductase-thymidylate synthase gene of *Plasmodium falciparum* Involved in cycloguanil resistance differ from those involved in pyrimethamine resistance. Proc Natl Acad Sci USA 87:3014–3017; [c] Peterson DS, Milhous WK, Wellems TE (1990) Molecular basis of differential resistance to cycloguanil and pyrimethamine in *Plasmodium falciparum* Malaria. Proc Natl Acad Sci USA 87:3018–3022; [d] Peterson DS, Walliker D, Wellems TE (1988) Evidence that a point mutation in dihydrofolate reductase-thymidylate synthase confers resistance to pyrimethamine in falciparum malaria. Proc Natl Acad Sci USA 85:9114–9118; [e] Plowe CV (2009) The evolution of drug-resistant malaria. Trans R Soc Trop Med Hyg 103:S11–S14
7. Reickmann KH (1973) Chemotherapy of malaria and resistance to antimalarials. World Health Organization technical report, vol 529. World Health Organisation, Geneva
8. Yuthavong Y, Tarnchompoo B et al (2012) Malarial dihydrofolate reductase as a paradigm for drug development against a resistance-compromised target. Proc Natl Acad Sci U S A 109: 16823–16828
9. [a] Adane L, Bharatam PV, Sharma V (2010) A common feature-based 3D-pharmacophore model generation and virtual screening: identification of potential *Pf*DHFR inhibitors. J Enzyme Inhib Med Chem 25:635–645; [b] Adane L, Patel DS, Bharatam PV (2010) Shape- and chemical feature-based 3D-pharmacophore model generation and virtual screening: identification of potential leads for *P. falciparum* DHFR enzyme inhibition. Chem Biol Drug Des 75:115–126
10. Mehdi A, Adane L, Patel DS, Bharatam PV (2010) Electronic structure and reactivity of guanylthiourea: a quantum chemical study. J Comput Chem 31:1259–1267
11. Abbat S, Jain V, Bharatam PV (2015) Origins of the specificity of inhibitor P218 toward wild-type and mutant *Pf*DHFR: a molecular dynamics analysis. J Biomol Struct Dyn 33:1913–1928
12. [a] Adane L, Bhagat S et al. (2014) Design and synthesis of guanylthiourea derivatives as potential inhibitors of *Plasmodium falciparum* dihydrofolate reductase enzyme. Bioorg Med Chem Lett 24:613–617; [b] Bhagat S, Arfeen M et al. (2017) Guanylthiourea derivatives as potential antimalarial agents: synthesis, *in vivo* and molecular modelling studies. Eur J Med Chem 135:339–348
13. [a] Taipale M, Jarosz DF, Lindquist S (2010) HSP90 at the hub of protein homeostasis: emerging mechanistic insights. Nat Rev Mol Cell Bio 11:515; [b] Wang T, Bisson WH et al. (2014) Differences in conformational dynamics between *Plasmodium falciparum* and human Hsp90 orthologues enable the structure-based discovery of pathogen-selective inhibitors. J Med Chem 57:2524–2535
14. Corbett KD, Berger JM (2010) Structure of the ATP-binding domain of *Plasmodium falciparum* Hsp90. Proteins 78:2738–2744
15. Shahinas D, Liang M, Datti A, Pillai DR (2010) A repurposing strategy identifies novel synergistic inhibitors of *Plasmodium falciparum* heat shock protein 90. J Med Chem 53:3552–3557
16. Kumar R, Musiyenko A, Barik S (2003) The heat shock protein 90 of *Plasmodium falciparum* and antimalarial activity of its inhibitor, geldanamycin. Malar J 2:30
17. Posfai D, Eubanks AL et al. (2018) Identification of Hsp90 inhibitors with anti-plasmodium activity. Antimicrob Agents Chemother 62:e01799–01717
18. Krüger T, Sanchez CP, Lanzer M (2010) Complementation of *Saccharomyces cerevisiae* $pik1^{ts}$ by a phosphatidylinositol 4-kinase from *Plasmodium falciparum*. Mol Biochem Parasitol 172:149–151
19. Rajkhowa S, Borah SM, Jha AN, Deka RC (2017) Design of *Plasmodium falciparum* PI(4) KIIIβ inhibitor using molecular dynamics and molecular docking methods. ChemistrySelect 2:1783–1792

20. McNamara CW, Lee MC et al (2013) Targeting *Plasmodium* PI(4)K to eliminate malaria. Nature 504:248–253
21. Achieng AO, Rawat M et al (2017) Antimalarials: molecular drug targets and mechanism of action. Curr Top Med Chem 17:2114–2128
22. Melo AM, Bandeiras TM, Teixeira M (2004) New insights into type II NAD (P) H: quinone oxidoreductases. Microbiol Mol Biol Rev 68:603–616
23. Biagini GA, Viriyavejakul P et al (2006) Functional characterization and target validation of alternative complex I of *Plasmodium falciparum* mitochondria. Antimicrob Agents Chemother 50:1841–1851
24. Pidathala C, Amewu R et al (2012) Identification, design and biological evaluation of bisaryl quinolones targeting *Plasmodium falciparum* type II NADH: quinone oxidoreductase (*Pf*NDH2). J Med Chem 55:1831–1843
25. Yang Y, Yu Y et al (2017) Target elucidation by cocrystal structures of NADH-ubiquinone oxidoreductase of *Plasmodium falciparum* (*Pf*NDH2) with small molecule to eliminate drug-resistant malaria. J Med Chem 60:1994–2005
26. Rodrigues T, Lopes F, Moreira R (2010) Inhibitors of the mitochondrial electron transport chain and *de novo* pyrimidine biosynthesis as antimalarials: the present status. Curr Med Chem 17:929–956
27. Alnabulsi S, Santina E et al (2016) Non-symmetrical furan-amidines as novel leads for the treatment of cancer and malaria. Eur J Med Chem 111:33–45
28. Banerjee AK, Arora N, Murty USN (2012) Aspartate carbamoyltransferase of *Plasmodium falciparum* as a potential drug target for designing anti-malarial chemotherapeutic agents. Med Chem Res 21:2480–2493
29. [a] Lunev S, Bosch SS et al. (2016) Crystal structure of truncated aspartate transcarbamoylase from *Plasmodium falciparum*. Acta Crystallogr F 72:523–533; [b] Lunev S, Bosch SS et al. (2018) Identification of a non-competitive Inhibitor of *Plasmodium falciparum* aspartate transcarbamoylase. Biochem Biophys Res Commun 497:835–842
30. Fritz-Wolf K, Jortzik E et al (2013) Crystal structure of the *Plasmodium falciparum* thioredoxin reductase-thioredoxin complex. J Mol Biol 425:3446–3460
31. McMillan PJ, Arscott LD et al (2006) Identification of acid-base catalytic residues of high-M_r thioredoxin reductase from *Plasmodium falciparum*. J Biol Chem 281:32967–32977
32. Boumis G, Giardina G et al (2012) Crystal structure of *Plasmodium falciparum* thioredoxin reductase, a validated drug target. Biochem Biophys Res Commun 425:806–811
33. McCarty SE, Schellenberger A et al (2015) *Plasmodium falciparum* thioredoxin reductase (*Pf*TrxR) and its role as a target for new antimalarial discovery. Molecules 20:11459–11473
34. Munigunti R, Gathiaka S et al (2013) Characterization of *Pf*TrxR inhibitors using antimalarial assays and in silico techniques. Chem Cent J 7:175
35. Munigunti R, Gathiaka S et al (2014) Determination of antiplasmodial activity and binding affinity of curcumin and demethoxycurcumin towards *Pf*TrxR. Nat Prod Res 28:359–364
36. Winkler M, Maynadier M et al (2015) Uncovering new structural insights for antimalarial activity from cost-effective aculeatin-like derivatives. Org Biomol Chem 13:2064–2077
37. Chaal BK, Gupta AP et al (2010) Histone deacetylases play a major role in the transcriptional regulation of the *Plasmodium falciparum* life cycle. PLoS Path 6:e1000737
38. [a] Gupta AP, Bozdech Z (2017) Epigenetic landscapes underlining global patterns of gene expression in the human malaria parasite, *Plasmodium falciparum*. Int J Parasitol 47:399–407; [b] Sumanadasa SDM, Goodman CD et al. (2012) Antimalarial activity of the anticancer histone deacetylase inhibitor SB939. Antimicrob Agents Chemother 56:3849–3856
39. Mukherjee P, Pradhan A et al (2008) Structural insights into the *Plasmodium falciparum* histone deacetylase 1 (*Pf*HDAC-1): a novel target for the development of antimalarial therapy. Bioorg Med Chem 16:5254–5265
40. Hansen FK, Sumanadasa SDM et al. Discovery of HDAC inhibitors with potent activity against multiple malaria parasite life cycle stages. Eur J Med Chem 82:204–213
41. Darkin-Rattray SJ, Gurnett AM et al (1996) Apicidin: a novel antiprotozoal agent that inhibits parasite histone deacetylase. Proc Natl Acad Sci USA 93:13143–13147

42. Harwaldt P, Rahlfs S, Becker K (2002) Glutathione S-transferase of the malarial parasite *Plasmodium falciparum*: characterization of a potential drug target. Biol Chem 383:821–830
43. Hiller N, Fritz-Wolf K et al (2006) *Plasmodium falciparum* glutathione S-transferase—structural and mechanistic studies on ligand binding and enzyme inhibition. Protein Sci 15:281–289
44. Fritz-Wolf K, Becker A et al (2003) X-ray structure of glutathione S-transferase from the malarial parasite *Plasmodium falciparum*. Proc Natl Acad Sci USA 100:13821–13826
45. Perbandt M, Eberle R et al (2015) High resolution structures of *Plasmodium falciparum* GST complexes provide novel insights into the dimer-tetramer transition and a novel ligand-binding site. J Struct Biol 191:365–375
46. Ahmad R, Srivastava AK (2008) Inhibition of glutathione-S-transferase from *Plasmodium yoelii* by protoporphyrin IX, cibacron blue and menadione: implications and therapeutic benefits. Parasitol Res 102:805–807
47. Miller RW, Kerr CT, Curry JR (1968) Mammalian dihydroorotate—ubiquinone reductase complex. Can J Biochem 46:1099–1106
48. Chen JJ, Jones ME (1976) The cellular location of dihydroorotate dehydrogenase: relation to de novo biosynthesis of pyrimidines. Arch Biochem Biophys 176:82–90
49. [a] Larsen JN, Jensen KF (1985) Nucleotide sequence of the pyrD gene of *Escherichia coli* and characterization of the flavoprotein dihydroorotate dehydrogenase. Eur J Biochem 151:59–65; [b] LeBlanc SB, Wilson CM (1993) The dihydroorotate dehydrogenase gene homologue of *Plasmodium falciparum*. Mol Biochem Parasitol 60:349–351
50. [a] Gardner MJ, Hall N et al. (2002) Genome sequence of the human malaria parasite *Plasmodium falciparum*. Nature 419:498–511; [b] K. Vyas V, Ghate M (2011) Recent developments in the medicinal chemistry and therapeutic potential of dihydroorotate dehydrogenase (DHODH) inhibitors. Mini-Rev Med Chem 11:1039–1055
51. Krungkrai J (1995) Purification, characterization and localization of mitochondrial dihydroorotate dehydrogenase in *Plasmodium falciparum*, human malaria parasite. Biochim Biophys Acta Gen Subj 1243:351–360
52. McRobert L, McConkey GA (2002) RNA Interference (RNAi) inhibits growth of *Plasmodium falciparum*. Mol Biochem Parasitol 119:273–278
53. Baldwin J, Farajallah AM et al (2002) Malarial dihydroorotate dehydrogenase: substrate and inhibitor specificity. J Biol Chem 277:41827–41834
54. Boa AN, Canavan SP et al (2005) Synthesis of brequinar analogue inhibitors of malaria parasite dihydroorotate dehydrogenase. Bioorg Med Chem 13:1945–1967
55. Baldwin J, Michnoff CH et al (2005) High-throughput screening for potent and selective inhibitors of *Plasmodium falciparum* dihydroorotate dehydrogenase. J Biol Chem 280:21847–21853
56. Heikkilä T, Thirumalairajan S et al (2006) The first de novo designed inhibitors of *Plasmodium falciparum* dihydroorotate dehydrogenase. Bioorg Med Chem Lett 16:88–92
57. Phillips MA, Rathod PK (2010) *Plasmodium* dihydroorotate dehydrogenase: A promising target for novel anti-malarial chemotherapy. Infect Disord Drug Targets 10:226–239
58. Löffler M, Fairbanks LD et al (2005) Pyrimidine pathways in health and disease. Trends Mol Med 11:430–437
59. Fagan RL, Nelson MN, Pagano PM, Palfey BA (2006) Mechanism of flavin reduction in class 2 dihydroorotate dehydrogenases. Biochemistry 45:14926–14932
60. Liu S, Neidhardt EA et al (2000) Structures of human dihydroorotate dehydrogenase in complex with antiproliferative agents. Structure 8:25–33
61. Fagan RL, Palfey BA (2009) Roles in binding and chemistry for conserved active site residues in the class 2 dihydroorotate dehydrogenase from *Escherichia coli*. Biochemistry 48:7169–7178
62. Hurt DE, Widom J, Clardy J (2006) Structure of *Plasmodium falciparum* dihydroorotate dehydrogenase with a bound inhibitor. Acta Crystallogr Sect D: Biol Crystallogr D62:312–323
63. Drozdetskiy A, Cole C, Procter J, Barton GJ (2015) JPred4: a protein secondary structure prediction server. Nucleic Acids Res 43:W389–W394

64. Deng X, Gujjar R et al (2009) Structural plasticity of malaria dihydroorotate dehydrogenase allows selective binding of diverse chemical scaffolds. J Biol Chem 284:26999–27009
65. Booker ML, Bastos CM et al (2010) Novel inhibitors of *Plasmodium falciparum* dihydroorotate dehydrogenase with anti-malarial activity in the mouse model. J Biol Chem 285:33054–33064
66. Coteron JM, Marco M et al (2011) Structure-guided lead optimization of triazolopyrimidine-ring substituents identifies potent *Plasmodium falciparum* dihydroorotate dehydrogenase inhibitors with clinical candidate potential. J Med Chem 54:5540–5561
67. Ross LS, Gamo FJ et al (2014) In vitro resistance selections for *Plasmodium falciparum* dihydroorotate dehydrogenase inhibitors give mutants with multiple point mutations in the drug-binding site and altered growth. J Biol Chem 289:17980–17995
68. Deng X, Kokkonda S et al (2014) Fluorine modulates species selectivity in the triazolopyrimidine class of *Plasmodium falciparum* dihydroorotate dehydrogenase inhibitors. J Med Chem 57:5381–5394
69. Phillips MA, Lotharius J et al (2015) A long-duration dihydroorotate dehydrogenase inhibitor (DSM265) for prevention and treatment of malaria. Sci Transl Med 7:296ra111
70. Deng X, Matthews D, Rathod PK, Phillips MA (2015) The X-ray structure of *Plasmodium falciparum* dihydroorotate dehydrogenase bound to a potent and selective *N*-phenylbenzamide inhibitor reveals novel binding-site interactions. Acta Crystallogr Sec F 71:553–559
71. Kokkonda S, Deng X et al (2016) Tetrahydro-2-naphthyl and 2-indanyl triazolopyrimidines targeting *Plasmodium falciparum* dihydroorotate dehydrogenase display potent and selective antimalarial activity. J Med Chem 59:5416–5431
72. Phillips MA, White KL et al (2016) A triazolopyrimidine-based dihydroorotate dehydrogenase inhibitor with improved drug-like properties for treatment and prevention of malaria. ACS Infect Dis 2:945–957
73. Malmquist NA, Gujjar R, Rathod PK, Phillips MA (2008) Analysis of flavin oxidation and electron-transfer inhibition in *Plasmodium falciparum* dihydroorotate dehydrogenase. Biochemistry 47:2466–2475
74. [a] Norager S, Jensen KF, Björnberg O, Larsen S (2002) *E. coli* Dihydroorotate dehydrogenase reveals structural and functional distinctions between different classes of dihydroorotate dehydrogenases. structure 10:1211–1223; [b] Rowland P, Bjornberg O et al. (1998) The crystal structure of *Lactococcus lactis* dihydroorotate dehydrogenase A complexed with the enzyme reaction product throws light on its enzymatic function. Protein Sci 7:1269–1279; [c] Rowland P, Nielsen FS, Jensen KF, Larsen S (1997) The crystal structure of the flavin containing enzyme dihydroorotate dehydrogenase A from *Lactococcus lactis*. Structure 5:239–252; [d] Sørensen PG, Dandanell G (2002) A new type of dihydroorotate dehydrogenase, type 1S, from the thermoacidophilic archaeon *Sulfolobus solfataricus*. Extremophiles 6:245–251
75. Bedingfield PT, Cowen D et al (2012) Factors influencing the specificity of inhibitor binding to the human and malaria parasite dihydroorotate dehydrogenases. J Med Chem 55:5841–5850
76. Copeland RA, Davis JP et al (1995) Recombinant human dihydroorotate dehydrogenase: expression, purification, and characterization of a catalytically functional truncated enzyme. Arch Biochem Biophys 323:79–86
77. [a] Löffler M, Knecht W et al. (2002) Drosophila melanogaster dihydroorotate dehydrogenase: the N-terminus is important for biological function in vivo but not for catalytic properties in vitro. Insect Biochem Mol Biol 32:1159–1169; [b] Rawls J, Knecht W et al. (2000) Requirements for the mitochondrial import and localization of dihydroorotate dehydrogenase. Eur J Biochem 267:2079–2087
78. Phillips MA, Gujjar R et al (2008) Triazolopyrimidine-based dihydroorotate dehydrogenase inhibitors with potent and selective activity against the malaria parasite *Plasmodium falciparum*. J Med Chem 51:3649–3653
79. Heikkilä T, Ramsey C et al (2007) Design and synthesis of potent inhibitors of the malaria parasite dihydroorotate dehydrogenase. J Med Chem 50:186–191

80. [a] Patel V, Booker M et al. (2008) Identification and characterization of small molecule inhibitors of *Plasmodium falciparum* dihydroorotate dehydrogenase. J Biol Chem 283:35078–35085; [b] Skerlj RT, Bastos CM et al. (2011) Optimization of potent inhibitors of *P. falciparum* dihydroorotate dehydrogenase for the treatment of malaria. ACS Med Chem Lett 2:708–713
81. Fritzson I, Bedingfield PTP et al (2011) N-substituted salicylamides as selective malaria parasite dihydroorotate dehydrogenase inhibitors. Med Chem Comm 2:895–898
82. Zhu J, Han L et al (2015) Design, synthesis, X-ray crystallographic analysis, and biological evaluation of thiazole derivatives as potent and selective inhibitors of human dihydroorotate dehydrogenase. J Med Chem 58:1123–1139
83. Azeredo LFSP, Coutinho JP et al (2017) Evaluation of 7-arylaminopyrazolo [1,5-*a*] pyrimidines as anti-*Plasmodium falciparum*, antimalarial, and *Pf* dihydroorotate dehydrogenase inhibitors. Eur J Med Chem 126:72–83
84. Xu M, Zhu J et al (2013) Novel selective and potent inhibitors of malaria parasite dihydroorotate dehydrogenase: discovery and optimization of dihydrothiophenone derivatives. J Med Chem 56:7911–7924
85. Schneider G, Fechner U (2005) Computer-based de novo design of drug-like molecules. Nat Rev Drug Discov 4:649
86. Gillet VJ, Newell W et al (1994) SPROUT: recent developments in the de novo design of molecules. J Chem Inf Comput Sci 34:207–217
87. Gujjar R, Marwaha A et al (2009) Identification of a metabolically stable triazolopyrimidine-based dihydroorotate dehydrogenase inhibitor with antimalarial activity in mice. J Med Chem 52:1864–1872
88. Gujjar R, El Mazouni F et al (2011) Lead optimization of aryl and aralkyl amine-based triazolopyrimidine inhibitors of *Plasmodium Falciparum* dihydroorotate dehydrogenase with antimalarial activity in mice. J Med Chem 54:3935–3949
89. Ojha PK, Roy K (2010) Chemometric modeling, docking and in silico design of triazolopyrimidine-based dihydroorotate dehydrogenase inhibitors as antimalarials. Eur J Med Chem 45:4645–4656
90. Shah P, Kumar S, Tiwari S, Siddiqi MI (2012) 3D-QSAR studies of triazolopyrimidine derivatives of *Plasmodium falciparum* dihydroorotate dehydrogenase inhibitors using a combination of molecular dynamics, docking, and genetic algorithm-based methods. J Chem Biol 5:91–103
91. Desai KR, Shaikh MS, Coutinho EC (2011) Molecular modeling studies, synthesis and biological evaluation of derivatives of *N*-phenylbenzamide as *Plasmodium falciparum* dihydroorotate dehydrogenase (*Pf*DHODH) inhibitors. Med Chem Res 20:321–332
92. Vyas VK, Parikh H, Ghate M (2013) 3D QSAR studies on 5-(2-methylbenzimidazol-1-yl)-*N*-alkylthiophene-2-carboxamide derivatives as *P. falciparum* dihydroorotate dehydrogenase (*Pf*DHODH) inhibitors. Med Chem Res 22:2235–2243
93. Wadood A, Zaheer-ulhaqb (2013) *In silico* identification of novel inhibitors against *Plasmodium falciparum* dihydroorate dehydrogenase. J Mol Graphics Model 40:40–47
94. Tseng TS, Lee YC et al (2016) Comparative study between 3D-QSAR and docking-based pharmacophore models for potent *Plasomodium falciparum* dihydroorotate dehydrogenase inhibitors. Bioorg Med Chem Lett 26:265–271
95. Hou X, Chen X, Zhang M, Yan A (2016) QSAR study on the antimalarial activity of *Plasmodium falciparum* dihydroorotate dehydrogenase (*Pf* DHODH) inhibitors. SAR QSAR Environ Res 27:101–124
96. Pavadai E, El Mazouni F et al (2016) Identification of new human malaria parasite *Plasmodium falciparum* dihydroorotate dehydrogenase inhibitors by pharmacophore and structure-based virtual screening. J Chem Inf Model 56:548–562

Recent Advancements in Computing Reliable Binding Free Energies in Drug Discovery Projects

N. Arul Murugan, Vasanthanathan Poongavanam and U. Deva Priyakumar

Abstract In recent times, our healthcare system is being challenged by many drug-resistant microorganisms and ageing-associated diseases for which we do not have any drugs or drugs with poor therapeutic profile. With pharmaceutical technological advancements, increasing computational power and growth of related biomedical fields, there have been dramatic increase in the number of drugs approved in general, but still way behind in drug discovery for certain class of diseases. Now, we have access to bigger genomics database, better biophysical methods, and knowledge about chemical space with which we should be able to easily explore and predict synthetically feasible compounds for the lead optimization process. In this chapter, we discuss the limitations and highlights of currently available computational methods used for protein–ligand binding affinities estimation and this includes force-field, ab initio electronic structure theory and machine learning approaches. Since the electronic structure-based approach cannot be applied to systems of larger length scale, the free energy methods based on this employ certain approximations, and these have been discussed in detail in this chapter. Recently, the methods based on electronic structure theory and machine learning approaches also are successfully being used to compute protein–ligand binding affinities and other pharmacokinetic and pharmacodynamic properties and so have greater potential to take forward computer-aided drug discovery to newer heights.

N. A. Murugan (✉)
Department of Theoretical Chemistry and Biology, School of Engineering Sciences in Chemistry, Biotechnology and Health, Royal Institute of Technology, Stockholm, Sweden
e-mail: murugan@kth.se

V. Poongavanam
Department of Physics, Chemistry, Pharmacy, University of Southern Denmark, Campusvej 55, DK-5230 Odense M, Denmark

U. D. Priyakumar
CCNSB, International Institute of Information Technology, Gachibowli 500032, Hyderabad, India

C. G. Mohan (ed.), *Structural Bioinformatics: Applications in Preclinical Drug Discovery Process*, Challenges and Advances in Computational Chemistry and Physics 27, https://doi.org/10.1007/978-3-030-05282-9_7

Keywords Computational drug discovery · Free energy of binding
Hybrid QM/MM · QM fragmentation · Binding affinity · Pharmacokinetic
(PK) properties · Machine learning approach

Abbreviations

FMO	Fragment molecular orbital
MAO-B	Monoamine oxidase B
MM-GBSA	Molecular mechanics–Generalized Born Surface Area
MM-PBSA	Molecular mechanics–Poisson–Boltzmann Surface Area
PD	Pharmacodynamic
PK	Pharmacokinetic
QM/MM	Quantum mechanics/molecular mechanics

1 Introduction: Drugs and Targets

Disease can be defined as an abnormal condition that alters the function or behaviour of an organism and this can be caused by different factors, i.e. internally e.g. due to the presence of disease-causing genes or due to external factors. Externally, disease may be caused due to malnutrition or subjecting a human to severe external conditions such as exposing to radiation or pollution or microbial infections or severe physiological conditions which leads to damage or malfunctioning of body machineries. Thanks to genomic analysis of normal and diseased persons, we know that the protein profile appears quite different in these two cases and by targeting the biomacromolecules expressed in the diseased state, and we can develop methods to arrest the progress of the disease. By comparative protein profiling of normal and diseased persons or by comparing the genomes of human and pathogenic micro-organisms, [1–3] we already know the information about the potential targets, but then the problem lies in identifying whether the aberrant expression of a certain biomacromolecule is the main cause of disease or may be a side product of another key process. Once the key target (protein or enzyme) is identified, primay task is to design small molecules that can modulate the target (this can be either of inhibitor, substrate, inducer). Subsequently, the active compound (also called hit molecule) is further optimized to pre-clinical candidate. The aim of lead optimization is not only to increase the potency, but also to reduce any off-target binding. In this chapter, we will discuss how to use computational approaches not only to identify small molecules that can inhibit or modulate the catalytic process of a key enzyme that is connected with disease, but also to understand the fundamental process of biomolecular recognition which assists in the lead optimization process in the drug discovery and development projects. The properties of the ligand to be optimized are binding affinity and specificity towards a key target biomacromolecule. These target molecules can be located within microorganism or within the host organism depending upon whether the disease belongs to infectious or autoimmune category disease, respectively.

2 Optimization of Drug-Likeness

In addition to binding affinity and specificity, there are certain other properties which are to be optimized for an effective drug i.e. low toxic with improved potency and orally bioavailable for conventional dosage forms. These properties are absorption (A), distribution (D), metabolism (M), excretion (E) and toxicity (T), and they collectively are called ADMET or pharmacokinetic (PK) properties. The properties in general refer to kinetic behaviour of drugs within body and give information about the timescale required for the drug to reach the potential target and lifetime within host organism before removal through excretion (this can be shortly described as "what the body does to a drug?"). The optimization of potency (binding affinity) and then the subsequent optimization of pharmacokinetic behaviour have been the major contributing factors for the failures at the phase II and phase III clinical trials [4–6]. So, it is necessary simultaneously to optimize the potency along with ADMET properties [7]. There are also other properties that are essential for oral bioavailability such as solubility and transport properties like membrane permeability (both cellular and across blood-brain barrier).

Overall, it is apparent that drug design is challenging as we need to optimize several properties at the same time [3, 6]. In certain cases, optimizing one property may lead to unexpected changes in another property making this optimization very complex , and in those cases we need to compromise on certain properties and try to balance different properties for better PD and PK profile. For example, if the potency of a drug is superior/outstanding but then if it has very poor PK and PD properties, then one can use suitable drug delivery systems such that the drug is delivered to its target biomacromolecule. Given that drug design is an optimization process, it is inevitable to avoid the use of computers as they can be used effectively to speed up the overall process. But the only requirement is that we need to have accurately enough methods that can be written in a numerically solvable form and can reliably describe these processes involved in the drug association with a biological target and its pharmacokinetics [8]. In general, quantum mechanics is the fundamental theory which can be used to describe any atomic and molecular systems and their association process and their response to any external variables like heat, pressure and fields and to any change in physiological conditions such as pH and ionic strength. However, the complexity of the mathematics involved in solving the Schrödinger equation grows with powers of n which is number of basis functions used to describe one electron orbital used to build wave function of the molecule that describes its energy and all other properties. For example, the computational demand is at the power of three in the case density functional theory and can go to power of 5–8 for the theories which can treat the electron correlation more accurately. When the system size is comparably larger than the wavelength of light and when we are not interested in processes where the matter interacts with light or laser field, it is pragmatic to use classical mechanics to describe the molecular systems which involves relatively simple mathematics, i.e. solving Newton's equation of motion to describe the interaction within system and their association with other systems and to model their time evolution and their response to external thermodynamic variables like temperature and pressure. As per classical mechanics, once we have the force-field information for a system, its entire future and past can be predicted by solving equation of motion. Force-fields can be developed by using various available structure databases and thermodynamics data. In this chapter, We briefly cover available force-field methods for computing the binding affinity in order to rank protein–ligand complexes in drug discovery and design. In addition, briefly discuss their limitations and also present the recent advancement in computational modelling approaches based on quantum mechanical theory and machine learning algorithm in a way suitable for drug discovery applications.

3 Free Energies Relevant to Describe Potency and Pharmacokinetics

Free energy is the key variable that dictates the structure of biomacromolecular complexes (protein–ligand protein, membrane–ligand, DNA–ligand etc.) and controls various molecular association and ligand transport processes. When there are many structures possible, the one with least free energy is the most stable one. Moreover, biomacromolecular or molecular association processes such as drug binding to receptor, protein–protein binding and drug transport involve the minimization of Gibbs free energy (**Δ**G). Any process that involves lowering of Gibbs free energy can proceed spontaneously. By calculating the free energy change, we can predict whether an association process is feasible or not. In the case of a drug, the most relevant aspect is to understand its binding affinity or potency towards a target biomacromolecule and its association with transport proteins like albumins [9] and metabolizing enzymes such as cytochrome P450s (CYP) [10–12] and with glycoproteins responsible for absorption. The ligand binding to target biomacromolecule, transport protein and metabolizing enzymes is dictated by the change in free energy of the ligand bound to these targets when compared to that in aqueous solution. Further, it is also necessary to understand whether the compounds will pass through certain cell membranes and also how well it will dissociate once it is taken through oral dose which is dependent on the physicochemical properties like lipophilicity [13] and aqueous solubility [14]. Schematic representation of various PK computational modeling is shown in Fig. 1, and free energy of relevance is provided in Table 1.

Computing the free energy of binding of the drug with a biological target and other targets (such as glycoproteins, albumins, cell membranes) that mediate the drug transport across the body to relevant target area is the main goal of any computational approaches. All these drug association-related processes and

Table 1 Computing various PK and PD properties and potency as a difference in free energy of the ligand in different environments

Property	Initial medium	Final medium	Free energyof relevance
Inhibition constant/binding affinity	Water	Target enzyme	$G_{enzyme} - G_{water}$
Absorption/distribution	Water	Glycoproteins/ albumin	$G_{albumin} - G_{water}$
Metabolism	Water	Cytochromes P450	$G_{P450} - G_{water}$
Permeability	Water	Membrane	$G_{membrane} - G_{water}$
Solubility	Crystalline	Water	$G_{water} - G_{crystal}$
Off-target binding	Water	Off-target (e.g. hERG)	$G_{offtarget} - G_{water}$

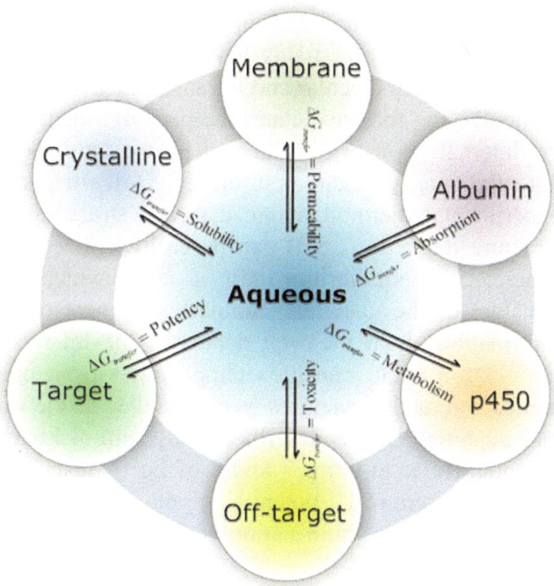

Fig. 1 Potency, pharmacodynamic property (such as solubility, permeability) and a few pharmacokinetic properties (such as drug absorption, distribution, metabolism and toxicity) are related to ΔG transfer, which is a free energy difference needed for driving a ligand from one environment to another

transport processes involve optimization (minimization in particular) of free energies, and the free energies associated with potency, bioavailability, drug absorption, distribution, metabolism, toxicity, solubility are listed in Table 1.

If we can calculate the free energy change for the drug to transfer from one medium to another, then we can predict how spontaneously this process will occur. As a conclusion, it can be deduced that the drug design involves calculations of free energy changes in two different media and the currently available methods are based on either force-fields or semi-empirical methods or electronic structure theory or combination of these. In this chapter, we will provide a brief outline of various computational methods available for computing free energy of binding of a ligand to a receptor which in turn can be used to optimize drug potency, absorption, distribution and metabolism. These methods are so general and also can be used to compute absolute free energies of a ligand or drug in crystalline or in different solvent environments making it feasible to predict other PD and PK properties like bioavailability, permeability and solubility [15, 16]. Very recently, machine learning approaches are also contributing to the computation of interaction energies of protein–ligand and protein–protein complexes and the progress in the use of such approaches for drug discovery projects will be discussed at the end.

4 Force-Field-Based Free Energies of Drug Target Binding

A force-field describes how the atomic and molecular systems behave at finite temperature, pressure and in a specific physiological condition or under any external fields. Force-fields have potential energy functions to explain the interactions of intermolecular and intramolecular degrees of freedom within a system. In particular, the former dictates the packing, relative orientation of molecules, while the internal geometry is dictated by the latter potentials. The currently available potential energy functions in different force-fields were parameterized using many experimental thermodynamic data and structural data [17]. For example, crystal structure database (CSD) can be easily utilized to get the information about the characteristic equilibrium radii of atoms which then dictate the size and overall structure of the materials. Similarly, the heat of vaporization can be used to get information about the well depths of the interaction potential which then gives information about transition temperatures from one phase to another. For convenience, interaction energies were modelled as the sum over pair potential where the pair potential itself is described using sum of Lennard-Jones (LJ) and electrostatic potential (refer to terms 7 and 8 of Eq. 1, respectively). As mentioned above, the two parameters of LJ potential, sigma and epsilon can be parameterized by using available structure information and thermodynamic data.

$$\begin{aligned} U_{\text{total}} = & \sum_{\text{bonds}} K_b(b-b_0)^2 + \sum_{\text{angles}} K_\theta(\theta-\theta_0)^2 + \sum_{\text{UB}} K_{\text{UB}}(S-S_0)^2 \\ & + \sum_{\text{dihedral}} K_x(1+\cos(cX-\delta)) + \sum_{\text{impropers}} K_{\text{impr}}(\varphi-\varphi_0)^2 \\ & + \sum_{\text{LJ}_{i\neq j}} \varepsilon_{ij}\left[\left(\frac{R_{\min_{ij}}}{r_{ij}}\right)^{12} - 2\left(\frac{R_{\min_{ij}}}{r_{ij}}\right)^{6}\right] + \sum_{\text{coulomb}} \frac{q_i q_j}{\varepsilon_i r'_{ij}} \end{aligned} \quad (1)$$

The LJ and electrostatic potentials describe only the intermolecular interactions, while it is not accounting for the structural changes in the molecule in the vicinities of other molecules. To describe such structural changes, we need to have as well the potential associated with changes in intramolecular structure. Usually, such a potential has terms to describe variation in bond length (bond length potential), angle (bond angle potential), dihedral angle (improper and proper dihedral angle potential) (refer to first five terms in Eq. 1). Since the classical force-field cannot describe the bond-breaking processes, a harmonic potential is in general used to describe structural changes associated with bond lengths and bond angles. However, note that the dihedral angle motion is not local and can describe conformational changes in molecules and in case of peptides these contribute to changes in the secondary structure. The total potential including both intra- and intermolecular interactions for a biomacromolecule alone or in solution or in combination with other molecules can be described by the Eq. 1.

The force constants and equilibrium values for bond length and bond angle are obtained from spectroscopic data and from the structural data, respectively. Also, electronic structure theory-based calculations can be employed to get these parameters.

In general, the binding of a ligand to a protein can be described as the equilibrium between the protein–ligand complex and the protein and the ligand (Eq. 2). The change in free energy/the binding free energy (ΔG_{Bind}) can thus be calculated as the difference between the free energies of the ligand and protein in free and bound form (Eq. 3) which is then compared to experimental binding affinity (inhibition constant or IC_{50}).

$$P_{\mathrm{Aq}} + L_{\mathrm{Aq}} \leftrightarrow PL_{\mathrm{Aq}} \tag{2}$$

$$\Delta G_{\mathrm{Bind}} = G(PL)_{\mathrm{Aq}} - G(P)_{\mathrm{Aq}} - G(L)_{\mathrm{Aq}} \tag{3}$$

All these free energies can be computed using explicit solvent models, namely SPC, TIP3P, TIP4P, TIP5P, but it is computationally very demanding. Alternatively, one can use the implicit solvent models, and then the free energies of the three systems, namely complex, receptor and ligand, can be computed with less computational effort [18]. This involves calculation of solvation free energy of a subsystem in a solvent media described with a dielectric constant which is a macroscopic parameter specific to solvent and describes its ability to polarize the solute [18]. The electrostatic interaction between the solute and the solvent is solved using generalized born (as in the MM-GBSA) or Poisson–Boltzmann (as in the MM-PBSA) to get the polar part of the solvation free energies [19]. The non-polar part of the solvation free energies is computed from the solvent-accessible surface area of the solute. In the implicit solvent model, the only used solvent parameter is dielectric constant, usually the solvent coordinates are removed from the molecular dynamics or Monte Carlo trajectories, and only the protein–ligand coordinates are used.

Force-field methods for calculating free energies (e.g. MM-GBSA or MM-PBSA) with implicit solvent models forsolvation part achieved considerable success in explaining the drug binding to a number of receptors or biomacromolecular targets [20–22]. In particular, MM-PBSA method was successfully used to predict the binding affinities of many antibacterial, antiviral benchmark datasets [22, 23]. Further, the MM-PBSA method has been extensively used to understand the interaction of various substrates with the targets that are relevant in the treatment of various neurodegenerative diseases. A detailed account of this can be found in the reference [22].

4.1 Molecular Docking

Molecular docking is the most simplistic method available for computing the protein–ligand binding affinities and for finding the most stable binding mode *(pose)* for a ligand within the binding site of the protein. The scoring functions are used to decide on the binders from non-binders and their least energy binding mode and pose which can be knowledge-based, empirical and force-field-based [24–26]. In the force-field-based scoring function, the free energy of binding dictates the drug potency. The interaction energies are calculated as a sum of polar and non-polar interactions such as van der Waals and electrostatic. The change in intramolecular energies of the ligand due to conformational change is also added to the total energies just to make sure ligand conformations with unusually high energies are avoided in the search. The entropic contributions due to conformational degree of freedom are included in a simple mean; i.e. each flexible bond is associated with 0.3 kcal/mol.

The working equation to compute the interaction energy between the protein and the ligand is as given below which is a sum of van der Waals (E_{vdw}), electrostatic (E_{elec}), hydrogen bonding (E_{Hbond}) and internal energies (E_{int}). The last term refers to the change in intramolecular energy of the ligand due to binding to receptor. In the gas phase, the ligand adopts geometry where the internal energy is assumed to be zero. But when it binds to a receptor, it undergoes certain structural changes (or conformational changes) and this increase in energy is contributing to internal energy. Such contributions are usually positive to the total binding energies; however, the other contributions are dominantly negative in magnitude making the protein–ligand association to happen instead of destabilization due to increase in internal energies.

$$E_{\text{dock}} = E_{\text{vdw}} + E_{\text{H-bond}} + E_{\text{electrostatic}} + E_{\text{internal}} \tag{4}$$

$$\begin{aligned} &= \sum_{\text{Protein}} \sum_{\text{ligand}} \left(\frac{A_{ij}}{d_{ij}^{12}} - \frac{B_{ij}}{d_{ij}^{6}} \right) + \sum_{\text{Protein}} \sum_{\text{ligand}} E(t) \\ &\quad \times \left(\frac{C_{ij}}{d_{ij}^{12}} - \frac{D_{ij}}{d_{ij}^{10}} \right) + \sum_{\text{Protein}} \sum_{\text{ligand}} 332.0 \frac{q_i q_j}{\varepsilon(d_{ij}) d_{ij}} \\ &\quad + \left\{ \sum_{\text{ligand}} \left(\frac{A_{ij}}{d_{ij}^{12}} - \frac{B_{ij}}{d_{ij}^{6}} \right) + \sum_{\textit{ligand}} E(t) \left(\frac{C_{ij}}{d_{ij}^{12}} - \frac{D_{ij}}{d_{ij}^{10}} \right) + \sum_{\text{ligand}} 332.0 \frac{q_i q_j}{4 d_{ij} d_{ij}} \right\} \end{aligned} \tag{5}$$

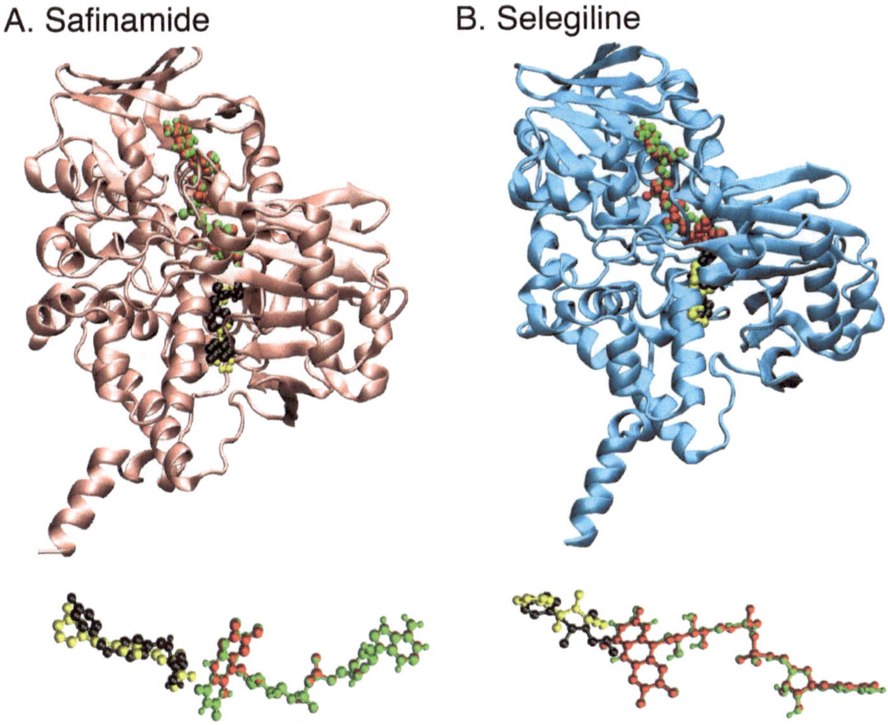

Fig. 2 Overlap of binding mode obtained from molecular docking with the experimental crystal structure

Due to the ease in doing calculations and lower computational demand, molecular docking methods are routinely used to rank the compounds according to their binding affinity or using other scoring. The pharmaceutical companies use this approach very efficiently to screen the chemical database containing millions of compounds against a potential target in the early virtual screening process before they can be synthezised as lead series. An elaborate list on use of molecular docking-based screening to design candidate drug molecules for various targets, namely G protein-coupled receptors, enzymes, ion channels, can be found in this reference [27].

Further, the binding mode and pose for number of ligands in their biological targets were predicted successfully using molecular docking tool. For example, there was a good overlap between the binding modes predicted from molecular docking and experimental crystal structure in the case of safinamide, a reversible inhibitor in monoamine oxidase B (MAO-B) (refer to Fig. 2A) [28, 29]. Interestingly, even in the case of a irreversible inhibitor such as selegiline

(L-deprenyl), where there is formation of covalent bond between the inhibitor and FAD cofactor, the predicted binding mode from molecular docking has reasonable overlap with the one from crystal structure (refer to Fig. 2B). Figure 2 shows the overlap of binding mode obtained from molecular docking with the experimental crystal structure. The target is monoamine oxidase B, and two inhibitors were considered, namely safinamide and selegiline. The former one is reversible MAO-B inhibitor, while the latter one is irreversible inhibitor which covalently bonding to the FAD cofactor.

4.2 Success Stories of Force-Field-Based Methods in Drug Discovery Projects

Drug discovery for a new disease is a complex project which requires knowledge from different domains, namely protein profiling (genomics), bioinformatics (for doing comparative genomics for target discovery), structural biology (for structure elucidation of the target), cheminformatics (chemical space), synthetic medicinal chemistry (design and synthesis of molecules), toxicology, pharmacology, pharmacokinetic property estimation, binding assay experiments, clinical studies. Computational approaches can be employed to speed up many of the intermediate steps involved in the drug discovery such as target discovery (computational comparative genomics), structure elucidation for a target (homology modelling) and lead compound prediction (using cheminformatics, virtual screening and de nova design) and ADMET property prediction (for screening the lead compounds with appropriate pharmacokinetic properties) and toxicity prediction (by studying the interaction of ligands with potential known off-targets). The chemical space consists of billions of small molecules [30], and huge genomics database suggests that there are thousands of targets and off-targets for studying the drug potency and its toxicity which makes the computational approaches as irreplaceable workhorses for the drug discovery projects. Thanks to such approaches, there are many drugs which are in the clinical trial phase as well as some of them are approved by FDA [31]. The lists include various drug compounds, namely Captopril, Dorzolamide, Saquinavir, Zanamivir, Oseltamivir, Aliskiren, Boceprevir, Nolatrexed, TMI-005, LY-517717, Rupintrivir and NVP-AUY922. In particular, the compounds Captopril and Aliskiren are used for treating heart disease, hypertension, and Saquinavir, Zanamivir, Oseltamivir, Rupintrivir are potential antiviral compounds (for HIV type I and type II, influenza virus and human rhinovirus).

4.3 *Limitations of Force-Field Methods and Need for an Alternative Approach*

In many occasions, force-field-based approaches were successful in explaining the ligand binding to receptors, in predicting the relative binding affinities of structurally similar ligands and in predicting the binding affinities towards various mutants of same receptors. However, many failures of these methods go unnoticed as these are not reported in general. We have noticed that the MM-GBSA and MM-PBSA methods cannot explain the relative binding affinities of indole-Substituted benzothiazoles and benzoxazoles compounds towards monoamine oxidase B and their binding specificity towards MOA-B when compared to MOA-A [32]. We have also reported that in the case of thiabendazole-based compounds the correlation between the experimental and computed binding affinities towards amyloid beta fibril using molecular docking and MM-GBSA approach when compared to quantum mechanics-based cluster model was not impressive [33].

The main reason behind is that force-field methods cannot account for the changes in the electronic structure of ligands when they are bound to the target. Usually, the charges for ligands are the same for the ligand in water as well as in the binding site of target. This is not true, the electronic structure and molecular dipole moment of the ligand can vary significantly depending upon the microenvironment [34, 35], and such polarization due to environment should be accounted for in the free energy calculations. Such a requirement automatically leads to the need for the description of the ligand using a quantum mechanical theory where the electronic degrees of freedom are treated explicitly and so the environment-specific changes in electronic structure and molecular structure can be accounted accurately [36, 37]. However, electronic structure theory is not suitable to describe protein–ligand complex systems as the number of electronic degrees of freedom is too many. So, many approximations are employed to treat the interactions between the protein–ligands in a quantum mechanical way.

5 Ab Initio Methods in Free Energy Calculations

It should be possible to calculate binding free energies using ab initio methods; however, calculation of the free energy is difficult and even intractable for large systems and an approximation is often invoked where only the energy is calculated (Eq. 6) and the temperature is assumed to be 0 K.

$$\Delta E = E_{\mathrm{Complex}} - E_{\mathrm{protein}} - E_{\mathrm{ligand}} \tag{6}$$

In this section, we briefly describe some of the known and recent developments in QM-based approaches which have been used for free energy-based drug development projects.

5.1 QM Cluster Model

In this model, the ligand and binding site residues are extracted and treated using electronic structure theory. Since the binding/catalytic site residues are mostly dictating the binding energies with ligand and the rest of the residues only play supportive role and are contributing to retain the structure of the enzyme, in particular the binding site conformation, this is reasonable approximation. Since not all the amino acids of enzyme are included in the calculation, certain approximations need to be applied. To avoid spurious charge accumulation in dangling bonds which might alter the energetics of the whole protein–ligand systems, the cut bonds are capped with hydrogens. Since the rigidity of the binding site was mostly stabilized by the rest of the protein, the free optimization of cluster might lead to unrealistic distortions in the binding site geometry. So, certain terminal residues are fixed in the space, only partial optimizations are performed, and the energies are computed for these geometries. In certain cases, the QM cluster is placed in continuum solvent to mimic the protein-like environment and the dielectric constant for the medium is chosen to be 4 [38]. There was the use of more than one quantum mechanical theory in some cluster calculations. For example as in the case of 8-Cl *TIBO* bound to human immunodeficiency virus reverse transcriptase, authors employed two-layer and three-layer ONIOM (in particular [MP2/6-31G(d), B3LYP/6-31G(d,p) and PM3]) approach to estimate the interaction energy. The residues closer to ligand are described using the high-level theory (like MP2) as these contribute to total interaction energy dominantly, while the residues far away from the ligand can be described using low-level theory as these contributions will not be very significant [39]. There are not many studies which employ this methodology to compute ligand binding energies or interaction energies with receptor [38, 40–42]. However, for modelling a number of enzymatic reactions, this method has been used successfully. In particular, the study on the enzymatic reaction of acetylene hydratase to produce vinyl alcohol using two different approaches, namely QM cluster model and QM/MM model, is worth recalling [43].

5.2 Hybrid QM/MM Approach

This approach combines the best of the two worlds, namely force-fields and electronic structure theory-based approach. Even though the receptor–ligand complex system is too large in length scale, most of the time the region of relevance to us is the ligand and certain residues that are in direct contact with the ligand. So, it is a smarter idea to split the system into two regions and use the more accurate level of theory (here, it is electronic structure theory) to describe the region of relevance and to use a relatively less accurate but cheaper (here, it is force-field) approach to describe the rest of the region. However, the harder part is the description of the interaction between these two regions or subsystems. If there is no charge transfer between these two subsystems, then one can add electrostatic and van der Waals terms to account for such interaction, and in this way the polarization of the quantum mechanical subsystem due to the system described using force-field is accounted for. The implementation is straightforward when the ligand alone is described by quantum mechanics and receptor and solvents are described using the force-fields. However, when certain residues of the receptors are to be included in the QM region, then the description of the chemical bonds between the receptor parts in QM region and MM region is a bit challenging. Methods such as hydrogen capping are developed to describe such regions, and it has become routine to use QM/MM methods for computing the protein–ligand interaction energies and free energies. Another main problem is due to the over-polarization of the terminal bonds in QM region due to atomic charges in the immediate MM region. Usually, the properties of certain atoms or groups closer to the interfacial region are moved further away into MM region so that such over-polarization is not a problem any more. Other option is to use damping function in the calculation of electrostatic contribution to deal with such effect in a mathematical way. It is also worth mentioning that the QM/MM methods in addition to energetics can be used to model the enzymatic catalytic reaction and can also be used to model the optical (linear and nonlinear) and magnetic properties of ligands when they are bound to receptors.

5.3 *Fragment Molecular Orbital*

A computationally viable strategy to evaluate the energy of an entire protein or protein–ligand complex is the fragment molecular orbital (FMO) method [44]. In the FMO method, the entire system is divided into several fragments and their energy is evaluated in the presence of all other fragments. This is known as the one-body FMO (FMO1) method. Usually, a single fragment consists of a single residue. To further enhance the quality of the calculation and include important QM effects, all pairs of fragments are evaluated in the presence of the rest of the fragments. This is known as the two-body FMO (FMO2) method. The total energy for an FMO2 calculation is given as in Eq. 7

$$E = \sum_{I}^{N} E_I + \sum_{I>J}^{N} \Delta E_{IJ} \tag{7}$$

$$\Delta E_{IJ} = E_{IJ} - E_I - E_J \tag{8}$$

where E_I is the energy of a monomer in the electrostatic potential (ESP) of all other monomers. ΔE_{IJ} is the interaction energy of fragment I and J evaluated as in Eq. 8.

5.3.1 Case study: HIV-1 RT RNase H Inhibition Screening

Many drugs or inhibitors potentially bind with metal ions in the catalytic site of enzyme or receptors in order to exhibit the therapeutic effect, e.g. magnesium ions containing enzymes such as HIV-1 integrase and RNase H [45, 46]. Thus, a good scoring function should be able to accurately calculate the metal–inhibitor interaction which impacts an overall binding affinity of individual compounds. Although the metal-binding term in the docking-scoring function is included (e.g. glide score), it considers only the anionic or highly polar interactions; therefore, ranking of actives is not appropriately achieved. On the other hand, it has been reported previously that magnesium ions in the HIV-1 reverse transcriptase-associated ribonuclease H (RNase H or RNH) play an essential role in binding and positioning of RNA–DNA duplex (natural substrate) during digestion in the viral genome reverse transcription process. Inhibition of this enzyme by chelation of magnesium ions (active site binder) is provided as an attractive approach in anti-HIV RT inhibition based drug design and discovery projects.

It is well known that the active site binder mechanism of inhibition is primarily through chelation with magnesium ions; thus, binding affinity prediction model was improved through the use of QM-based calculations by primarily considering the chelation mechanism of inhibitors with the catalytically active magnesium ions. This could be useful as a high-throughput filter in the virtual screening process.

The simplest possible model (**scenario 1**) to describe the binding of the ligand is to only describe the chelation process between the magnesium ions and the ligand in solution yielding the following approximation to Eq. 8. To further refine scenario 1, we consider in **scenario 2** geometry optimization of the protein–ligand complexes using the Qsite module (version 5.0) of the Schrödinger suite. Here, the magnesium ions and inhibitors were considered in the QM region (optimized with B3LYP and the 6-31G(d) basis set). The rest of the protein and water molecules were considered in the MM region (evaluated using the OPLS 2005 force-field) and kept frozen.

In general, docking methods could also be used for ranking compounds; however, the correlation between scoring functions and experimental values for binding free energies is rather poor in this case, and one reason is the lack of protein flexibility in the majority of the docking experiments. The correlation between molecular docking using glide score and experimental activity was quite low (n=7; $R^2 = 0.098$). However, when the atomic coordinates were used for chelation energy calculations using the DFT-B3LYP method (scenario 2; n=7; $R^2 = 0.93$) and FMO methods($R^2 = 0.80$-0.94). [47].

As we have discussed above that an effective virtual screening of RNase H inhibitors from large chemical databases could be achieved using the combination of docking and QM-based refinement calculations. In order to identify a novel chemotype for RNase H inhibition and to validate previously developed computational methods, the best models were used to screen the Specs database (containing 277,325 drug-like compounds for purchase) for HIV-1 RNase H inhibition screening (Fig. 3). A set of 1205 compounds was obtained at the end of the docking-based virtual screening, and these compounds were subsequently used for

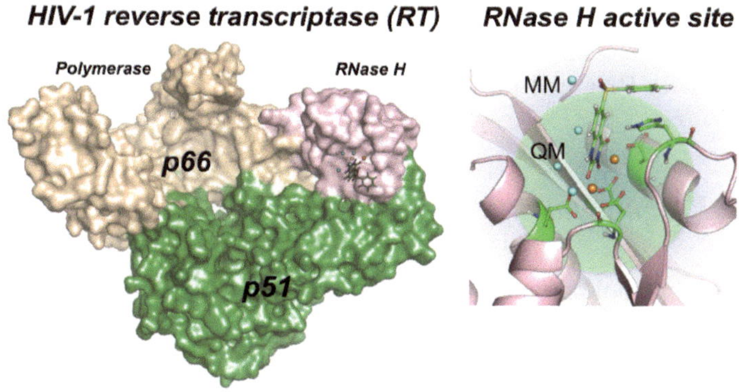

Fig. 3 Structure of HIV-1 reverse transcriptase enzyme and its active site. The components of the QM and MM region for geometry optimization are shown

QM-based refinement calculation based on density functional theory (DFT) calculations as described above (Eq. 8). The best-ranked 180 compounds from the screening were sorted for further inspection. To select a diverse set of structures for the biochemical assays, these compounds were clustered according to the structural similarity. Of the 50 structural clusters, 25 structurally diverse compounds, with the best scores, were chosen and purchased from the chemical vendor (www.specs.net) to be tested against the HIV RT-associated RNase H function in enzymatic assays. The overall workflow of the virtual screening process is shown in Fig. 4. Out of 25 compounds tested, 3 compounds inhibited the RNase H activity

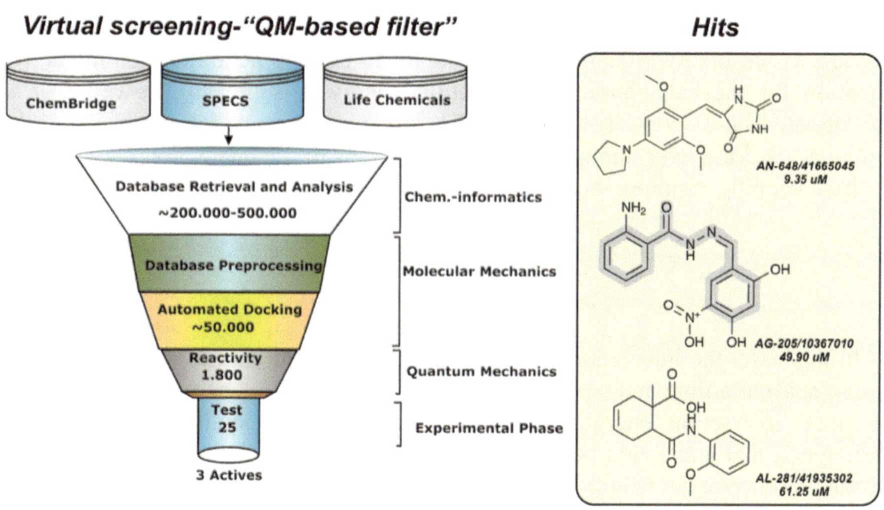

Fig. 4 Overall workflow of structure-based virtual screening strategies applied. Initial hit molecules from the first screening are provided and one of the compound is highlighted in regions where it shares a common structural pattern with known RNase H inhibitor BHMP07

below an IC_{50} value of 100 μM and compound AN-648/41665045 showed an IC_{50} value of 9.35 μM. Notably, none of these compounds has previously been reported as an inhibitor for RNase H [48]. (Fig. 4).

5.4 QM Fragmentation Approach

In this approach, whole protein is fragmented into individual amino acids, and the fragment-wise interaction energies with ligand are calculated and added together to get the total interaction energies as in the equation below. Since the whole protein is broken into individual fragments, the size of the protein is not a problem any more and a high-level electronic structure theory such as Møller–Plesset perturbation theory or coupled cluster method that accounts for electronic correlation explicitly can be used to compute the subsystem interaction energies [49].

$$\Delta E = \sum_{i=1}^{n} \Delta E_{(\mathrm{Ai-ligand})} \tag{9}$$

where Ai is the ith amino acid in a receptor, n is the total number of amino acids, and Ai-ligand refers to the ith residue–ligand complex. The ΔE is interaction energy between the ith residue and the ligand, which itself is computed as below:

$$\Delta E_{\mathrm{Ai-ligand}} = E_{\mathrm{Ai-ligand}} - E_{\mathrm{Ai}} - E_{\mathrm{ligand}} \tag{10}$$

The amino acids are cut along the peptide bond and capped either with hydrogens or with other functional groups to mimic protein-like environment around the residue. When the hydrogen atoms are employed as capping atom, then the above equation for the calculation of interaction energy is sufficient. However, it is appropriate to use $-NHCH_3$ and $-CO-CH_3$ as capping groups for either side of the amino acids. Moreover, the additional contributions to the interaction energies due to these capping residues should be removed as below:

$$E_{p-L} = \sum_{k=1}^{N-2} E_{F_k-L} - \sum_{k=1}^{N-3} E_{CC_k-L} - \sum_{k=1}^{N-2} E_{Fk} + \sum_{k=1}^{N-3} E_{CC_k} - E_L \tag{11}$$

In this case, the interactions are due to two molecular entities (a ligand and an amino acid) at a time and so we completely ignore the three-body contributions to the total interaction energies. In other words, this is similar to making an assumption that interaction between an amino acid and the ligand is not modulated by the presence of the neighbouring amino acids (or fragments). However, by doing additional calculations for estimating the interaction energies of dipeptide (or in units of two amino acids) and ligand at a time, such three-body contributions can be included. The expression for interaction energy is now a bit more complicated and

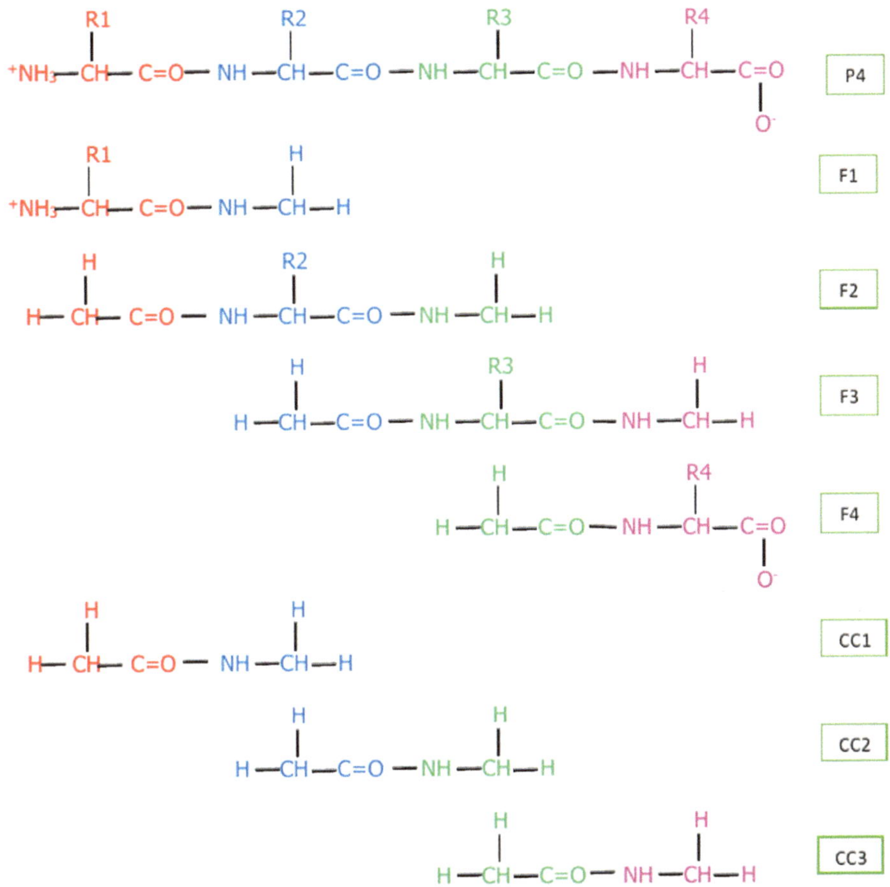

Scheme 1 Construction of various capped fragments for a peptide made of four amino acids (and so three peptide bonds). As can be seen, there are eventually four fragments (referred to F1, F2, F3 and F4). Each peptide bond can be capped with three pairs of –CO–CH_3 and –NH–CH_3 groups, so there are three conjugate caps (referred to CC1, CC2 and CC3), and the interactions of these with ligands should be removed as these are counted twice. (Note the positive sign for these contributions in the equation above.) It can be seen for a peptide with n amino acids there can be $n - 1$ fragments formed and $n - 2$ conjugate gaps possible if we fragment them using a scheme shown above

involves the calculation of trimer (two residues and a ligand), dimer (one residue and ligand) and monomer energies.

QM fragmentation energies can be further made sophisticated by computing the individual monomer, dimer, trimer energies with an embedding scheme which allows the interaction between these fragments with the rest of the protein through an effective Hamiltonian. This part is methodologically very similar to the above-discussed QM/MM approach where the QM system interacts with MM subsystem through electrostatic and van der Waals interaction. However, care

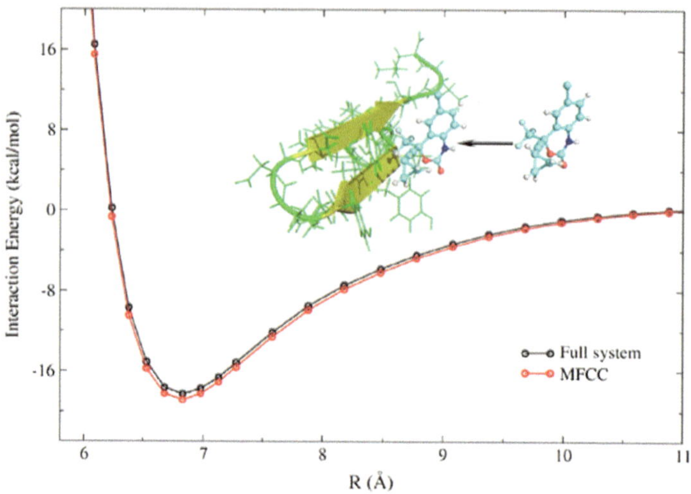

Fig. 5 Interaction energy calculated using M062X/6-311G** for Efavirenz with a fragment of HIV-1 reverse transcriptase containing residues in the range from Asn175 to Leu193 of chain A. Reprinted (adapted) with permission from (Acc. Chem. Res., **2014**, 47 (9), pp 2748–2757). Copyright (2014) American Chemical Society

should be taken to make sure that certain subsystem interactions are not double counted or in general over-counted (Scheme 1).

QM fragmentation scheme has been employed successfully to compute the interaction of ligands with various drug targets. Interestingly, certain studies showed that the computed interaction energy is comparable to that of QM cluster model. Figure 5 compares the interaction energy of Efavirenz with HIV NNRT target based on the two approaches, namely QM cluster and QM fragmentation schemes [50]. As can be seen, the interaction energy as obtained from QM fragmentation scheme agrees well with the full QM model suggesting that the former scheme is accurate as well as quite inexpensive.

Recently, it has been shown [51] that QM fragmentation method was able to correctly reproduce the relatively larger binding affinity of a tracer, FDDNP towards tau fibril when compared to amyloid beta fibril. In contrary, the MM-GBSA-based method predicted that FDDNP has a larger binding affinity towards amyloid beta fibril which is not in agreement with experimental binding affinity data. As shown in Fig. 6, it is necessary to include interaction energy of the ligand with water (within a cut-off of 15 Å) with that of its interaction with protein residues to correctly reproduce the experimental binding affinity data.

QM fragmentation scheme has been applied to compute not only protein–ligand interaction energies but also solvation energy, molecular electrostatic potential and properties such as NMR chemical shifts of the ligands in solvent and bio-environment [50]. Even the electron density of the whole biomolecule can be obtained using this approach.

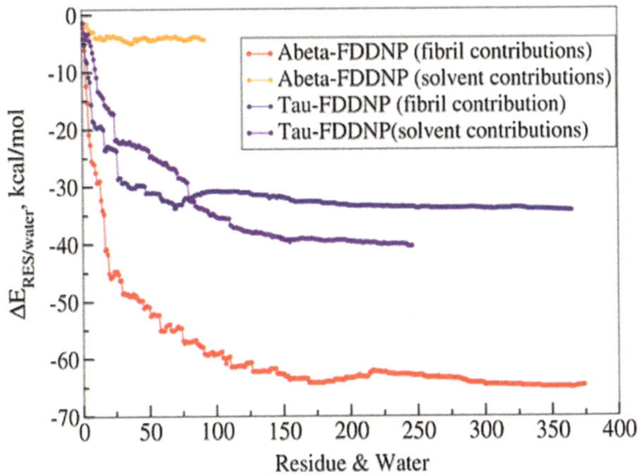

Fig. 6 Total interaction energy between the FDDNP tracer and amyloid and tau fibrils with increasing number of residues (related to increased cut-off). Also, the interaction energy of tracer with solvents located near the binding site is shown with increasing number of solvent. The residues and water solvents were first arranged with increased distance from the tracer centre of mass, and their contributions were computed and added to the total interaction energy. As can be seen with inclusion of around 125 residues, the major part of interaction energy with amyloid and tau fibril is retrieved. The figure has been reproduced with peermission from (ACS Chem. Neurosci., **2018**, 9 (7), pp 1757–1767). Copyright (2018) American Chemical Society

6 Entropic Contributions in the QM-Based Free Energy Calculations

So far in all the electronic structure theory-based approaches, we have only seen how to compute the interaction energies between a receptor and a ligand. However, the quantity of interest is the free energy of binding and not the interaction energy. For this, we also need to add the entropic contributions. The translational, rotational and vibrational contributions to the entropies are computed from the translational, rotational and vibrational partition functions as given in the reference by Yu et al. [52]. The translational and rotational contributions to the protein–ligand association are usually positive, while the vibrational contributions favour the association process. The vibrational contribution has been often reported to be much smaller in quantity when compared to the translational and rotational contributions. In some cases, we have noticed that the addition of translational and rotational contributions to total interaction energy yielded positive binding free energies. The computation of absolute free energy (including all these different entropic contributions) still remains as a challenge as there are no detailed benchmarking studies on the estimation of the translational and rotational contributions and their relative contributions to binding free energies.

7 Machine Learning (ML)-Based Approaches for Drug Discovery

Application of machine learning (ML) methods to problems in chemistry, biology, materials, etc., has taken a huge leap during the last few years. Specifically, a number of problems related to accurate intermolecular potentials, [53] drug design, [54] protein–protein interaction, [55] viable retrosynthetic pathways, [56] stability of solids, [57] potential energy surfaces [58], etc., are being addressed [59]. Advances that are being made in this space in terms of tackling problems in a way that was not thought about even few years are rapid, and the number of papers that are being published in this area is increasing exponentially. Unlike in the most research areas of science and technology, traditional ML methods such as single-layer neural networks or random forest have been applied in the area of computer-aided drug design long time ago. However, modern deep learning methods within ML are expected to make significant contributions to the area of drug design in the coming days [54, 60, 61]. Given that the last fifteen years have witnessed prolific generation of experimental data in terms of synthesizable compounds, their pharmacodynamic and pharmacokinetic properties, application of data-driven methods is likely to advance the field significantly. The following sections give a brief account of ML and some of the recent successes in application of ML methods in areas relevant to various drug design projects, including off-targets [62].

There are two fundamentally different methods in ML: supervised and unsupervised learning. Given a large data of inputs and outputs, supervised learning methods try to learn a function so that given a set of inputs, output may be predicted. Supervised learning methods such as the artificial neural networks (ANNs) are pertinent in quite a few drug discovery applications. On the other hand, unsupervised learning methods learn structure within the data when only inputs are available, which are typically applied dimensional reduction, pattern recognition, etc. Most of the ML methods are based on ANNs that connect the input and the output layers via an interconnected neural network (hidden layer(s)). The ANNs consist of a number of layers with each containing a number of neurons. Output of one of the layers is taken as input of the next layer, and the output values are calculated using an activation function. Fully connected deep neural network (DNN), recurrent neural network (RNN), convolutional neural network (CNN) and autoencoders are some of the variants of ANNs that are very successful as efficient methods for statistical modelling in a variety of fields. For a detailed account of different machine learning methods relevant to drug discovery, the readers may refer to the review by Lavecchia [63].

7.1 Structure-Based ML Approaches in Drug Design

As explained in previous sections, molecular recognition is a fundamental phenomenon behind all biological processes and in drug binding. While the number of drug-like molecules that could be synthesized is estimated to be around 10^{60}, the current experimental techniques cannot possibly screen all of these within reasonable time and expense. Computational methods such as docking calculations address this to some extent; however, the accuracy of the scoring functions behind these algorithms is still not good enough to efficiently narrow the search space that can be explored by experiments. Recently, it has been shown that machine learning (ML)-based scoring functions can predict binding affinities better than the classical scoring functions that are primarily used in computer-aided drug design [64]. Wojcikowski et al. recently reported a systematic study on the performance of ML-based scoring functions and compared it to well-known established methods [65]. They proposed a scoring function based on the random forest method (RF-Score-VS) that was trained on about 15,000 active and 900,000 inactive molecules against about 100 different drug targets . However, the authors do indicate that use of better molecular representations and descriptors will further increase the success of machine learning scoring functions. In addition, Kinnings et al. also showed that the support vector machines (SVMs) can be used to improve the performance of scoring functions. They constructed two prediction models; one is a regression model to predict the IC_{50} values, and the second is a classification model that was shown to perform very well across the entire data set [66]. Ragoza et al. have proposed a CNN-based model for scoring functions that can be used in structure-based drug design [67]. The model used the existing three-dimensional structures of protein–ligand complexes to train a model that predicts the binding affinity corresponding to any protein and ligand. The model was systematically trained by including a series of structural and binding variations such as high affinity binders, low affinity binders, correct binding pose and incorrect poses. They found that the scoring function obtained based on the CNN algorithm performs significantly better than AutoDock Vina in terms of predicting both binding poses and affinities. Recently, Dror and co-workers proposed a method named Siamese Atomic Surfacelet Network (SASNet) that applies CNN to predict protein–protein binding interfaces with high accuracy compared to the previously available knowledge-based and ML-based methods [58]. Interestingly, the training was done on a biased data where binding-driven conformational changes are not part of the data set; however, the model was shown to perform very well suggesting that the model has possibly learned the inherent structural and dynamic properties of proteins in general. In addition to the traditional neural network-based algorithms, there are other deep learning methods such as reinforcement learning that have been found to be very effective in drug design. Reinforcement learning is based on two neural networks, namely the generative and predictive neural networks. Popova et al. have recently proposed, Reinforcement Learning for Structural Evolution (ReLeaSE), a de novo method based on reinforcement learning [68]. Initially, the generative and predictive networks are trained individually using one of the supervised learning methods followed by training of both models together. This allows for predicting new chemical structures with desired biological activities. They have shown test cases by generating libraries of molecules with desired melting points, hydrophobicity and biological activities. They propose that it is possible to use a similar approach for optimization of multiple properties such as biological

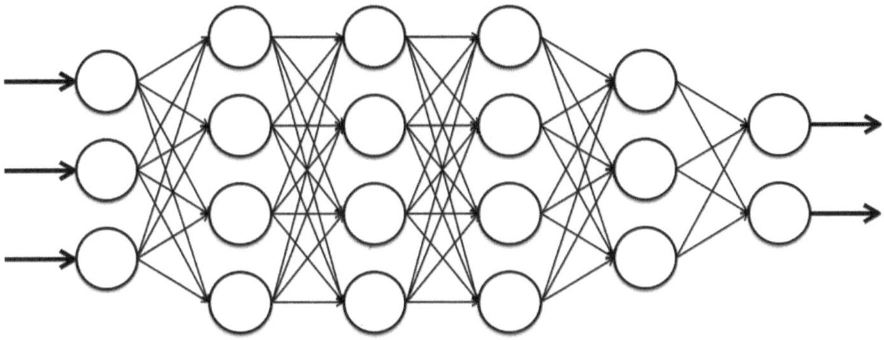

Fig. 7 Schematic representation of a multi-layer feed forward ANN

activity and different ADMET properties simultaneously to identify small molecules with desired pharmacodynamic and pharmacokinetic properties at the same time. Machine learning methods are also being used in ligand-based drug design projects . The main goal of ligand-based drug design activities is to predict how a chemical structure can be modified to achieve desired biological activity and/or ADMET properties. The aim of QSAR, one of the major methodologies in ligand-based drug design, is to generate a predictive regression model that gives a relationship between biological activity (or any other property) and a set of molecular descriptors. Such an exercise is inherently very suitable for traditional machine learning algorithms, and hence it has been adopted very early [69]. Supervised learning algorithms such as neural networks, random forest, SVMs and k-nearest neighbour have been used in QSAR [55, 60, 61]. Similarly, application of unsupervised methods such as clustering methods, principal component analysis and independent component analysis has been successful. Schematic representation of ANN workflow is shown in Fig. 7.

7.2 *Future Prospects of AI-ML in Drug Design*

Machine learning methods have been used in ligand-based drug design for a long time and have been reasonably successful. During the last five years, applications of deep learning algorithms have showed a lot of promise in terms of their superior performance compared to traditional ML methods used in drug design. The rate of advance of computational methodologies that are traditionally applied to drug design seems to be far lower than the advances that are being made by machine learning-based methods. The availability of high-quality data, improved biophysical experimental techniques, increasing computational resources/power and faster evolution of machine learning methods such as deep learning are further pushing the drug design efforts in right direction. Although the efforts seem to be fragmented at this point of time, further involvement of research groups with varied backgrounds and availability of clean data are expected to inspire emergence of more efficient workflows in drug design that combine traditional methods and machine learning methods.

8 Conclusions

The current drug discovery projects can be benefited a lot from advancements in the structure elucidation methods such as cryogenic electron microscopy, NMR spectroscopy, X-ray crystallograpy and from computational free energy calculation methods. This chapter presents various computational approaches available for estimating the free energies of drugs in different environments which surrogate various components of fate of drugs in the biosystem and this not only limited to drug binding to its targets but also other interactions and its relevant properties e.g. ADMET. We present various free energy calculation methods which use force-field, semi-empirical and ab initio electronic structure theory-based methods. Until a decade ago, using the electronic structure theory method for studying the structure and energetics of biomacromolecule was formidable. Thanks to the fragmentation and effective Hamiltonian approaches, it is possible to employ these methods for computing the interaction energy between ligand and biomacromolecular fragments reliably. Here, various working principles of these approaches along with key illustrative examples are presented. Even though the methods appear very promising for computing the free energy of the ligands in solvent or in biomacromolecular (such as enzyme, membrane, fibril, DNA and RNA) environment, we need to systematically study various receptor–ligand systems and test for their ability to reproduce experimental binding affinity and other pharmacokinetic parameters before employing them as lead compounds in drug discovery projects. While physics-based methods such as those mentioned above are important and unavoidable, alternative approaches based on machine learning algorithms that exploit existing experimental/computational data are emerging to be powerful tools for drug design. We expect that elegant combination of traditional physics-based methods, better computational power and more sophisticated machine learning algorithms will enable efficient and accurate quantification of protein–ligand binding affinities for improved lead identification/optimization processes in the drug design and discovery projects.

References

1. Rask-Andersen M, Almen MS, Schioth HB (2011) Trends in the exploitation of novel drug targets. Nat Rev Drug Discov 10:579–590
2. Lenz GR, Nash HM, Jindal S (2000) Chemical ligands, genomics and drug discovery. Drug Discov Today 5(4):145–156
3. Knowles J, Gromo G (2003) Target selection in drug discovery. Nat Rev Drug Discov 2: 63–69

4. Kubinyi H (2003) Drug research: myths, hype and reality. Nat Rev Drug Discov 2(8):665–668
5. Hodgson John (2001) ADMET-turning chemicals into drugs. Nat Biotechnol 19(8):722
6. Caldwell GW (2000) Compound optimization in early-and late-phase drug discovery: acceptable pharmacokinetic properties utilizing combined physicochemical, in vitro and in vivo screens. Curr Opin Drug Discov Devel 3(1):30–41
7. Zamora I, Oprea T, Cruciani G, Pastor M, Ungell AL (2003) Surface descriptors for protein — ligand affinity prediction. J Med Chem 46(1):25–33
8. De Waterbeemd Van, Han Eric Gifford (2003) ADMET in silico modelling: towards prediction paradise. Nat Rev Drug Discov 2(3):192–204
9. Colmenarejo G (2003) Insilico prediction of drug-binding strengths to human serum albumin. Med Res Rev 23(3):275–301
10. Guengerich FP (2006) Cytochrome P450 s and other enzymes in drug metabolism and toxicity. AAPS J 8(1):E101–E111
11. Vasanthanathan P, Hritz J, Taboureau O, Olsen L, Jorgensen FS, Vermeulen NPE, Oostenbrink C (2009) Virtual screening and prediction of site of metabolism for cytochrome P450 1A2 ligands. J Chem Inf Model 49:43–52
12. Vasanthanathan P, Olsen L, Jorgensen FS, Vermeulen NPE, Oostenbrink C (2010) Calculation of Binding Free Energy for CYP1A2 Ligands by Using Empirical Free Energy Method. Drug Metab Dispos 38:1347–1354
13. Leung SS, Mijalkovic J, Borrelli K, Jacobson MP (2012) Testing physical models of passive membrane permeation. J Chem Inf Model 52(6):1621–1636
14. Westergren J, Lindfors L, Höglund T, Lüder K, Nordholm S, Kjellander R (2007) In silico prediction of drug solubility: 1. Free energy of hydration. J Phys Chem 111(7):1872–1882
15. Rossi Sebastiano M, Doak BC, Backlund M, Poongavanam V, Over B, Ermondi G, Caron G, Matsson P, Kihlberg J (2018) Impact of dynamically exposed polarity on permeability and solubility of chameleonic drugs beyond the rule of 5. J Med Chem 61(9):4189–4202
16. Wan J, Zhang L, Yang GF, Zhan CG (2004) Quantitative structure-activity relationship for cyclic imide derivatives of protoporphyrinogen oxidase inhibitors: a study of quantum chemical descriptors from density functional theory. J Chem Inf Comput Sci 44:20
17. Hopfinger AJ, Pearlstein RA (1984) Molecular mechanics force-field parameterization procedures. J Comput Chem 5(5):486–99.99–2105
18. Kollman PA, Massova I, Reyes C, Kuhn B, Huo SH, Chong L, Lee M, Lee T, Duan Y, Wang W, Donini O, Cieplak P, Srinivasan J, Case DA, Cheatham TE (2000) Calculating structures and free energies of complex molecules: Combining molecular mechanics and continuum models. Acc Chem Res 33:889–897
19. Genheden S, Ryde U (2015) The MM/PBSA and MM/GBSA methods to estimate ligand-binding affinities. Expert Opin Drug Discov 10:449–461
20. Wang W, Donini O, Reyes CM, Kollman PA (2001) Biomolecular simulations: Recent developments in force fields, simulations of enzyme catalysis, protein-ligand, protein-protein, and protein-nucleic acid noncovalent interactions. Annu Rev Biophys Biomol Struct 30:211–243
21. Massova I, Kollman PA (2000) Combined molecular mechanical and continuum solvent approach (MM-PBSA/GBSA) to predict ligand binding. Perspect Drug Discov 18:113–135
22. Wang C, Greene DA, Xiao L, Qi R, Luo R (2018) Recent developments and applications of the MMPBSA method. Front Mol Biosci 10(4):87
23. Wang J, Morin P, Wang W, Kollman PA (2001) Use of MM-PBSA in reproducing the binding free energies to HIV-1 RT of TIBO derivatives and predicting the binding mode to HIV-1 RT of efavirenz by docking and MM-PBSA. J Am Chem Soc 123(22):5221–5230
24. Meng XY, Zhang HX, Mezei M, Cui M (2011) Molecular docking: a powerful approach for structure-based drug discovery. Curr Comput Aided Drug Des 7(2):146–157
25. Jain AN (2006) Scoring functions for protein-ligand docking. Curr Protein Pept Sci 7:407–420
26. Stahl M, Rarey M (2001) Detailed analysis of scoring functions for virtual screening. J Med Chem 44:1035–1042

27. Kellogg GE. (2006) In: Ekins S (ed) Computer applications in pharmaceutical research and development. Wiley, Hoboken, NJ
28. Binda C, Wang J, Pisani L, Caccia C, Carotti A, Salvati P, Edmondson DE, Mattevi A (2007) Structures of human monoamine oxidase B complexes with selective noncovalent inhibitors: safinamide and coumarinanalogs. J Med Chem 50(23):5848–5852
29. De Colibus L, Li M, Binda C, Lustig A, Edmondson DE, Mattevi A (2005) Three-dimensional structure of human monoamine oxidase A (MAO A): relation to the structures of rat MAO A and human MAO B. Proc Natl Acad Sci 102(36):12684–12689
30. Ruddigkeit L, Van Deursen R, Blum LC, Reymond JL (2012) Enumeration of 166 billion organic small molecules in the chemical universe database GDB-17. J Chem Inf Model 52 (11):2864–2875
31. Talele TT, Khedkar SA, Rigby AC (2010) Successful applications of computer aided drug discovery: moving drugs from concept to the clinic. Curr Top Med Chem 10(1):127–141
32. Nam MH, Park M, Park H, Kim Y, Yoon S, Sawant VS, Choi JW, Park JH, Park KD, Min SJ, Lee CJ (2017) Indole-substituted benzothiazoles and benzoxazoles as selective and reversible MAO-B inhibitors for treatment of Parkinson's disease. ACS Chem Neurosci 8(7):1519–1529
33. Balamurugan K, Murugan NA, Ågren H (2016) Multistep modeling strategy to improve the binding affinity prediction of PET tracers to Aβ42: case study with styrylbenzoxazole derivatives. ACS Chem Neurosci 7(12):1698–1705
34. Murugan NA, Aidas K, Kongsted J, Rinkevicius Z, Agren H (2012) NMR spin-spin coupling constants in polymethine dyes as polarity indicators. Chem Eur J 18:11677–11684
35. Murugan NA, Kongsted J, Rinkevicius Z, Agren H (2012) Color modeling of protein optical probes. Phys Chem Chem Phys 14:1107–1112
36. Ryde U, Soderhjelm P (2016) Ligand-binding affinity estimates supported by quantum-mechanical methods. Chem Rev 116:5520–5566
37. Cavalli A, Carloni P, Recanatini M (2006) Target-related applications of first principles quantum chemical methods in drug design. Chem Rev 106:3497–3519
38. Nikitina E, Sulimov V, Zayets V, Zaitseva N (2004) Semiempirical calculations of binding enthalpy for protein—ligand complexes. Int J Quantum Chem 97:747–763
39. Saen-oon S, Kuno M, Hannongbua S (2005) Binding energy analysis for wild-type and Y181C mutant HIV-1 RT/8-Cl TIBO complex structures: Quantum chemical calculations based on the ONIOM method. Proteins Struct Funct Bioinf 61(4):859–869
40. Perakyla M, Pakkanen TA (1994) Quantum mechanical model assembly study on the energetics of binding of arabinose, fucose, and galactose to L-arabinose-binding protein. Proteins Struct Funct Genet 20:367–372
41. Perakyla M, Pakkanen TA (1995) Model assembly study of the ligand binding by p-hydroxybenzoate hydroxylase: correlation between the calculated binding energies and the experimental dissociation constants. Proteins Struct Funct Genet 21:22–29
42. Nikitina E, Sulimov V, Grigoriev F, Kondakova O, Luschekina S (2006) Mixed implicit/explicit solvation models in quantum mechanical calculations of binding enthalpy for protein—ligand complexes. Int J Quantum Chem 106:1943–1963
43. Liao RZ, Thiel W (2012) Comparison of QM-only and QM/MM models for the mechanism of tungsten-dependent acetylene hydratase. J Chem Theory Comput 8(10):3793–3803
44. Fedorov DG, Kitaura K (2007) Extending the power of quantum chemistry to large systems with the fragment molecular orbital method. J Phys Chem 111:6904–6914
45. Klumpp K, Hang JQ, Rajendran S, Yang Y, Derosier A et al (2003) Two metal ion mechanism of RNA cleavage by HIV RNase H and mechanism-based design of selective HIV RNase H inhibitors. Nucleic Acids Res 31:6852–6859
46. Budihas SR, Gorshkova I, Gaidamakov S, Wamiru A, Bona MK et al (2005) Selective inhibition of HIV-1 reverse transcriptase-associated ribonuclease H activity by hydroxylatedtropolones. Nucleic Acids Res 33:1249–1256
47. Poongavanam V, Steinmann C, Kongsted J (2014) Inhibitor ranking through QM based chelation calculations for virtual screening of HIV-1 RNase H inhibition. PLoS ONE 9(6): e98659

48. Poongavanam V, Corona A, Steinmann C, Scipione L, Grandi N, Pandolfi F, Santo RD, Esposito F, Tramontano E, Kongsted J (2018) Structure-guided approach identifies a novel class of HIV-1 ribonuclease H inhibitors: binding mode insights through magnesium complexation and site-directed mutagenesis studies. Med Chem Comm 9:562–575
49. Zhang Lei, Li Wei, Fang Tao, Li Shuhua (2017) accurate relative energies and binding energies of large ice-liquid water clusters and periodic structures. J Phys Chem 121(20):4030–4038
50. He X, Zhu T, Wang X, Liu J, Zhang JZ (2014) Fragment quantum mechanical calculation of proteins and its applications. Acc Chem Res 47(9):2748–2757
51. Murugan NA, Nordberg A, Ågren H (2018) Different positron emission tomography tau tracers bind to multiple binding sites on the tau fibril: insight from computational modeling. ACS Chem Neurosci 9 (7):1757–1767
52. Yu YB, Privalov PL, Hodges RS (2001) Contribution of translational and rotational motions to molecular association in aqueous solution. Biophys J 81(3):1632–1642
53. von Lilienfeld OA (2018) Quantum machine learning in chemical compound space. Angew Chem Int Ed 57(16):4164–4169
54. Chen H, Engkvist O, Wang Y, Olivecrona M, Blaschke T (2018) The rise of deep learning in drug discovery. Drug Disco Today 23(6):1241–1250
55. Townshend RJ, Bedi R, Dror RO (2018) Generalizable protein interface prediction with end-to-end learning. arXiv preprint arXiv:1807.01297
56. Segler MH, Preuss M, Waller MP (2018) Planning chemical syntheses with deep neural networks and symbolic AI. Nature 555(7698):604
57. Butler KT, Davies DW, Cartwright H, Isayev O, Walsh A (2018) Machine learning for molecular and materials science. Nature 559(7715):547
58. Smith JS, Isayev O, Roitberg AE (2017) ANI-1: an extensible neural network potential with DFT accuracy at force field computational cost. Chem Sci 8(4):3192–3203
59. Chattopadhyay A, Zheng M, Waller MP, Priyakumar UD (2018) A probabilistic framework for constructing temporal relations in replica exchange molecular trajectories. J Chem Theory Comput 14(7):3365–3380
60. Klepsch F, Vasanthanathan P, Ecker GF (2014) Ligand and structure-based classification models for Prediction of P-glycoprotein inhibitors. J Chem Inf Model 54:218–229
61. Poongavanam V, Kongsted J (2013) Virtual screening models for prediction of HIV-1 RT associated RNase H inhibition. PLoS ONE 16(8):e73478. https://doi.org/10.1371/journal.pone.0073478
62. Vasanthanathan P, Lastdrager J, Oostenbrink C, Commandeur JNM, Vermeulen NPE, Jørgensen FS, Olsen Lars (2011) Identification of CYP1A2 ligands by structure-based and ligand-based virtual screening. Med Chem Comm 2:853–859
63. Lavecchia A (2015) Machine-learning approaches in drug discovery: methods and applications. Drug Disco Today 20(3):318–331
64. Colwell LJ (2018) Statistical and machine learning approaches to predicting protein-ligand interactions. Curr Opin Struct Biol 49:123–128
65. Wójcikowski M, Ballester PJ, Siedlecki P (2017) Performance of machine-learning scoring functions in structure-based virtual screening. Sci Rep 7:46710
66. Kinnings SL, Liu N, Tonge PJ, Jackson RM, Xie L, Bourne PE (2011) A machine learning-based method to improve docking scoring functions and its application to drug repurposing. J Chem Inf Model 51(2):408–419
67. Ragoza M, Hochuli J, Idrobo E, Sunseri J, Koes DR (2017) Protein-ligand scoring with convolutional neural networks. J Chem Inf Model 57(4):942–957
68. Popova M, Isayev O, Tropsha A (2018) Deep reinforcement learning for de novo drug design. Sci Adv 4(7):eaap7885
69. Lo YC, Rensi SE, Torng W, Altman RB (2018) Machine learning in chemoinformatics and drug discovery. Drug Disco Today 23(8):1538–1546.M

Integrated Chemoinformatics Approaches Toward Epigenetic Drug Discovery

Saurabh Loharch, Vikrant Karmahapatra, Pawan Gupta, Rethi Madathil and Raman Parkesh

Abstract Epigenetics has become an important field of research in drug discovery. Epigenetic mechanisms are dynamic in nature and play a fundamental role in cellular processes. Dysregulation of epigenetic events, including cross-talk between DNA methylation and histone modifications, not only affects gene expression but also causes pathophysiological effects leading to cancer, aging, cardiovascular, neurological, and metabolic disorders. Epigenetic targets have captured the attention of researchers from diverse backgrounds to identify potential drugs for various diseases. However, drug development is a complex, time-consuming process and challenged by the high attrition rate. As with many chemotherapeutics, it is pertinent to avoid possible risk factors in epigenetic drug discovery. In this context, computational approaches can rationally guide the search for active compounds by utilizing the accumulated epigenetics knowledge base. In this chapter, we have described the chemoinformatic strategies that can be applied to facilitate the early-stage lead discovery in epigenetics, based on current best practices.

Keywords Chemoinformatics · Epigenetics · Drug discovery · Screening library Scaffolds · Chemical space

Abbreviations

ATP	Adenosine triphosphate
CTCL	Cutaneous T cell lymphoma
DNA	Deoxyribonucleic acid
DNMT	DNA methyltransferase
FDA	Food and Drug Administration
HDAC	Histone deacetylases
miRNA	MicroRNA
RNA	Ribonucleic acid

S. Loharch · V. Karmahapatra · P. Gupta · R. Madathil · R. Parkesh (✉)
Advanced Protein Science Building, Institute of Microbial Technology,
Chandigarh 160036, India
e-mail: rparkesh@imtech.res.in

C. G. Mohan (ed.), *Structural Bioinformatics: Applications in Preclinical Drug Discovery Process*, Challenges and Advances in Computational Chemistry and Physics 27, https://doi.org/10.1007/978-3-030-05282-9_8

1 Introduction

Discovery and development of new drugs is a complex, expensive, and time-consuming expedition that requires high R&D costs and extensive clinical testing. Developing a drug may take a span of 10–15 years, with a cost of several hundred million dollars [1]. Hitherto 90% of lead molecules end up in failure [2] owing to a number of factors including problematic functionalities, solubility at relevant concentrations, toxicity, off-target effects, etc. Epigenetic drug discovery is emerging as a promising therapeutics for small-molecule modulation of various metabolic, neurodegenerative, and cardiovascular diseases. The field of epigenetics is relatively new and growing quickly. The word epigenetics is derived from Greek prefix: 'ἐπί-' meaning 'above,' and thereby, the term 'epigenetics' literally describes the regulation at a level above genetic mechanisms. While all cells in an organism share identical genome, they are able to maintain unique physical characteristics and biological functions. The characteristics of a cell are determined by DNA sequence as well as gene expression pattern. Since DNA sequence remains the same in all cell types, the key role in cell fate is determined by epigenetics. Many non-specific external stimuli such as temperature, viral infections, bacteria, and diet can affect the DNA packing [3, 4] and as a result influence epigenetic states, impacting cellular phenotype without disrupting nucleotide sequence. Epigenetic changes are responsible for the cellular plasticity and enable cellular reprogramming and environmental responses. Epigenetic mechanisms play the critical role in diseases related to diet, lifestyle, environmental exposures to chemicals, toxins, etc., and, thus, offer a robust platform to explore therapeutic potential in various diseases (Table 1). The epigenetic states of a cell are dynamic in nature and can be manipulated by targeting the molecular factors associated with a disease. Chromatin remodeling via ATP-dependent processes, regulation by noncoding RNAs, DNA methylation, histone acetylation, and histone methylation are some of the key mechanisms involved in epigenetic gene regulation [5]. The recent advances in epigenetic mechanisms, gene expression control, and cellular functions have prompted the researchers to develop small-molecule inhibitors to target these processes. The successful approval of many epigenetic drugs is timely and promising. Pharmaceutical companies endorse this growing interest with huge investments in order to explore new epigenetic drugs. According to recent market reports, epigenetic drugs and diagnostic technology market are estimated to be worth US$5.7 billion by 2018 [6]. Many existing drugs significantly affect the epigenetic events [7, 8] during its disease-modifying action and further validate its uniqueness as drug target. A recent study has shown that of all the FDA-approved drugs, 1%, show significant epigenetic activity [9].

Despite the successful approval of epi-drugs, the druggability of many of the epigenetic modulators remains challenging. Many factors such as selectivity, poly-pharmacology, drug combination, toxicity, and target 'confidence' need to be addressed. The new generation of epi-drugs is expected to be more selective and specific with defined drug targets. Considering the increasing rate of failures in drug

Table 1 Epigenetic control in different diseases

Disease	Target[a]	Alteration	References
Cancer	DNMT1, DNMT3, TET, GCN5, p300/CBP, MYST, HDAC1, HDAC2, HDAC3, HDAC6, HDAC9, SIRT1, SIRT2, SIRT3, SIRT6, SIRT7, EZH2, DOT1L, SETDB1, NSD1, LSD1, JMJD2C, and JMJD3	Global DNA hypomethylation, CpG island hypermethylation, hypoacetylation, and hypermethylation	[10–12]
Aging	DNMT1, TET, SET2, p300/CBP, SUV39H1, SIRT1, SIRT6, HDAC4, HDAC5, HDAC7, KDM2, and KDM7	Global DNA hypomethylation, CpG island hypermethylation, aberrant histone acetylation and methylation	[13, 14]
Diabetes	DNMT1, p300/CBP SET7/9	DNA hypermethylation of the PPARGC1A[b] promoter in pancreatic islets Insulin gene displays hyperacetylation and hypermethylation in β cells	[15]
Parkinson's disease	DNMT1, HDAC1, SIRT2	Global DNA hypomethylation, aberrant histone acetylation	[10, 16, 17]
Alzheimer's disease	DNMT3B, HDAC2, HDAC6	CpG island hypermethylation, aberrant histone acetylation	[10, 16, 17]
Huntington's disease	p300/CBP, ESET	Aberrant histone acetylation and methylation	[10, 18]
Multiple sclerosis	DNMT1, TET2, HDAC1, HDAC2	CpG island hypomethylation	[10, 19]
Systemic lupus erythematosus	DNMT1, DNMT3a, DNMT3b, HDAC2, HDAC7, MAPK	CpG island hypomethylation, aberrant acetylation and aberrant phosphorylation	[10, 20–22]
Rheumatoid arthritis	DNMT1, HDAC4	Global DNA hypomethylation, CpG island hypermethylation	[10, 23]

[a]DNMT1 (DNA methyltransferase 1), DNMT3 (DNA methyltransferase 3), DNMT3a (DNA methyltransferase 3a), DNMT3b (DNA methyltransferase 3b), DOT1L (disruptor of telomeric silencing-1 like), EZH2 (enhancer of zeste homolog 2), ESET (ERG-associated protein with SET domain), GCN5 (general control nonderepressible 5), HDAC1 (histone deacetylase 1), HDAC2 (histone deacetylase 2), HDAC3 (histone deacetylase 3), HDAC4 (histone deacetylase 4), HDAC5 (histone deacetylase 5), HDAC6 (histone deacetylase 6), HDAC7 (histone deacetylase 7), HDAC9 (histone deacetylase 9), JMJD2C (Jumonji domain-containing 2C), JMJD3 (Jumonji domain-containing 3), KDM2 (lysine demethylase 2), KDM7 (lysine demethylase 7), LSD1 (lysine-specific histone demethylase 1), MAPK (mitogen-activated protein kinase), MYST (Moz, Ybf2/Sas3, Sas2, Tip60), NSD1 (nuclear receptor-Binding SET domain protein 1), p300/CBP (CREB-binding protein), SET2 (SET domain-containing 2), SET7/9 (SET domain-containing 7/9), SETDB1 (SET domain, bifurcated 1), SIRT1 (silent mating-type information regulation 2 homolog 1), SIRT2 (silent mating-type information regulation 2 homolog 2), SIRT3 (silent mating-type information regulation 2 homolog 3), SIRT6 (silent mating-type information regulation 2 homolog 6), SIRT7 (silent mating-type information regulation 2 homolog 7), SUV39H1 (suppressor Of variegation 3-9 homolog 1), TET (ten-eleven translocation), and TET2 (ten-eleven translocation 2)
[b]PPARGC1A (PPARG coactivator 1 Alpha)

discovery in past decades, it is advisable to make upfront choices before selecting and acquiring chemical libraries for a challenging target like epigenetics. This chapter offers a set of chemoinformatic tools to identify possible risk factors and pitfalls, and explores the synergy with experimental methods to advance the discovery of small molecules that modulate various epigenetic targets.

2 Overview: Epigenetic Control of Gene Expression

Chromatin organization

Multicellular organisms have the large genome, and it must maintain the integrity and specialized functions of different cell types. Therefore, eukaryotes have evolved a dynamic and efficient packaging system, where chromosomal DNA is organized inside the nucleus of the cell with the help of histone proteins (Fig. 1). The condensed DNA and associated proteins together constitute the nucleosome, the basic structural unit of chromatin [24]. The nucleosome consists of a histone octamer core which is wrapped around by 146 base pairs of dsDNA. Each octamer contains two copies of each of four histone proteins—H2A, H2B, H3, and H4. Further, these nucleosomes align along the DNA with the help of a fifth type of histone (H1 and isoforms) which acts as a linker. Histone H1, its isoforms, RNAs, and other non-histone proteins together contribute to the next level of condensation forming the 30 nm fiber. Further, scaffold proteins and chromatin-remodeling complexes help to

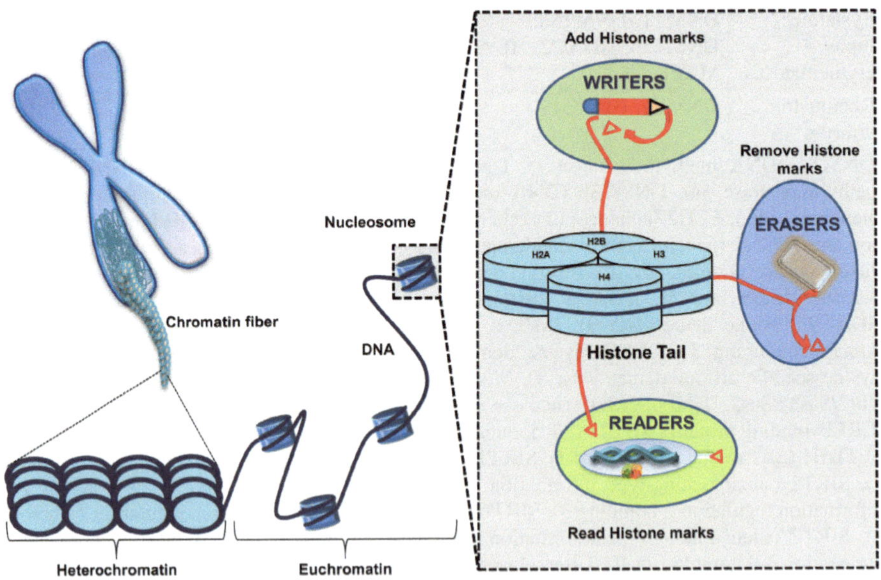

Fig. 1 Chromatin organization and main categories of epigenetic protein family

maintain the chromatin structure and allow dynamic movements between euchromatin and heterochromatin [4] (Fig. 1). Euchromatin is loosely packed, thereby accessible to transcription machinery and genetically active. In contrast, the heterochromatin contains highly compact DNA in which case it is difficult to access by transcription machinery and, consequently, is genetically inactive.

Epigenetic modifications

Alterations in chromatin structure and subsequent change in the gene expression are achieved by two main mechanisms: (1) DNA methylation at cytidine residues and (2) histone modifications. Histone modifications may include acetylation, methylation, phosphorylation, ubiquitylation, sumoylation and siRNA-controlled gene expression. These modifications lead to multiple chromatin states, creating combinatorial patterns of DNA and histone variant mark areas of distinct genome functions [25]. Such patterns both reflect and influence chromatin-related processes, mainly DNA replication, DNA repair, transcription, and chromosomal segregation. The epigenetic proteins that facilitate such modifications can be divided into three major categories based on their broad functions: (1) writers that embed epigenetic marks on DNA or histones, (2) erasers that remove such marks, and (3) readers that identify these marks (Fig. 1).

3 The First and Second Generation of Epi-Drugs

Research interest has grown exponentially within academia and industries to develop therapeutic agents for cancer, aging, diabetes, neurodegenerative, and cardiovascular disorders. This has led to the development of five successful epi-drugs that have been approved by the FDA (Food and Drug Administration), representing the first generation of epigenetic drugs (Table 2):

(1) **5-azacytidine** (also known as Vidaza®): 5-Azacytidine is an analog of the cytidine and was first synthesized about 40 years ago. Azacitidine inhibits DNA methylation by stoichiometrically binding to DNMT1 (DNA methyltransferase 1). Azacitidine has been used primarily in the treatment of leukemia and MDS (myelodysplastic syndrome).
(2) **5-aza-2′deoxycytidine** (also known as decitabine or trade name: Dacogen®): 5-aza-2′deoxycytidine is also an analog of cytidine. Similar to Azacitidine, it also inhibits DNMT1 (DNA methyltransferase 1) and exhibits clinical utility in MDS (myelodysplastic syndrome) and leukemia.
(3) **Vorinostat** (chemical name: suberoylanilide hydroxamic acid, commercially known as Zolinza®): Vorinostat is an inhibitor of class I and II HDACs (histone deacetylases). Vorinostat has been approved for the treatment of CTCL (cutaneous T cell lymphoma).

Table 2 The first generation of epi-drugs

Drug name	Chemical structure	Target[a]	Use[b]
Azacytidine	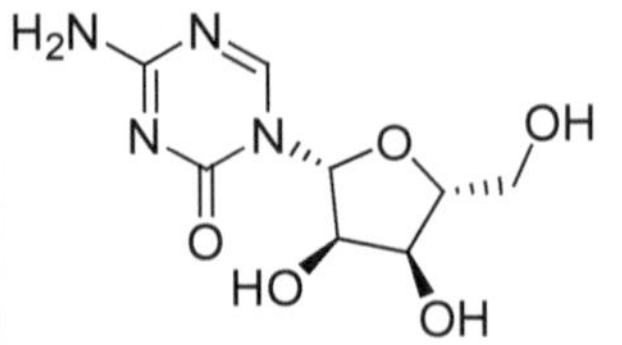	DNMT1	Approved for the treatment of MDS [26] and relapsed AML [27]
Decitabine	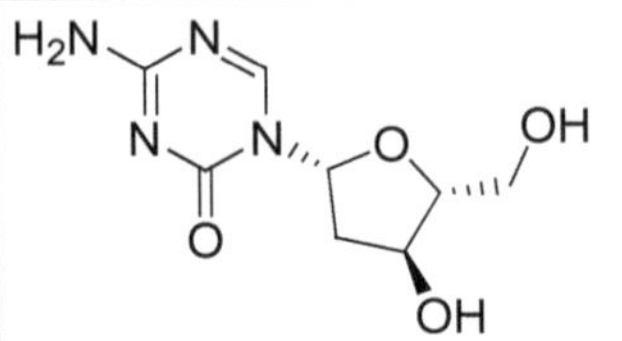	DNMT1	Approved for MDS and for treatment of elderly AML patients [28]
Vorinostat	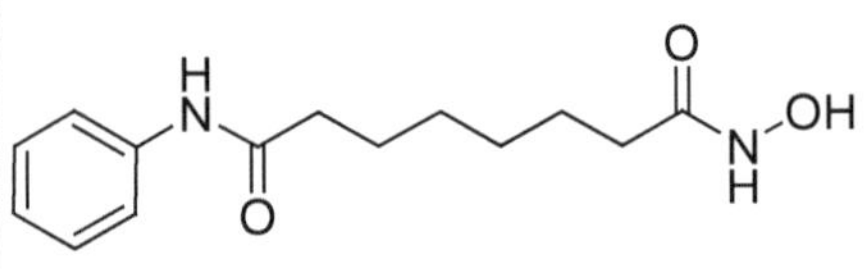	Class I & II HDACs	Approved for CTCL [29]. Phase II trial in recurrent glioblastoma and unresectable gastric cancer [30]
Romidepsin	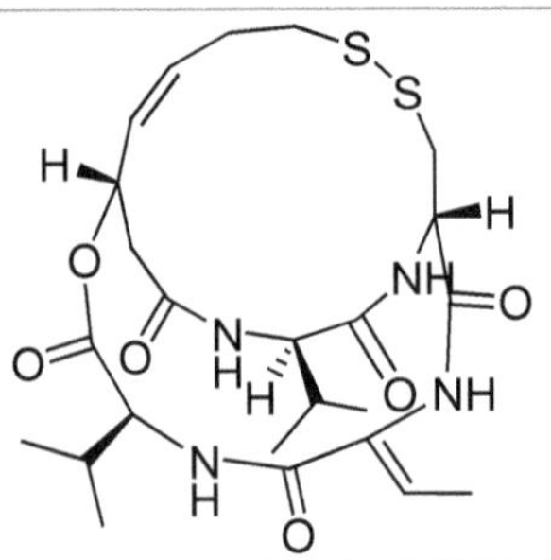	Class I HDACs	Approved for CTCL [31] and relapsed PTCL [32] Phase I/II trial in recurrent high-grade glioma [33]
Belinostat	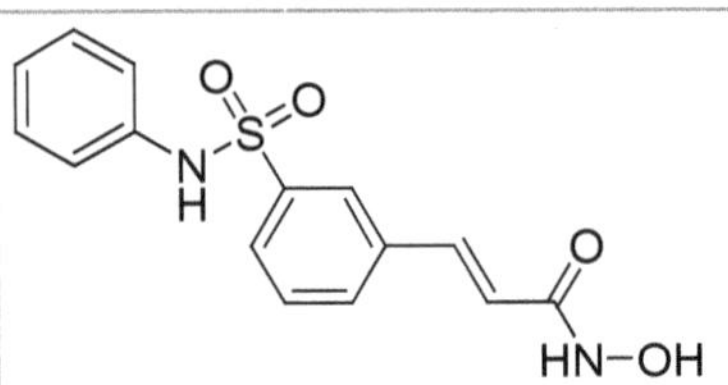	pan-HDAC	Approved for relapsed/ refractory PTCL [34]

[a]DNMT1 (DNA methyltransferase 1), HDACs (histone deacetylases)
[b]MDS (myelodysplastic syndrome), AML (acute myeloid leukemia), CTCL (cutaneous T cell lymphoma), PTCL (peripheral T cell lymphoma)

(4) **Romidepsin** (also known as Istodax®): Romidepsin is a selective inhibitor of HDAC (histone deacetylase) that was discovered from cultures of *Chromobacterium violaceum*. Romidepsin was too approved for valuable treatment of CTCL (cutaneous T cell lymphoma).

Table 3 The second generation of epi-drugs

Drug name	Chemical structure	Target[a]	Disease condition[b]	Reference
GSK126	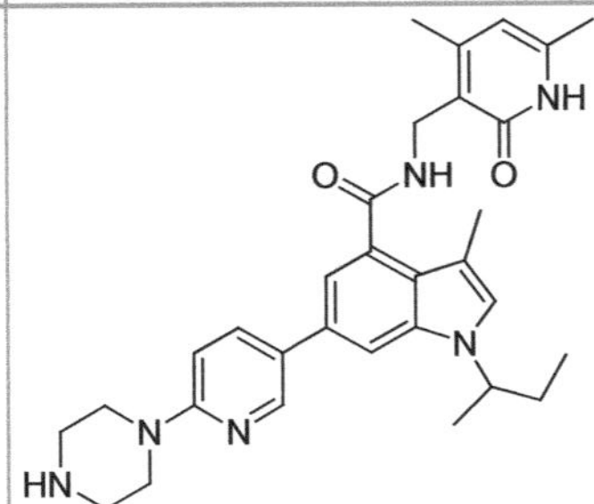	EZH2	Relapsed/refractory DLBCL, transformed follicular lymphoma, other non-Hodgkin's lymphomas, solid tumors, and multiple myeloma	[35–37]
C646	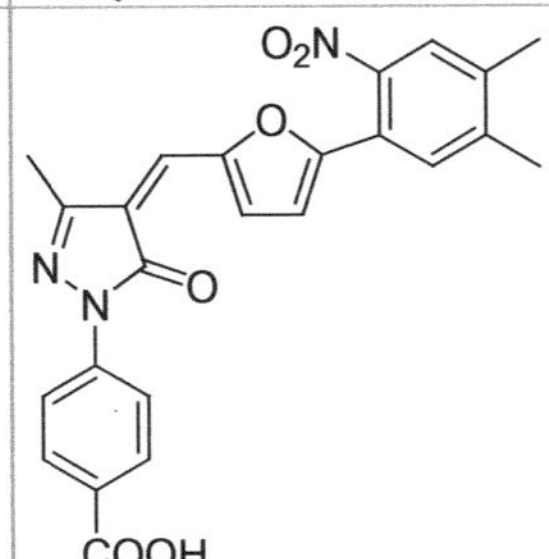	p300	Abrogates TSA-induced acetylation in cells, radiosensitizes NSCLC, potent anti-influenza virus candidate	[38–40]
EPZ004777	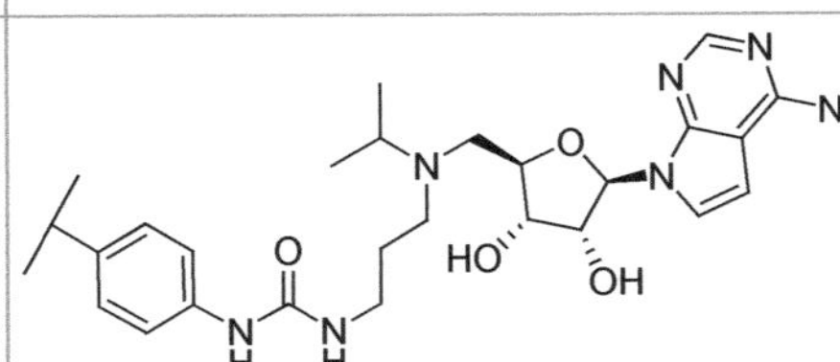	DOT1L	Induces apoptosis, cell cycle arrest, and terminal differentiation in DNMT3A-mutant human AML cells, induces apoptosis in MLL-rearranged leukemia cells	[41, 42]

(continued)

Table 3 (continued)

Drug name	Chemical structure	Target[a]	Disease condition[b]	Reference
JQ1	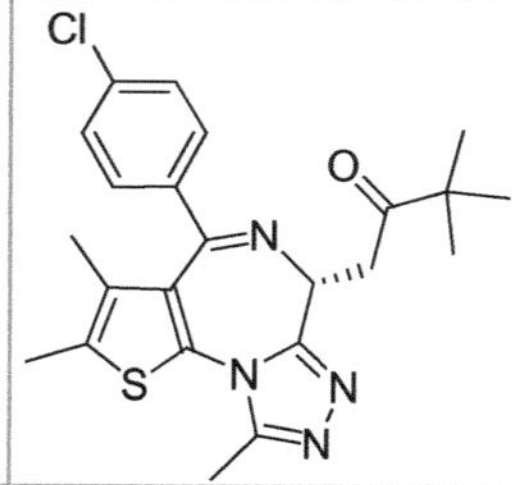	BRD4	Suppresses cell growth and invasion in oral squamous cell carcinoma, killing of MLL cells	[43, 44]
UNC669	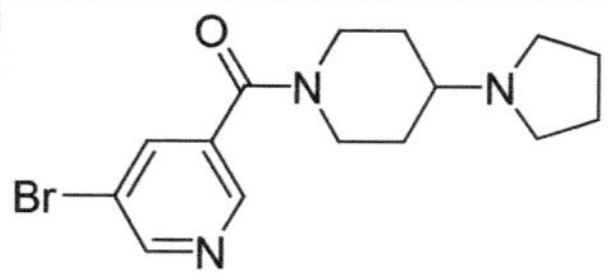	L3MBTL1	Effective and selective MBT inhibitor	[45]

[a]EZH2 (enhancer of zeste homolog 2), p300 (P300 acetyltransferase), BRD4 (bromodomain-containing 4), DOT1L (disruptor of telomeric silencing-1 like), L3MBTL1 (lethal(3)malignant brain tumor-like protein 1), MBT (malignant brain tumor)

[b]DLBCL (diffuse large B cell lymphoma), NSCLC (non-small cell lung carcinoma cells), AML (acute myeloid leukemia), MLL (mixed lineage leukemia)

(5) **Belinostat** (also known as PXD-101 or Beleodaq®): Belinostat is another histone deacetylase (HDAC) inhibitor that gained accelerated approval by the FDA for the treatment of patients with relapsed or refractory PTCL (peripheral T cell lymphoma).

Currently, the second generation of epi-drugs is entering clinical trials and holds more promise because of their greater intrinsic selectivity for the molecular targets (Table 3):

(1) **GSK126**: GSK126 has been validated as an histone methyltransferase (HMT) inhibitor of EZH2 (enhancer of zeste homolog 2). It has shown promising results in inhibition of DLBCL (diffuse large B cell lymphoma).
(2) **C646**: C646 is a potent and selective (HAT—histone acetyltransferase) inhibitor of p300 enzyme, which has proven to be effective against cancer cell growth.
(3) **EPZ004777**: EPZ004777 is a selective inhibitor of DOT1L (disruptor of telomeric silencing 1-like), a protein methyltransferase. It has been demonstrated to kill MLL (mixed lineage leukemia) cells in vitro and prolongs survival in an MLL xenograft mouse model.
(4) **JQ1**: JQ1 is a bromodomain inhibitor for BRD4 (bromodomain-containing 4) protein. It promotes differentiation, tumor regression, and prolonged survival in murine models.
(5) **UNC669**: UNC669 is a potent and selective MBT (malignant brain tumor) inhibitor for L3MBTL1 (lethal(3)malignant brain tumor-like protein 1).

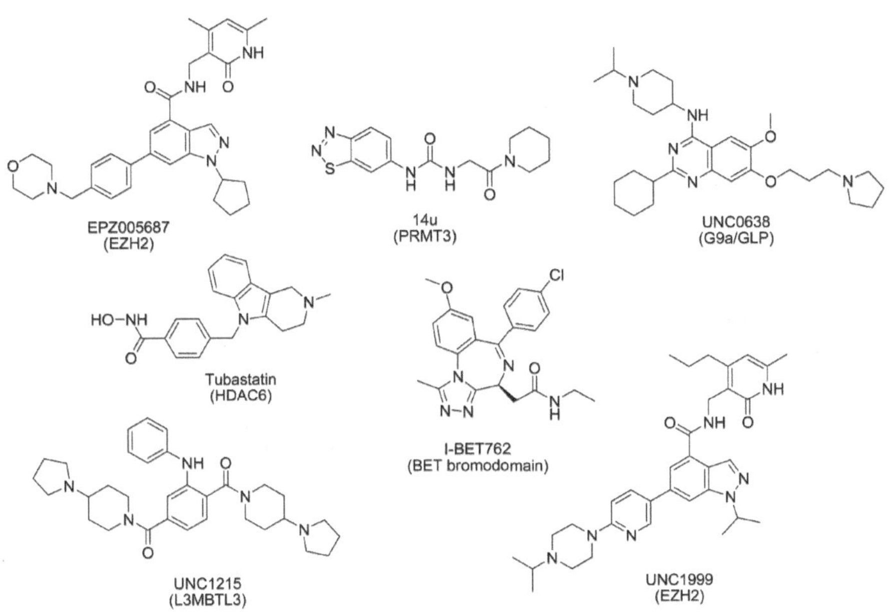

Fig. 2 Illustrative examples of pharmacological inhibitors of epigenetic proteins. The names of target proteins are written in parentheses

Apart from the above-mentioned drugs, several other inhibitors of epigenetic targets are discovered. Some of these inhibitors are depicted in Fig. 2.

4 Chemoinformatics Study on Epigenetic Modulators

This section briefly describes the key strategies to select and assess a small-molecule library based on current best practices and theoretical guidelines. Once a target is determined, one needs to design and select a chemical compound library. Recently, we have developed a powerful and curated database to assist the researchers in epigenetic drug discovery (www.epidbase.org). Our database has diverse molecules and scaffolds that are found to be active against various epigenetic proteins. Detailed methodology and analysis can be availed from our previously published study [46]. There are numerous chemoinformatics tools available to support the synthetic and medicinal chemists in vetting the promising libraries [47–49]. These tools are available in various forms to select building blocks, enumerate compounds, and also calculate chemical descriptors and fingerprints to sketch the library. In order to utilize these tools, an important prerequisite is to store and sort the molecules using one of the data formatting systems such as SMILES (simplified molecular input line entry specification format) or SDF (structure-data file format). Many software packages including Schrodinger, Tripos, and Pipeline Pilot are powered with structural, physicochemical, ADME, and diversity filtering tools.

Strategy 1. Analyze the chemical space and scaffold diversity

The term 'scaffold' represents the core structures of bioactive compounds and is often used interchangeably with terms like 'framework,' 'substructure,' or 'fragment.' As proposed by Bemis and Murcko [50], the framework can be obtained by trimming the side chain atoms which are not positioned in the connecting path between two rings. The concept of chemical scaffold diversity is widely applied in drug discovery. The chemical space in relevance to drug-like molecules is estimated to be of 10^{60} molecules (i.e., between 300 and 500 Da of molecular weight). Practically, chemical space is infinite and meagerly populated and it is impossible to mine out a drug. In drug discovery, a chemical library is not of significant use if the biological and chemical space does not overlap [51]. Therefore, scaffolds should be focused upon as the core of small chemical libraries to be synthesized or acquired. There is a significant number of studies that have applied large substructure similarity to analyze the diversity based on ring systems in the structures [52–57]. Using such a strategy enables the medicinal chemist to select structures which belong to the same chemical family and have a common molecular framework. For example, in our database, to narrow down the chemical space and select the scaffolds, we searched the literature for known modulators of epigenetic proteins. Despite the importance of epigenetic proteins in therapeutics, there were no resources or platforms available to comprehensively analyze the epigenetic modulators for drug discovery. Databases such as NCBI Epigenomics, HEMD, ChEpimod, ChEMBL, and ChromoHub provide access only to epigenetic proteins and their phylogenetic associations. However, their chemical libraries are mainly based on virtual inhibitors of epigenetic proteins (e.g., PubChem bioassay data) [58]. To retain only validated molecular scaffolds that actually interact with epigenetic targets, we collected data manually from the literature for experimentally verified inhibitors. We designed EpiDBase, to view, explore, search, and analyze the small-molecule modulators targeting various epigenetic protein families. EpiDBase is manually curated to identify unique molecular scaffolds and includes text search, structural editor, and chemical fingerprint search, for powerful browsing [46]. Molecules from EpiDBase can be selected and further derivatized to obtain a potent lead toward an epigenetic protein in drug discovery.

Strategy 2. Weed out the problematic compounds

One of the key aspects in the prioritization of chemical matter is to weed out the 'problematic' or 'risky' compounds. Such compounds called 'PAINS' are 'frequent hitters' and belong to a subset of chemical substructures that interact non-specifically with proteins in various unrelated bioassays. PAINS can readout as false positives due to non-selective binding with proteins [59], fluorescence [60], redox activity [61], cysteine oxidation [62], and aggregation [63]. More than 450 structural classes [64] have been identified as PAINS, and a typical academic screening library may consist of 5–12% of such compounds [65]. The most recurring chemotypes are rhodanines, phenol-sulfonamides, toxoflavins, isothiazolones, enones, curcumin, hydroxyphenylhydrazones, quinones, and catechol. For example, rhodanines are reported as promising bioactive compounds, but they may undergo light-induced reactions and modify some proteins covalently [66]. Likewise, phenol-sulfonamides are unstable compounds and can alter the redox cycle and covalently modify the target proteins. Unfortunately, several articles and patents in the literature include PAINS as potential bioactive compounds [67–70]. A medicinal chemists' precise look at the structure can help the biologist to exclude such compounds. Recently, chemical substructure filters and rules (e.g., PAINS, REOS, and others [49, 71–74]) have been introduced to identify these problematic moieties. In EpiDBase, Eli Lilly MedChem regular rules were applied to weed out such promiscuous compounds [46]. The Eli Lilly MedChem rules are defined by a set of 275 rules and are capable of identifying compounds that may interfere with biological assays. We processed a total of 5401 molecules, out of which 1664 molecules were rejected by the filter rules. The remaining 3737 molecules can be further exploited in epigenetic drug discovery and can prove promising.

Strategy 3. Perform the computational ADME/toxicity prediction

Majority of drug candidates fail in clinical trials owing to their unfavorable absorption, distribution, metabolism, excretion, and toxicity (ADMET) profile (Fig. 3) [75–77]. A recent study [2] showed that only 32% of Phase II drug candidates are able to make it to Phase III clinical trials and overall just ~10% of drugs reach to market. Unfortunately, terminating a lead molecule suffers from the loss that increases exponentially as it moves further down the pipeline. For that reason, it is important to adhere to conventional rules for drug-likeness and oral bioavailability of drug candidates. These rules are defined by more than 3300 molecular descriptors [78] including physicochemical, geometrical, topological, electropological, quantum chemical, and molecular fingerprints. Drug-likeness of a molecule can be evaluated based on statistical rules (e.g., 'rule of five') or its physicochemical properties (e.g., solubility, lipophilicity, rotatable bonds, polar surface area). It is well established that prediction of ADMET properties at the earliest stages prevents from depletion of scarce resources on bad leads and expensive clinical trials. This allows allocation of drug development resources on fewer but much promising drug leads. Various software packages including DEREK, METEOR, Discovery Studio are available to predict the ADMET properties. Though not very accurate, these software packages may provide key insights into a drug's safety and efficacy profile to curtail the high cost of failures in clinical trials. In EpiDBase, we performed ADMET analysis using FAF-Drugs2 [79], to filter out the toxic, unstable molecules, and/or functional groups. Further, we used ZINC property filter to assess the drug-likeness of 3737 molecules using various physicochemical descriptors such as MW, hydrogen bond acceptor (HBA), hydrogen bond donor (HBD), rotatable bonds, polar surface area. The filtered molecules represent a set of epigenetic ligands that possess drug-like properties and can be further explored by virtual screening and docking experiments to find potential ligands for various epigenetic proteins.

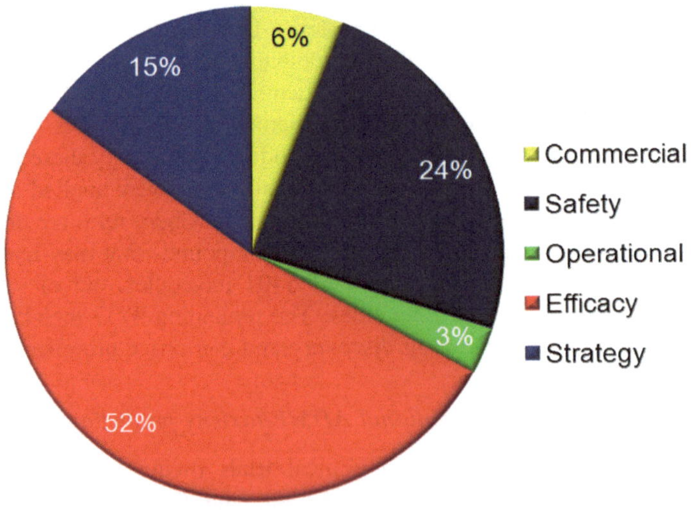

Fig. 3 Reasons for drug failures during 2013–2015 [77]

Strategy 4. Generate 3D conformers

Drugs act through physical interaction with specific biological targets. Such interactions are determined primarily by complementarity of shape and properties between the interacting molecules. On that account, the biological activity of a drug depends on its three-dimensional structure. In solution, most drug molecules are flexible and exist as an ensemble of low-energy conformations (shapes) in equilibrium with one another. The biologically active conformation (target-bound) can either be similar to the conformations present in solution or can be induced by target binding [80]. Also, different target proteins or cellular environments can induce different conformations of the same ligand. Thus, conformational adaptation is an important aspect in pharmacophore modeling, rigid docking, shape-based screening, 3D-QSAR, and virtual screening and must be considered carefully during lead optimization. There are various small-molecule conformer generation tools available such as BALLOON, CONFAB, FROG2, RDKIT, and OMEGA. These tools use diverse algorithms which may be based on approaches including knowledge-based rule sets [81, 82], random coordinate changes [83], random torsional angle changes [84, 85], and/or distance geometry [86, 87]. Numerous studies have reported the use of conformer generating tools for predicting bioactive conformations [88, 89]. In view of this, we generated a multi-conformer database of 5447 compounds using Universal Force Field (UFF) to facilitate epigenetic ligands with maximum coverage of conformational space [46]. Conformers were generated using RDKit, an open-source toolkit that comes under the permissive Berkeley Software Distribution (BSD) license and employs the distance geometry approach [90]. For each compound, we generated 50 conformers with RMSD cutoff of 0.5. For some compounds, the number of conformers generated was less than 50 as a result of RMSD criteria. In total, 269,052 conformers were generated which provide the structural information about the conformational states of all epigenetic ligands in the database and can be exploited further for in silico drug designing.

Strategy 5. Perform clustering analysis

Clustering is a powerful tool to identify homogeneous subsets within a heterogeneous compound dataset using structural characteristics [76, 91, 92]. For example, it can be utilized to explore a large dataset to correlate compounds based on their biological activity/property and scaffold hopping [93]. Clustering tools are applied widely in drug discovery for chemical diversity, compound selection, and data reduction in libraries. In times, it is advantageous to cluster small subsets of compounds together to perform assays where it is not feasible to perform the high-throughput screening. An ideal clustering process creates a series of clusters from a larger library of compounds. Each cluster consists of compounds with similar datapoints clubbed together as per the chosen criteria for similarity [94]. JKlustor, ChemMine Tools, and PKOM are some examples of the available clustering tools. In EpiDBase, ChemMine Tools were utilized to perform clustering analysis. ChemMine has an online workbench that provides three important clustering methods including hierarchical clustering, multidimensional scaling (MDS), and binning clustering. ChemMine tool first calculates the similarity matrix based on atom pair descriptors for each compound using the Tanimoto coefficient. The Tanimoto coefficient may range between 0 and 1, indicating '1' as the highest similarity and '0' as no similarity. The similarity matrix is then converted into a distance matrix by deducting the similarity values from 1. The hierarchical clustering arranges the similar compounds in a tree with branch representation, where branch lengths are proportional to the similarity between compounds. However, MDS represents compounds as a scatter plot. The binning cluster displays the results as a table where similar compounds are grouped together for a user-definable cutoff. In EpiDBase, binning clustering was performed using a similarity cutoff of 0.6 (Tanimoto coefficient) to define the chemical diversity and space coverage of epigenetic ligands. For example, 622 ligands of SIRT2 were clustered using ChemMine to classify similar compounds and scaffolds into groups. Out of 108 clusters formed, 65 were single molecule clusters represented by unique ligands. Such clusters provide a rare opportunity for medicinal chemists to populate them using rational design and structure-activity relationship (SAR) studies to identify potential therapeutics.

Strategy 6. Optimization of fragment-based library

Fragment-based drug design has emerged as a powerful technique in lead discovery paradigm which is used as an alternative or often complementary to traditional high-throughput screening (HTS). Fragments are 'atom-efficient' binders that can be further expanded into high-affinity lead compounds [95]. In comparison with compounds, fragments exhibit weak affinity toward the target and need high-end instruments for their detection. Furthermore, fragment screening needs a high concentration of protein and fragments. These challenges can be overcome by using a wide variety of computational approaches to identify potential fragments and binding sites. In EpiDBase, Retrosynthetic Combinatorial Analysis Procedure (RECAP) [96] was used to generate the fragment database for SIRT2 modulators. RECAP cleaves along the bonds using chemical knowledge and generates a collection of fragments suitable for combinatorial library synthesis. Such fragments can be further clustered to generate effective libraries for epigenetic drug discovery.

Strategy 7. Optimization of commercial library for HTS

High-throughput screening (HTS) is a robust approach to drug discovery that allows the assaying of a large number of small molecules against a validated target. The aim of HTS is to accelerate drug discovery by identifying active chemical series. It is essential to enrich small-molecule libraries with high chemical diversity to increase the hit rate. The compound enrichment can be done using a multitude of techniques including scaffold tree classification, virtual docking or general structure and property filters (Fig. 4) [97, 98]. In our earlier study [99], various commercially available chemical libraries were analyzed for their exclusiveness, drug-likeness, and scaffold similarities (using asymmetrical metrics). The study demonstrates that larger libraries have optimal chemo-diversity and unique scaffolds with drug-like properties for HTS. Accordingly, this method can be implemented in epi-drug discovery to identify the unique scaffolds that can be further optimized to design high-affinity lead molecules with enhanced activity (Fig. 4).

Strategy 8. Structure-based drug discovery in epigenetics

Structure-based drug design (SBDD, also known as rational drug design) is an essential tool in drug discovery and has delivered many successful drugs [100–105]. In contrast to conventional ways of drug discovery (which are mostly hit and trials), SBDD is more efficient since it incorporates the 3D structural information of biological targets to understand their functional role in a disease [106]. SBDD methods are now often applied much earlier in the drug discovery to save resources and time during preclinical and early clinical stages.

SBDD begins with the selection of a potent biological target for a given therapeutic need. Once a target is selected, its structure must be determined to identify potential ligand binding site using techniques such as X-ray crystallography, cryo-EM (cryo-electron microscopy), and NMR (nuclear magnetic resonance spectroscopy). The ligand binding site is ideally defined by a variety of hydrogen bond donors and acceptors, hydrophobic residues, and molecular surface area. In cases, where the structure of the target protein is not available, a homology model can be deduced using computational tools such as SWISS-MODEL [107], Modeller [108], Phyre2 [109]. Finally, a lead molecule is designed to interact with the target

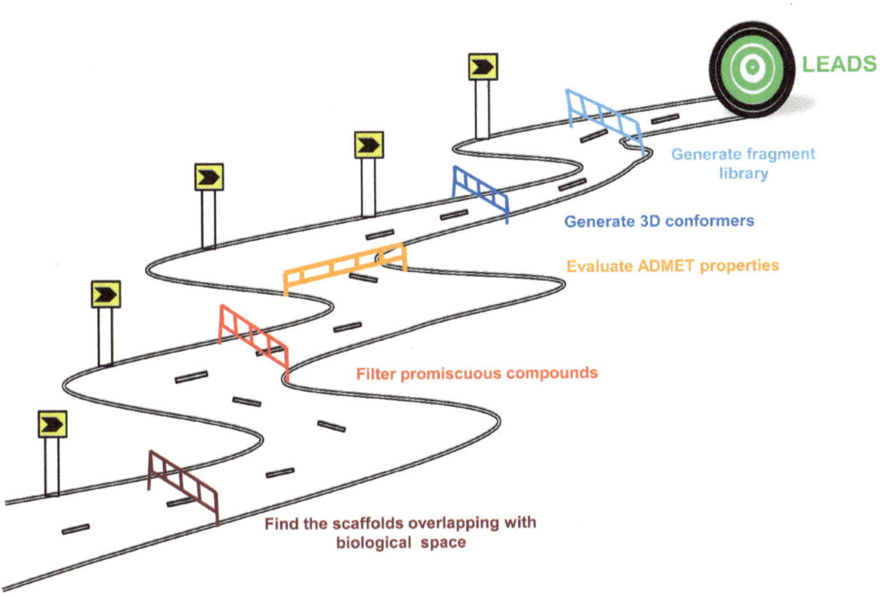

Fig. 4 Barriers in drug discovery

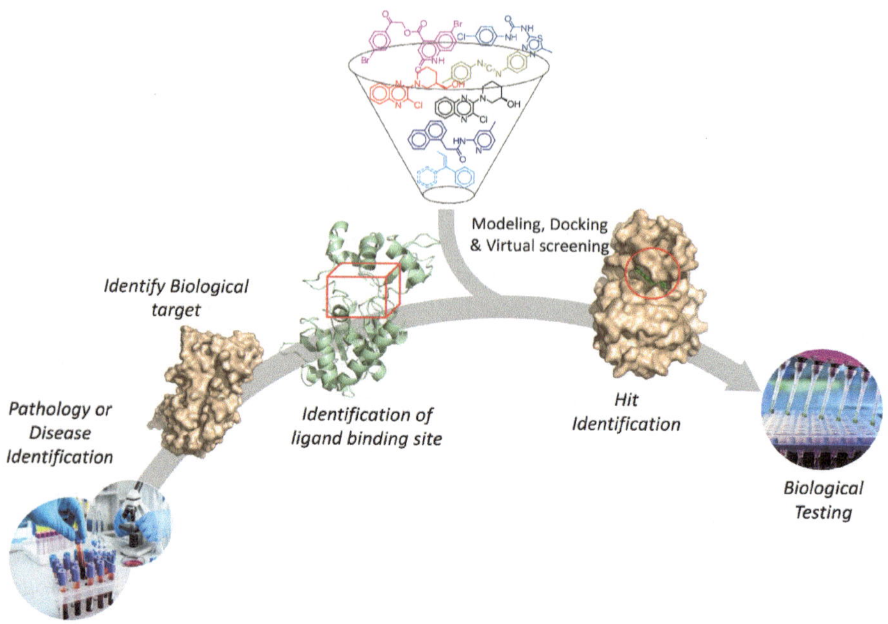

Fig. 5 General strategy of structure-based drug design (SBDD)

protein to modulate its biological activity. The general strategy of structure-based drug discovery is outlined in Fig. 5.

With the advent of technologies such as high-throughput X-ray crystallography [110], cryo-EM [111], NMR [112], and homology modeling [113], the proteomic and structural information of new biological targets is flaring up and has further opened up new opportunities for future lead discovery. Subsequently, the current information on epigenetic targets in the form of available structural data and drug design perspectives is evolving at an impressive rate. Recently, Shao et al. have discovered novel inhibitors targeting DNMT3A by utilizing the structure-based virtual screening in addition to biological assays [114]. Two of compounds, 40 and 40_3, showed low micromolar inhibitory activity through binding to S-adenosyl-l-methionine and may further serve as scaffolds for drug optimization. In another study, co-crystal structures of PRMT2 and PRMT4 with S-adenosyl-L-homocysteine or other compounds (including Cp1, a synthetic inhibitor of PRMT2) were investigated [115]. The comparison of inhibitor interactions with two proteins revealed that compound Cp1 is efficient at inhibiting PRMT2 [115]. The study represents an initiative toward a better understanding of PRMT2 substrate recognition. It may provide further insights into structure-based drug design of PRMT2 inhibitors. Similarly, Siedlecki et al. predicted homology model of DNMT1 [116], performed structure-based virtual screening with ~2000 compounds, and identified RG108 inhibitor [117, 118]. Bowers et al. discovered C646 compound as cofactor-competitive and cofactor-selective inhibitor of p300 [38].

The examples cited above represent the significance of SBDD approaches in epigenetic drug discovery. However, despite the success of SBDD and availability of

X-ray crystal structures of many epigenetic proteins, the SBDD approaches are not competent in all circumstances. For example, the thermodynamics of ligand-receptor association cannot be predicted precisely because the current methodologies do not take into account the factors such as receptor flexibility, solvation, entropy, and dynamic inclusion of water molecules [119, 120]. Additionally, some tricky epigenetic targets such as HATs and other complexes lack potent chemical probes and are unexplored [121]. Nevertheless, epigenetic drug discovery is evolving and it is necessary to have access to experimental data related to bioassays and 3D structures of epigenetic enzymes. To support SBDD for epigenetic targets, EpiDBase [46] provides information regarding available structures of epigenetic proteins with the cross-reference to Protein Data Bank, resolution of crystal structures, the method of obtaining crystal structure, and the information about ligand present in crystal structure, if any. EpiDBase can assist in homology modeling, docking, virtual screening and can prove to be resourceful to accelerate structure-based drug discovery.

5 Conclusion and Future Perspectives

Transforming lead molecules into successful drugs is still a challenging task, in spite of tremendous advances in pharmacology and medicinal chemistry [122]. In fact, 99% of drug discovery projects fail and most leads are unable to make it through the later stages of drug development (clinical trials). The advancement in the knowledge of epigenetic changes associated with specific disease has encouraged the research efforts in the field of epigenetic drug discovery.

Box 1. Questions to ask before obtaining a chemical library

Questions?	Why ask?
Do the scaffolds overlap with the biological space of epigenetic target binding site?	The selection of scaffolds for a particular target limits the chemical space and provides the key building blocks while designing the drugs
Do the scaffolds have synthetic accessibility?	The synthesis of various derivatives can be achieved for scaffolds, which demonstrate synthetic tractability. This will probably lower the cost of synthesis and save time
Do the compounds exhibit favorable physicochemical properties?	All chemical compounds are not drugs. Drug-likeness of a compound depends on various physical and chemical properties. Compounds with favorable physicochemical properties are more promising at later stages of drug development

(continued)

(continued)

Questions?	Why ask?
Do the compounds pass through the filters for promiscuity and common sources of assay artifacts?	Certain chemical moieties can interact non-specifically with proteins in multiple assays. It is crucial to weed out such artifacts in order to avoid expenses on bad lead molecules
What is the library size?	The library should fit as per the needs of the project and should not be populated unnecessarily. It is better to keep fewer but useful compounds in the library. However, larger libraries offer chemical diversity for HTS
What is the clustering density of compounds?	Clustering density can reveal interesting patterns among the target specific compounds synthesized over many decades. The underexplored clusters can be scaled up to synthesize the derivatives

The progress made so far in terms of epi-drugs is only the beginning of a revolution. Implementing computational strategies in the selection and screening of chemical libraries at an early stage of epigenetic drug discovery is vital for identifying promising lead molecules. Furthermore, epigenetic specific databases such as EpiDBase, NCBI Epigenomics, ChEMBL, and HEMD are powerful tools. For example, EpiDBase can facilitate interactive exploring of epigenetic proteins, their curated ligands, SAR studies, statistical analysis, and fragment-based drug design. The database can be employed to study epigenetic ligands for their experimental IC_{50} values, structural data, toxicological, and chemoinformatic information. We have attempted to provide an overview of the myriad of considerations to all researchers engaged in epigenetic drug discovery for selecting a small-molecule screen library (Box 1). We have highlighted the strategies to design a chemical library based on current best practices and theoretical considerations (Fig. 4). Nevertheless, it is hoped that this study would be beneficial for the design and discovery of modulators to influence epigenetic states of various diseases. Ultimately, it shall help in guiding compounds through all the potential pitfalls to determine the success of an epi-drug discovery campaign.

References

1. IFPMA. *The pharmaceutical industry and global health: Facts and Figures 2017*. 2017; Available from: https://www.ifpma.org/wp-content/uploads/2017/02/IFPMA-Facts-And-Figures-2017.pdf
2. Hay M et al (2014) Clinical development success rates for investigational drugs. Nat Biotechnol 32(1):40–51

3. Abraham AL et al (2012) Genetic modifiers of chromatin acetylation antagonize the reprogramming of epi-polymorphisms. PLoS Genet 8(9):e1002958
4. Bierne H, Hamon M, Cossart P (2012) Epigenetics and bacterial infections. Cold Spring Harb Perspect Med 2(12):a010272
5. Brookes E, Shi Y (2014) Diverse epigenetic mechanisms of human disease. Annu Rev Genet 48:237–268
6. Epigenetics Drugs and Diagnostic Technologies Market - Global Industry Analysis, Size, Share, Growth, Trends and Forecast, 2012–2018
7. Raynal NJ et al (2017) Repositioning FDA-approved drugs in combination with epigenetic drugs to reprogram colon cancer epigenome. Mol Cancer Ther 16(2):397–407
8. Mendez-Lucio O et al (2014) Toward drug repurposing in epigenetics: olsalazine as a hypomethylating compound active in a cellular context. ChemMedChem 9(3):560–565
9. Raynal NJ et al. (2014) Discovery of new epigenetic drugs among FDA-approved drug libraries. Cancer Res. 74:19
10. Portela A, Esteller M (2010) Epigenetic modifications and human disease. Nat Biotechnol 28(10):1057–1068
11. Bhattacharjee D, Shenoy S, Bairy KL (2016) DNA methylation and chromatin remodeling: the blueprint of cancer epigenetics. Scientifica (Cairo) 2016:6072357
12. Virani S et al (2012) Cancer epigenetics: a brief review. ILAR J 53(3–4):359–369
13. Sen P et al (2016) Epigenetic mechanisms of longevity and aging. Cell 166(4):822–839
14. Benayoun BA, Pollina EA, Brunet A (2015) Epigenetic regulation of ageing: linking environmental inputs to genomic stability. Nat Rev Mol Cell Biol 16(10):593–610
15. Ling C, Groop L (2009) Epigenetics: a molecular link between environmental factors and type 2 diabetes. Diabetes 58(12):2718–2725
16. Landgrave-Gomez J, Mercado-Gomez O, Guevara-Guzman R (2015) Epigenetic mechanisms in neurological and neurodegenerative diseases. Front Cell Neurosci 9:58
17. Coppede F (2014) The potential of epigenetic therapies in neurodegenerative diseases. Front Genet 5:220
18. Lee J et al (2013) Epigenetic mechanisms of neurodegeneration in Huntington's disease. Neurotherapeutics 10(4):664–676
19. Peedicayil J (2016) Epigenetic drugs for multiple sclerosis. Curr Neuropharmacol 14(1):3–9
20. Relle M, Foehr B, Schwarting A (2015) Epigenetic aspects of systemic lupus erythematosus. Rheumatol Ther 2(1):33–46
21. Wu H et al (2015) The real culprit in systemic lupus erythematosus: abnormal epigenetic regulation. Int J Mol Sci 16(5):11013–11033
22. Hedrich CM (2017) Epigenetics in SLE. Curr Rheumatol Rep 19(9):58
23. Klein K, Ospelt C, Gay S (2012) Epigenetic contributions in the development of rheumatoid arthritis. Arthritis Res Ther 14(6):227
24. Bernstein BE, Meissner A, Lander ES (2007) The mammalian epigenome. Cell 128(4): 669–681
25. Wang Z et al (2008) Combinatorial patterns of histone acetylations and methylations in the human genome. Nat Genet 40(7):897–903
26. Kaminskas E et al (2005) FDA drug approval summary: azacitidine (5-azacytidine, Vidaza) for injectable suspension. Oncologist 10(3):176–182
27. Von Hoff DD, Slavik M, Muggia FM (1976) 5-Azacytidine. A new anticancer drug with effectiveness in acute myelogenous leukemia. Ann Intern Med 85(2):237–245
28. Joeckel TE, Lubbert M (2012) Clinical results with the DNA hypomethylating agent 5-aza-2'-deoxycytidine (decitabine) in patients with myelodysplastic syndromes: an update. Semin Hematol 49(4):330–341
29. Mann BS et al (2007) FDA approval summary: vorinostat for treatment of advanced primary cutaneous T-cell lymphoma. Oncologist 12(10):1247–1252
30. Galanis E et al (2009) Phase II trial of vorinostat in recurrent glioblastoma multiforme: a north central cancer treatment group study. J Clin Oncol 27(12):2052–2058

31. Prince HM, Dickinson M (2012) Romidepsin for cutaneous T-cell lymphoma. Clin Cancer Res 18(13):3509–3515
32. Iyer SP, Foss FF (2015) Romidepsin for the treatment of peripheral T-cell lymphoma. Oncologist 20(9):1084–1091
33. Iwamoto FM et al (2011) A phase I/II trial of the histone deacetylase inhibitor romidepsin for adults with recurrent malignant glioma: North American Brain Tumor Consortium Study 03-03. Neuro Oncol 13(5):509–516
34. Rashidi A, Cashen AF (2015) Belinostat for the treatment of relapsed or refractory peripheral T-cell lymphoma. Future Oncol 11(11):1659–1664
35. McCabe MT et al (2012) EZH2 inhibition as a therapeutic strategy for lymphoma with EZH2-activating mutations. Nature 492(7427):108–112
36. Chen YT et al (2016) The novel EZH2 inhibitor, GSK126, suppresses cell migration and angiogenesis via down-regulating VEGF-A. Cancer Chemother Pharmacol 77(4):757–765
37. Zeng D, Liu M, Pan J (2017) Blocking EZH2 methylation transferase activity by GSK126 decreases stem cell-like myeloma cells. Oncotarget 8(2):3396–3411
38. Bowers EM et al (2010) Virtual ligand screening of the p300/CBP histone acetyltransferase: identification of a selective small molecule inhibitor. Chem Biol 17(5):471–482
39. Oike T et al (2014) C646, a selective small molecule inhibitor of histone acetyltransferase p300, radiosensitizes lung cancer cells by enhancing mitotic catastrophe. Radiother Oncol 111(2):222–227
40. Zhao D et al (2015) C646, a novel p300/CREB-binding protein-specific inhibitor of histone acetyltransferase, attenuates influenza A virus infection. Antimicrob Agents Chemother 60(3):1902–1906
41. Rau RE et al (2016) DOT1L as a therapeutic target for the treatment of DNMT3A-mutant acute myeloid leukemia. Blood 128(7):971–981
42. Wong M, Polly P, Liu T (2015) The histone methyltransferase DOT1L: regulatory functions and a cancer therapy target. Am J Cancer Res 5(9):2823–2837
43. Wang L et al (2016) JQ1, a small molecule inhibitor of BRD4, suppresses cell growth and invasion in oral squamous cell carcinoma. Oncol Rep 36(4):1989–1996
44. Daigle SR et al (2011) Selective killing of mixed lineage leukemia cells by a potent small-molecule DOT1L inhibitor. Cancer Cell 20(1):53–65
45. Herold JM et al (2011) Small-molecule ligands of methyl-lysine binding proteins. J Med Chem 54(7):2504–2511
46. Loharch S, et al (2015) EpiDBase: a manually curated database for small molecule modulators of epigenetic landscape. Database (Oxford), 2015
47. Huggins DJ, Venkitaraman AR, Spring DR (2011) Rational methods for the selection of diverse screening compounds. ACS Chem Biol 6(3):208–217
48. Walters WP, Namchuk M (2003) Designing screens: how to make your hits a hit. Nat Rev Drug Discov 2(4):259–266
49. Brenk R et al (2008) Lessons learnt from assembling screening libraries for drug discovery for neglected diseases. ChemMedChem 3(3):435–444
50. Bemis GW, Murcko MA (1996) The properties of known drugs. 1. Molecular frameworks. J Med Chem 39(15):2887–2893
51. McMillan M, Kahn M (2005) Investigating Wnt signaling: a chemogenomic safari. Drug Discov Today 10(21):1467–1474
52. Nilakantan R, Bauman N, Haraki KS (1997) Database diversity assessment: new ideas, concepts, and tools. J Comput Aided Mol Des 11(5):447–452
53. Lee ML, Schneider G (2001) Scaffold architecture and pharmacophoric properties of natural products and trade drugs: application in the design of natural product-based combinatorial libraries. J Comb Chem 3(3):284–289
54. Lewell XQ et al (2003) Drug rings database with web interface. A tool for identifying alternative chemical rings in lead discovery programs. J Med Chem 46(15):3257–3274
55. Kho R et al (2005) Ring systems in mutagenicity databases. J Med Chem 48(21):6671–6678

56. Lameijer EW et al (2006) Mining a chemical database for fragment co-occurrence: discovery of "chemical cliches". J Chem Inf Model 46(2):553–562
57. Ertl P, et al (2006) Quest for the rings. In silico exploration of ring universe to identify novel bioactive heteroaromatic scaffolds. J Med Chem 49(15):4568–4573
58. Xie XQ (2010) Exploiting pubchem for virtual screening. Expert Opin Drug Discov 5(12): 1205–1220
59. Huth JR et al (2005) ALARM NMR: a rapid and robust experimental method to detect reactive false positives in biochemical screens. J Am Chem Soc 127(1):217–224
60. Gul S, Gribbon P (2010) Exemplification of the challenges associated with utilising fluorescence intensity based assays in discovery. Expert Opin Drug Discov 5(7):681–690
61. Soares KM et al (2010) Profiling the NIH small molecule repository for compounds that generate H2O2 by redox cycling in reducing environments. Assay Drug Dev Technol 8 (2):152–174
62. Crowe A et al (2013) Aminothienopyridazines and methylene blue affect Tau fibrillization via cysteine oxidation. J Biol Chem 288(16):11024–11037
63. Feng BY et al (2007) A high-throughput screen for aggregation-based inhibition in a large compound library. J Med Chem 50(10):2385–2390
64. Jasial S, Hu Y, Bajorath J (2017) How frequently are pan-assay interference compounds active? Large-scale analysis of screening data reveals diverse activity profiles, low global hit frequency, and many consistently inactive compounds. J Med Chem 60(9):3879–3886
65. Baell J, Walters MA (2014) Chemistry: chemical con artists foil drug discovery. Nature 513(7519):481–483
66. Tomasic T, Peterlin Masic L (2012) Rhodanine as a scaffold in drug discovery: a critical review of its biological activities and mechanisms of target modulation. Expert Opin Drug Discov 7(7):549–560
67. Ge Y et al (2012) Discovery and synthesis of hydronaphthoquinones as novel proteasome inhibitors. J Med Chem 55(5):1978–1998
68. Priyadarsini KI (2013) Chemical and structural features influencing the biological activity of curcumin. Curr Pharm Des 19(11):2093–2100
69. Qin J et al (2012) Identification of a novel family of BRAF(V600E) inhibitors. J Med Chem 55(11):5220–5230
70. Rai D et al (2008) Curcumin inhibits FtsZ assembly: an attractive mechanism for its antibacterial activity. Biochem J 410(1):147–155
71. Baell JB (2010) Observations on screening-based research and some concerning trends in the literature. Future Med Chem 2(10):1529–1546
72. Habig M et al (2009) Efficient elimination of nonstoichiometric enzyme inhibitors from HTS hit lists. J Biomol Screen 14(6):679–689
73. Jadhav A et al (2010) Quantitative analyses of aggregation, autofluorescence, and reactivity artifacts in a screen for inhibitors of a thiol protease. J Med Chem 53(1):37–51
74. Bruns RF, Watson IA (2012) Rules for identifying potentially reactive or promiscuous compounds. J Med Chem 55(22):9763–9772
75. Kennedy T (1997) Managing the drug discovery/development interface. Drug Discovery Today 2(10):436–444
76. Downs GM, Barnard JM (2002) Clustering methods and their uses in computational chemistry. In: Lipkowitz KB, Boyd DB (eds) Reviews in computational chemistry. Wiley, New York, pp 1–40
77. Harrison RK (2016) Phase II and phase III failures: 2013–2015. Nat Rev Drug Discovery 15:817
78. Todeschini R, Consonni V (eds) (2009) Molecular descriptors for chemoinformatics. Wiley-VCH Verlag GmbH & Co. KGaA, Weinheim, Germany, pp I–XLI
79. Lagorce D et al (2011) The FAF-Drugs2 server: a multistep engine to prepare electronic chemical compound collections. Bioinformatics 27(14):2018–2020

80. Perola E, Charifson PS (2004) Conformational analysis of drug-like molecules bound to proteins: an extensive study of ligand reorganization upon binding. J Med Chem 47(10): 2499–2510
81. Bostrom J (2001) Reproducing the conformations of protein-bound ligands: a critical evaluation of several popular conformational searching tools. J Comput Aided Mol Des 15(12):1137–1152
82. Chen IJ, Foloppe N (2008) Conformational sampling of druglike molecules with MOE and catalyst: implications for pharmacophore modeling and virtual screening. J Chem Inf Model 48(9):1773–1791
83. Stahura FL, Bajorath J (2005) New methodologies for ligand-based virtual screening. Curr Pharm Des 11(9):1189–1202
84. Lorber DM, Shoichet BK (1998) Flexible ligand docking using conformational ensembles. Protein Sci 7(4):938–950
85. Lyne PD (2002) Structure-based virtual screening: an overview. Drug Discov Today 7 (20):1047–1055
86. Sadowski J, Gasteiger J, Klebe G (1994) Comparison of automatic three-dimensional model builders using 639 X-ray structures. J Chem Inf Comput Sci 34(4):1000–1008
87. Kirchmair J et al (2006) Comparative performance assessment of the conformational model generators omega and catalyst: a large-scale survey on the retrieval of protein-bound ligand conformations. J Chem Inf Model 46(4):1848–1861
88. Vainio MJ, Johnson MS (2007) Generating conformer ensembles using a multiobjective genetic algorithm. J Chem Inf Model 47(6):2462–2474
89. Liu X et al (2009) Cyndi: a multi-objective evolution algorithm based method for bioactive molecular conformational generation. BMC Bioinformatics 10:101
90. Blaney JM, Dixon JS (2007) Distance geometry in molecular modeling. In: Lipkowitz KB, Boyd DB (eds) Reviews in computational chemistry. pp 299–335
91. Wild, DJ, Blankley CJ (1999) VisualiSAR: a web-based application for clustering, structure browsing, and structure-activity relationship study. J Mol Graph Model 17(2):85–89, 120–125
92. Nicholls A et al (2010) Molecular shape and medicinal chemistry: a perspective. J Med Chem 53(10):3862–3886
93. Bohm HJ, Flohr A, Stahl M (2004) Scaffold hopping. Drug Discov Today Technol 1(3): 217–224
94. Willett P (1987) A review of chemical structure retrieval systems. J Chemom 1(3):139–155
95. Scott DE et al (2012) Fragment-based approaches in drug discovery and chemical biology. Biochemistry 51(25):4990–5003
96. Lewell XQ et al (1998) RECAP–retrosynthetic combinatorial analysis procedure: a powerful new technique for identifying privileged molecular fragments with useful applications in combinatorial chemistry. J Chem Inf Comput Sci 38(3):511–522
97. Varin T et al (2010) Compound set enrichment: a novel approach to analysis of primary HTS data. J Chem Inf Model 50(12):2067–2078
98. Dandapani S et al (2012) Selecting, acquiring, and using small molecule libraries for high-throughput screening. Curr Protoc Chem Biol 4:177–191
99. Petrova T et al (2012) Structural enrichment of HTS compounds from available commercial libraries. MedChemComm 3(5):571–579
100. Kaldor SW et al (1997) Viracept (nelfinavir mesylate, AG1343): a potent, orally bioavailable inhibitor of HIV-1 protease. J Med Chem 40(24):3979–3985
101. Schindler T et al (2000) Structural mechanism for STI-571 inhibition of abelson tyrosine kinase. Science 289(5486):1938–1942
102. Varghese JN (1999) Development of neuraminidase inhibitors as anti-influenza virus drugs. Drug Dev Res 46(3–4):176–196
103. Rutenber EE, Stroud RM (1996) Binding of the anticancer drug ZD1694 to E. coli thymidylate synthase: assessing specificity and affinity. Structure 4(11):1317–1324
104. Filikov AV et al (2000) Identification of ligands for RNA targets via structure-based virtual screening: HIV-1 TAR. J Comput Aided Mol Des 14(6):593–610

105. Lind KE et al (2002) Structure-based computational database screening, in vitro assay, and NMR assessment of compounds that target TAR RNA. Chem Biol 9(2):185–193
106. Lionta E et al (2014) Structure-based virtual screening for drug discovery: principles, applications and recent advances. Curr Top Med Chem 14(16):1923–1938
107. Schwede T et al (2003) SWISS-MODEL: An automated protein homology-modeling server. Nucleic Acids Res 31(13):3381–3385
108. Eswar N, et al (2006) Comparative protein structure modeling using Modeller. Curr Protoc Bioinformatics Chapter 5: p. Unit-5 6
109. Kelley LA et al (2015) The Phyre2 web portal for protein modeling, prediction and analysis. Nat Protoc 10(6):845–858
110. Blundell TL, Patel S (2004) High-throughput X-ray crystallography for drug discovery. Curr Opin Pharmacol 4(5):490–496
111. Boland A, Chang L, Barford D (2017) The potential of cryo-electron microscopy for structure-based drug design. Essays Biochem 61(5):543–560
112. Sugiki, T, et al (2018) Current NMR techniques for structure-based drug discovery. Molecules, 23(1)
113. Vyas VK et al (2012) Homology modeling a fast tool for drug discovery: current perspectives. Indian J Pharm Sci 74(1):1–17
114. Shao Z et al (2017) Discovery of novel DNA methyltransferase 3A inhibitors via structure-based virtual screening and biological assays. Bioorg Med Chem Lett 27(2):342–346
115. Cura V et al (2017) Structural studies of protein arginine methyltransferase 2 reveal its interactions with potential substrates and inhibitors. FEBS J 284(1):77–96
116. Siedlecki P et al (2003) Establishment and functional validation of a structural homology model for human DNA methyltransferase 1. Biochem Biophys Res Commun 306(2): 558–563
117. Siedlecki P et al (2006) Discovery of two novel, small-molecule inhibitors of DNA methylation. J Med Chem 49(2):678–683
118. Brueckner B et al (2005) Epigenetic reactivation of tumor suppressor genes by a novel small-molecule inhibitor of human DNA methyltransferases. Cancer Res 65(14):6305–6311
119. Clark DE (2008) What has virtual screening ever done for drug discovery? Expert Opin Drug Discov 3(8):841–851
120. Lavecchia A, Di Giovanni C (2013) Virtual screening strategies in drug discovery: a critical review. Curr Med Chem 20(23):2839–2860
121. Wapenaar H, Dekker FJ (2016) Histone acetyltransferases: challenges in targeting bi-substrate enzymes. Clin Epigenetics 8:59
122. Kannt A, Wieland T (2016) Managing risks in drug discovery: reproducibility of published findings. Naunyn Schmiedebergs Arch Pharmacol 389(4):353–360

Structure-Based Drug Design with a Special Emphasis on Herbal Extracts

D. Velmurugan, N. H. V. Kutumbarao, V. Viswanathan and Atanu Bhattacharjee

Abstract Structure-based drug design (SBDD) is a computational analysis of identifying ligands which can potentially inhibit the target. SBDD is a cluster of methods and modules which reduces the cost and time spent on experimental procedures. SBDD plays a crucial role in preclinical drug development procedures. There is a vast development in techniques and methods related to theoretical physics and chemistry, computers processers, and pharmacokinetic analysis which helps in elucidating the biological role of ligands and their receptors. Here, the general theoretical backgrounds of various SBDD and simulation approaches employed are discussed. These methods are also discussed with respect to the identification of potential drug-like molecules from natural sources to control human ailments.

Keywords Docking · Molecular simulations · Pharmacophore
Force field · Crystallography · Natural products

1 Introduction

Drug discovery involves computation in major ways. Structure-based methods involve discovery of lead compounds, their refinement, and re-engineering to overcome resistance. As the number of protein structures available in the Protein Data Bank (PDB) has crossed 1.3 lakhs, SBDD effort with potent targets has progressed well. Compounder model uses many advances in the visualization of molecular structures. Insight II, Quanta, Cerius 2 [1], Sybyl [2], and CAChe [3] are

D. Velmurugan (✉) · N. H. V. Kutumbarao · V. Viswanathan
CAS in Crystallography and Biophysics, University of Madras,
Guindy Campus, Chennai 600025, Tamil Nadu, India
e-mail: shirai2011@gmail.com

A. Bhattacharjee
Department of Biotechnology & Bioinformatics, North-Eastern
Hill University, Umshing Mawkynroh, Shillong 793022, India

C. G. Mohan (ed.), *Structural Bioinformatics: Applications in Preclinical Drug Discovery Process*, Challenges and Advances in Computational Chemistry and Physics 27, https://doi.org/10.1007/978-3-030-05282-9_9

some examples of the commercial programs, and Macromodel [4], Grasp [5], and PyMOL [6] are some of the academic programs available. In all the stages, the macromolecular design process depends heavily on the molecular structure and visualizing it. X-ray structures of macromolecular targets/their inhibitor complexes make it simple to visualize the active site of the enzyme, etc., where water coordination having hydrogen bonds with the backbone of amino acids can be seen. During the design process, the active site water molecule can be replaced by the suitable inhibitor which can bind with the catalytic residues. The above type of molecular visualization was carried out with HIV-1 protease, and the crystallographic study of its complex confirmed this. There are many examples in the literature like the above case using which inhibitors resulting as drugs have come out. Molecular hydrophobicity maps are useful to understand the binding mechanism of inhibitors to the active site. Free energy perturbation calculations are used in the above.

Lipinski's "rule of five" is employed in the selection and filtering of compounds with enhanced oral administration property. Four parameters are used as yardsticks, such as hydrogen bond donors less than or equal to 5, hydrogen bond acceptors less than or equal to 10, molecular weight should be under 500, and CLogP less than 5.

The lead molecules are tested for absorption (A), distribution (D), metabolism (M), excretion (E), and toxicity (T) profile, also known as ADMET properties. These physical and chemical properties have a greater influence on the biological effect of the lead molecule, which determines how drug molecules can sustain in plasma and can determine concentration for oral administration, and also indicates its effects.

In the discovery or in the refinement of new leads, thousands of compounds from various databases are docked with the known target structures. During the docking, the energies of the above complexes are evaluated and the one with the lowest energy is selected as the possible lead compound. Many reviews deal with docking approach [7–11]. During the initial calculation of docking, hundreds of inhibitors may sometime be used. In this situation, receptor site is kept rigid and flexibility is provided with the ligand. This is called rigid docking. After the analysis of score and binding energy, few ligands will be selected for docking and now flexibility will be given to both the receptor site and ligand for a better fit at the active site. This is called induced fit docking (IFD). In this way, docking analysis can lead to the selection of ligands in short time. There are many ligand databases available.

The binding constant of the compounds is related to the free energy of the binding. Many computational tools provide rapid estimates of these free energy changes. For very simple solutes, above calculations are possible using molecular dynamics or Monte Carlo simulations [12]. Direct calculation of free energies of binding is not possible for solutes of complex nature. In this situation, an indirect way of calculating relative free energies of binding is possible [13]. In recent years, computational alchemy methods have been applied to modify the binding of the compounds. For increasing the solubility (and bioavailability), an aromatic H is usually modified by OH and NH_2. For enzymes like thymidylate synthase [14], acetylcholinesterase [15], adenosine deaminase [16], and elastase [17], sets of related inhibitors have been designed in the above way.

Computational alchemy in drug discovery involves potential energy function, and the development of potential functions still continues to tackle various situations. Knowledge of structural waters is useful in inhibitor design.

In structure-based drug design (SBDD), the three-dimensional structure of bioactive agents and their targets are the bases. There are many approved drugs, like AIDS medications Crixivan and Viracept, the flu drug Tamiflu, the leukemia therapy Gleevec, and the cancer agent Tarceva, which have come out through SBDD. X-ray crystallography and NMR spectroscopy are very much useful in knowing the three-dimensional molecular structures which play a crucial role in SBDD. Important antibiotics with the targets lipid II and ribosome 50's subunit have come out using SBDD. Using modeling programs like Analog and QikeProp, many antibiotics are in human clinical trials. X-ray crystal structure of Human Immunodeficiency Virus Reverse Transcriptase (HIV-RT) and the complexes with various antiviral drugs, diarylprimidines, have been discovered as promising inhibitors and some of these are in phase II and phase III trials. In structural-based vaccine design, at Rutgers University, the first structure of a virus that infects animals, the cold-causing human rhinovirus, has been obtained, and its modified form in complex with an anti-HIV antibody was used as the basis [18]. For Tumor necrosis factor-Alpha-Converting Enzyme (TACE), novel hydroxamates were developed using X-ray crystallography in combination with structure–activity relationship (SAR). TACE inhibitors are potential anti-inflammatory agents. X-ray crystallography along with molecular modeling has been used to identify potent and selective inhibitors of human beta-secretase-1 (BASE–I). Alzheimer treatment and selective oral chymase inhibitors for asthma and dermatitis have entered clinical trials where SBDD was used. Since 25% of genes code for membrane proteins like G Protein-Coupled Receptors (GPCRs), these are highly important targets for SBDD. Unfortunately, these membrane proteins are hard to isolate and crystallize. Due to the difficulties in crystallization and not many three-dimensional structures are available, molecular modeling has helped to develop potent inhibitors of glycogen phosphorylase, the target for designing anti-diabetic drugs. Using SBDD approach, compounds were designed with 3- to 14-fold better potency than the lead compound. Recently, fragment-based approaches (FAPs) are used in SBDD. Using the combination of fragment screening, computational chemistry, and structural biology for fragment-based drug discovery, a compound was synthesized and went to clinical trial approval in just 14 months and went to the phase I clinical trial treatment of refractory solid tumors. Like the above, there are many examples showing how many potent drugs have come out using SBDD approach.

2 Role of X-Ray Crystallography in SBDD and Medicine

Crystallography as a science and as a diffraction technique had started and grown exponentially after the famous Laue experiment. The diffraction experiment elucidates the structure and arrangement of atoms/molecules, which form a perfectly ordered crystal. Apart from many biomedical applications, X-rays are majorly used in the elucidation of three-dimensional structures of molecule(s) from the days where X-rays were produced from gas tubes invented by Coolidge to the synchrotron(s) where monochromatic waves are produced and used for diffraction whose energy may vary depending upon the source. The wavelength of X-rays ranges from 0.8 to 2.3 Å, suitable for protein crystallography. The low wavelength and high-energy X-rays help in analyzing the atomic details of proteins and viruses which largely help in the structure-based drug designing. The prerequisite of this diffraction experiment is growth of good quality single crystals. Materials in nature (atom, molecules) possess the property to form crystalline solid, where the basic constituents of the material tend to arrange themselves in a highly ordered manner. This arrangement is called "crystalline lattice," and the solid material is crystal. Crystals are very regular in shape and highly ordered and symmetrical in nature, which can be understood from the external examination. The periodic arrangement of atom and ions in three-dimensional space is "crystal lattice." The smallest repetitive unit in three dimensions, which on translation gives the entire crystal structure, is called "unit cell." Depending upon the number of atoms in the molecule which is crystallized, there are two categories, namely small molecular crystallography and macromolecular crystallography.

The structure determination of molecules which are in the range of several hundred Daltons is named as small molecular crystallography. These structures usually contain inorganic, organic, metallo-organic, and material structures. Cambridge Structural Database, CSD, is the world's repository for small molecular organic and organo-metallic crystal structures. The CCDC plays a major role in the collection of nearly 9, 50,000 updated entries by 2018, and this information is made available to all scientists across the world.

The importance of enzymes, their characterization, and crystallization was stated by Sumner, Northrup, Kunitz, Herriott and their colleagues [19–21]. These investigations proved to be an important tool in identification of the properties and nature of catalytic mechanism of macromolecules and their nature to form crystals. There too, many factors control the crystal growth of macromolecules. The structures of macromolecules are deposited in Protein Data Bank (PDB) which is an open-source database. The structure and also other experimental and relevant information are incorporated for comparative studies. The total number of entries in PDB as of August 2018 is 1,43,392.

Importance of protein crystallography in medicine

The availability of structural information of small molecules assisted scientists to exploit and alter the molecule for biomedical benefits, which was evident to crystallographers and also further in extending this method for proteins as well, since such application has greater potential with greater degree of medical implications. The idea of structure–function relationship met with greater effect with examples such as function of oxygen binding and affinity toward hemoglobin, insulin function. There is a greater report on the function of proteases from different viruses, especially, protease from HIV, where structures of apo as well as ligand bound forms have provided greater insight into functioning and inhibitions [22]. Crystallography has helped in the structural characterization of many mutations from HIV protease, where these mutations alter the active site and these mutants reduce the drug efficacy. In case of HIV-RT, the structures helped in proposing three mechanisms of drug resistance because of mutations, where these mutations caused the alteration of the binding sites for the nucleoside analogue or non-nucleoside inhibitors, mutations at the template DNA-binding site, and finally mutations at active site region influencing the conformation of the enzyme [23–27].

Apart from individual laboratories trying to solve protein structures, there are different structural consortium projects undertaken at higher level, which have also been facilitated by the determination of genome sequences of most microorganism and also humans. This has provided us with a great wealth of protein structures which are crucial for pathogen survival and replication, thus chosen as targets in designing new pharmaceutical importance. The role of crystallography in elucidating the advent of many genetic disorders is commendable and successful. For instance, the understanding of sickle cell anemia, thalassemias, and other deficiencies of hemoglobin [28] is a well-known fact that has come out as consequence of structure–function relationships.

The success of crystallography in the identification of inhibitors can be observed in case of viral diseases, especially in the case of influenza virus by neuraminidase. Numerous structures which are crucial drug targets of various bacteria have been solved, and their function was understood using X-ray crystallography. In addition, structural studies of a wide spectrum of target proteins from protozoa species such as Plasmodium falciparum, Trypanosoma cruzi, and different Leishmania species are carried out. Other phenomena such as detoxification, mutation, and enzyme replacement mechanisms which lead to resistance are explained in molecular detail which is possible largely by employing crystallography.

Role of synchrotron radiation in SBDD

As discussed earlier, crystallography helps in understanding the mode of binding of ligand with target protein, and it provides detailed pattern of interactions of ligand with which it inhibits or promotes the function of proteins. These interactions are used as stepping stones in designing new class of ligands with better efficiency, improving their specificity, physicochemical properties, reducing interactions which might induce cross-interaction with protein leading to side effects [29].

The bottleneck problem arising at this situation is the availability of high-throughput and pipelined techniques at different stages, of which high-energy synchrotron radiation plays a crucial role. The high energetic, physically tunable wavelength X-ray radiation helps in obtaining phases (using anomalous dispersion) relatively easily than any other methods. Advent of improved detectors (CCD and pixel array detectors) and advanced goniometers also play major roles in more structures being determined [30].

Pipelining helps in saving time and increased precision by least human intervention; they provide users with workflow to solve structures starting with data collection, processing, structure determination, and finally with refinement with no time and least intervention. This continuous workflow is important in understanding the structure–activity relationship of numerous ligands toward single target. Crystallography offers us glimpse into the stable and global minima structure and the atomic details of catalytic domains. Thus, implementing a high-throughput screening of bound ligands and analyzing their ability to bind crucial residues and identification of different modes of binding for different fragment scaffolds of ligands are major outcomes that play crucial roles in SBDD. Identification of such varied mechanisms in ligand binding, if exists, can throw light into aspects of improving the ligand efficacy to accommodate key interactions. With speedy data collection strategies and pipeline in structure solution and analysis, synchrotron beamlines help in carrying out ligand-binding experiment in a high-throughput manner. One such automation process is called DIMPLE [31]. High-throughput workflow helps in employing the screening and analyzing in the crystallographic readout for the presence of ligands and understanding their binding with the protein domain. These structural screening methods have helped in identification of fragment-based ligand scaffolds which are carried forward for further clinical testing. Automation of these modules has been completed and can be executed remotely without being at site of diffraction. For the identification of lead molecules, computational analysis is very much useful [32, 33]. Docking plays a major role in this [34].

3 Docking

Molecular docking is a method of identification of binding mode of a small molecule in the active site of the protein with more stability. In docking studies, the score and energy associated with each binding pose are related to the activity. One can refer a review article for molecular docking-related terminologies [35].

Docking aims to predict an accurate enzyme-inhibitor (*EI*) complex under equilibrium conditions. The equilibrium depends on the factors such as desolvation, rational entropy, and translational entropy.

$$[EI]\text{aq} \leftrightarrow [E]\text{aq} + [I]\text{aq} \tag{1}$$

$$[E]\text{aq} + [I]\ \underrightarrow{\text{aq}\Delta G_{\text{bind}}}\ [E + I]\text{aq} \tag{2}$$

Following equation relates the binding affinity and binding free energy

$$\Delta G = -RT\text{In}K_{\text{A}} \tag{3}$$

$$K_{\text{A}} = K_{\text{i}}^{-1} = \frac{[EI]}{[E][I]} \tag{4}$$

where [E], [I], and [EI] are concentration of enzyme, inhibitor, and enzyme-inhibitor complex, respectively. The solvation term (aq), association constant (K_{A}), and the inhibitory constant (K_i) are related using the above expressions. In the process of identification of correct poses, the ranking is based on correlation with corresponding K_A values of the training compounds.

Steric, electrostatic, hydrogen bonding, inhibitor strain, and enzyme strain are some of the important factors affecting the predictive accuracy of *EI* complex.

The binding energy is calculated from electrostatic (E_{coul}) and van der Waal's (E_{vdW}) interaction energies (Eqs. 5 and 6).

$$E_{coul}(r) = \sum_{i=1}^{N_A} \sum_{j=1}^{N_B} \frac{q_i q_j}{4\pi\varepsilon_0 r_{ij}} \tag{5}$$

where N is the number of atoms in the molecules A and B, q is the charge on each atom, r is the distance separating the two point charges, and ε_0 is vacuum permittivity.

The other contributing factor to the potential energy calculation is the van der Waal's contribution. The treatment of the non-bonded interactions is done by implementing Lennard-Jones 12-6 function.

$$E_{vdW}(r) = \sum_{j=1}^{N_A} \sum_{i=1}^{N_B} 4\varepsilon \left[\left(\frac{\sigma_{ij}}{r_{ij}} \right)^{12} - \left(\frac{\sigma_{ij}}{r_{ij}} \right)^{6} \right] \tag{6}$$

where

ε is the well depth of potential energy.
σ is the collision diameter of the atoms i and j, respectively (Fig. 1).

Generally, molecules will be represented as a function of potential energy. Other two ways of representation are surface and grid methods, of which surface representation was pioneered by Connolly [36, 37]. This surface-based docking is more suitable for protein–protein docking to a good extent. Grid representation is also a standard method for protein–protein docking [38].

There are three methods of treating the conformation freedom during docking.

(1) Systematic methods.
(2) Random or stochastic methods.
(3) Simulation methods.

Systematic method is a stepwise or incremental search. The algorithm tries to explore all the degrees of freedom. The search can be done in two ways. First, fragments of molecule are docked into the active site region and then joined to each other. Second, the ligand is divided into rigid (core fragment) and the flexible part (side chain). The rigid portion of the ligand is first docked into the grid region to which the flexible regions are attached in an orderly manner.

In random search method, the search was accomplished by implementing random changes to either ligand or a cluster of ligands. The pre-defined probability

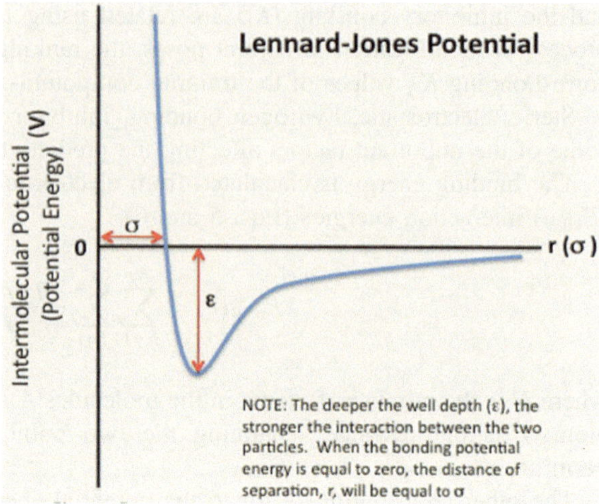

Fig. 1 Pictorial representation of the Lennard-Jones equation (adopted from picturesquephysics)

function is used to evaluate the newly obtained ligand. Monte Carlo and genetic algorithms are the two most widely used random search algorithms. AutoDock uses a variant of Monte Carlo algorithm, and DOCK and GOLD use genetic algorithms.

Simulation method is a widely implemented approach with the only limitation for crossing high-energy barrier within the stipulated time. This allows the ligands which are trapped in local minima to cross the barrier. Monte Carlo is also added at times to compliment the simulation method. DOCK and Glide use this method. Some of the important and widely used search methods are listed in Table 1.

Flexibility of Protein: The methodology applied to introduce flexibility of ligand is well parameterized when compared with the flexibility of protein. There are algorithms which make a part of the protein flexible, during the docking process. Monte Carlo simulations, rotamer libraries, and protein ensemble grids help in this regard [48]. One of the approaches is to generate average potential energy grid for the ensemble, and the other one is to map different receptors to each grid point and then score ligand with each of the receptor possible.

Scoring: Generating ligand conformation is achieved, but sorting and ranking the predicted conformation are more appropriate and the most vital role in choosing the best ligand. Separating correct pose from the incorrect poses is the crucial step of docking as this process helps in identification of the reliable ligand which can

Table 1 Flexible ligand search methods

Random/stochastic	Systematic	Simulation
AutoDock (MC) [39]	DOCK (incremental) [43]	DOCK
MOE-Dock (MC,TS) [40]	FlexX (incremental) [44]	Glide
GOLD (GA) [41]	Glide (incremental) [45]	MOE-Dock
PRO_LEADS (TS) [42]	Hammerhead (incremental) [46]	AutoDock
	FLOG (database) [47]	Hammerhead

Table 2 Types of scoring functions

Force-field-based	Empirical-based	Knowledge-based
D-Score [49]	LUDI [52, 53]	PMF [59–61]
G-Score [49]	F-Score [44]	DrugScore [62]
GOLD [50]	ChemScore [54]	SMoG [63]
AutoDock [51]	SCORE [55, 56]	
DOCK [43]	Fresno [57]	
	X-SCORE [58]	

Note PMF—Potential of the mean force, GOLD—Genetic optimization for ligand docking

bind with the target molecule. Good scoring functions are important for docking procedure. There are three different kinds of scoring functions which are widely used. Different software which use a particular scoring function are listed in Table 2. The scoring functions are

1. Force-field-based scoring.
2. Empirical scoring.
3. Knowledge-based scoring.

4 Protein Simulation and Drug Designing

Molecular dynamics (MD) and Monte Carlo (MC) simulations are widely used to understand the structure–function relationship of proteins. A large spectrum of studies can be studied ranging from ligand binding, enzyme mechanism, folding–unfolding, etc., and the method started at its infancy 25 years before. In these analyses, the simulations are understood in terms of energy as the function of atomic coordinates. The series of frames of structure generated at thermal equilibrium (trajectory) are function of low potential energy, and forces on individual atoms are related to the gradient of this function, which is commonly referred as "force field."

The Born–Oppenheimer ground-state energy is the energy surface. It is assumed that the atoms moving on potential energy surface obey it. The calculation of such energy state directly is possible by means of quantum mechanics calculations; however, it is quite difficult to do such calculations for macromolecules with number of atoms more than 150–200, depending on the use of appropriate basis set and method, and the available computation facility. In most practical simulations, simple classical energy functions are used. Most of the force fields which are employed nowadays are developed in the early 1990s.

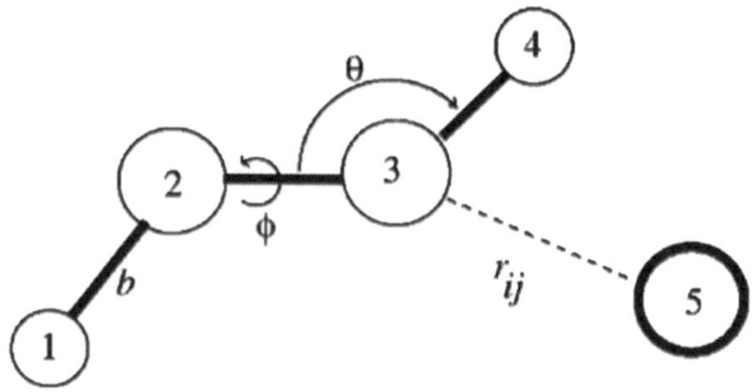

Fig. 2 Representation of bonded (atoms 1–4) and non-bonded (including atom 5) terms of force field

The widely used force field has the potential energy function:

$$V(r) = \sum_{\text{bonds}} k_b(b-b_0)^2 + \sum_{\text{angles}} k_\theta(\theta-\theta_0)^2 + \sum_{\text{torsions}} k_\phi[\cos(n\phi+\delta)+1] + \sum_{\substack{\text{nonbonded}\\ \text{pairs}}} \left[\frac{q_i q_j}{r_{ij}} + \frac{A_{ij}}{r_{ij}^{12}} - \frac{C_{ij}}{r_{ij}^{6}}\right] \quad (7)$$

First three summations denote the bond, angle, and torsional term, respectively (Fig. 2).

The final summation excludes 1–2 and 1–3 interactions and often uses separate parameters for 1–4 interactions. The equation explains electrostatics which uses partial charges q_i on every atom which are in interaction bound by Coulomb's law. The Lennard-Jones 6-12 potential represents the combination of dispersion and exchange of repulsion forces. This function is usually called "van der Waals" term. This equation helps in exploring the basic aspects of potential energy landscapes in atomic detail. The combination of potential energy function with different parameters $(k_b, b_0, k_\theta, \theta_0)$ helps in constructing it and is labeled as "Force field."

The "force field" history dates back to the year 1980, when the simulation technique was started. Building of force field for protein simulation started with a template, which is from the force fields of organic chemistry. Few of the potentials are ECEPP potential by Scheraga and workers [64, 65] and CFF [66–68]. The popular force fields that we employ are given below.

AMBER force field

The key input in the early stages of AMBER development is the charges that were derived from quantum chemistry calculations fitting partial atomic charges to the quantum electrostatic potential, which are generally called "electrostatic potential," ESP charges. The polar hydrogen was explicitly represented, but hydrogen atoms bonded to carbon were combined with united atoms. The van der Waal's (vdW) terms were derived from crystal data by Lifson's group [67, 68] and from the liquid simulations by Jorgensen [69]. The force constants, idealized bond lengths, and angles taken from crystal structure and normal mode frequencies are used for a number of peptide fragments. In due time, the studies led to employing all-atom force fields. Further, most of the force fields are extended to their all-atom versions with reference to the work done by Weiner [70]. The advent of high-performance computers has prompted the development of new form of force field which is popularly known as ff94 force field [71]. The algorithms and automation for force field beyond protein molecule were achieved by the introduction of antechamber program, which completely automates the creation of AMBER-like force field for molecule. The use of fitted charges at the HF/6-31G level has shown a general way to develop charges for all 20 amino acids in the way which is roughly consistent with the water molecule. The implementation of the above method for the development of charges has two major complications: one being the underdetermination of effective charge in more buried atoms and the other being the procedure of implementing Restrained Electrostatic Potential Fit (RESP) where the charges depend on the molecular conformations. To overcome these problems, a more complex term is required, and this is overcome in ff94 by fitting the charges simultaneously to several conformations, hoping to achieve optimal average behavior. Torsion angle parameters for the ϕ and ψ backbone angles affect largely the energies of helices and sheet in proteins. This was done by the ff94 by fitting representative points on the dipeptide maps for glycine and alanine and computed at MP2 level with TZP basic set. In recent years, work is still actively undertaken to test the potentials with the experimental values of short peptides. From the works of Damm, Mitsutake and Garcia [72–74], there are modifications done to the ff94, and at least two modifications have been proposed based on large-scale short peptide simulations [75, 76].

CHARMM force field

Chemistry at Harvard using Molecular Mechanics (CHARMM) program [77] was developed in 1983. There were rather many versions of CHARMM, namely CHARMM19, CHARMM22, and CHARMM27, with different charge-deriving methodologies incorporated. The same difficulty of torsion angle parameterization exits in CHARMM as observed in case of ff94. The comparison of AMBER and CHARMM force fields in respect to protein potential shows that the peptide carbonyl group is less polar in CHARMM22 than with ff94, and NH dipole is less polar in the case of AMBER. We can observe the similarities in the charge model between the two force fields prominently than the differences between them.

OPLS force field

This is the other force field developed during the 1980s by Jorgensen and co-workers to simulate liquid state. This force field called optimized potential for liquid simulations (OPLS) plays a great importance to non-bonded interactions by comparison with liquid-state thermodynamics [69]. For proteins, polar hydrogen model only was developed initially with the atoms type and valance (bond, angle, dihedral) parameters form AMBER, and later, all-atoms force field was developed (OPLS-AA).

Importance of protein flexibility

SBDD works by the identification of sites on the protein surfaces, named “hot spots,” where the chemical motifs will bind the receptor residues. The methods which are generally employed in most docking methods are devoid of important aspects, protein flexibility, and are not calibrated enough to understand the difference in surface recognition by ligands and water. The target surface potential is roughed, and in addition, local minima are identified which are false. Research work carried by Lexa et al., analyzed the effect of complete flexible protein and solvent on simulation and drug identification. The results point out that only when protein is allowed to be fully flexible the true local minima were identified and also the accurate identification of ligand-binding regions (hot spots). A protocol with mixed-solvent molecular dynamics (MixMD) was proposed to map hot spots which matched with that identified using an experimental method MSCS. MixMD simulations help in finding true binding minima and hot spots, thus retaining the importance of incorporating flexibility to protein receptors [78].

Importance of protein–ligand interactions in drug design

Enzymatic reaction and ligand binding are the key steps, and the detailed understanding of interaction between small molecules and protein may form the basis for a rational drug design [79–82] to address the major pathologies such as cancers [83, 84] and cardiovascular [85–87] diseases. Docking studies have been useful to identify new lead compounds as anti-infection agents for Mycobacterium tuberculosis or Plasmodium falciparum, the two pathogens involved in the development of TB and malaria, respectively [88].

Protein–protein docking approach

Protein–protein interactions (PPIs) play a vital role in all biological/cellular processes. This results in a formation of the macromolecular assembly crucial for different cellular functions. Initially, the protein–protein docking was introduced in 1978 itself [89] and was later extended to the interaction between the macro- and small molecules [34]. Currently, protein–protein docking algorithms have been developed in light of the critical assessment of prediction of interaction (CAPRI) rule which has accelerated the development of more efficient protein docking methods [90]. Prediction of protein complex interface is a major driving factor of the accurate outcome [91–94]. Incorporation of global and local flexibility in the docking algorithms provides invaluable information in mutagenesis studies and to steer drug design applications [95–98]. Residue interaction networks (RINs) are small-world networks, and their topological analyses have been used in particular to study protein–protein interfaces [99] and to optimize scoring functions for the evaluation of docking poses [100, 101]. The docking prediction can be used in combination with homology-based methodologies and integrated into PPI networks to enhance the structural information [102, 103]. Several disease-causing mutations were located at the interface of protein; these key elements could be targeted by drugs. It was predicted that, on an average, a drug binds to six different targets, including both the primary target and additional "of targets" [104]. Using the idea, reverse docking can be performed where one single molecule is screened against multiple receptors instead of screening multiple small molecules against several receptors [105, 106].

Solvent effect
Catalytic water molecules are also crucial in enzyme-substrate recognition and enzymatic reaction [107, 108]. The electrostatic screening performed by water molecules helps in identification of ligand binding to the protein [109]. In a living system, proteins are active and move in solvent environment with a dielectric constant of around 80, where waters are arranged around the protein with constant motion [110]. There are many types of water models, and their effect in docking and simulations are well studied. In both explicit and implicit models, incorporation the effect of salvation is available, and choice of the model depends on the computational resource available with the user.

Modeling solvent molecules
Explicit water models
The first model for liquid was proposed by Bernal and Flower [111]. Then, ST2 model of water proposed by Stillinger and Rahman [112] was widely used during initial stages of development of the protein force fields. The SPC [113] and TIP3P [114] are similar to each other in terms of atomic point charge. In both the models, three-site rigid water models are parameterized to produce structure which is in bulk phase. Thermodynamics of liquid water is taken care. There are other advances in recent times, and newly developed solvent models such as TIP4P and TIP5P have the most agreement with the experimentally calculated internal energy [115]. There are different continuum models, majorly COSMO model [116], Poisson–Boltzmann (PB) models, and the most commonly used Generalized Born (GB) model [117, 118].

The increase in the computation power as a result of evolution of core processors and GPU computations helps in the betterment of the force field development and matches with the experimental data. The design of good potential approximation used for simulation analysis helps in prediction of ligand binding, protein structure prediction, and drug designing accurately. Hence, the potential functions and their approximation are crucial. The development of force field will lead to the better chemical accuracy that has to be reached for the best simulation of the biological molecules aiding in studying their properties.

This development of the force field, simulation methods, and algorithms that make the calculations automatic and varied potential energy descriptors has a great impact on the aspect of computer-aided drug design (CADD). Molecular optimization and free energy calculations are important to a major extent that will aid in the understanding of a molecule binding with protein. In drug designing, it should be noted that simple and fast methods should be used to screen large number of molecules. There are different model systems available with different properties that can be used to analyze the ligand binding to the protein and its recognition. Different models are briefly stated below.

Fixed conformation models
Force field calculations
Most of the molecular mechanics-based assessments of ligand binding with the receptor follow the below expression

$$\Delta E = E_{\text{complex}} - \left(E_{\text{protein}} \mp E_{\text{ligand}}\right) \quad (8)$$

where $E_{\text{complex}}, E_{\text{protein}}, E_{\text{ligand}}$ are the potential energies of the complex, protein, and ligand, respectively. In which, the hydrophobic contributions have the function

$$G_{\phi} = \gamma \text{SA} + c \quad (9)$$

where γ is a microscopic surface tension, SA is the solvent accessible surface area of the solute, and c is a constant.

Poisson model

This model provides good description of electrostatic properties of molecules using the following function

$$-\nabla \cdot \varepsilon(\vec{r})\nabla\phi(\vec{r}) = \rho(\vec{r}) \quad (10)$$

where $\varepsilon(\vec{r})$ is the dielectric function, $\phi(\vec{r})$ is the electrostatic potential, and $\rho(\vec{r})$ is the charge density at $\vec{r}$.

The analytical solution for the simple system is quite possible, but for calculation of complicated models, the finite difference methods can be used for obtaining numerical solution of the Poisson model. Delphi [119] and UHBD [120] are the two packages for execution of these calculations.

Poisson–Boltzmann model

The salt effect on the solution is incorporated into the Poisson equation to generalize the method to account for the experimental conditions. The density of ith type of ion n_i at different points in space can be explained using the relation,

$$n_i = n_i^0 \exp\left(-\frac{q_i \phi}{RT}\right) \quad (11)$$

where n_i^0 is the number of density if ion of ith type in pure salt solution, R is the gas constant, and T is the absolute temperature. The atomic point charge is:

$$\rho_i = n_i q_i \quad (12)$$

where q_i is the charge of type of ion.

When this charge is added to the Poisson equation, it gives the Poisson–Boltzmann equation, where the sum is over all types of mobile ions.

$$-\nabla \cdot \varepsilon(\vec{r})\nabla\phi(\vec{r}) = \rho(\vec{r}) + \sum_i q_i n_i^0 \exp\left(-\frac{q_i \phi}{RT}\right) \quad (13)$$

The function given above and its linearized variant which is used for low univalent salt concentrations and moderately charged systems help in analyzing the effect of salt which is considered in a mean-field sense. This approximation is more appropriate for many drug design applications.

Generalized Born model
Born method is relatively simpler and faster in execution [117]. Different variants are available, of which Qiu et al. [121] use the function given below to estimate the electrostatic contribution to the solvation energy

$$-\frac{1}{2}\left(1-\frac{1}{\epsilon}\right)\sum_{i=1}^{N}\sum_{j=1}^{N}\frac{q_i q_j}{\sqrt{r_{ij}^2+a_{ij}^2 \mathrm{e}^{-D}}} \tag{14}$$

where $a_{ij}^2 = a_i a_j$ and $D = r_{ij}^2/(2a_{ij})^2$

a_i is the Born radius of atom i, and r_{ij} is the inter-atomic distance between the atoms i and j.

At the end of molecular dynamics simulations, one usually interprets the following graphs: Root Mean Squared Deviation (RMSD) as a function of simulation time, Root Mean Squared Fluctuation (RMSF) for each amino acid residues in the protein of interest, Radius of Gyration (Rg) as a function of simulation time (if there are large-scale conformational changes like open-to-close conformational transition, domain movement, α-to-β transition in protein aggregation, etc.), and frequency/occupancy of H-bond interactions (to assess the lifetime of H-bond throughout the simulation time).

5 Different Approaches in Drug Designing

High-throughput screening
Virtual screening is one of the commonly used approaches in lead identification step and is seen as a complementary approach to experimental high-throughput screening (HTS) to improve the speed and efficiency of the drug discovery and development process [122]. This involves explicit molecular docking (process to predict binding mode) of each ligand to the binding site of the target and scoring (process to measure binding affinity). The compounds in the databases screened are ranked to select and experimentally test a small subset for biological activity, considered to be appropriate for a given receptor. Many successful applications have been reported in the field of molecular docking-based virtual screening. Although the energy calculations involved are crude, the compounds in the library are readily available, making experimental testing easy and false positives tolerable [123].

Fragment-based screening
Optimization of lead molecules which were identified from crystallographic analysis, structure-based analysis of target, ligand screening using HTS, and other

computational methods may only provide a core structure of a drug molecule. This basic molecule can be altered to reduce its side effects by increasing the specificity, efficiency in bonding pattern, change in molecule weight, and so on. In overcoming the hurdles during high-throughput screening process in identification of good lead compound, shaping and building the core molecule architecture through step-by-step substructure improvement was introduced. This process is called fragment-based drug development. This method of sculpturing of molecule for increased efficiency was first developed by Fesik group around 1996 [124]. A similar initiative was introduced well ahead by Hol et al., during 1990 [125]. This approach has come a long way in various procedures specifically adopted in identification of less complex, more specific small molecule with accepted molecular weight which binds to both protein and DNA molecules. The fragment-based drug identification is complemented with many structural characterization methods such as fragment-based approach which has also a crucial role in identification of lead molecules for unconventional targets [126–128], to identify chemical probes of biological systems [129, 130].

A two-way approach exists in implementing the fragment-based method. Here, the first case screening will be carried out to analyze a small set of compounds with lower molecular weight to understand the binding mode against a protein site. The molecular weight range of these compounds should be manageably good enough to possess interaction and not large to overcome unfavorable interactions. In the second phase, these small substructures are optimized to lead compound by addition or inclusion of properties of individual molecules so as to obtain a better-optimized ligand. Different steps or doorways are present in developing lead using fragment method, starting with library, method in identifying the interacting fragment with protein target, structural analysis of such bound fragment, choosing best fragment, and building the fragment into lead compound. This method largely draws its structural and interaction information from X-ray crystallography, NMR, surface plasma resonance, and other biophysical techniques.

The greater advantage of using the fragment-based method in drug discovery strategies is its success in varied targets where regular high-throughput screening fails. The targets may include large multimeric proteins, protein–protein complexes, ubiquitin-specific proteases, etc. [131].

With a few slight variations in the workflow, the basic approach for Fragment-Based Method (FBM) was well derived by the wake of the twenty-first century [132–134]. The increased use of surface plasma resonance (SPR) has boosted the fragment approach to new heights with rigorous and better fragment binding understanding [135, 136]. In addition, many developments in implementation of other techniques such as immobilization of the protein and its reorganization by ligand using optical, NMR, mass spectrum, and other fluorescent-based techniques [137–140] had happened.

FBM has helped in the identification and development of an FDA-approved drug [141], and much more are in clinical trials [142]. Integration of FBM with SBDD in identification of appropriate scaffold can significantly increase the result rate. It should be noted that the importance and significance of small units (lower molecular weight), which in a combination would work to a good molecule, would have been screened out in regular HTS workflow; this method can show and exploit the potential of different functional groups and core structures in harboring key interactions with the active site pocket.

Pharmacophore modeling
Pharmacophore is the method to identify and understand the structural features of the active site or ligand-binding site of receptor which is responsible for the biological activity. Recent years had witnessed the extensive use of pharmacophore individually or in combination with various SBDD methods for better results [143–146]. The structural aspects and the model are derived from the sets of ligand molecule which are substrates or reported binders toward the target protein. This model which is derived projects the requirements which are important for ligands to bind to receptor for their recognition, leading to biological response. The pharmacophore model can be established as a ligand-based or in a structure-based manner.

Ligand-based pharmacophore modeling
Common chemical features are extracted from the three-dimensional structures of a set of known ligands. During the pharmacophore generation, the conformational space for each ligand in the training set is created which represents the conformational flexibility of ligands. Some of the commercially available software packages for pharmacophore modeling are HipHop, Hypogen (http://accelrys.com/services/training/life-science/pharmacophore-modeling.html), DISCO, GASP (http://pharmacophore.org/), Phase (https://www.schrodinger.com/), molecular operating environment (MOE, https://www.chemcomp.com), etc. There are also several academic programs available. In these programs, algorithms are used for handling the flexibility of ligands [147, 148] and for the alignment of molecules.

Quantitative structure–activity relationship (QSAR)
For the correlation of biological activity with the molecular properties (descriptors) calculated from the two-dimensional or three-dimensional structures of ligands [149], QSAR is used. For forming a linear equation of this type, a set of ligands with the known biological activities called the training set is essential. Unknown activity of novel compounds can then be predicted using the relationship: $v = f(p)$, where v is the biological activity and p is a set of descriptor properties. Most of the software packages listed for pharmacophore modeling are having QSAR modules. Most widely used methods are kNN, PLS, MLR, etc., and these functionalities are available with different software packages such as VLife sciences, Schrodinger, and MOE.

Peptide-based drugs
Drugs which are available can be categorized into smaller molecules which are usually <500 Da, and others are larger molecules with molecular weight >5000 Da. The lower molecular weight drugs can be orally administered, and larger ones are delivered through injections.

The disadvantage of small molecule drugs is that they have poor target selectivity which results in a broad range of side effects. The larger molecule-based drugs have an advantage of target specificity, majorly based on their biological characteristics, but the disadvantage of these molecules is their lesser bioavailability and metabolism. There is a great deal of research in the search of peptide drugs which lie between the extreme end of the molecular weight spectrum and possess the advantages of both.

Small molecules are identified from screening, ligand-based, structure-based, or receptor-based design, and such small molecules were majorly subject of research with major successes in the treatment of diseases. With advent of molecular biology, purification, and biophysical characterization, new classes of specific molecules were discovered which are named "Biologics." These are usually antibiotics, growth factors, insulin substitute, etc., yet these are intravenously administered. Molecules which are presently in market such as infliximab to treat arthritis, bevacizumab in treating colorectal cancer, and trastuzumab for breast cancer, insulin for diabetes, Epogen and Avonex are few peptide-based molecules which have a high pharmaceutical importance and a financial market. The market for peptide-based drugs is around $40 billion yearly.

The development of sequencing techniques and analyzing the large datasets of genes, proteins, and also transcription products had resulted in the identification of many molecules which are proteins in nature that exhibit specificity toward receptors and drug targets. Majority of these molecules do not come under the spectrum of rule-of-five. In addition to this deviation, these peptide-based molecules are metabolized easily by the proteases present in the body rapidly. They have large limitation to cross the membrane because of the size and others. In spite of these prominent disadvantages, they possess high efficiency in selective binding and are strong and natural binders of multiple targets. They are found to accumulate less compared to synthetic compounds and finally less toxic. Most of the peptides used in pharmaceuticals belong mostly to size lesser than 10 amino acids, with few exceptions. Some of the peptides which are at the high end of molecular spectrum are 32-residue-long calcitonin, 34- and 36-residue-long teriparatide and Fuzeon, respectively. Food and Drug Administration (FDA) has approved about 60 peptide drugs into market, about 140 are under clinical trials, and around 500 are in the stage of preclinical trials [150].

It is interesting to note that recent peptide molecules out in the industry are exenatide and ziconotide. Exenatide is a 39-residue peptide employed in the diabetes treatment, isolated from saliva of lizard [151]. Ziconotide is isolated from marine cone snail and targets pain relief. In addition, other low molecular peptides, such as captopril and tirofiban (3 aa) and eptifibatide (7 aa), isolated from venom of animals are marketed. Different organisms, namely reptiles, marine fishes, marine plants, are rich with peptides of moderate size, and they play crucial role in binding with many neural and inflammatory-related targets. In addition, venom of them constitutes of various membrane-binding and pore-forming peptides that can act as anti-bacterial and antifungal agents. Design of peptide with important functional scaffold is an important approach which helps in producing bioactive mimicking sequences. Many such scaffolds like knottins, cyclotides [152–155], conotoxins from marine snail [156, 157] were identified. The advent of proteomic data analysis with Mass Spectroscopy (MS) and Next-Generation Sequencing (NGS) helps in identification of such potential peptides which are conserved evolutionary. Proper screening of such peptides can be helpful in the identification of peptides which might be closer to lower end of molecular weight spectrum (500 Da) and far from the higher end (5000 Da) and understanding structure–function relationship of bioactive sequences. Such systematic search can yield side effect-free peptide or peptide-like molecules with potential pharmaceutical application.

Other approaches in exploration of potential lead molecule employing different approaches and revisiting old ones with new perspective will help in SBDD, and one such aspect is drug repurposing. Drug repurposing or drug repositioning is a process in which new uses are identified and attributed for a drug which already exists [158]. With a huge set of drug molecules in the clinical and preclinical stages, trials will not be carried out for next level because of side effects or toxicity, etc. Researchers have started to understand the toxic effect of the drug-like molecule and maintain the database with these prediction results [159]. The concept of drug repurposing was introduced to understand the properties of existing drugs for being safe [160]. Drug repurposing helps in avoiding adverse effects and also in saving time in drug development pipeline [161], and this is achieved by analyzing the similarities of drugs. Such similarities provide with a situation, where they inhibit the same target and result in the same function. The profiling of pharmacological properties of drugs which are presently available was done by Cheng et al., by comparing the chemical similarity of drug molecules and their phenotypic effects [162]. The larger information available deals not only with the properties of drugs but also regarding targets, and this has resulted in "big data" analysis, network analysis, machine learning by Bayesian statistics [163], deep learning, multi-task learning [164], and ligand-based chemogenomic analysis [163, 165, 166], which are advanced and recent algorithms, and routines are employed in drug designing approaches. Yet, a lot of scopes still persist and also large information is to be explored for a better profiling and understanding of the drugs and their physiological roles. Out of many aspects, drug–drug interaction (DDI) is studied in recent years. This will help in understanding the effect of one drug on the other and their interaction. Networking analysis among drugs can help in grouping drugs which might manifest similar physiological effect. Different parameters can be computed with benchmark parameter with which the dataset can be classified and even be validated using training and test datasets. Different statistical methods are employed. For instance, Tanimoto coefficient [167] was employed in understanding the DDI among 6711 drugs collected from the DrugBank [168]. Similarity analysis, association networking for scoring the drugs, and target prediction to predict the numbers of targets each drug binds and target prediction with chemical structure of drug molecules were performed. This approach will provide us with parameters, based on which two drugs interact, and the results project that the DDI effect can manifest due to pharmacokinetic parameters. Even though DDI can help in predicting targets, proposing new targets, and providing us with drug–drug synergetic effect on same target, there are still many hurdles posed for employing this method.

This situation basically arises from the fact that in vitro or physiological mapping of the molecule is a prerequisite for DDI.

One other approach aiding SBDD is de novo *method,* which means "from the beginning." Active site of drug targets when characterized from a structural point of view will shed light on its binding features. This information of active site composition and the orientation of various amino acids at the binding site can be used to design ligands specific to that particular target. Computational tools that can analyze protein active site and suggest potential compounds are extensively used for de novo design methods. Many promising approaches with the goal of ligand design have been reported.

Gane and Dean [169] reported that various de novo methods, especially whole molecule methods like docking, have become integrated within disciplines that include chemistry, pharmacology, molecular biology, and computer modeling. Electrostatic and solvation terms critical for evaluating correct binding energies are difficult and slow to calculate. Advances in algorithm sophistication are providing better approximations for these parameters.

Advancement in different concepts and parameters that are crucial in SBDD is to be employed to improve the outcome of ligand search. For instance, SBDD techniques range from simple docking process to proteins considered to be rigid where complex calculations involving the water mediator interactions are involved. Flexible docking approach, first introduced by Totrov and Abagyan [170], and the use of water molecules in docking calculation, introduced by Lengauer [171], are two of the several milestones achieved in structure-based drug designing approaches. There are wide varieties of docking programs available for users in both public and private. One of the advancements is found in combining the aspects of protein flexibility and displaced water molecule in a docking processor, named FITTED for better binding calculations. The success of the program can be seen in the identification of molecules in various drug discovery programs and collaborations [172].

Complementarity and ranking are calculated by a scoring function that is either based on empirically fit descriptors [44, 54, 173–175], knowledge-based potential functions [61, 62, 176], or physics-based terms [51, 93, 177–179]. Physics-based scoring functions borrow force field derived terms, such as van der Waals (vdW) and electrostatics, to calculate the protein–ligand interaction energy [180, 181]. A new scoring function was introduced by Micheal et al., where scoring term is dependent on context-dependent ligand desolvation. Here, every ligand atom's Born radii is related to fractional desolvation. This fraction is employed in scaling an atom-by-atom decomposition of the full transfer free energy [182]. This is scored across the grid, and the new method improves docking performance. This method is effective in discriminating ligands and that of other charged molecules compared to others. With the advantage of calculating the context-dependent ligand desolvation beforehand, the scoring function can enhance the docking consistency without costing much time. Such advancements in different aspects of programming and analysis, both in experimental and also theoretical understanding, will greatly influence the outcome of our quest in identification of drug molecules in subsiding various human ailments and infections.

6 Applications of SBDD Using Natural Products

The molecular modeling, docking, and SBDD modules were executed using various modules available in Schrodinger suite (USA) [183]. Glide docking followed by flexible docking using induced fit docking protocol was used in all the cases. Molecular simulations were carried to analyze the stability of ligand–protein complex over time. Simulations were carried out using AMBER. Table 3 summarizes the results described in the following sections.

6.1 Toward Antidotes with PLA_2 as Target

The snake venom is largely composed of melittin, which is a stimulant of PLA_2. The arachidonic acid is released excessively from the phospholipid membrane due to the increased presence and activity of PLA_2 resulting from a snakebite [184]. There are some crystal structures available for PLA_2 inhibitor complexes. The *Indigofera tinctoria Linn (Neeli), Cocculus hirsutus (Linn) Diels (Kattukodi),*

Table 3 Different targets and the compounds identified

Disease	Plant	Compound	Target
Snakebite	*Indigofera tinctoria Linn* *Cocculus hirsutus (Linn) Diels* *Andrographis paniculata* *Vitex negundo* *Acalypha indica* *Corallocarpus epigaeus* *Leucas aspera Spreng* *Tinospora cordifolia*	Tris(2,4-di-*tert*-butylphenyl) phosphate Octadecanoic acid Vitamin E β-amyrin	2B17 3H1X 1DB5 1KVO
Cancer	*Sphaeranthus Amaranthoides* *Stephania hernandifolia*	Tetrahydropalmatine Decahydro-6-(iminomethyl)- 4a-methylnaphthalen-2-ol Diethylstilbestrol Ethyl oleate	3UE4 1W84 1XJD 5T97
Diabetics		ZINC00187322 ZINC00754305 ZINC00754341 ZINC00754234	2FZD
Dengue virus	*Azadirachta indica* *Aegle marmelos* *Murraya koenigii* *Heliopsis scabra* *Taiwania cryptomerioides* *Calophyllum* *Carica papaya* *Fishes and crab*	Oleic acid Stearic acid Palmitic acid Gly-His-Met-Ser (GHMS) Ser-Met-His-Gly (SMHG) FB10251 FB08615	2FOM 2M9P

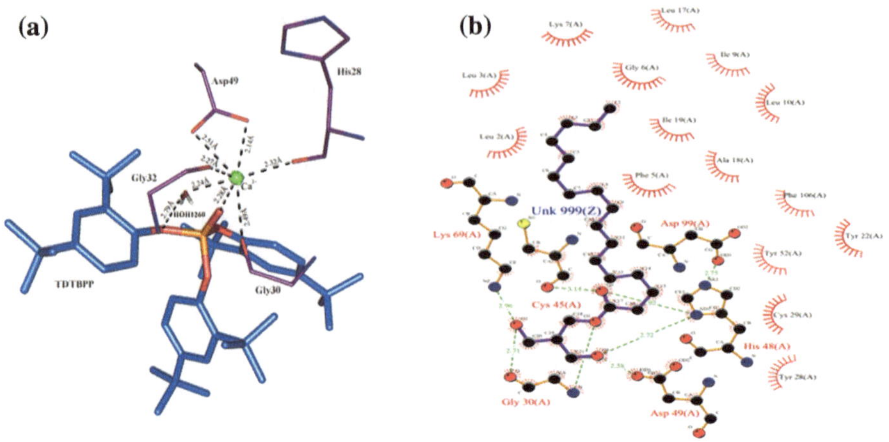

Fig. 3 **a** Tris(2,4-di-*tert*-butylphenyl) phosphate, **b** octadecanoic acid

Andrographis paniculata (Siriyanangai), Vitex negundo (Nochi), Acalypha indica (Kuppameni), Corallocarpus epigaeus (Akasakarudan), Leucas aspera Spreng (Thumbi), and Tinospora cordifolia (Shindilkodi) plants are reported to have medicinal value, especially with antidote property. The extracts were collected and GC-MS analysis was carried out. From these eight medicinal plants, we have identified 100 different compounds from the methanolic extract. Glide SP was used to screen these compounds with the four different targets (PDB ID: 2B17 and 3H1X [185, 186], PDB ID: 1DB5 and 1KVO [187, 188]), followed by induced fit docking (IFD), resulting in 20 compounds. The Tris(2,4-di-tert-butylphenyl) phosphate and octadecanoic acid were the best antidote compounds, based on the docking score, glide energy, and interactions with the active site residues (Fig. 3a, b).

6.2 Toward Anticancer Compounds with Various Targets

Different types of cancer targets were chosen such as Abl Tyrosine Kinase, p38alpha MAP Kinase, Protein Kinase Cθ and BCL-2 [189–192]. *Sphaeranthus amaranthoides* (Sivakaranthai) and *Stephania hernandifolia* (Jabung), the two medicinal plants from south- and northeastern regions of India, which are being used by herbalists for cancer treatment, are considered. The *S. Amaranthoides* plant sample was collected from Palani hills, and *S. hernandifolia* plant (rhizome) was collected from northeast region of India. From these two medicinal plants, 40 different compounds were subjected to Glide SP, resulting in five favorable compounds which were analyzed using induced fit docking (IFD). The tetrahydropalmatine and decahydro-6-(iminomethyl)-4a-methylnaphthalen-2-ol compounds were the best anticancer compounds, based on the docking score, glide energy, and interaction with the active site residues (Fig. 4a, b). The cell line studies carried out

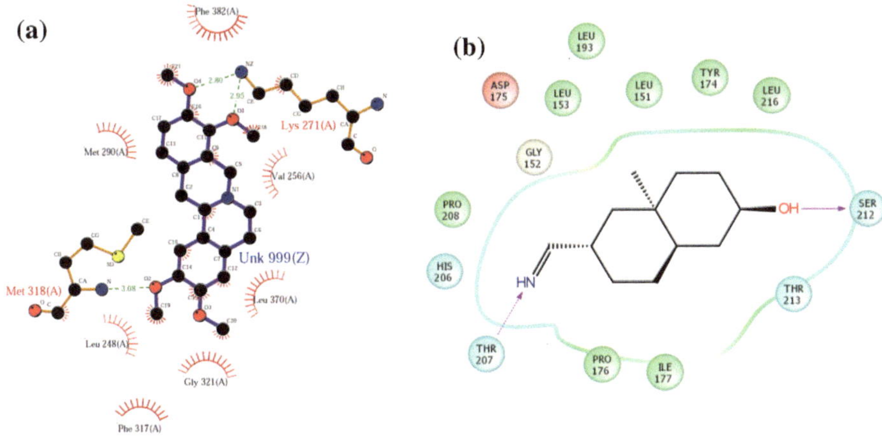

Fig. 4 **a** Tetrahydropalmatine, **b** decahydro-6-(iminomethyl)-4a-methylnaphthalen-2-ol

also confirm the inhibition activity of tetrahydropalmatine (from Jabung) toward cancerous cell lines HCT-116, MCF-7, and A-549 with an IC50 of 17, 10.5, and 27 μg/ml, respectively.

6.3 With Diabetic Target

Aldose reductase (ALR2) is one of the key enzymes involved in the pathogenesis of many diabetic complications [193, 194]. ALR2 catalyzes rate-limiting step of polyol pathway of glucose metabolism [195, 196]. Multistage filtering approach was employed to filter 16,000 compounds to 300 compounds. Two compounds (ZINC00754341 and ZINC00754234) bind with the hydrophobic pocket (specificity pocket) and also with the catalytic residues (Fig. 5a, b).

6.4 Toward Dengue Virus

Dengue is one of the life-threatening diseases. The NS2B/NS3 protease (PDB ID: 2FOM, 2M9P) is a crucial component in the replication cycle of the virus. The ligands undertaken for the modeling studies range from compounds belonging to plants such as Neem (*Azadirachta indica*), Bael (*Aegle marmelos*), *Murraya koenigii*, *Heliopsis scabra*, *Taiwania cryptomerioides*, *Calophyllum,* and also peptides from edible fishes and crab. Some of the phytochemicals from the above herbs bind well with the active site [197, 198]. Fatty acids isolated from the leaves of *Carica papaya* showed good binding at the active site [199]. The peptides GHMS and SMHG isolated from marine edible fishes and two compounds taken from FooDB

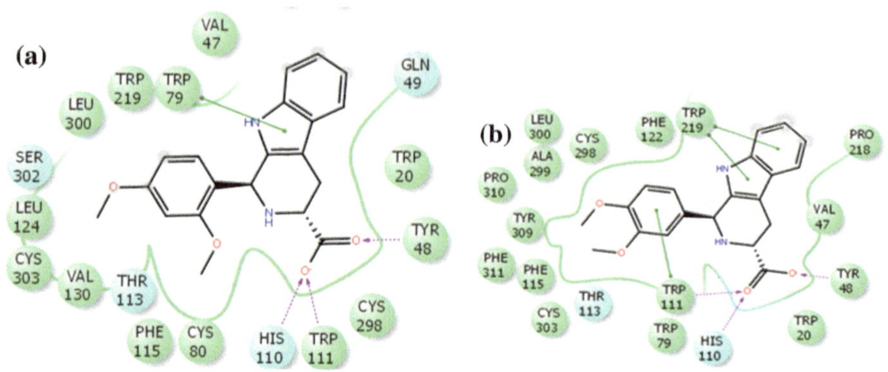

Fig. 5 **a** ZINC00754341, **b** ZINC00754234

(FDB008615, FDB010251) identified from docking studies were further analyzed for binding using molecular simulation studies [200]. Flavone- and chalcone-based inhibitors bind favorably at the active site of protease. Acridone and xanthone prefer to bind an alternative site (the tunnel-like pocket) understood from blind docking. Flexible docking, MD simulation, and binding free energy calculations confirm that acridone and flavone show favorable binding suggesting that these two compounds will have synergistic inhibitory activity against the proteolytic activity of NS2B/NS3pro (Fig. 6a, b). Binding studies for the dual inhibition sites have not been reported so far [201].

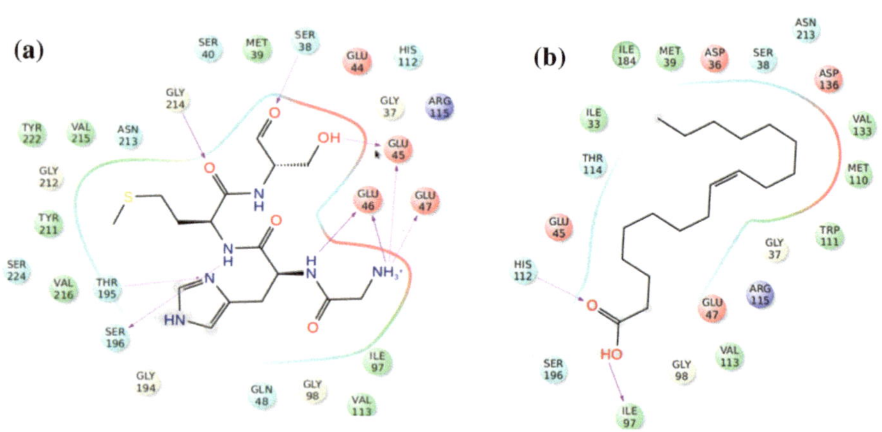

Fig. 6 **a** GHMS. **b** Oleic acid

7 Conclusion

Computer-aided drug design leads to many successful discoveries of structure-based drugs. Combination of molecular and quantum mechanics finds potential use in investigation of enzymatic mechanism. Protein–protein interactions are found to be relevant in drug design. Due to the sequencing of the human genome, many therapeutic targets are now available for structure-based drug design. Advancement in many aspects of crystallography and NMR methods has contributed to the high-resolution structures of many protein–protein–ligand complexes. Since each computational method in the field of SBDD has its own field of applicability, drawbacks, and limitations, they should be used in combination to come out with potential drugs. Since potential links exist between drugs and diseases, drug repurposing creates faster way in the field of drug discovery. In the design of protein–protein interaction inhibitors, the challenge is to discover druggable pockets in the interfaces of proteins engaged in transient interactions. A thorough study of successive compounds binding the same target assists in understanding structure–activity relationships, binding modes, and conformational changes. As more and more human protein structures are getting solved, structure-based drug design will have more impact in the discovery of new drugs to combat various diseases as structural biology is a major partner in drug development.

Acknowledgements The authors would like to thank Dr. C. Ramakrishnan, Postdoctoral Fellow, Department of Biotechnology, Indian Institute of Technology Madras, Chennai, for his kind help in the revising of the manuscript.

References

1. Quanta/InsightII/Cerius2. Molecular Simulations Inc., 9685 Scranton Road, San Diego, CA 92121-3752
2. Sybyl. Tripos, Inc., 1699 South Hanley Road, St. Louis, MO 63144-2913
3. CAChe. CAChe Scientific, Inc., P.O. Box 4003, Beaverton, OR 97076
4. MacroModel. Department of Chemistry, Columbia University, New York, NY 10032
5. Weiner SJ, Kollman PA, Case DA, Singh UC, Ghio C et al (1984) A new force field for molecular mechanical simulation of nucleic acids and proteins. J Am Chem Soc 106:765–784
6. DeLano WL (2002) The PyMOL molecular graphics system. http://pymol.org
7. Rosenfeld R, Vajda S, DeLisi C (1995) Flexible docking and design. Annu Rev Biophys Biomol Struct 24:677–700
8. Lybrand TP (1995) Ligand-protein docking and rational drug design. Curr Opin Struct Biol 5:224–228
9. Jones G, Willett P (1995) Docking smallmolecule ligands into active sites. Curr Opin Biotechnol 6:652–656
10. Kuntz ID, Meng EC, Shoichet BK (1994) Structure-based molecular design. Acc Chem Res 27:117–123
11. Schoichet BK, Kuntz ID (1996) Predicting the structure of protein complexes: a step in the right direction. Chem Biol 3:151–156

12. Straatsma TP, McCammon JA (1992) Computational alchemy. Annu Rev Phys Chem 43:407–435
13. Tembe BL, McCammon JA (1984) Ligand-receptor interactions. Comput Chem 8:281–283
14. Rami Reddy M, Bacquet RJ, Zichi D, Matthews DA, Welsh KM et al (1992) Calculation of solvation and binding free energy differences for folate-based inhibitors of the enzyme thymidylate synthase. J Am Chem Soc 114:10117–10122
15. Wlodek ST, Antosiewicz J, McCammon JA, Straatsma TP, Gilson MK et al (1996) Binding of tacrine and 6-chlorotacrine by acetylcholinesterase. Biopolymers 38:109–117
16. Marrone TJ, Straatsma TP, Briggs JM, Wilson DK, Quiocho FA, McCammon JA (1996) Theoretical study of inhibition of adenosine deaminase by 8Rcoformycin and 8R-deoxycoformycin. J Med Chem 39:277–284
17. Damewood JR Jr (1996) Peptide mimetic design with the aid of computational chemistry. Reviews in computational chemistry, In: Lipkowitz KB, Boyd DB (ed), vol 9. VCH Publishers, New York, pp 1–79
18. Borman S (2005) Drug by design. C&EN, Washington, DC
19. Sumner JB (1926) The isolation and crystallization of the enzyme urease preliminary paper. J Biol Chem 69(2):435–441
20. Sumner JB, Dounce AL (1937) Crystalline catalase. J Biol Chem 121(2):417–424
21. Northrop JH, Kunitz M, Herriott RM (1948) Crystalline Enzymes, Columbia Biological Series, No. 12. Columbia University Press, New York, 16, p 305
22. Wlodawer A, Vondrasek J (1998) Inhibitors of HIV-1 protease: a major success of structure-assisted drug design. Annu Rev Biophys Biomol Struct 27(1):249–284
23. Das K, Ding J, Hsiou Y, Clark AD, Moereels H, Koymans L, Andries K, Pauwels R, Janssen PA, Boyer PL, Smith RH Jr, Kroeger Smith MB, Michejda CJ, Hughes SH, Arnold E, Clark P (1996) Crystal structures of 8-Cl and 9-Cl TIBO complexed with wild-type HIV-1 RT and 8-Cl TIBO complexed with the Tyr181Cys HIV-1 RT drug-resistant mutant. J Mol Biol 264(5):1085–1100
24. Hsiou Y, Das K, Ding J, Clark AD, Kleim JP, Rösner M, Winkler I, Riess G, Hughes SH, Arnold E (1998). Structures of Tyr188Leu mutant and wild-type HIV-1 reverse transcriptase complexed with the non-nucleoside inhibitor HBY 097: inhibitor flexibility is a useful design feature for reducing drug resistance. J Mol Biol 284(2):313–323
25. Huang H, Chopra R, Verdine GL, Harrison SC (1998) Structure of a covalently trapped catalytic complex of HIV-1 reverse transcriptase: implications for drug resistance. Science 282(5394):1669–1675
26. Ren J, Esnouf RM, Hopkins AL, Jones EY, Kirby I, Keeling J, Ross CK, Larder BA, Stuart DI, Stammers DK (1998) 3′-Azido-3′-deoxythymidine drug resistance mutations in HIV-1 reverse transcriptase can induce long range conformational changes. Proc Natl Acad Sci 95(16):9518–9523
27. Sarafianos SG, Das K, Clark AD, Ding J, Boyer PL, Hughes SH, Arnold E (1999) Lamivudine (3TC) resistance in HIV-1 reverse transcriptase involves steric hindrance with ß-branched amino acids. Proc Natl Acad Sci 96(18):10027–10032
28. Dickerson RE, Geis I (1983) Hemoglobin: structure, function, evolution, and pathology, vol 1983. Benjamin-Cummings Publishing Company
29. Williams DP, Naisbitt DJ (2002) Toxicophores: groups and metabolic routes associated with increased safety risk. Curr Opin Drug Discov Devel 5(1):104–115
30. Aishima J, Owen RL, Axford D, Shepherd E, Winter G, Levik K, Gibbons P, Ashton A, Evans G (2010) High-speed crystal detection and characterization using a fast-readout detector. Acta Crystallogr Sect D: Biol Crystallogr 66(9):1032–1035
31. Winn MD, Ballard CC, Cowtan KD, Dodson EJ, Emsley P, Evans PR, Keegan RM, Krissinel EB, Leslie AG, McCoy A, McNicholas SJ, Murshudov GN, Pannu NS, Potterton EA, Powell HR, Read RJ, Vagin A, Wilson KS (2011) Overview of the CCP4 suite and current developments. Acta Crystallogr Sect D: Biol Crystallogr 67(4):235–242
32. Walters WP, Stahl MT, Murcko MA (1998) Virtual screening—an overview. Drug Discov Today 3(4):160–178

33. Bajorath J (2002) Integration of virtual and high-throughput screening. Nat Rev Drug Discov 1(11):882–894
34. Kuntz ID, Blaney JM, Oatley SJ, Langridge R, Ferrin TE (1982) A geometric approach to macromolecule-ligand interactions. J Mol Biol 161(2):269–288
35. Kitchen DB, Decornez H, Furr JR, Bajorath J (2004) Docking and scoring in virtual screening for drug discovery: methods and applications. Nat Rev Drug Discov 3(11): 935–949
36. Connolly ML (1983) Solvent-accessible surfaces of proteins and nucleic acids. Science 221 (4612):709–713
37. Connolly ML (1983) Analytical molecular surface calculation. J Appl Crystallogr 16 (5):548–558
38. Goodford PJ (1985) A computational procedure for determining energetically favorable binding sites on biologically important macromolecules. J Med Chem 28(7):849–857
39. Goodsell DS, Lauble H, Stout CD, Olson AJ (1993) Automated docking in crystallography: analysis of the substrates of aconitase. Proteins: Struct, Funct, Bioinf 17(1):1–10
40. Chemical Computing Group. MOE (2003) Montreal. Quebec, Canada
41. GOLD Version 1.2. [online] 2003. http://www.ccdc.cam.ac.uk/products/life_sciences/gold/
42. Westhead DR, Clark DE, Murray CW (1997) A comparison of heuristic search algorithms for molecular docking. J Comput Aided Mol Des 11(3):209–228
43. Ewing TJ, Makino S, Skillman AG, Kuntz ID (2001) DOCK 4.0: search strategies for automated molecular docking of flexible molecule databases. J Comput Aided Mol Des 15 (5):411–428
44. Rarey M, Kramer B, Lengauer T, Klebe G (1996) A fast flexible docking method using an incremental construction algorithm. J Mol Biol 261(3):470–489
45. Friesner RA, Banks JL, Murphy RB, Halgren TA, Klicic JJ, Mainz DT, Repasky MP, Knoll EH, Shelley M, Perry JK, Shaw DE, Francis P, Shenkin PS (2004) Glide: a new approach for rapid, accurate docking and scoring. 1. Method and assessment of docking accuracy. J Med Chem 47(7):1739–1749
46. Welch W, Ruppert J, Jain AN (1996) Hammerhead: fast, fully automated docking of flexible ligands to protein binding sites. Chem Biol 3(6):449–462
47. Kearsley SK, Underwood DJ, Sheridan RP, Miller MD (1994) Flexibases: a way to enhance the use of molecular docking methods. J Comput Aided Mol Des 8(5):565–582
48. Knegtel RM, Kuntz ID, Oshiro CM (1997) Molecular docking to ensembles of protein structures. J Mol Biol 266(2):424–440
49. Kramer B, Rarey M, Lengauer T (1999) Evaluation of the FLEXX incremental construction algorithm for protein–ligand docking. Proteins: Struct, Funct, Bioinf 37(2):228–241
50. Verdonk ML, Cole JC, Hartshorn MJ, Murray CW, Taylor RD (2003) Improved protein–ligand docking using GOLD. Proteins: Struct, Funct, Bioinf 52(4):609–623
51. Morris GM, Goodsell DS, Halliday RS, Huey R, Hart WE, Belew RK, Olson AJ (1998) Automated docking using a Lamarckian genetic algorithm and an empirical binding free energy function. J Comput Chem 19(14):1639–1662
52. Böhm HJ (1994) The development of a simple empirical scoring function to estimate the binding constant for a protein-ligand complex of known three-dimensional structure. J Comput Aided Mol Des 8(3):243–256
53. Böhm HJ (1998) Prediction of binding constants of protein ligands: a fast method for the prioritization of hits obtained from de novo design or 3D database search programs. J Comput Aided Mol Des 12(4):309–323
54. Eldridge MD, Murray CW, Auton TR, Paolini GV, Mee RP (1997) Empirical scoring functions: I. The development of a fast empirical scoring function to estimate the binding affinity of ligands in receptor complexes. J Comput Aided Mol Des 11(5):425–445
55. Wang R, Liu L, Lai L, Tang Y (1998) SCORE: a new empirical method for estimating the binding affinity of a protein-ligand complex. J Mol Model 4(12):379–394
56. Tao P, Lai L (2001) Protein ligand docking based on empirical method for binding affinity estimation. J Comput Aided Mol Des 15(5):429–446

57. Rognan D, Lauemøller SL, Holm A, Buus S, Tschinke V (1999) Predicting binding affinities of protein ligands from three-dimensional models: application to peptide binding to class I major histocompatibility proteins. J Med Chem 42(22):4650–4658
58. Wang R, Lai L, Wang S (2002) Further development and validation of empirical scoring functions for structure-based binding affinity prediction. J Comput Aided Mol Des 16(1): 11–26
59. Muegge I (2000) A knowledge-based scoring function for protein-ligand interactions: probing the reference state. Perspect Drug Discov Des 20(1):99–114
60. Muegge I (2001) Effect of ligand volume correction on PMF scoring. J Comput Chem 22(4): 418–425
61. Muegge I, Martin YC (1999) A general and fast scoring function for protein-ligand interactions: a simplified potential approach. J Med Chem 42(5):791–804
62. Gohlke H, Hendlich M, Klebe G (2000) Knowledge-based scoring function to predict protein-ligand interactions. J Mol Biol 295(2):337–356
63. DeWitte RS, Shakhnovich EI (1996) SMoG: de novo design method based on simple, fast, and accurate free energy estimates. 1. Methodology and supporting evidence. J Am Chem Soc 118(47):11733–11744
64. Momany FA, McGuire RF, Burgess AW, Scheraga HA (1975) Energy parameters in polypeptides. VII. Geometric parameters, partial atomic charges, nonbonded interactions, hydrogen bond interactions, and intrinsic torsional potentials for the naturally occurring amino acids. The J Phys Chem 79(22):2361–2381
65. Nemethy G, Pottle MS, Scheraga HA (1983) Energy parameters in polypeptides. 9. Updating of geometrical parameters, nonbonded interactions, and hydrogen bond interactions for the naturally occurring amino acids. The J Phys Chem 87(11):1883–1887
66. Lifson S, Warshel A (1968) Consistent force field for calculations of conformations, vibrational spectra, and enthalpies of cycloalkane and n-alkane molecules. J Chem Phys 49 (11):5116–5129
67. Hagler AT, Huler E, Lifson S (1974) Energy functions for peptides and proteins. I. Derivation of a consistent force field including the hydrogen bond from amide crystals. J Am Chem Soc 96(17):5319–5327
68. Hagler AT, Lifson S (1974) Energy functions for peptides and proteins. II. Amide hydrogen bond and calculation of amide crystal properties. J Am Chem Soc 96(17):5327–5335
69. Jorgensen WL (1981) Quantum and statistical mechanical studies of liquids. 10. Transferable intermolecular potential functions for water, alcohols, and ethers. Application to liquid water. J Am Chem Soc 103(2):335–340
70. Weiner SJ, Kollman PA, Case DA, Singh UC, Ghio C, Alagona G, Profeta S, Weiner P (1984) A new force field for molecular mechanical simulation of nucleic acids and proteins. J Am Chem Soc 106(3):765–784
71. Cornell WD, Cieplak P, Bayly CI, Gould IR, Merz KM, Ferguson DM, Spellmeyer DC, Fox T, Caldwell JW, Kollman PA (1995) A second generation force field for the simulation of proteins, nucleic acids, and organic molecules. J Am Chem Soc 117(19):5179–5197
72. Damm W, van Gunsteren WF (2000) Reversible peptide folding: dependence on molecular force field used. J Comput Chem 21(9):774–787
73. García AE, Sanbonmatsu KY (2001) Exploring the energy landscape of a ß hairpin in explicit solvent. Proteins: Struct, Funct, Bioinf 42(3):345–354
74. Mitsutake A, Sugita Y, Okamoto Y (2001) Generalized-ensemble algorithms for molecular simulations of biopolymers. Pept Sci 60(2):96–123
75. García AE, Sanbonmatsu KY (2002) a-Helical stabilization by side chain shielding of backbone hydrogen bonds. Proc Natl Acad Sci 99(5):2782–2787
76. Simmerling C, Strockbine B, Roitberg AE (2002) All-atom structure prediction and folding simulations of a stable protein. J Am Chem Soc 124(38):11258–11259
77. Brooks BR, Bruccoleri RE, Olafson BD, States DJ, Swaminathan SA, Karplus M (1983) CHARMM: a program for macromolecular energy, minimization, and dynamics calculations. J Comput Chem 4(2):187–217

78. Lexa KW, Carlson HA (2010) Full protein flexibility is essential for proper hot-spot mapping. J Am Chem Soc 133(2):200–202
79. Jazayeri A, Dias JM, Marshall FH (2015) From G protein-coupled receptor structure resolution to rational drug design. J Biol Chem 290(32):19489–19495
80. Singh R, Singh S, Nath Pandey P (2016) In-silico analysis of Sirt2 from Schistosoma monsoni:structures, conformations and interactions with inhibitors. J Biomol Struct Dyn 34(5):1042–1051. https://doi.org/10.1080/07391102.2015.1065205
81. Sliwoski G, Kothiwale S, Meiler J, Lowe EW Jr (2014) Computational methods in drug discovery. Pharmacol Rev 66(1):334–395
82. Alvarez Dorta D, Sivignon A, Chalopin T et al (2016) The antiadhesive strategy in Crohn's disease: orally active mannosides to decolonize pathogenic *Escherichia coli* from the gut. ChemBioChem 17(10):936–952
83. Amin KM, Anwar MM, Kamel MM, Kassem EM, Syam YM, Elseginy SA (2013) Synthesis, cytotoxic evaluation and molecular docking study of novel quinazoline derivatives as PARP-1 inhibitors. Acta Pol Pharm 70(5):833–849
84. Sabbah DA, Saada M, Khalaf RA et al (2015) Molecular modeling based approach, synthesis, and cytotoxic activity of novel benzoin derivatives targeting phosphoinostide 3-kinase (PI3Kalpha). Bioorg Med Chem Lett 25(16):3120–3124
85. Frederick R, Robert S, Charlier C et al (2005) 3,6-disubstituted coumarins as mechanism-based inhibitors of thrombin and factor Xa. J Med Chem 48(24):7592–7603
86. Dong MH, Chen HF, Ren YJ, Shao FM (2016) Molecular modeling studies, synthesis and biological evaluation of dabigatran analogues as thrombin inhibitors. Bioorg Med Chem 24 (2):73–84
87. Mena-Ulecia K, Tiznado W, Caballero J (2015) Study of the differential activity of thrombin inhibitors using docking, QSAR, molecular dynamics, and MM-GBSA. PLoS ONE 10(11): e0142774
88. Vitoria M, Granich R, Gilks CF et al (2009) The global fight against HIV/AIDS, tuberculosis, and malaria: current status and future perspectives. Am J Clin Pathol 131(6): 844–848
89. Wodak SJ, Janin J (1978) Computer analysis of protein-protein interaction. J Mol Biol 124 (2):323–342
90. Janin J, Henrick K, Moult J et al (2003) CAPRI: a critical assessment of predicted interactions. Proteins. 52(1):2–9
91. Lensink MF, Wodak SJ (2010) Blind predictions of protein interfaces by docking calculations in CAPRI. Proteins 78(15):3085–3095
92. Fleishman SJ, Whitehead TA, Strauch EM et al (2011) Community-wide assessment of protein-interface modeling suggests improvements to design methodology. J Mol Biol 414 (2):289–302
93. Dominguez C, Boelens R, Bonvin AM (2003) HADDOCK: a protein-protein docking approach based on biochemical or biophysical information. J Am Chem Soc 125(7):1731–1737
94. May A, Zacharias M (2007) Protein-protein docking in CAPRI using ATTRACT to account for global and local flexibility. Proteins 69(4):774–780
95. Sable R, Jois S (2015) Surfing the protein-protein interaction surface using docking methods: application to the design of PPI inhibitors. Molecules 20(6):11569–11603
96. Bier D, Thiel P, Briels J, Ottmann C (2015) Stabilization of protein-protein interactions in chemical biology and drug discovery. Prog Biophys Mol Biol 119(1):10–19
97. Persico M, Di Dato A, Orteca N et al (2015) From protein communication to drug discovery. Curr Top Med Chem 15(20):2019–2031
98. Kuenemann MA, Sperandio O, Labbe CM, Lagorce D, Miteva MA, Villoutreix BO (2015) In silico design of low molecular weight protein-protein interaction inhibitors: overall concept and recent advances. Prog Biophys Mol Biol 119(1):20–32
99. Pons C, Glaser F, Fernandez-Recio J (2011) Prediction of protein-binding areas by small-world residue networks and application to docking. BMC Bioinform 12:378

100. Chang S, Jiao X, Li CH, Gong XQ, Chen WZ, Wang CX (2008) Amino acid network and its scoring application in protein-protein docking. Biophys Chem 134(3):111–118
101. del Sol A, O'Meara P (2005) Small-world network approach to identify key residues in protein-protein interaction. Proteins 58(3):672–682
102. Hammad N, Jingdong J (2013) Structure-based protein-protein interaction networks and drug design. Quant Biol 1(Issue 3):183–191
103. Wang X, Wei X, Thijssen B, Das J, Lipkin SM, Yu H (2012) Three-dimensional reconstruction of protein networks provides insight into human genetic disease. Nat Biotechnol 30(2):159–164
104. Mestres J, Gregori-Puigjane E, Valverde S, Sole RV (2009) The topology of drug-target interaction networks: implicit dependence on drug properties and target families. Mol BioSyst 5(9):1051–1057
105. Xie L, Xie L, Bourne PE (2011) Structure-based systems biology for analyzing off-target binding. Curr Opin Struct Biol 21(2):189–199
106. Yang L, Chen J, He L (2009) Harvesting candidate genes responsible for serious adverse drug reactions from a chemical-protein interactome. PLoS Comput Biol 5(7):e1000441
107. Ben-Naim A (2002) Molecular recognition—viewed through the eyes of the solvent. Biophys Chem 10(1–102):309–319
108. Bienstock RJ (2015) Solvation methods for protein-ligand docking. Methods Mol Biol 1289:3–12
109. Zhang L, Yang Y, Kao YT, Wang L, Zhong D (2009) Protein hydration dynamics and molecular mechanism of coupled water-protein fluctuations. J Am Chem Soc 131(30): 10677–10691
110. Schutz CN, Warshel A (2001) What are the dielectric "constants" of proteins and how to validate electrostatic models? Proteins 44(4):400–417
111. Bernal JD, Fowler RH (1933) A theory of water and ionic solution, with particular reference to hydrogen and hydroxyl ions. J Chem Phys 1(8):515–548
112. Stillinger FH, Rahman A (1974) Improved simulation of liquid water by molecular dynamics. J Chem Phys 60(4):1545–1557
113. Berendsen HJ, Postma JP, van Gunsteren WF, Hermans J (1981) Interaction models for water in relation to protein hydration. In: Intermolecular forces. Springer, Netherlands, pp 331–342
114. Jorgensen WL, Chandrasekhar J, Madura JD, Impey RW, Klein ML (1983) Comparison of simple potential functions for simulating liquid water. J Chem Phys 79(2):926–935
115. Mahoney MW, Jorgensen WL (2000) A five-site model for liquid water and the reproduction of the density anomaly by rigid, nonpolarizable potential functions. J Chem Phys 112(20): 8910–8922
116. Klamt A, Schüürmann GJGJ (1993) COSMO: a new approach to dielectric screening in solvents with explicit expressions for the screening energy and its gradient. J Chem Soc, Perkin Trans 2(5):799–805
117. Still WC, Tempczyk A, Hawley RC, Hendrickson T (1990) Semianalytical treatment of solvation for molecular mechanics and dynamics. J Am Chem Soc 112(16):6127–6129
118. Bashford D, Case DA (2000) Generalized born models of macromolecular solvation effects. Annu Rev Phys Chem 51(1):129–152
119. Rocchia W, Sridharan S, Nicholls A, Alexov E, Chiabrera A, Honig B (2002) Rapid grid-based construction of the molecular surface and the use of induced surface charge to calculate reaction field energies: applications to the molecular systems and geometric objects. J Comput Chem 23(1):128–137
120. Madura JD, Briggs JM, Wade RC, Davis ME, Luty BA, Ilin A, Antosiewicz J, Gilson MK, Bagheri B, Scott LR, McCammon JA (1995) Electrostatics and diffusion of molecules in solution: simulations with the University of Houston Brownian Dynamics program. Comput Phys Commun 91(1–3):57–95

121. Qiu D, Shenkin PS, Hollinger FP, Still WC (1997) The GB/SA continuum model for solvation. A fast analytical method for the calculation of approximate Born radii. The J Phys Chem A 101(16):3005–3014
122. Ghosh S, Nie A, An J, Huang Z (2006) Structure-based virtual screening of chemical libraries for drug discovery. Curr Opin Chem Biol 10(3):194–202
123. Shoichet BK (2004) Virtual screening of chemical libraries. Nature 432(7019):862–865
124. Shuker SB, Hajduk PJ, Meadows RP, Fesik SW (1996) Discovering high-affinity ligands for proteins: SAR by NMR. Science 274(5292):1531–1534
125. Verlinde CL, Fan E, Shibata S, Zhang Z, Sun Z, Deng W, Ross J, Kim J, Xiao L, Arakaki TL, Bosch J, Bosch J, Caruthers JM, Larson ET, Letrong I, Napuli A, Kelly A, Mueller N, Zucker F, Van Voorhis WC, Buckner FS, Merritt EA, Hol WG (2009) Fragment-based cocktail crystallography by the medical structural genomics of pathogenic protozoa consortium. Curr Top Med Chem 9(18):1678–1687
126. Maurer T, Garrenton LS, Oh A, Pitts K, Anderson DJ, Skelton NJ, Fauber BP, Pan B, Malek S, Stokoe D, Ludlam MJ, Bowman KK, Wu J, Giannetti AM, Starovasnik MA, Mellman I, Jackson PK, Rudolph J, Wang W, Fang G (2012) Smallmolecule ligands bind to a distinct pocket in Ras and inhibit SOS-mediated nucleotide exchange activity. In: Proceedings of the National Academy of Sciences of the United States of America, 109, pp 5299–5304
127. Sun Q, Burke JP, Phan J, Burns MC, Olejniczak ET, Waterson AG, Lee T, Rossanese OW, Fesik SW (2012) Discovery of small molecules that bind to K-Ras and inhibit Sos-mediated activation. Angewandte Chemie 124(25):6244–6247
128. Ostrem JM, Peters U, Sos ML, Wells JA, Shokat KM (2013) K-Ras (G12C) inhibitors allosterically control GTP affinity and effector interactions. Nature 503(7477):548–551
129. Yin Z, Whittell LR, Wang Y, Jergic S, Liu M, Harry EJ, Dixon NE, Beck JL, Kelso MJ, Oakley AJ (2014) Discovery of lead compounds targeting the bacterial sliding clamp using a fragment-based approach. J Med Chem 57(6):2799–2806
130. Darby JF, Landström J, Roth C, He Y, Davies GJ, Hubbard RE (2014) Discovery of Selective Small-Molecule Activators of a Bacterial Glycoside Hydrolase. Angew Chem Int Ed 53(49):13419–13423
131. Komander D, Rape M (2012) The ubiquitin code. Annu Rev Biochem 81:203–229
132. Lau WF, Withka JM, Hepworth D, Magee TV, Du YJ, Bakken GA, Miller MD, Hendsch ZS, Thanabal V, Kolodziej SA, Hu Q, Narasimhan LS, Love R, Charlton ME, Hughes S, van Hoorn WP, Mills JE, Xing L (2011) Design of a multi-purpose fragment screening library using molecular complexity and orthogonal diversity metrics. J Comput-Aided Mol Des 25(7):621
133. Doak BC, Morton CJ, Simpson JS, Scanlon MJ (2014) Design and evaluation of the performance of an NMR screening fragment library. Aust J Chem 66(12):1465–1472
134. Albert JS, Blomberg N, Breeze AL, Brown AJ, Burrows JN, Edwards PD, Folmer RH, Geschwindner S, Griffen EJ, Kenny PW, Nowak T, Olsson LL, Sanganee H, Shapiro AB (2007) An integrated approach to fragment-based lead generation: philosophy, strategy and case studies from AstraZeneca's drug discovery programmes. Curr Top Med Chem 7 (16):1600–1629
135. Giannetti AM (2011) 8 From Experimental Design to Validated Hits: A Comprehensive Walk-Through of Fragment Lead Identification Using Surface Plasmon Resonance. Methods Enzymol 493:169
136. Giannetti AM, Koch BD, Browner MF (2008) Surface plasmon resonance based assay for the detection and characterization of promiscuous inhibitors. J Med Chem 51(3):574–580
137. Rich RL, Quinn JG, Morton T, Stepp JD, Myszka DG (2010) Biosensor-based fragment screening using FastStep injections. Anal Biochem 407(2):270–277
138. Siegal G, Hollander JG (2009) Target immobilization and NMR screening of fragments in early drug discovery. Curr Top Med Chem 9(18):1736–1745

139. Meiby E, Simmonite H, le Strat L, Davis B, Matassova N, Moore JD, Mrosek M, Murray J, Hubbard RE, Ohlson S (2013) Fragment screening by weak affinity chromatography: comparison with established techniques for screening against HSP90. Anal Chem 85 (14):6756–6766
140. Pollack SJ, Beyer KS, Lock C, Müller I, Sheppard D, Lipkin M, Hardick D, Blurton P, Leonard PM, Hubbard PA, Todd D, Richardson CM, Ahrens T, Baader M, Hafenbradl DO, Hilyard K, Bürli RW (2011) A comparative study of fragment screening methods on the p38a kinase: new methods, new insights. J Comput Aided Mol Des 25(7):677–687
141. FDA FY 2011 Innovative Drug Approvals; 2011
142. Erlanson DA (2011) Introduction to fragment-based drug discovery. In: Fragment-based drug discovery and X-ray crystallography. Springer, Berlin, pp 1–32
143. Dror O, Shulman-Peleg A, Nussinov R, Wolfson HJ (2004) Predicting molecular interactions in silico: I. A guide to pharmacophore identification and its applications to drug design. Curr Med Chem 11(1):71–90
144. Guner OF (2002) History and evolution of the pharmacophore concept in computer-aided drug design. Curr Top Med Chem 2(12):1321–1332
145. Mason JS, Good AC, Martin EJ (2001) 3-D pharmacophores in drug discovery. Curr Pharm Des 7(7):567–597
146. Gund P (1977) Three-dimensional pharmacophoric pattern searching. Prog Mol Subcell Biol 5:117
147. Poptodorov K, Luu T, Hoffmann RD (2006) Pharmacophore model generation software tools. Methods Princ Med Chem 32:17
148. Wolber G, Seidel T, Bendix F, Langer T (2008) Molecule-pharmacophore superpositioning and pattern matching in computational drug design. Drug Discov Today 13(1):23–29
149. Potemkin V, Grishina M (2008) Principles for 3D/4D QSAR classification of drugs. Drug Discov Today 13(21):952–959
150. Fosgerau K, Hoffmann T (2015) Peptide therapeutics: current status and future directions. Drug Discov Today 20(1):122–128
151. Montanya E (2012) A comparison of currently available GLP-1 receptor agonists for the treatment of type 2 diabetes. Expert Opin Pharmacother 13(10):1451–1467
152. Heitz A, Avrutina O, Le-Nguyen D, Diederichsen U, Hernandez JF, Gracy J, Kolmar H, Chiche L (2008) Knottin cyclization: impact on structure and dynamics. BMC Struct Biol 8 (1):54
153. Gracy J, Le-Nguyen D, Gelly JC, Kaas Q, Heitz A, Chiche L (2007) KNOTTIN: the knottin or inhibitor cystine knot scaffold in 2007. Nucleic Acids Res 36(suppl_1):D314–D319
154. Gould A, Ji Y, Aboye TL, Camarero JA (2011) Cyclotides, a novel ultrastable polypeptide scaffold for drug discovery. Curr Pharm Des 17(38):4294–4307
155. Chan LY, Gunasekera S, Henriques ST, Worth NF, Le SJ, Clark RJ, Campbell JH, Craik DJ, Daly NL (2011) Engineering pro-angiogenic peptides using stable, disulfide-rich cyclic scaffolds. Blood 118(25):6709–6717
156. Terlau H, Olivera BM (2004) Conus venoms: a rich source of novel ion channel-targeted peptides. Physiol Rev 84(1):41–68
157. Adams DJ, Alewood PF, Craik DJ, Drinkwater RD, Lewis RJ (1999) Conotoxins and their potential pharmaceutical applications. Drug Dev Res 46(3–4 Special Issue: Biotechnology and Pharmacology in Australia):219–234
158. Novac N (2013) Challenges and opportunities of drug repositioning. Trends Pharmacol Sci 34(5):267–272
159. Combes RD (2011) Challenges for computational structure–activity modelling for predicting chemical toxicity: future improvements? Expert Opi Drug Metabolism Toxicol 7(9):1129–1140
160. Cavalla D (2013) Predictive methods in drug repurposing: gold mine or just a bigger haystack? Drug Discov Today 18(11):523–532
161. Kuhn M, Campillos M, González P, Jensen LJ, Bork P (2008) Large-scale prediction of drug–target relationships. FEBS Lett 582(8):1283–1290

162. Cheng F, Li W, Wu Z, Wang X, Zhang C, Li J, Liu G, Tang Y (2013). Prediction of polypharmacological profiles of drugs by the integration of chemical, side effect, and therapeutic space. J Chem Inf Model 53(4):753–762
163. Bender A, Scheiber J, Glick M, Davies JW, Azzaoui K, Hamon J, Urban L, Whitebread S, Jenkins JL (2007) Analysis of pharmacology data and the prediction of adverse drug reactions and off-target effects from chemical structure. ChemMedChem 2(6):861–873
164. Unterthiner T, Mayr A, Klambauer G, Steijaert M, Wegner JK, Ceulemans H, Hochreiter S (2014) Deep learning as an opportunity in virtual screening. In: Proceedings of the deep learning workshop at NIPS
165. Mestres J, Martín-Couce L, Gregori-Puigjané E, Cases M, Boyer S (2006) Ligand-based approach to in silico pharmacology: nuclear receptor profiling. J Chem Inf Model 46 (6):2725–2736
166. Gregori-Puigjané E, Mestres J (2008) A ligand-based approach to mining the chemogenomic space of drugs. Comb Chem High Throughput Screening 11(8):669–676
167. Willett P, Barnard JM, Downs GM (1998) Chemical similarity searching. J Chem Inf Comput Sci 38(6):983–996
168. Zhou B, Wang R, Wu P, Kong DX (2015) Drug repurposing based on drug-drug interaction. Chem Biol Drug Des 85(2):137–144
169. Gane PJ, Dean PM (2000) Recent advances in structure-based rational drug design. Curr Opin Struct Biol 10(4):401–404
170. Alvarez JC (2004) High-throughput docking as a source of novel drug leads. Curr Opin Chem Biol 8(4):365–370
171. Rarey M, Kramer B, Lengauer T (1999) The particle concept: placing discrete water molecules during protein-ligand docking predictions. Proteins: Struct Funct Bioinf 34(1):17–28
172. Corbeil CR, Englebienne P, Moitessier N (2007) Docking ligands into flexible and solvated macromolecules. 1. Development and validation of FITTED 1.0. J Chem Inf Model 47 (2):435–449
173. Jones G, Willett P, Glen RC, Leach AR, Taylor R (1997) Development and validation of a genetic algorithm for flexible docking. J Mol Biol 267(3):727–748
174. Friesner RA, Banks JL, Murphy RB, Halgren TA, Klicic JJ, Mainz DT, Repasky MP, Knoll EH, Shelley M, Perry JK, Shaw DE, Francis P, Shenkin PS (2004) Glide: a new approach for rapid, accurate docking and scoring. 1. Method and assessment of docking accuracy. J Med Chem 47(7):1739–1749
175. Jain AN (2003) Surflex: fully automatic flexible molecular docking using a molecular similarity-based search engine. J Med Chem 46(4):499–511
176. Velec HF, Gohlke H, Klebe G (2005) DrugScoreCSD knowledge-based scoring function derived from small molecule crystal data with superior recognition rate of near-native ligand poses and better affinity prediction. J Med Chem 48(20):6296–6303
177. Meng EC, Shoichet BK, Kuntz ID (1992) Automated docking with grid-based energy evaluation. J Comput Chem 13(4):505–524
178. Abagyan R, Totrov M, Kuznetsov D (1994) ICM—a new method for protein modeling and design: applications to docking and structure prediction from the distorted native conformation. J Comput Chem 15(5):488–506
179. McMartin C, Bohacek RS (1997) QXP: powerful, rapid computer algorithms for structure-based drug design. J Comput Aided Mol Des 11(4):333–344
180. Moitessier N, Englebienne P, Lee D, Lawandi J, Corbeil AC (2008) Towards the development of universal, fast and highly accurate docking/scoring methods: a long way to go. Br J Pharm 153(S1)
181. Brooijmans N, Kuntz ID (2003) Molecular recognition and docking algorithms. Annu Rev Biophys Biomol Struct 32(1):335–373
182. Mysinger MM, Shoichet BK (2010) Rapid context-dependent ligand desolvation in molecular docking. J Chem Inf Model 50(9):1561–1573
183. Schrödinger L (2009) Schrödinger Suite 2009. LLC, New York, NY

184. Murakami M, Masuda S, Ichiro KUDO (2003) Arachidonate release and prostaglandin production by group IVC phospholipase A2 (cytosolic phospholipase A2gamma). Biochem J 372(3):695–702
185. Singh N, Jabeen T, Sharma S, Somvanshi RK, Dey S, Srinivasan A, Singh TP (2006) Specific binding of non-steroidal anti-inflammatory drugs (NSAIDs) to phospholipase A2: structure of the complex formed between phospholipase A2 and diclofenac at 2.7 Å resolution. Acta Crystallogr D Biol Crystallogr 62(4):410–416
186. Singh N, Kumar RP, Kumar S, Sharma S, Mir R, Kaur P, Srinivasan A, Singh TP (2009) Simultaneous inhibition of anti-coagulation and inflammation: crystal structure of phospholipase A2 complexed with indomethacin at 1.4 Å resolution reveals the presence of the new common ligand-binding site. J Mol Recognit 22(6):437–445
187. Schevitz RW, Bach NJ, Carlson DG, Chirgadze NY, Clawson DK, Dillard RD, Draheim SE, Hartley LW, Jones ND, Mihelich ED, Olkowski JL (1995) Structure-based design of the first potent and selective inhibitor of human non-pancreatic secretory phospholipase A2. Nat Struct Mol Biol 2(6):458–465
188. Cha SS, Lee D, Adams J, Kurdyla JT, Jones CS, Marshall LA, Bolognese B, Abdel-Meguid SS, Oh BH (1996) High-resolution X-ray crystallography reveals precise binding interactions between human nonpancreatic secreted phospholipase A2 and a highly potent inhibitor (FPL67047XX). J Med Chem 39(20):3878–3881
189. Levinson NM, Boxer SG (2012) Structural and spectroscopic analysis of the kinase inhibitor bosutinib and an isomer of bosutinib binding to the Abl tyrosine kinase domain. PLoS ONE 7(4):e29828
190. Gill AL, Frederickson M, Cleasby A, Woodhead SJ, Carr MG, Woodhead AJ, Walker MT, Congreve MS, Devine LA, Tisi D, O'Reilly M, Seavers LC, Davis DJ, Curry J, Anthony R, Padova A, Murray CW, Carr RA, Jhoti H (2005) Identification of novel p38a MAP kinase inhibitors using fragment-based lead generation. J Med Chem 48(2):414–426
191. Xu ZB, Chaudhary D, Olland S, Wolfrom S, Czerwinski R, Malakian K, Lin L, Stahl ML, Joseph-McCarthy D, Benander C, Fitz L, Greco R, Somers WS, Mosyak L (2004) Catalytic domain crystal structure of protein kinase C-? (PKC?). J Biol Chem 279(48):50401–50409
192. Burks HE, Abrams T, Kirby CA, Baird J, Fekete A, Hamann LG, Kim S, Lombardo F, Loo A, Lubicka D, Macchi K, McDonnell DP, Mishina Y, Norris JD, Nunez J, Saran C, Sun Y, Thomsen NM, Wang C, Wang J, Peukert S (2017) Discovery of an acrylic acid based tetrahydroisoquinoline as an orally bioavailable selective estrogen receptor degrader for ERa + breast cancer. J Med Chem 60(7):2790–2818
193. Kinoshita JH, Nishimura C (1988) The involvement of aldose reductase in diabetic complications. Diabetes/Metabolism Res Rev 4(4):323–337
194. Pugliese G, Tilton RG, Williamson JR (1991) Glucose-induced metabolic imbalances in the pathogenesis of diabetic vascular disease. Diabetes/Metabolism Res Rev 7(1):35–59
195. Wilson DK, Bohren KM, Gabbay KH, Quiocho FA (1992) An unlikely sugar substrate site in the 1.65 A. Science 257:81
196. Bohren KM, Grimshaw CE, Lai CJ, Harrison DH, Ringe D, Petsko GA, Gabbay KH (1994) Tyrosine-48 is the proton donor and histidine-110 directs substrate stereochemical selectivity in the reduction reaction of human aldose reductase: enzyme kinetics and crystal structure of the Y48H mutant enzyme. Biochemistry 33(8):2021–2032
197. Velmurugan D, Malar Selvi U, Mythily U, Rao K, Rajarajeshwari R (2012) Structure-based discovery of anti-viral compounds for hepatitis B & C, human immunodeficiency, and dengue viruses. Curr Bioinform 7(2):187–211
198. Velmurugan D, Mythily U, Rao K (2014) Design and docking studies of peptide inhibitors as potential antiviral drugs for dengue virus ns2b/ns3 protease. Protein Pept Lett 21(8):815–827
199. Kutumbarao NHV, Velmurugan D (2016) Structural analysis and molecular modeling studies of fatty acids and peptides binding with NS2B/NS3 dengue protease. J Emerg Dis Virol 2(4); 2473-1846

200. Velmurugan D (2017) Designing, molecular docking and simulation studies of dengue protease inhibitors. Res J Med Allied Sci (RJMAS) 1(1)
201. Kutumbarao NHV, Ramakrishnan C, Balasubramanian K, Velmurugan D (2016) Computational assessment of inhibitory activity of acridone, xanthone and flavone derivatives against NS2B/NS3pro of Dengue Virus Type 2. J Emerg Dis Virol 2(4); 2473-1846

Impact of Target-Based Drug Design in Anti-bacterial Drug Discovery for the Treatment of Tuberculosis

Anju Choorakottayil Pushkaran, Raja Biswas and C. Gopi Mohan

Abstract Tuberculosis (TB) is an infectious disease caused by *Mycobacterium tuberculosis* (*Mtb*) and is a major public health concern. According to the 2017 WHO report, global burden of TB infection was 10.4 million people causing the mortality rate of $\sim$1.6 million. The rapid emergence of multidrug-resistant (MDR) and extensively drug-resistant (XDR) TB is of major concern in anti-TB drug discovery. There are different druggable targets and its pathways involved in the virulence, which include *Mtb* cell wall, replication and transcription, regulatory, protein synthesis, membrane transport, and energy production which need to be explored for efficient killing of the bacteria. The ability of the tubercle bacilli to remain within the host intracellular compartment is of other major concern in TB therapy. Thus, to tackle the TB drug resistance, potent inhibitors with novel mechanism of action of different *Mtb* druggable targets need to be discovered. Three-dimensional structure of different *Mtb* target was solved for structure-based drug design. The current chapter focuses on some of the key druggable targets in *Mtb* and also the recent advances in target-based drug designing in the area of anti-tubercular drug discovery.

Keywords *Mycobacterium tuberculosis* · TB therapy · Drug targets
Structural biology · Chemoinformatics · Drug resistance

Abbreviations

3D	Three-dimensional
ag85	Antigen 85
AMPK	5′adenosine monophosphate-activated protein kinase
AspS	Aspartyl tRNA synthetase
BCG	Bacille Calmette-Guérin

A. C. Pushkaran · R. Biswas · C. G. Mohan (✉)
Center for Nanosciences and Molecular Medicine,
Amrita Institute of Medical Sciences and Research Centre,
Amrita Vishwa Vidyapeetham, Ponekkara, Kochi 682041, Kerala, India
e-mail: cgmohan@aims.amrita.edu

C. G. Mohan (ed.), *Structural Bioinformatics: Applications in Preclinical Drug Discovery Process*, Challenges and Advances in Computational Chemistry and Physics 27, https://doi.org/10.1007/978-3-030-05282-9_10

ClpP	Caseinolytic peptidase P
CmaA1	Cyclopropane synthase
D-Ala	D-Alanine
D-Glu	D-Glutamic acid
DprE1	Decaprenylphosphoryl-β-D-ribofuranose 2′-oxidase
FtsZ	Filamenting temperature-sensitive protein Z
GlcB	Malate synthase
GlcNAc	*N*-acetylglucosamine
GyrB	DNA gyrase subunit B
HTS	High-throughput screening
L-Ala	L-Alanine
Ldt	L,D-transpeptidase
LeuRS	Leucyl-tRNA synthetase
L-Lys	L-Lysine
Lpd	Lipoamide dehydrogenase
MDR	Multidrug-resistant
MEPS	Molecular electrostatic potential surface
meso-DAP	*meso*-diaminopimelic acid
MIC	Minimum inhibitory concentration
MSA	Multiple sequence alignment
Mtb	*Mycobacterium tuberculosis*
MurNGlyc	*N*-glycolylmuramic acid
NMR	Nuclear magnetic resonance
PDB	Protein data bank
PDF	Peptide deformylase
PG	Peptidoglycan
PtpA	Tyrosine phosphatase A
PtpB	Tyrosine phosphatase B
Qcrb	Cytochrome bc1 complex
QSAR	Quantitative structure activity relationship
RNAP	RNA polymerase enzyme
ROS	Reactive oxygen species
TB	Tuberculosis
TCA	Tricarboxylic acid
VS	Virtual screening
WHO	World Health Organization
XDR	Extensively drug-resistant

1 Introduction

Tuberculosis (TB) is a highly infectious disease caused by *Mycobacterium tuberculosis* (*Mtb*). The global incidence of TB disease can be controlled by effective chemotherapy; however, these are moderately protective, so novel effective anti-tuberculosis agents are required to fight against TB. According to World Health Organization (WHO), ~10.4 million people are infected with the *Mtb* and mortality was about 1.6 million people, shown in Fig. 1 [1]. Further, one-third of the world's population is latently infected with *Mt*b.

Robert Koch isolated first *Mtb* pathogen in 1882. Further, Paul Ehrlich a German doctor and bacteriologist developed a staining method for the identification of *Mtb* bacteria and provided the basis to develop Ziehl-Neelsen acid-fast staining to detect the bacterium and which is still prevalent as a tool in TB diagnosis [2, 3]. In the current scenario, novel compounds and its druggable targets for the disease therapy need to be discovered by taking into account the drug resistance, safety, and duration of TB treatment. *Mtb* cell wall having the mycolic acid forms an unusual waxy coating on its surface making the TB drug penetration a major bottleneck. TB chemotherapy revolution begins with Streptomycin in 1944, which was the first antibiotics effective against *Mtb*. Eight years later, Isoniazid, the first orally administered anti-TB drug, was introduced, which drastically reduces the TB mortality rate [4].

Ethambutol and Rifampicin drugs were later introduced in the 1970s as the first-line TB drugs. The current anti-TB therapy involves a standard two-month course of first-line anti-TB drugs, Rifampicin, Isoniazid, Ethambutol, and Pyrazinamide followed by a four-month course of Rifampicin and Isoniazid drugs (Fig. 2). Treatment of MDR-TB involves two-year duration with the combination of at least five second-line expensive drugs—Capreomycin, Ethionamide, Bedaquiline, Moxifloxacin, and Streptomycin as well as first-line drugs shown in Table 1.

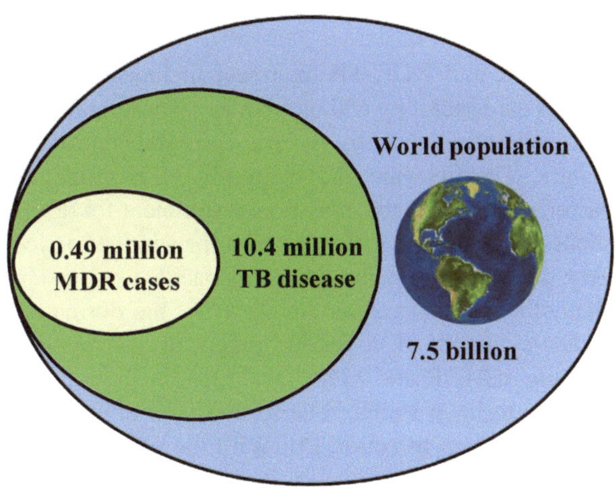

Fig. 1 Tuberculosis infection worldwide statistics 2017

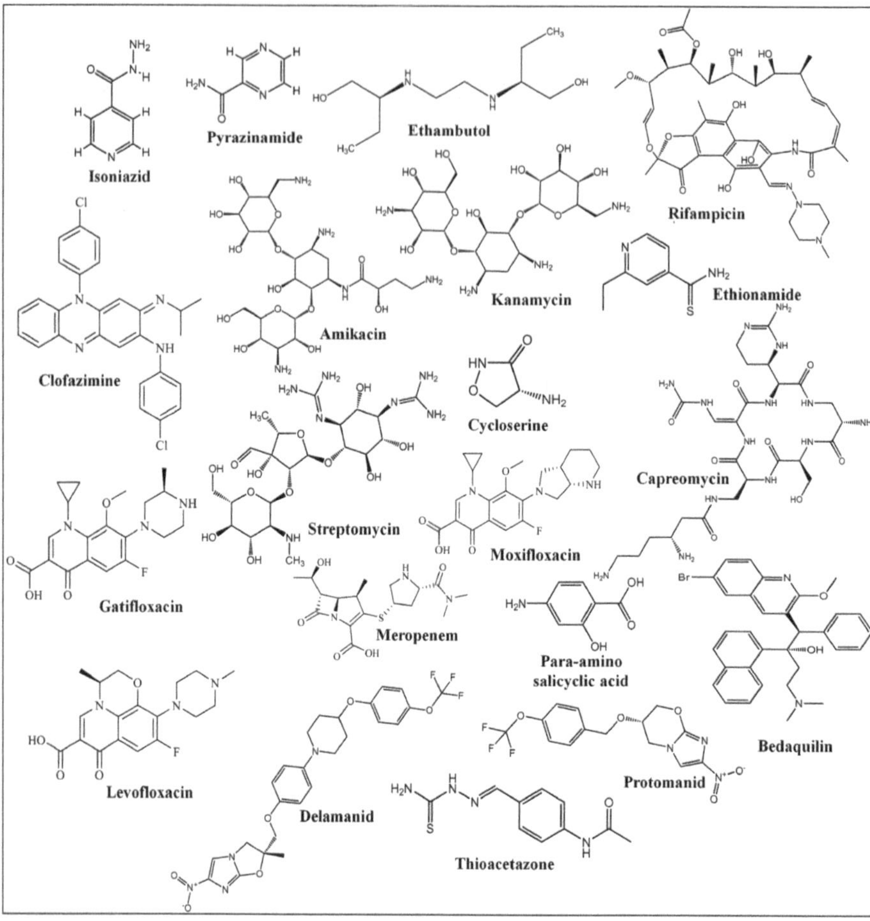

Fig. 2 Structure of first-line and second-line TB drugs

The MDR and XDR-TB treatment includes six months of daily injections having serious cardiotoxicity and ototoxicity issues. Another challenge in TB management is the cyclic reinfection by the persistent tubercle bacillus present in the host macrophages. The first-line and second-line TB drug treatment generally kills growing bacteria. However, this may not be sufficient for dormant *Mtb*, which is metabolically silent and persistent within the host for a longer period of time. Thus, TB treatment is very difficult due to the nature of bacilli to withstand the host immune attack and chemotherapy, as well as the ability of the dormant *Mtb* to survive for decades.

Development of novel and potent drugs for better TB treatments, especially in the case of MDR and XDR-TB infection, should be taken up in war-footing manner in order to have stable, relapse-free, and effective sterilization of the diverse populations of *Mtb* infection [5]. With the advent of the modern technology, the drug

Table 1 First- and second-line drugs for the treatment of sensitive and drug-resistant TB disease

S. no.	Drugs	Targets involved in drug binding	Molecular mechanism of drug action
First line of drugs for drug-sensitive TB cases			
1	Isoniazid	Enoyl-acyl-carrier protein reductase InhA	Inhibition of mycolic acid synthesis
2	Pyrazinamide	Multiple targets: FAS I, QAPRTase, RpsA, PanD, Rv2783 of *Mtb*	Membrane energetic disruption, fatty acid and ribosomal protein synthesis inhibition
3	Rifampicin	RNA polymerase β subunit	Inhibits transcription
4	Ethambutol	Arabinosyltransferase	Inhibits arabinogalactan biosynthesis
Second line of drugs for drug-resistant TB cases			
1	Streptomycin Amikacin	16S rRNA subunit of *Mtb* ribosome	Inhibits protein synthesis
2	Kanamycin Capreomycin	30S rRNA subunit of *Mtb* ribosome	Inhibits protein synthesis
3	Gatifloxacin Levofloxacin Moxifloxacin	DNA gyrase and topoisomerase	Inhibits DNA synthesis
4	Linezolid	23S rRNA of 50S subunit	Inhibits protein synthesis
5	Clofazimine	Bacterial DNA	Inhibits template function of DNA
6	Ethionamide	Enoyl-acyl-carrier protein reductase InhA by activating catalase peroxidase of *Mtb*	Inhibits mycolic acid synthesis
7	Cycloserine	Alanine racemase and D-alanine: D-alanine ligase	Inhibits peptidoglycan synthesis
8	Para-amino salicylic acid	Dihydrofolate reductase	Inhibits folate biosynthesis
Other drugs for TB chemotherapy (WHO Group D Add on drugs)			
1	Delamanid Pretomanid	Not exactly known	Inhibits mycolic acid synthesis
2	Bedaquiline	ATP synthase subunit c encoded by atpE gene	Inhibits ATP production
3	Thioacetazone	Mycolic acid cyclopropane synthases	Inhibits mycolic acid synthesis
4	Meropenem	D,D-transpeptidase and L,D-transpeptidase	Inhibits peptidoglycan biosynthesis

resistance issue of the first-line and second-line TB drugs can be addressed more effectively by understanding the novel mechanisms of action of the target concerned and also the pharmacokinetics/pharmacodynamic properties of the drug variability arising in different patient populations. The application of computational power to streamline the drug discovery and development process is gaining interest. In past few years, several resources are being reported including *Mtb* druggable target databases and even software for predicting molecular targets. This will further facilitate the identification and development of novel anti-bacterial compounds. One of the most challenging parts of computer-assisted modeling and biochemical screening toward antibiotic development is that the compounds may lack biological activity in the later stages of clinical studies.

2 *Mtb* Druggable Target Identification and Validation

The whole genome sequencing of the *Mtb* H37Rv strain in 1998 has revealed several crucial genes which are necessary for growth, survival, and virulence of the bacteria. This led to the target-based design of new anti-tubercular molecules. Further, the drug should be orally effective, cell wall permeability, metabolically stable, and target vulnerability by addressing the drug resistance [6]. Successful integration of computational and wet-laboratory studies is reported for different *Mtb* druggable targets (without mammalian counterpart), which includes the proteins involved in the pathway of cell wall biosynthesis, metabolism, energy production, and regulatory processes [7–10]. A pictorial representation of these different druggable *Mtb* targets is presented below as Fig. 3.

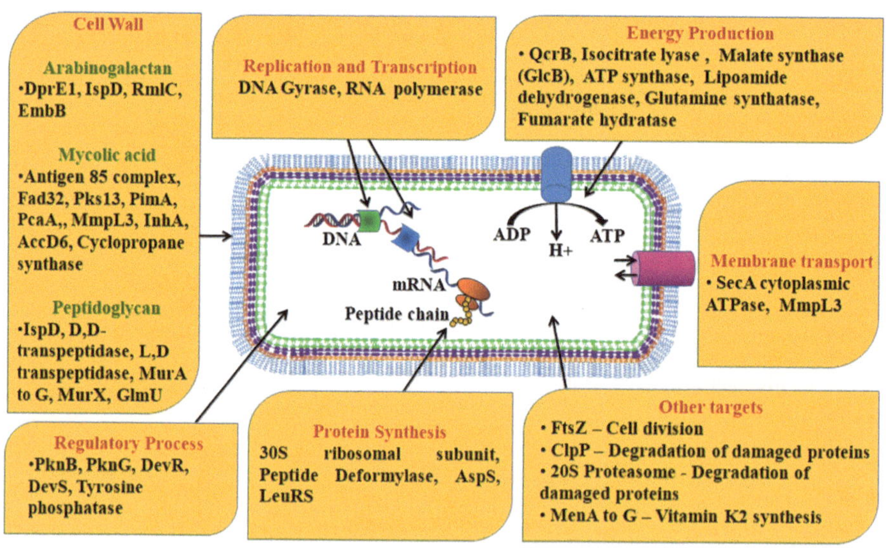

Fig. 3 Different druggable protein targets in *Mtb* for drug discovery program

2.1 Molecular Targets Involved in Cell Wall Biosynthesis and Its Inhibitors

The *Mtb* cell wall is unique, and its low permeability makes the drug difficult to penetrate the cell wall, which in turn cause bacterial survival in the host [11]. The primary targets involved in anti-TB drug discovery process are associated with *Mtb* cell wall biosynthesis pathway. The *Mtb* cell wall has covalently linked arabinogalactan, mycolic acid, and peptidoglycan (PG) layers. The enzymes associated with the biosynthesis of these layers are excellent druggable targets and are depicted in Fig. 3. In order to understand the structure–function relationships and biological mechanism of action of these targets for discovering small-molecule inhibitors, different research groups worldwide used structural biology and computational techniques [12]. Some of the best-known *Mtb* druggable targets and its inhibitors are explained below to understand its molecular mechanism of inhibition.

Decaprenylphosphoryl-β-D-ribofuranose 2′-oxidase (DprE1) and arabinosyltransferase encoding **EmbB** protein are the main druggable targets in arabinogalactan biosynthesis pathway. DprE1 is a well-validated druggable target for anti-bacterial drug designing. Benzothiazinones chemical class of compounds are covalently bind and inhibit the enzyme activity of DprE1 [13]. BTZ043 is a potent inhibitor of DprE1 which is in clinical trials. The X-ray crystal structure of DprE1 in complex with BTZ043 has already been solved, which revealed the key molecular mechanism of enzyme inhibition. BTZ043 forms a covalent bond with active site Cys 387 and other interacting residues are Gly 117, Lys 134, Ser 228, Leu 317, Leu 363, Val 365, Lys 367, Phe 369, Asn 35, and Lys 418 shown in Fig. 4.

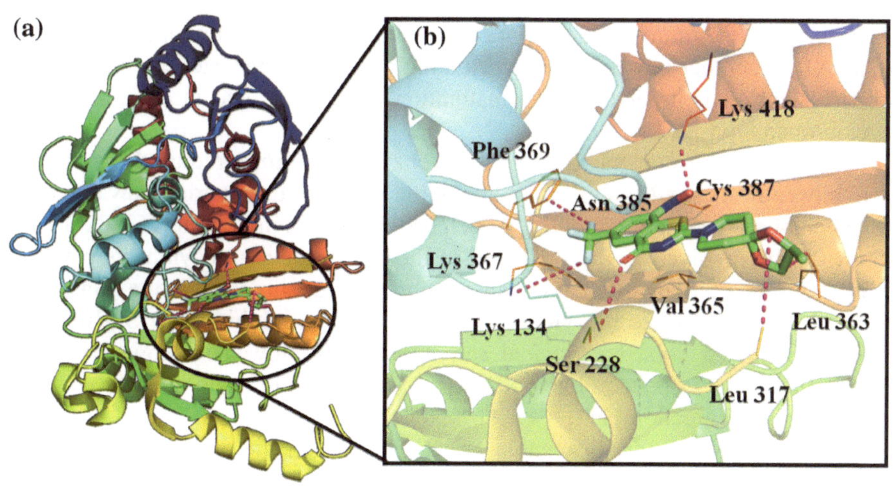

Fig. 4 Crystal structure of DprE1 in complex with BTZ-043 (PDB ID: 6HEZ [15]). **(a)** The binding site of BTZ-043 (sticks) and **(b)** the molecular interactions with the active site amino acid residues (lines)

The EmbB is another validated mycobacterial drug target. Ethambutol which is a first-line TB drug inhibits EmbB target, and the mechanism of drug action is well studied [13, 14] (Table 1; Fig. 3).

IspD a key enzyme of methylerythritol phosphate pathway and **RmlC** enzyme of rhamnose pathway form a link toward biosynthesis of arabinogalactan and PG, which is essential for the *Mtb* cell wall integrity and growth [16, 17]. Both these chemotherapeutic targets are essential for *Mtb* growth and are absent in mammals. In order to carry out structure-based drug discovery, both IspD and RmlC crystal structures are available in complex with small-molecule inhibitors and its molecular mechanism of action was clearly elucidated [18].

2.1.1 Mycolic Acid Biosynthesis Pathway Targets

InhA is an NADH-dependent key enzyme of enoyl-acyl-carrier protein reductase of fatty acid synthase-II involved in the biosynthesis of the mycolic acid pathway which is targeted by Isoniazid, one of the first-line anti-TB drug and a second-line drug Ethionamide [19, 20] shown in Table 1 and Fig. 2. Hence, mycobacterial InhA serves as a validated target for TB therapy. Isoniazid requires activation by KatG enzyme before binding to InhA. Most of the resistance toward Isoniazid confers to the mutations in *KatG* gene. So designing of inhibitors that directly target InhA has been of interest by several research groups. Different InhA inhibitors are already being identified using various techniques consisting of high-throughput screening (HTS), encoded library technology, and in silico drug design techniques [21–23]. However, most of the identified inhibitors lack good pharmacokinetic profile. In 2016, Martínez-Hoyos et al. identified GSK693, a direct oral InhA inhibitor with potent anti-tubercular activity against MDR and XDR clinical isolates and also in TB murine models [24]. Mycobacterial **cyclopropane synthase** (CmaA1) is another key enzyme contributing toward the persistence and virulence of *Mtb*. CmaA1 is involved in the maturation of mycolic acid in a process called cyclopropanation [25].

Wilson et al. demonstrated that **Pks13** enzyme of *Mtb* is required for mycolic acid biosynthesis and is an essential druggable target by discovering new classes of thiophene-based compounds acting as cell wall synthesis inhibitors [26]. Pks13 was known to be involved in the final step of the mycolic acid biosynthesis pathway. **MmpL3** is another key druggable transmembrane target involved in the transport of trehalose monomycolate biosynthetic pathway, and its inhibition by SQ109 small molecule showed good bactericidal activity. SQ109 showed synergism with the anti-tubercular drug bedaquiline and was very effective in acute and chronic mice models of *Mtb* infection [27].

Another potential *Mtb* druggable target involved in the cell wall assembly is **Antigen 85 (Ag85) complexes** consisting of three proteins (Ag85A, B, and C). These proteins exhibit mycolyltransferase activities through disruption of cord factor biosynthesis by the biogenesis of trehalose dimycolate and are also useful in controlling MDR and XDR-TB. Crystal structure of Ag85C protein was solved at

1.5 Å resolutions in complex with a covalent inhibitor. This structure revealed an alpha-/beta-hydrolase polypeptide fold and a catalytic triad formed by three amino acid residues, Ser 124, Glu 228, and His 260 [28, 29]. Kovac et al. discovered few sulfonate inhibitors against Ag85C protein based on the catalytic mechanism of action elucidated by its crystal structure and potent compound discovered having promising activity of IC_{50} = 4.3 μM using the mycolyltransferase inhibition assay [30]. Ag85B in complex with trehalose crystal structure was solved by Anderson et al. to study the structure-based anti-tubercular drug design aspects of understanding its interface mechanisms [31].

The building block of mycolic acid biosynthesis in *Mtb* is controlled by an essential carboxyltransferase enzyme, **AccD6**, which is involved in the synthesis of malonyl-CoA by the catalysis of acetyl-CoA [32]. The crystal structure of *Mtb* AccD6 in complex with haloxyfop-R, a herbicide targeting plant acetyl-CoA carboxylases, revealed its molecular basis of inhibitor binding leading to the development of novel herbicides with better ADMET profile as *Mtb* AccD6 inhibitors [33]. Other druggable target of *Mtb* cellular growth includes **PimA**, and the target is proven by different research groups using in vitro and in vivo techniques [34]. PimA takes part in the biosynthesis of phosphatidyl-myoinositol mannosides. Thus, it could be used as potent in silico/in vitro target-based HTS program for the TB therapy [34].

2.1.2 Peptidoglycan Biosynthesis Pathway Targets

Gram-positive and Gram-negative bacteria cell walls have unique biopolymer PG, which is necessary for keeping its cellular integrity. The *Mtb* cell wall possesses key druggable Mur ligases—**MurA**, **MurC**, **MurD**, **MurE,** and **MurF**, which are essential for biosynthesis of bacterial PG and not present in the mammalian system [35]. Mur proteins are conserved among different bacterial species and also possess a common three-dimensional (3D) structural motif [38]. Structure–function relationship studies on *Mtb*-MurB oxidoreductase enzyme were performed in our laboratory by an integrated approach involving multiple sequence alignment (MSA), homology modeling, molecular dynamics, molecular electrostatic potential surface (MEPS) mapping, and molecular docking studies. In order to understand the sequence conservation, MSA among different mycobacterium *Mtb*, *Escherichia coli* and *Staphylococcus aureus* MurB proteins showed that Tyr122, Gly123, Arg156, Arg218, and Ser237 residues are conserved among these microbes (Fig. 5a).

The binding analysis of the natural ligand naphthyl tetronic acid toward *Mtb* and *E. coli*-MurB is presented in Fig. 5b–d. Molecular docking studies using different chemical classes of well-known 28 MurB inhibitors belonging to 3,5-dioxopyrazolidine derivatives showed hydrogen bonding and other week interactions with these *Mtb*-MurB residues for the development of broad-spectrum anti-bacterial drug. Further, our computational binding affinity showed good correlation of 0.83 with its experimental IC_{50} value. This binding study with most potent compound 10a also supported the experimental site-directed mutational studies on key functional *Mtb*-MurB residues and is presented in Table 2 [37].

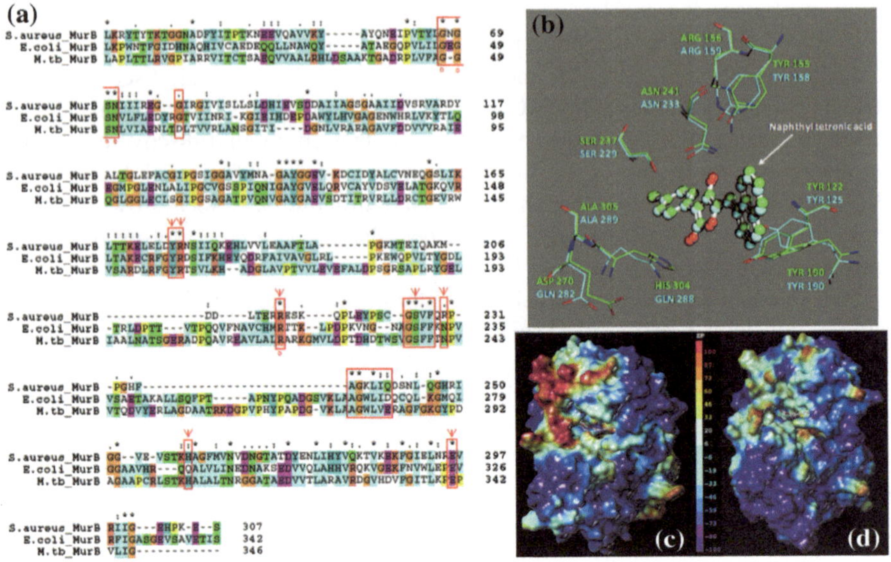

Fig. 5 Sequence alignment of *Mtb*-MurB with template (*E. coli*)-MurB. Red boxes show highly conserved amino acids among MurB of different bacterial species, and arrows above the residues indicate the amino acids in *E. coli* and *Mtb*-MurB investigated for mutational study in *Staphylococcus aureus*. Colors indicate amino acids with their similar characteristics, and stars, colons, and dots represent the identical amino acids, similar amino acids, and almost similar amino acids, respectively. The red circles indicate flavin adenine dinucleotide binding residues (**a**), superimposition of active site residues within 5.5 Å region surrounding naphthyl tetronic acid inhibitor in *Mtb*-MurB (green) and (*E. coli*)- MurB (cyan) (**b**); molecular electrostatic potential surface (MEPS) map of *Mtb*-MurB (**c**) and (*E. coli*)- MurB (**d**) along with the MEPS color ramp from (+100 to −100) kcal/mol. Most positive potential regions are shown by red color, while most negative potential regions with blue color, on the same potential scale for comparison. The ligand naphthyl tetronic acid is shown in ball-and-stick model (copyright permission, Springer)

We also adopted a similar theoretical strategy for MurD and uridine monophosphate kinase as druggable *Mtb* targets for curing TB disease [35, 36]. Recently, another key target in this family *Mtb* **phospho-MurNAc-pentapeptide translocase** (also termed as MurX) involved in the PG biosynthesis was identified and a sansanmycin uridyl peptide natural product analogue was discovered to be selective and potent inhibitor by in vitro and intracellular anti-mycobacterial activity assay [39]. The biggest challenge in the development of inhibitors targeting Mur ligase enzymes will be the chemical structure optimization of the compounds to enable the passage through the cell wall of *Mtb* and its stability in the cytoplasm. Thus, *in vivo* and clinically successful compounds are yet to be discovered based on the MurC ligases [40, 41].

Next key targets in almost all Gram-positive and Gram-negative bacterial species are involved in the PG biosynthesis. *Mtb* PG is composed of a repeating disaccharide sugar units consisting of *N*-glycolylmuramic acid (MurNGlyc) linked to

Table 2 *Mtb*-MurB, *S.aureus*-MurB, and *E. coli* MurB structure-based molecular docking studies to understand the changes in the binding affinity due to point mutations in comparison with its experimental kinetic studies (copyright permission, Springer)

Mtb-MurB	Compound structure (**10a**)	Autodock[a]		(*S. aureus*)-MurB	(*E. coli*)-MurB
		Binding energy (kcal/mol)	Inhibition constant (μM)	K_d (μM)	K_d (μM)
Y155F (mutant)	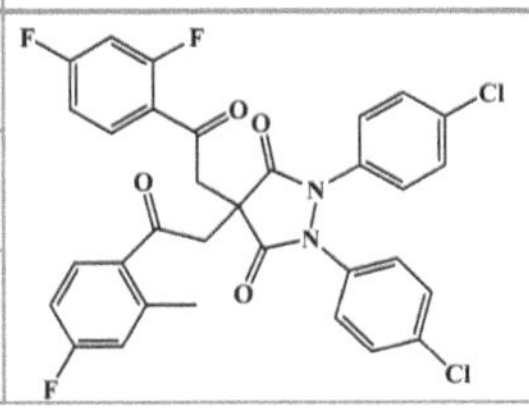	−8.48	0.61	173 (Y175F)	
S237A (mutant)		−7.45	3.44	180 (S226A)	7.3
Wild type		−8.73	0.39	41	4.1

[a]Autodock analysis was carried out using the most potent inhibitor (**10a**) from the series

N-acetylglucosamine (GlcNAc) through a β-1,4-glycosidic bond. The MurNGlyc unit is linked to a linear stem peptide chain consisting of amino acids L-Alanine (L-Ala), D-Glutamic acid (D-Glu), *meso*-diaminopimelic acid (*meso*-DAP) or L-Lysine (L-Lys) and D-Alanine (D-Ala). In Gram-positive bacterial PG, the adjacent stem peptides are highly cross-linked and the cross-linking takes place between D-Ala^4 and *meso*-DAP^3 of neighboring stem peptides (D-Ala^4 → *meso*-DAP^3), called a classical type of PG cross-link mediated by D,D-transpeptidase enzyme [42]. **D,D-transpeptidase** is the molecular target of β-lactam antibiotics. However, *Mtb* is resistant to β-lactam antibiotics because of the production of chromosomally encoded Ambler class A β-lactamase enzyme, which cleaves the β-lactam ring of the antibiotic and also due to the presence of non-classical type PG cross-link between neighboring *meso*-DAP^3–*meso*-DAP^3 residues. The non-classical type of PG cross-linking is catalyzed by **L,D-transpeptidase** (Ldt) enzymes [43]. In *Mtb*, two functional Ldt paralogs are present, namely $\mathrm{Ldt_{Mt1}}$ and $\mathrm{Ldt_{Mt2}}$; among them, $\mathrm{Ldt_{Mt2}}$ is predominantly expressed than $\mathrm{Ldt_{Mt1}}$ and both these enzymes signify an important druggable target [44, 45]. Recently, the crystal structure of Ldt in complex with Meropenem, Biapenem, Tebipenem drugs are reported, which is very useful for structure-based anti-bacterial drug discovery against this target [46, 47].

2.2 *Target-Based Drug Design Toward* Mtb *Regulatory Process*

The literature reports suggest that the integrated signaling networks in mycobacteria play significant roles in host–pathogen interactions as well as in intracellular survival strategies [48]. The response regulators **DevR** and **DevS** form a well-characterized two-component regulatory system of *Mtb*, which is important for the adaptation and dormancy of mycobacteria within the host tissues in response to hypoxia, nitric oxide, carbon monoxide, and ascorbic acid [49–52]. These regulators represent compelling targets for anti-bacterial drug design towards *Mtb*. Kaur et al. identified two DevR mimetic peptides which specifically inhibit the DevR-dependent transcriptional activity, thereby blocking the survival of *Mtb* under hypoxic conditions [53]. Gupta et al. developed a homology model of DevR and used it for structure-based screening of 2.5 million ZINC compounds. They identified a potential compound and validated its sterilizing activity against tubercle bacilli by *in vitro* techniques [54]. Other *Mtb* target involved in the regulatory process is **PknB**. It is a transmembrane serine/threonine-protein kinase B and plays a crucial role in a number of signal transduction through phosphorylation of protein and also regulates cell division and differentiation [55]. Lougheed et al. performed an HTS of ~54,000 compounds against PknB target and identified number of inhibitors with anti-mycobacterial activities in the micromolar range [56]. **PknG** is another serine/threonine-protein kinase critical in signal transduction pathway of *Mtb*, and a few PknG target-based inhibitors are reported. By adopting a sequential pharmacophore-based virtual screening method and threefold docking using different search algorithms followed by molecular dynamic simulations, Singh et al. identified few inhibitors against PknG target. The in vitro validation resulted in three of these compounds with significant inhibitory activity against *Mtb* PknG. Further, the *Mtb* survival studies within the infected THP-1 macrophage cells demonstrated that NRB04248 compound inhibited the growth of *Mtb bovis* BCG [57].

The pathogenicity of *Mtb* is based on the bacilli's ability to inhibit phagosome acidification and maturation processes after endocytosized by macrophages. Tyrosine phosphatase is enzyme which dephosphorylates the host proteins in human and which is involved in the signaling pathways leading to the prevention of the initiation of host defense mechanisms, including phagosome acidification. Two types of tyrosine phosphatase enzymes are present in *Mtb,* namely **tyrosine phosphatase A** (PtpA) and **tyrosine phosphatase B** (PtpB) [58–60]. Inhibition of these enzymes leads to the decrease in the proliferation of *Mtb* in host macrophages and thus represents a key druggable target for TB therapy.

2.3 *Druggable Targets Involved in* Mtb *Protein Synthesis*

In bacterial protein biosynthesis, **peptide deformylase** (PDF), a metalloprotease enzyme, plays a pivotal role in the maturation of nascent polypeptides. Hence, PDF represents a potential druggable target for the TB therapy [61]. In human, PDF homologue has been identified; however, there is a notable difference among the *Mtb* and human PDF proteins. The main difference between them lies in the fingerprint active site motif pattern, in which leucine residue is mutated into glutamic acid in human PDF [62]. The sequence alignment of both *Mtb* and human PDF showed very less identity in the active site region, suggesting the specific residues involved in both the species for its function [62]. Hence, these major differences between them enable the design and discovery of novel inhibitors with selective inhibitory activity toward *Mtb* PDF. Also, the *Mtb* PDF contains highly conserved motifs, EGCLS and QHEXXH (where X is any hydrophobic amino acid), which are important for the metal ion coordination and thus necessary for its enzyme activity. A number of *Mtb* PDF inhibitors are reported till date by knowing the 3D structure-based mechanism or HTS techniques [63–65].

Another important druggable target in *Mtb* protein synthesis mechanism is **aspartyl tRNA synthetase** (AspS), which represents a key molecular target for designing anti-TB agents [66]. Soto et al. recently identified a couple of inhibitors, GSK97C (spiro-oxazolidin-2-one), GSK93A (2-amino-1,3-thiazole), GSK85A, and GSK92A (enamides) targeting *Mtb* AspS using a whole cell-target-based approach [67]. The X-ray crystal structure of *Mycobacterium smegmatis* AspS is already being reported and the catalytic site consisting of three important amino acid residues, Asp 174, Phe521, and Thr565 which will further aid in structure-based drug discovery process [66]. **Leucyl–tRNA synthetase** (LeuRS) which functions in protein synthesis represents another potent druggable target in *Mtb*. The compound, GSK-070, an oxaborole derivative, is demonstrated to inhibit LeuRS and consequently blocks the protein synthesis by forming a complex with tRNA and trapping the enzyme–tRNA complex on the editing site. GSK-070 has recently completed Phase I clinical trial [68, 69].

2.4 Molecular Targets in Mtb *Energy Production and Metabolism*

Key druggable target in energy metabolism includes **Cytochrome b** subunit of the **cytochrome bc1 complex (Qcrb)** of *Mtb* which is mainly involved in the respiratory transport chain for ATP synthesis [70]. HTS campaign identified novel imidazo [1,2-a] pyridine inhibitors and Q203 compound as a potential clinical candidate for *Mtb* therapy [71]. However, this compound is bacteriostatic and does not kill dormant bacteria. Further, homology modeling and docking study of Qcrb with Q203 inhibitor was performed by Choi and Ko to elucidate the drug–target molecular mechanism of action. Parish et al. discovered benzimidazoles of phenyl alkyl groups showing good anti-mycobacterial intracellular activity in the nanomolar range targeting Qcrb protein, with promising sterilizing activity and low cytotoxicity against eukaryotic cells. This study further paved way for the molecular mechanism of action of small-molecule inhibitor against Qcrb inhibition [72]. **Isocitrate lyase** is the first enzyme of the glyoxylate shunt pathway involved in the *Mtb* carbon metabolism by cleaving isocitrate to succinate and glyoxylate. This enzyme is critical in *Mtb* for its replication and persistence, shown in *in vivo* mouse model [73, 74].

Malate synthase (GlcB) is the second enzyme of the *Mtb* carbon metabolism and anaplerosis with catalytic conversion from glyoxylate to malate and acetyl-coenzyme A (acetyl-CoA) into CoA, respectively. Crystal structure of GlcB of *Mtb* was solved with several inhibitors by fragment-based screening and HTS methods. These structures revealed conformational flexibility at the active sites during catalysis in order to screen preferred chemotypes and its molecular mode of binding. Further, in vivo mouse model of acute infection showed reduced bacterial load using GlcB inhibitors [75, 76]. **ATP synthase** is another key target in anti-mycobacterial drug discovery. This is primarily involved in cellular energy production in all microorganisms, plants, and animals. The structural and functional activity of the ATP synthase enzyme is same in all organisms. Bedaquiline, a second-line TB drug, targets ATP synthase subunit ε and inhibits *Mtb* ATP production (Table 1; Fig. 2). It was approved by FDA as MDR-TB drug [77, 78]. An imidazopyridine amide compound (Q203) was discovered by the whole cell screen in infected macrophages. Q203 is a potent inhibitor by disrupting the electron transport chain of ATP synthesis and successfully entered in Phase I clinical trials [79]. **Lipoamide dehydrogenase** (Lpd), another key target member of three multienzyme, complexes the energy metabolism pathway for *Mtb* virulence and can be potential for anti-TB drug discovery program. *Mtb* of Lpd is crucial for the metabolism of branched chain amino acids and thus essential for its pathogenicity. The small-molecule inhibitor triazospirodimethoxybenzoyl which inhibits the *Mtb* Lpd enzyme selectively without affecting the human Lpd enzyme can be successfully used for TB therapy [80, 81].

Another important druggable target in *Mtb* is **glutamine synthetase** required for both its nitrogen metabolism and cell wall biosynthesis. Glutamine synthetase is essential for *Mtb* virulence [82], and the inhibition of the enzyme resulted in reduced growth of *Mtb* [83]. The tricarboxylic acid (TCA) cycle plays a key role in the metabolism of almost all pathogens including mycobacteria. The **fumarate hydratase** enzyme has been identified as one of the important enzymes, which catalyzes the reversible conversion of fumarate to (L) malate in the TCA cycle. The presence of the human homologue of fumarate hydratase implicated the inhibitor designing toward this target. The first selective small-molecule inhibitor for *Mtb* fumarate hydratase is reported in 2016, in which the inhibitor is binding to an allosteric site consisting of amino acid residues which are different between human and *Mtb* [84].

2.5 Protein Membrane Transport Targets for Mtb *Inhibition*

Mtb protein export pathway system comprising **SecA cytoplasmic ATPase** is required for bacterial virulence. The SecA is an important *Mtb* target, since no eukaryotic homolog exists suggesting for safe development of anti-TB drug design and discovery. Crystal structure of SecA was solved, which was further used by Li et al. to discover new inhibitors by structure-based virtual screening (VS) [85]. Further, Chen et al. developed biochemical ATPase assay and validated these inhibitors at the micromolar range for TB therapy [86].

Pyrazinamide is a first-line TB drug, and its possible targets include FAS I, QAPRTase, RpsA, PanD, Rv2783 (Table 1; Fig. 2). Pyrazinamide is a prodrug which is converted by pyrazinamidase into the active form pyrazinoic acid [87, 88]. This drug inhibits the transport mechanism and energy pathway in *Mtb,* thereby disrupting its growth. Pyrazinoic acid is able to inhibit the enzyme, fatty acid synthase, in *Mtb,* which in turn inhibits the production of fatty acids. There is also evidence of this inhibitor in disrupting the membrane potential and energy production, which are essential for *Mtb* survival [89].

2.6 Mycobacterial Drug Targets Involved in Replication and Transcription

DNA gyrase is an essential ATP-dependent bacterial enzyme that acts by creating a transient double-stranded DNA break, which relieves the strain caused by unwinding of double-stranded DNA by helicase enzyme. DNA gyrase is found in all the bacterial species and essential for DNA replication, transcription, and recombination processes (Fig. 3). The enzyme exists as a heterotetramer consisting of two A subunits and two B subunits (A2B2) [90]. In *Mtb,* it is proven that the inhibition of DNA gyrase results in high anti-mycobacterial activity toward both actively replicating and non-replicating, dormant bacilli, which is necessary for shortening the TB treatment [91]. The synthetic antimicrobial class, Fluoroquinolones, has been demonstrated to have anti-tubercular activity by inhibiting to the mycobacterial DNA gyrase subunit A [92]. Fluoroquinolones are currently being used as second-line drugs for TB, and also these drugs have been used in TB treatment to hamper the development of XDR-TB from MDR-TB (Table 1; Fig. 2). Though the emergence of fluoroquinolone resistance affects its use as second-line drugs [93], this has driven the interest in targeting the DNA gyrase subunit B (GyrB). An approved anti-bacterial agent novobiocin was the only approved GyrB inhibitor; however, it was withdrawn from the market due to safety concerns [94]. A new class of compounds, aminobenzimidazole antibiotics targeting the ATP-binding site of GyrB, has also been reported [95]. Chopra et al. evaluated the biological activity of both aminobenzimidazole and novobiocin. The drugs are demonstrated to be active against *Mtb* with minimum inhibitory concentration (MIC) of 1 and 4 mg/ml, respectively. Only aminobenzimidazole compound exhibited a time-dependant mycobactericidal activity against both drug-sensitive and drug-resistant bacilli, which also showed potent activity against non-replicating persistent *Mtb*. Further, this compound significantly reduced the lung colony forming unit counts in the mice TB model [95].

Shirude et al. identified aminopyrazinamides as novel and specific GyrB inhibitors that kill replicating and non-replicating *Mtb* bacilli through HTS of compounds. They solved the X-ray crystal structure of GyrB of *M. smegmatis* in complex with one of the aminopyrazinamides. The aminopyrazinamides also showed significant anti-mycobacterial activity under *in vitro*, intracellular, and hypoxic conditions. The interactions with the hydrophobic pockets consisting of Val 49, Val 123, Val 128, and Ile 171 contribute to the specificity of the compound as compared to other known GyrB inhibitors. Another, notable amino acid residues at the active site regions include Asp 79, Arg 82, and Arg 141 obtained from GyrB crystal structure [96]. Later, a couple of studies also reported novel classes of compounds targeting *Mtb* GyrB through molecular hybridization studies followed by protein–ligand molecular docking [97–99]. More recently, VS studies have been conducted for a group of flavonoid compounds to identify a dual inhibitor for DNA gyrase and isoleucyl-tRNA synthetase enzymes. The binding of the high-ranked flavonoid, taxifolin, to both these enzymes was validated through molecular dynamics simulation. Further, anti-mycobacterial activity of taxifolin was evaluated using a cell viability assay resulted in MIC $\leq$ 12.5 μg/ml against *Mtb* [100].

Another notable *Mtb* target is **RNA polymerase enzyme** (RNAP), which is required for the transcription process. RNAP is the target for the first-line anti-TB drug Rifampicin. An MDR and XDR strain of *Mtb* is resistant to Rifampicin due to the mutations in the rpoB gene encoding RNAP. Rifampicin-resistant *Mtb* strains show mutation in the 81 bp hotspot region of rpoB gene, stretching from codons 507 to 533 [101]. Researchers are underway to identify derivatives of Rifampicin which are not affected by rpoB mutations in the binding site of Rifampicin. Lin et al. recently solved the X-ray crystal structure of RNAP in its ligand free form as well as in complex with Rifampicin at 3.8–4.4 Å resolution. They identified novel compounds *N*α-aroyl-*N*-aryl-phenylalaninamides (AAPs) by HTS and solved the crystal structure of RNAP-AAPs complexes. Further, they showed binding to a different site than Rifampicin in the mycobacterial RNAP. Hence, mutations at the Rifampicin binding site will not hamper AAPs activity. Also, AAPs showed addictive anti-mycobacterial activity in combination with Rifampicin [102].

More recently, new anti-bacterial target in *Mtb* RNAP has been reported by Wang et al. They constructed phylogenetic trees for 17 genes important for the functioning of RNAP enzyme in 13 different mycobacterial species and identified positive selection sites or conserved regions. They modeled the 3D structure of RNAP and performed molecular docking calculations with anti-bacterial drugs. By comparing the positive selection site and as well as molecular interaction, they proposed a putative drug binding site near Cys 933 and His 935 residues on the rpoB subunit [103]. Several research groups are underway to identify novel inhibitors against *Mtb* RNAP as well as to identify potential Rifampicin analogues effective against MDR and XDR strains of TB [104, 105].

2.7 Other Druggable Targets of Mtb

Filamenting temperature-sensitive protein Z (FtsZ) is a cytoskeletal protein involved in the *Mtb* cell division. The protein forms a contractile ring structure (Z ring) at the site of cell division. FtsZ also functions to recruit cell division proteins to the septum required for the formation of new cell wall between dividing cells. FtsZ is a prokaryotic homologue to the eukaryotic protein tubulin. It contains a GGGTGTG motif like in tubulin for guanine binding in FtsZ and also shows GTPase activity [106]. So, blocking of FtsZ leads to cell division arrest leading to bacterial death [107]. Das et al. recently identified phytochemicals as FtsZ inhibitor through molecular docking-based VS; however, no experimental validation was performed by that group [108].

Caseinolytic peptidase P (ClpP) is another important molecular target in *Mtb*. ClpP acts in association with ATPases to perform energy-dependant degradation of damaged proteins within the cell. *Mtb* encodes two ClpP homologues, ClpP1 and ClpP2, and forms a mixed protein called ClpP1P2 [109]. It was experimentally validated that both ClpP1 and ClpP2 are required for the protein degradation and depletion of either of the protein results in bacterial death [110]. A novel natural product, lassomycin, is found to inhibit ClpP [111]. Schmitz et al. solved the X-ray structure of ClpP1P2 and elucidated the molecular mechanism of association with ATPases [112]. This will aid in designing of novel and more potent inhibitors of ClpP protein. *Mtb* produces 20S proteasome, which is essential for the survival of the bacteria within the host and helps in defending the bacilli against nitrosative stress [109, 113]. Gandotra et al. demonstrated that prcBA genes encoding mycobacterial proteasome are essential for the *Mtb* survival in chronic phase of infection in mice [114]. Hence, **20S proteasome** serves as an important mycobacterial target for inhibitor design.

The enzymes involved in menaquinone (Vitamin K2) synthesis, **menA**, **menB**, **menC**, **menD**, **menE**, **menF**, and **ubiE** (**menG**), serve as an important druggable target for anti-mycobacterial therapy [115, 116]. It has been proved that Menaquinone synthesis is important for maintaining the mycobacterial viability during the exponential growth phase and recovery from non-replicating persistence and also involved in electron transfer pathways.

3 Structure-Based Anti-TB Drug Design Approach and Its Molecular Mechanism of Action

Structure-based inhibitor design for the past two decades became significant due to the theoretical and experimental technological advancements which include: protein modeling, ab initio modeling, homology modeling, protein folding dynamics, molecular docking, pharmacophore modeling, virtual screening, quantitative structure activity relationship (QSAR), structural biology, nuclear magnetic resonance (NMR) studies toward the preclinical drug discovery program. 3D structural data present in the protein data bank (PDB) and other pharmaceutical databases contain many millions of datasets which fit into the big data domain. This vast amount of data can provide information about the key molecular mechanism of action at the atomic level. Nowadays, structure- and ligand-based drug design has also become a fundamental strategy in both lead generation and lead optimization.

The main challenge in the anti-TB drug discovery program is the different types of screening compounds using virtual, cell-based, and target-based *in vitro* methods lack the important *Mtb* physiology in which its cell wall permeability, metabolic stability, and druggable target's resistance are often neglected. While maintaining the compound efficacy, these unique properties are difficult to achieve for its success toward TB drug discovery. Thus, *in silico* and *in vitro* biochemical techniques should provide guidance for the compound to be selective and specific with respect to particular druggable targets under different physiologically appropriate conditions. This in turn can avoid compounds with unfavorable physicochemical as well pharmacokinetics and pharmacodynamic properties. Further, the study will progress by identifying new hits having novel scaffolds for drug development and new protein targets, showing promising mechanism of action from the existing knowledge on drugs and its resistance mechanisms.

The structure-guided inhibitor designing is performed based on the structure of the molecular targets. Due to the advances in X-ray crystallographic techniques, many mycobacterial protein targets have been crystallized, solved, and deposited in the PDB. If the crystal structure of a target protein is not solved experimentally using X-ray crystallography or NMR, homology modeling can be done to build predictive model of the protein using a crystal structure of related proteins with a good sequence identity. The 3D structures of the molecular targets provide an understating of protein folding, function and active site regions that are important aspects in the anti-tubercular drug discovery process. The druggable molecular target in *Mtb* for which the crystal structure is submitted in the PDB is given in Table 3. In addition to crystal coordinates of the target proteins, the interactions with known inhibitors are also crucial for the structure-based inhibitor designing. The binding mode of these compounds/drugs toward the well-characterized molecular target aids the effective drug discovery process. Another notable advance in structure-based drug designing is the drug repurposing/reprofiling strategy which is described below.

Conventionally, the discovery and development of anti-bacterial agents were based on the identification of novel compounds targeting various bacterial targets. This process of finding novel compounds is exceptionally expensive and takes an enormous amount of time. Only a few compounds pass the clinical trials with good safety profiles among thousands of compounds tested, which makes the drug discovery and development process very time-consuming and expensive [117]. In the last few years, a number of new anti-TB agents have been proposed and developed into effective therapeutics; some of them were obtained by modifying the already existing drugs or scaffolds, and many are developed by repurposing strategy [118].

Drug repurposing or repositioning is a discovery process, which takes drugs that have been approved for one disease and repositioning them for another disease [118–124]. The traditional drug discovery and development process takes enormous amount of time to reach into market. However, discovering new indications for already approved drugs can improve the drug safety and can lower the development cost and time. Many repurposed drugs are being used nowadays for the treatment of various diseases. Also drug repurposing is becoming very popular in

Table 3 Crystal structure of the key druggable targets in *Mtb*

Molecular target	Pathways involved	PDB ID
DprE1	Arabinogalactan biosynthesis	6G83, 5OEL, 5OEP, 5OEQ, 4CVY, 4P8C, 4P8 K, 4P8L, 4P8P, 4NCR
IspD	Arabinogalactan biosynthesis	3OKR, 3Q7U, 3Q80, 2XWN
Antigen 85 complex	Mycolic acid synthesis	5OCJ, 5VNS, 4QDO, 4QDT, 4QDU, 3HRH
Pks13	Mycolic acid synthesis	6D8I, 6D8 J, 5XUO, 5V3 W, 5V3X
PcaA	Mycolic acid synthesis	1L1E
InhA	Mycolic acid synthesis	
D,D-transpeptidase	Peptidoglycan biosynthesis	5CRF, 5CXW, 4RYE
L,D transpeptidase-1	Peptidoglycan biosynthesis	4JMN, 4JMX, 5E51, 5E5L
L,D transpeptidase-2	Peptidoglycan biosynthesis	5DZP, 5DVP, 5D7H, 4GSU, 4GSR, 4GSQ, 4HU2, 4HUC, 3VAE, 3U1P, 3TX4, 3TUR
GlmU	Peptidoglycan biosynthesis	2QKX
RmlC	Arabinogalactan biosynthesis	2IXC, 1PM7
PknB	Signal transduction	6B2P, 6B2Q 5U94, 5E0Y, 5E0Z, 5E10, 5E12
PknG	Signal transduction	2PZI, 4Y0X, 4Y12
DevS	Regulatory process	2W3H, 2W3G, 2W3F, 2W3E, 4YOF, 4YNR
AspS	Protein synthesis	5W25
Peptide Deformylase	Maturation of nascent polypeptides	3E3U
DNA Gyrase	DNA replication process	5BTN, 5BTL, 5BTI, 5BTG, 5BTF, 5BTD, 5BTC, 5BTA
RNA polymerase	Transcription process	5W36, 5W35, 5W33, 6C06, 6C05, 6C04
Isocitrate lyase	Carbon metabolism	6C4A, 6C4C, 5DQL
Malate synthase	Carbon metabolism	1N8I, 1N8 W, 4TVM
Lpd	Energy metabolism	4M52, 3II4
FtsZ	Cell division	1RQ7, 1RQ2, 1RLU, 2Q1X, 5V68
ClpP	Degradation of damaged protein	4U0H, 2CE3
20S proteasome	Protein processing	2FHG
Glutamine synthetase	Nitrogen metabolism and cell wall biosynthesis	1HTO, 1HTQ
Fumarate hydratase	Citric acid cycle	3NO9, 5F92, 5F91
MenB	Vitamin K2 synthesis	1Q52, 1Q51
MenD	Vitamin K2 synthesis	5ESD, 5ESO, 5ERY, 5ERX

TB drug discovery, and in past years, a number of repurposed TB drugs are being reported (Table 4; Fig. 6). Some of the drugs which are repurposed for TB treatment are given below.

Table 4 Repurposed drugs that were initially developed for the treatment of other diseases and now being evaluated for tuberculosis treatment

Drug	FDA Approval	Mechanism of action	Repurposed mechanism of action in *Mtb*
Amoxicillin/ Clavulanate	Bacterial infections including strep throat, pneumonia, skin infections, and urinary tract infections	Amoxicillin inhibit D, D-transpeptidase enzyme and inhibit bacterial cell wall synthesis; clavulanate inhibits β-lactamase enzyme	Inhibits D, D-transpeptidase enzyme and inhibit bacterial cell wall synthesis; clavulanate inhibits β-lactamase enzyme of *Mtb*
Meropenem Imipenem	Broad-spectrum bacterial infections	Inhibit D,D-transpeptidase enzyme and blocks bacterial cell wall synthesis	Inhibit both D,D-transpeptidase and L,D-transpeptidase enzymes for *Mtb* PG cross-linking
Clarithromycin	Bacterial infections affecting skin and respiratory tract	Binds to 23S rRNA of bacterial 50S ribosomal subunit and inhibits protein synthesis	Binds to 23S rRNA of bacterial 50S ribosomal subunit and inhibits specifically mycobacterial protein synthesis
Linezolid	Infections caused by Gram-positive bacteria: skin infections and pneumonia	Inhibit protein synthesis by binding to 23S rRNA of ribosomal 50S subunit	Inhibit protein synthesis by binding to 23S rRNA of ribosomal 50S subunit of *Mtb*
Levofloxacin	Mainly for acute bacterial sinusitis, pneumonia, urinary tract infections, chronic prostatitis, and some types of gastroenteritis	Inhibit bacterial DNA replication by binding to DNA gyrase and topoisomerase IV	Inhibit mycobacterial DNA replication by binding to DNA gyrase and topoisomerase IV
Moxifloxacin	Chronic bronchitis, acute bacterial sinusitis, pneumonia	Inhibit bacterial DNA synthesis by inhibiting DNA gyrase	Inhibit mycobacterial DNA replication by inhibiting DNA gyrase
Gatifloxacin	Respiratory tract infections	Inhibit bacterial DNA replication by binding to DNA gyrase and topoisomerase IV	Inhibit mycobacterial DNA replication by binding to DNA gyrase and topoisomerase IV

(continued)

Table 4 (continued)

Drug	FDA Approval	Mechanism of action	Repurposed mechanism of action in *Mtb*
Metronidazole	Antibiotic and anti-protozoal	Covalently binds to DNA, disrupt its helical structure, inhibit bacterial nucleic acid synthesis	Interfere with mycobacterial DNA
Chlorpromazine	Antipsychotic	Antagonist (blocking agent) on different postsynaptic receptors	Inhibit NADH: menaquinone oxidoreductase
Clofazimine	Anti-leprosy	Interference with template function of DNA in *M. leprae*; alteration of membrane structure and its transport function; disruption of mitochondrial electron transport chain	Interfere with mycobacterial DNA
Metformin	Anti-diabetic	Suppress hepatic gluconeogenesis by inhibition of the mitochondrial respiratory chain, activation of AMPK, inhibition of glucagon-induced elevation of cyclic adenosine monophosphate (cAMP)	Activates host AMPK leading to production of mitochondrial reactive oxygen species, and aids phagosome–lysosome fusion
Artemisinin	Antimalarial	React with heme and iron(II) oxide, results in the generation of free radicals that in turn damage susceptible parasitic proteins	Inhibit the establishment of dormancy by binding to heme molecule of the mycobacterial oxygen sensor, DosS/T which turns down its ability to detect oxygen levels

β-lactam antibiotics: They are developed primarily to treat Gram-positive bacterial infections, but have been repurposed to be used as an anti-TB agent. Enormous research is being carried out to study the efficacy of β-lactam antibiotics in combination with β-lactamase inhibitor. Among all the β-lactams, carbapenems are considered to be more promising as anti-TB agents, as they are more stable to

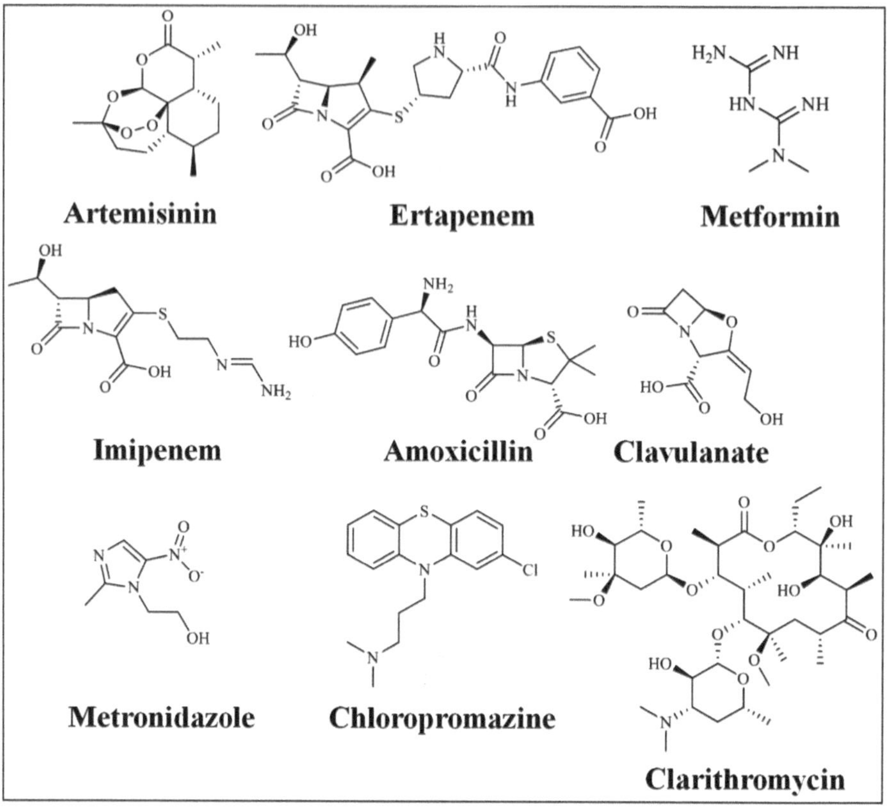

Fig. 6 Structure of repurposed drugs for TB

mycobacterial β-lactamase enzyme [125]. The anti-tubercular activity of carbapenems, i.e., Meropenem, Imipenem, has been of immense interest in recent years, and some of the carbapenem antibiotics are in clinical trials for the treatment of TB. These antibiotics targets covalently modify both D,D-transpeptidase enzyme and Ldt enzymes involved in PG cross-linking. The mechanism of action of carbapenems against Ldt enzymes are well studied, and crystal structure of some of these antibiotic compounds has been solved and deposited in PDB. The drug forms a covalent bond with the thiol group of Cys 354/226 present in the active site of both the Ldt enzymes (Ldt_{Mt1} & Ldt_{Mt2}). Other amino acid residues which stabilize the binding are His 208/336, Ser 209/337, Met 175/303, Tyr 190/318, His 224/352, Gly 225/353, and Asn 228/356 of Ldt_{Mt1} and Ldt_{Mt2}, respectively [46, 47, 126]. The molecular mechanism of binding of Meropenem, a carbapenem class of antibiotic which is in clinical trials for TB therapy toward Ldt_{Mt2}, is given in Fig. 7.

Artemisinin: It is an ancient Chinese medicine extracted from sweet wormwood *Artemisia annua* and is used to treat malaria. In 2016, Abramovitch et al. demonstrated the anti-tubercular activity of Artemisinin. The compound targets

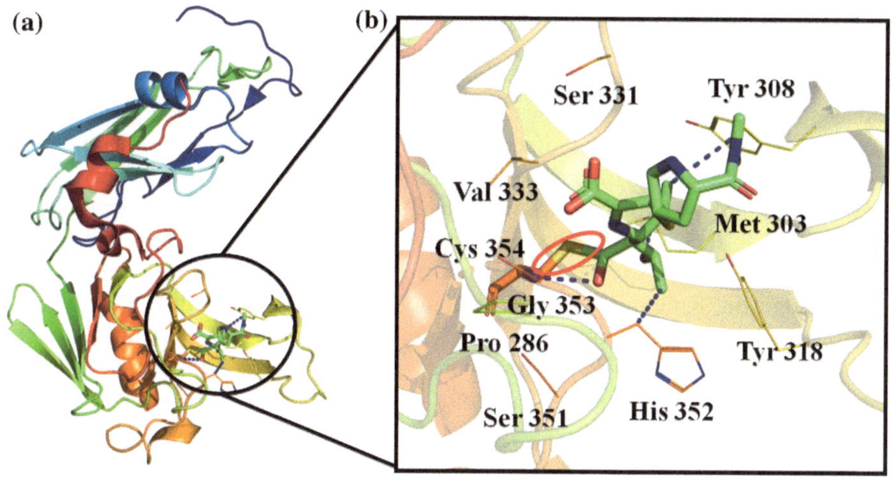

Fig. 7 Molecular interactions of Meropenem with the active of Ldt_{Mt2} (PDB ID: 4GSU [47]). **(a)** The binding site of Meropenem (sticks) and **(b)** the molecular interactions with the active site amino acid residues (lines); the catalytic Cys 354 is represented as orange sticks and the covalent bond with Meropenem is encircled

heme molecule present in the oxygen sensor called DosS/T present in *Mtb* and inhibits the bacilli to establish dormancy during hypoxic conditions. However, further studies are required to better understand the anti-mycobacterial activity of Artemisinin [127].

Metformin: It is an oral drug which is approved for the treatment of type 2 diabetes mellitus. The first report on Metformin as adjunct anti-tuberculosis therapy was reported in 2016 [128]. Instead of targeting the *Mtb* protein, Metformin is involved in the host-directed therapy, which involves targeting the harmful inflammation leading to tissue damage. Metformin is an activator of 5'adenosine monophosphate-activated protein kinase (AMPK), and this in turn increases the mitochondrial reactive oxygen species (ROS). The macrophages, which are exposed to Metformin, showed high *in vitro* anti-mycobacterial activity because of the increased level of ROS [128]. Briefly, Metformin is shown to promote phagocytosis, phagosome–lysosome fusion, and autophagy, which are the host cellular processes evaded by *Mtb* to survive inside macrophages. Singhal et al. further confirmed that diabetes mellitus patients on Metformin treatment had higher control of TB infection [128, 129].

Clofazimine: Clofazimine is a riminophenazine dye used in the leprosy treatment for decades. The efficacy of treating MDR and XDR-TB is already validated, and it is listed as a WHO-recommended second-line drug for the treatment of TB. Clofazimine is shown to decrease the TB treatment duration [130]. However, the main side effect of the drug is skin discoloration owing to its long half life and higher lipophilic nature [131, 132].

Clarithromycin: It is a semisynthetic macrolide antibiotic, which is approved to treat various skin as well as respiratory tract infections. The drug binds reversibly to the bacterial 23S rRNA of 50S ribosomal subunit and impedes the amino acid translocation and further protein assembly. Clarithromycin is also used to treat nontuberculous mycobacterial infections [133]. Several studies have been conducted from the mid-90s to study the effect of Clarithromycin on *Mtb*. Clarithromycin gets metabolized into 14-hydroxyclarithromycin; both of them act synergistically [134]. Cavalieri et al. reported that clarithromycin/14-hydroxyclarithromycin had considerably improved the in vitro anti-mycobacterial activities of the conventional TB drugs, Isoniazid, Rifampicin, and Ethambutol against MDR bacilli [135].

Fluoroquinolones: Fluoroquinolones are one of the broad-spectrum antibiotic classes effective against both Gram-negative and Gram-positive bacteria. They kill the bacteria by preventing the bacterial DNA replication. The third-generation Levofloxacin and the fourth-generation Moxifloxacin and Gatifloxacin are now used as second-line anti-TB drugs for treating drug-resistant strains of *Mtb* [136, 137].

Linezolid: It is a synthetic antibiotic belonging to the class of oxazolidinone. Linezolid is used to treat Gram-positive bacterial infections. Linezolid functions by binding to the peptidyl transferase center located in the 23S portion of 50S ribosome, thereby inhibiting protein synthesis. The bactericidal activity of Linezolid against drug-resistant *Mtb* has been well studied in the past few years, and now, it is being used as a second-line TB drug in MDR and XDR-TB treatment regimen [136]. But there were concerns against its safety and tolerability [138].

Chlorpromazine: It is an FDA-approved drug used to treat psychotic disorders such as schizophrenia. Chlorpromazine showed in vitro anti-tuberculous activity by inhibiting NADH:menaquinone oxidoreductase enzyme [139]. Also, it is shown to improve the efficacy of the first-line anti-TB drugs in combination with Chlorpromazine [140, 141].

4 Recent Computer-Aided Drug Design Approaches for Anti-TB Drug Discovery

The advances in the computational methods and techniques have a great impact on the drug discovery and development. The effectiveness of computational algorithms and software tools has tremendous impact on speeding up the conventional drug discovery. Now, computational techniques are inseparable part of drug discovery and development process. Traditionally, the computer-aided drug discovery process involves virtual screening (VS) of a set of small molecules against the X-ray crystal structures. If the crystal structure data is not available, ligand-based drug designing uses knowledge of existing active compounds against a particular target proteins. In ligand-based inhibitor designing, using the structural features of the known ligands of druggable molecular targets, new compound with improved potency can be developed. It majorly involves QSAR or pharmacophore-based modeling. In QSAR modeling, the relationship between physicochemical properties called molecular descriptors of the known ligands with the biological activity will be expressed as a regression equation. Further, using this QSAR equation the activities of new compounds will be predicted. QSAR is based on the assumption that the biological activity of a compound depends upon the molecular features present in the structure. With the advancement of powerful drug designing algorithms and software, the process of finding novel drugs has been improved a lot. Several successful drugs are already being in a market which is developed by computational chemistry and drug designing strategies. Some of the reported computer-aided drug designing studies toward the development of novel anti-tubercular agents are already being discussed in the above sections. With the use of high-end computational techniques, many research works are carried out by different research groups, and some of the recent advances in the in silico-based designing of anti-tubercular agents are discussed here.

Many anti-tubercular drug designing strategies are reported in past few years involving development of structural analogues to existing TB drugs using ligand-based strategies. Aragwal et al. performed VS using a ligand-based pharmacophore model and identified 95 compounds from small-molecule database. The identified hits were evaluated further using molecular docking calculations, and 15 short listed compounds were biological assays [139]. Anquetin et al. carried out QSAR studies on the quinolone series of compounds. Based on the QSAR results, they synthesized six fluoroquinolones, and four of them were found to be active against *Mtb* [142]. In another study, structure- and ligand-based computational models were developed to identify potential inhibitors for *Mtb* GlmU. The ligand-based method involved QSAR analysis of known GlmU inhibitors, and they identified lead compounds with potential anti-mycobacterial activity [143].

Recently, an in silico-guided polypharmacological approach for drug screening was reported. This approach considered three *Mtb* molecular targets, InhA, GlmU, and DapB with the aim of simultaneous inhibition of several potential targets for the treatment of drug-resistant TB. So, a combination of pharmacophore and QSAR-based VS strategy considering these three targets resulted in initial 784 hits from Asinex database small-molecule library. These hits were further subjected to molecular docking calculations with other 33 *Mtb* druggable targets. Finally, 110 potential polypharmacological hits identified; however, they have not tested the in vitro activity of those hits [144]. Choudhary et al. developed dynamics-based pharmacophore model for mycobacterial cyclopropane synthase (CmaA1) based on known inhibitors for the screening molecule library [145].

One of the most efficient computational methods was combinatorial design of small molecules and screening of these compounds by a ligand-based pharmacophore models. Nandi et al. adopted such a method and generated a combinatorial library of a series of 3850 fluoroquinolone and isothiazoloquinolone compounds, which was further VS based on QSAR model against mycobacterial DNA gyrase. The interactions of hits obtained were compared with the known ligands and 68 compounds, including 34 fluoroquinolones and 34 isothiazoloquinolones which were selected as potential leads [146]. Similarly, Singh et al. identified substituted hydrazine carbothioamide as potent anti-tubercular agents by using QSAR studies [147]. In other studies, *Mtb* DprE1 inhibitors are identified by employing molecular docking calculations [148, 149].

In order to find novel drug candidates, the understanding of pathogenesis of the underlying disease needs to be done. Kandasami et al. used bioinformatics techniques involving pharmacophore modeling to identify the catalytic residues of PknI, and those residues were experimentally validated by site-directed mutagenesis technique. This active site residue information was used in identifying an inhibitor specific to PknI, which was further validated by laboratory experiments [150]. Several other computer-aided drug designing studies are reported involving various in silico techniques comprising of homology modeling, QSAR, pharmacophore modeling, molecular docking-based VS, etc., against various *Mtb* drug targets [151–154].

Apart from the in silico drug screening techniques, computational approaches are also used to build databases of compounds for the ease of VS them toward various druggable targets. Prakash et al. developed an anti-tubercular compound database and a data mining procedure in the search for novel anti-tubercular agents and targets. They computed a minimum common bioactive substructure (MCBS) responsible for the activity of anti-tubercular agents by employing QSAR and pharmacophore modeling techniques [155]. This database of compounds will help to identify potential compounds with the structural feature similar to the known anti-tubercular drugs. Dalecki et al. developed an easy-to-use software solution for streamlining, processing, and analysis of biological screening data for *Mtb*. This software also offers a scaffold of compounds from screening data, which will further expand the scope of finding new compounds with improved activity [156]. A database, BioPhytMol, was designed to store and analyze the anti-mycobacterial phytomolecules and plant extracts, which would in future have immense potential as a drug discovery resource [157].

5 Novel TB Drugs in the Clinical Pipeline

Several new compounds as well as repurposed drugs for TB treatment are now in the clinical pipeline and are listed in Table 5. Auranofin is enrolled in the Phase 2 clinical trials for TB, which is an anti-rheumatic agent. Auranofin targets mycobacterial thioredoxin reductase, and in a cell-based screen, the drug exhibited anti-tubercular activity against non-replicating *Mtb* [158]. Nitazoxanide is another repurposed drug for TB, now in Phase 2 clinical trials. It is an FDA-approved anti-protozoal agent [159]. GSK-286 is a new chemical class having a novel mechanism of action against *Mtb*. It targets mycobacterial cholesterol catabolism, and the compound was found to penetrate into the necrotic lesions and kill intracellular *Mtb* with an MIC of $> 10\ \mu M$. GSK070 belongs to the oxaborole chemical class, shown to inhibit mycobacterial LeuRS enzyme. It has been shown to be active in in vivo murine TB model [160]. An oxazolidinone class of compounds Sutezolid, Delpazolid, TBI-223, and Contezolid are in the various phases of clinical trials for TB. These compounds inhibit protein synthesis in *Mtb* [161–163]. The benzothiazinone class of compounds, BTZ-043 and Macozinone, is demonstrated to inhibit the DprE1 enzyme of mycobacteria leading to the inhibition of arabinogalactan synthesis. Both these compounds are now in Phase 1 clinical trials.

The development of vaccines against *Mtb* has also been of interest by several research groups. H56 vaccine is in clinical trials, which is a multistage vaccination strategy consisting of a combination of early antigen Ag85B and early secretary antigen target (ESAT-6) with Rv2660c protein, which is associated with latency. The vaccine is demonstrated to promote T-cell response, and it is also controlled by the reactivation of the bacilli [164].

6 Future Directions to TB Drug Discovery Process

With the availability of validated *Mtb* molecular targets by the knowledge of the complete genome sequence of *Mtb* H37Rv, target-based discovery of new inhibitors is gaining interest in which the target modeling and chemoinformatics approaches have played a pivotal role. Host-directed therapies are also reported in recent years, which mainly focus on boosting the immune system of the host. Recently, pharmaceutical chemists have been pushing the discipline beyond computer-aided drug design in the field of chemical biology, to study and manipulate the biological systems at the system level. In the small-molecule drug discovery program, HTS campaigns using in silico and in vitro techniques are of paramount importance for the hit to lead identification. In addition, a recent trend in the preclinical campaign includes fragment-based drug discovery process, which includes focused or specific screening and iterative screening. This has been coupled with the speed and automation of a number of in silico (pipeline pilot mode) and biophysical techniques, which has the capacity to measure quantitatively the direct interaction between small molecule and druggable protein of interest. The application of computational and biophysical techniques demonstrates the direct target engagement with its hits or small molecules, which in turn increases the confidence in the HTS campaign.

During TB drug development program, different metrics were adopted in optimizing the hits to lead to compound discovery with an ultimate goal of decreasing the late-stage attrition in the clinical trials. The disease-driven biochemical pathways and sometimes the complex target space associated with this at the tissue level lead to new developments in the TB drug discovery process. This involves small-molecule library design to complex physiological models, which would in turn mimic the target tissue in the TB disease model system [165]. In the present scenario, the heterogeneous TB cellular models are studied thoroughly to understand variety of phenotypes and its 3D cellular imaging. These in turn occupy the bigger dataset. The informatics and algorithms for computing and analysis these big

Table 5 Newer TB drugs in the clinical pipeline

Compound	Chemical class	Target	Development stage	Sponsor/coordinator
GSK-286	–	Cholesterol catabolism of intracellular *Mtb*	Preclinical	GlaxoSmithKline, TB Drug Accelerator
TBAJ-587	Diarylquinoline	ATP Synthase	Preclinical	TB Alliance, University of Auckland
TBI-223	Oxazolidinone	Protein synthesis	Preclinical	TB Alliance, Institute of Materia Medica
Spectinamide 1810	Spectinamide	Protein synthesis	Preclinical	Microbiotix, Inc.
BTZ-043	Benzothiazinone	DprE1	Phase 1	University of Munich, Hans-Knöll Institute, Jena, German Center for Infection Research
GSK 070, GSK3036656	Oxaborole	LeuRS	Phase 1	GlaxoSmithKline
Contezolid (MRX-4)	Oxazolidinone	Protein synthesis (V region of 23S rRNA)	Phase 1	MicuRx Pharmaceuticals, Inc.
OPC-167832	3,4-dihydrocarbostyril derivative	DprE1	Phase 1	Otsuka Pharmaceutical Development & Commercial. Inc.
Macozinone (MCZ, PBTZ-169)	Benzothiazinone	DprE1	Phase 1	iM4 TB - Innovative Medicines for Tuberculosis, Bill & Melinda Gates Found
TBI-166	Riminophenazines	Mycobacterial DNA	Phase 1	Institute of Materia Medica, CAMS & PUMC
TBA-7371	Azaindole	DprE1	Phase 1	TB Alliance
Telacebec (Q203)	Imidazopyridine amide	qcrB subunit of the cytochrome bc1 complex	Phase 2	Qurient Co., Ltd, Qurient Co. Ltd./LLC "Infectex", a portfolio firm of Maxwell Biotech Venture Fund
Sutezolid	Oxazolidinone	Protein synthesis (bacterial 23S rRNA of the 50S subunit)	Phase 2	Sequella, Inc., TB Alliance
Delpazolid (LCB01-0371)	Oxazolidinone	Protein synthesis (V region of 23S rRNA)	Phase 2	LegoChem Biosciences, Inc.

(continued)

Table 5 (continued)

Compound	Chemical class	Target	Development stage	Sponsor/coordinator
SQ109	Ethylenediamine	MmpL3	Phase 2	Sequella, Inc.
Auranofin	Gold-containing thiosugar and triethyl phosphine	Thioredoxin reductase: trxB2	Phase 2	The Aurum Institute NPC, Calibr, The Scripps Research Institute
Nitazoxanide	Synthetic nitrothiazolyl-salicylamide derivative	–	Phase 2	Weill Medical College of Cornell University
Bedaquiline	Diarlyquinoline	ATP Synthase	Phase 3	Janssen Research & Development, LLC
Delamanid	Nitrodihydro-imidazo oxazole	Mycolic acid synthesis	Phase 3	Otsuka Pharmaceutical Development & Commercialization
Rifamycin (priftin)		RpoB Bacterial RNA polymerase	Phase 3	CDC TBTC, Sanofi

dataset is challenging and needs to be developed well for novel anti-TB drug discovery.

The traditional drug discovery and development process takes up to 10–15 years. It involves identification of right protein target in a disease, designing new inhibitors, optimizing the activity of the molecule and further preclinical and toxicity analyses. The enormous amount of scientific data is being reported, and analysis of this "big data" is one of the hurdles in the target identification process. In this data-driven process, the application of artificial intelligence (AI) plays a pivotal role nowadays. AI and machine learning have a crucial part in the first analysis of the massive scientific data to form essential new knowledge for the drug development process [166]. This digital electronic field is emerging and has an immense potential toward cost, speed, and efficiency in the drug discovery program. In drug discovery and clinical diagnostics, AI can outperform humans on certain tasks, and machine learning toward identifying the spot patterns and its relationship within big data provide guidance in this aspect, which is beyond our reach. There are different case studies of AI in the area of target identification and validation as well as in medicinal and synthetic chemistry [167–169]. These new data-driven technologies are proving to be tremendously promising when it reveals new mechanistic insights to disease, thereby helping to identify promising targets. In the context of computer-aided drug design, AI and machine learning techniques can process broader and varied chemical space in a much faster manner to identify the potential molecules from the bigger dataset for disease cure [170].

7 Conclusions

Current anti-TB drug regimens require better understanding of the drug–target relationships in order to decipher the structure–function relationships and its molecular mechanism of action with the drugs. Review of the literature surveys and studies on the basis of bioinformatics, structure-based TB drug discovery, computer-aided drug design, and drug repurposing study was promising for TB diagnostics, therapeutics, and molecular mechanism of action toward MDR-TB and XDR-TB. The proper selection of druggable TB target and its molecular mechanism of inhibition are essential to understand the TB drug resistance at the fundamental level in which the structure-based anti-TB drug design will play a pivotal role. The present crisis toward antibiotic resistance and the discovery of bedaquiline, delamanid, and recently eravacycline drugs has promised a sigh of relief for TB patients.

Drug repurposing or repositioning is an alternative step in the anti-TB drug discovery program addressing the drug resistance. The synergistic effect of the repurposed/combination drugs linezolid, clofazimine, benzoxaboroles, fluoroquinolones, trimethoprim, thioridazine, sulfamethoxazole, sulfadiazine, minocycline, amoxicillin/clavulanic acid, and carbapenems like Meropenem along with the new FDA-approved drugs bedaquiline, delamanid, and eravacycline can be successfully used in the treatment regimen for drug-resistant TB. This combinatorial chemotherapy toward MDR-TB, XDR-TB, and TDR-TB can definitely improve the life expectancy and reduce the mortality rate of TB patients to some extent. The main obstacle in the current anti-TB drug discovery program is the drug toxicity, mode of delivery, and duration of medication. This can surmount if new oral drugs with good pharmacokinetic profile and anti-TB activity at the nanomolar level can be discovered for patient compliance and safety.

Acknowledgements The author ACP acknowledges Kerala State Council for Science, Technology & Environment (KSCSTE) for awarding Junior Research Fellowship (Grant No:1132/2013/KSCSTE), India. The authors thank Indian Council for Medical Research (ICMR) and Department of Biotechnology (DBT; BT/PR5659/MED/29/564/2012), Government of India, New Delhi, India, for financial support. We also acknowledge gratefully Centre for Nanosciences and Molecular Medicine, Amrita Institute of Medical Sciences and Research Centre, Kochi, for the infrastructure support.

References

1. World Health Organization (2017) Global tuberculosis report 2017
2. Ducati RG, Ruffino-Netto A, Basso LA, Santos DS (2006) The resumption of consumption: a review on tuberculosis. Memórias do Instituto Oswaldo Cruz 101:697–714
3. Strebhardt K, Ullrich A (2008) Paul Ehrlich's magic bullet concept: 100 years of progress. Nat Rev Cancer 8:473–480
4. Bloom BR, Murray CJ (1992) Tuberculosis: commentary on a reemergent killer. Science 257:1055–1064
5. Zhang Y (2005) The magic bullets and tuberculosis drug targets. Annu Rev Pharmacol Toxicol 45:529–564
6. Cole S, Brosch R, Parkhill J, Garnier T, Churcher C, Harris D, Gordon S, Eiglmeier K, Gas S, Barry Iii C (1998) Deciphering the biology of *Mycobacterium tuberculosis* from the complete genome sequence. Nature 393:537–544
7. Chaitanya M, Babajan B, Anuradha C, Naveen M, Rajasekhar C, Madhusudana P, Kumar CS (2010) Exploring the molecular basis for selective binding of *Mycobacterium tuberculosis* Asp kinase toward its natural substrates and feedback inhibitors: a docking and molecular dynamics study. J Mol Model 16:1357–1367
8. da Cunha EF, Barbosa EF, Oliveira AA, Ramalho TC (2010) Molecular modeling of *Mycobacterium tuberculosis* DNA gyrase and its molecular docking study with gatifloxacin inhibitors. J Biomol Struct Dyn 27:619–625
9. Khedkar SA, Malde AK, Coutinho EC, Srivastava S (2007) Pharmacophore modeling in drug discovery and development: an overview. Med Chem 3:187–197
10. Yuan T, Sampson NS (2018) Hit generation in TB drug discovery: from genome to granuloma. Chem Rev 118:1887–1916
11. Jarlier V, Nikaido H (1994) Mycobacterial cell wall: structure and role in natural resistance to antibiotics. FEMS Microbiol Lett 123:11–18
12. Brennan PJ, Crick DC (2007) The cell-wall core of *Mycobacterium tuberculosis* in the context of drug discovery. Curr Top Med Chem 7:475–488
13. Trefzer C, Škovierová H, Buroni S, Bobovská A, Nenci S, Molteni E, Pojer F, Pasca MR, Makarov V, Cole ST (2011) Benzothiazinones are suicide inhibitors of mycobacterial decaprenylphosphoryl-β-D-ribofuranose 2′-oxidase DprE1. J Am Chem Soc 134:912–915

14. Mikusova K, Slayden RA, Besra GS, Brennan PJ (1995) Biogenesis of the mycobacterial cell wall and the site of action of ethambutol. Antimicrob Agents Chemother 39:2484–2489
15. Richter A, Rudolph I, Möllmann U, Voigt K, Chung C-W, Singh OM, Rees M, Mendoza-Losana A, Bates R, Ballell L (2018) Novel insight into the reaction of nitro, nitroso and hydroxylamino benzothiazinones and of benzoxacinones with *Mycobacterium tuberculosis* DprE1. Sci Rep 8:13473
16. Gao P, Yang Y, Xiao C, Liu Y, Gan M, Guan Y, Hao X, Meng J, Zhou S, Chen X (2012) Identification and validation of a novel lead compound targeting 4-diphosphocytidyl-2-C-methylerythritol synthetase (IspD) of mycobacteria. Eur J Pharmacol 694:45–52
17. Kantardjieff KA, Kim C-Y, Naranjo C, Waldo GS, Lekin T, Segelke BW, Zemla A, Park MS, Terwilliger TC, Rupp B (2004) *Mycobacterium tuberculosis* RmlC epimerase (Rv3465): a promising drug-target structure in the rhamnose pathway. Acta Crystallogr D Biol Crystallogr 60:895–902
18. Björkelid C, Bergfors T, Henriksson LM, Stern AL, Unge T, Mowbray SL, Jones TA (2011) Structural and functional studies of mycobacterial IspD enzymes. Acta Crystallogr D Biol Crystallogr 67:403–414
19. DeBarber AE, Mdluli K, Bosman M, Bekker L-G, Barry CE (2000) Ethionamide activation and sensitivity in multidrug-resistant *Mycobacterium tuberculosis*. Proc Natl Acad Sci 97:9677–9682
20. Johnsson K, King DS, Schultz PG (1995) Studies on the mechanism of action of isoniazid and ethionamide in the chemotherapy of tuberculosis. J Am Chem Soc 117:5009–5010
21. Manjunatha UH, Rao SP, Kondreddi RR, Noble CG, Camacho LR, Tan BH, Ng SH, Ng PS, Ma NL, Lakshminarayana SB (2015) Direct inhibitors of InhA are active against *Mycobacterium tuberculosis*. Sci Transl Med 7:269ra3
22. Pan P, Tonge JP (2012) Targeting InhA, the FASII enoyl-ACP reductase: SAR studies on novel inhibitor scaffolds. Curr Topics Med Chem 12:672–693
23. Šink R, Sosič I, Živec M, Fernandez-Menendez R, Turk S, Pajk S, Alvarez-Gomez D, Lopez-Roman EM, Gonzales-Cortez C, Rullas-Triconado J (2014) Design, synthesis, and evaluation of new thiadiazole-based direct inhibitors of enoyl acyl carrier protein reductase (InhA) for the treatment of tuberculosis. J Med Chem 58:613–624
24. Martínez-Hoyos M, Perez-Herran E, Gulten G, Encinas L, Álvarez-Gómez D, Alvarez E, Ferrer-Bazaga S, García-Pérez A, Ortega F, Angulo-Barturen I (2016) Antitubercular drugs for an old target: GSK693 as a promising InhA direct inhibitor. EBioMedicine 8:291–301
25. Barkan D, Liu Z, Sacchettini JC, Glickman MS (2009) Mycolic acid cyclopropanation is essential for viability, drug resistance, and cell wall integrity of *Mycobacterium tuberculosis*. Chem Biol 16:499–509
26. Wilson R, Kumar P, Parashar V, Vilchèze C, Veyron-Churlet R, Freundlich JS, Barnes SW, Walker JR, Szymonifka MJ, Marchiano E (2013) Antituberculosis thiophenes define a requirement for Pks13 in mycolic acid biosynthesis. Nat Chem Biol 9:499–506
27. Tahlan K, Wilson R, Kastrinsky DB, Arora K, Nair V, Fischer E, Barnes SW, Walker JR, Alland D, Barry CE (2012) SQ109 targets MmpL3, a membrane transporter of trehalose monomycolate involved in mycolic acid donation to the cell wall core of *Mycobacterium tuberculosis*. Antimicrob Agents Chemother (AAC) 05708-11
28. Ronning DR, Klabunde T, Besra GS, Vissa VD, Belisle JT, Sacchettini JC (2000) Crystal structure of the secreted form of antigen 85C reveals potential targets for mycobacterial drugs and vaccines. Nat Struct Mol Biol 7:141–146
29. Warrier T, Tropis M, Werngren J, Diehl A, Gengenbacher M, Schlegel B, Schade M, Oschkinat H, Daffe M, Hoffner S (2012) Antigen 85C inhibition restricts *Mycobacterium tuberculosis* growth through disruption of cord factor biosynthesis. Antimicrob Agents Chemother 1735–1743
30. Kovač A, Wilson RA, Besra GS, Filipič M, Kikelj D, Gobec S (2006) New lipophilic phthalimido-and 3-phenoxybenzyl sulfonates: inhibition of antigen 85C mycolyltransferase activity and cytotoxicity. J Enzyme Inhib Med Chem 21:391–397

31. Anderson DH, Harth G, Horwitz MA, Eisenberg D (2001) An interfacial mechanism and a class of inhibitors inferred from two crystal structures of the *Mycobacterium tuberculosis* 30 kda major secretory protein (antigen 85B), a mycolyl transferase1. J Mol Biol 307:671–681
32. Pawelczyk J, Brzostek A, Kremer L, Dziadek B, Rumijowska-Galewicz A, Fiolka M, Dziadek J (2011) AccD6, a key carboxyltransferase, essential for mycolic acid synthesis in *Mycobacterium tuberculosis*, is dispensable in a non-pathogenic strain. J Bacteriol JB:05638-11
33. Reddy MC, Breda A, Bruning JB, Sherekar M, Valluru S, Thurman C, Ehrenfeld H, Sacchettini JC (2014) Structure, activity, and inhibition of the carboxyltransferase β-subunit of acetyl-coa carboxylase (AccD6) from *Mycobacterium tuberculosis*. Antimicrob Agents Chemother 6122–6132
34. Boldrin F, Ventura M, Degiacomi G, Ravishankar S, Sala C, Svetlikova Z, Ambady A, Dhar N, Kordulakova J, Zhang M (2014) The phosphatidyl-myo-inositol mannosyltransferase PimA is essential for *Mycobacterium tuberculosis* growth in vitro and in vivo. J Bacteriol 3441–3451
35. Arvind A, Jain V, Saravanan P, Mohan CG (2013) Uridine monophosphate kinase as potential target for tuberculosis: from target to lead identification. Interdiscip Sci Comput Life Sci 5:296–311
36. Arvind A, Kumar V, Saravanan P, Mohan CG (2012) Homology modeling, molecular dynamics and inhibitor binding study on MurD ligase of *Mycobacterium tuberculosis*. Interdiscip Sci Comput Life Sci 4:223–238
37. Kumar V, Saravanan P, Arvind A, Mohan CG (2011) Identification of hotspot regions of MurB oxidoreductase enzyme using homology modeling, molecular dynamics and molecular docking techniques. J Mol Model 17:939–953
38. Mansour TS, Caufield CE, Rasmussen B, Chopra R, Krishnamurthy G, Morris KM, Svenson K, Bard J, Smeltzer C, Naughton S (2007) Naphthyl tetronic acids as multi-target inhibitors of bacterial peptidoglycan biosynthesis. ChemMedChem Chem Enabling Drug Discov 2:1414–1417
39. Tran AT, Watson EE, Pujari V, Conroy T, Dowman LJ, Giltrap AM, Pang A, Wong WR, Linington RG, Mahapatra S (2017) Sansanmycin natural product analogues as potent and selective anti-mycobacterials that inhibit lipid I biosynthesis. Nat Commun 8:14414
40. Silver LL (2003) Novel inhibitors of bacterial cell wall synthesis. Curr Opin Microbiol 6:431–438
41. Silver LL (2006) Does the cell wall of bacteria remain a viable source of targets for novel antibiotics? Biochem Pharmacol 71:996–1005
42. Vollmer W, Blanot D, De Pedro MA (2008) Peptidoglycan structure and architecture. FEMS Microbiol Rev 32:149–167
43. Lavollay M, Arthur M, Fourgeaud M, Dubost L, Marie A, Veziris N, Blanot D, Gutmann L, Mainardi J-L (2008) The peptidoglycan of stationary-phase *Mycobacterium tuberculosis* predominantly contains cross-links generated by L,D-transpeptidation. J Bacteriol 190:4360–4366
44. Gupta R, Lavollay M, Mainardi J-L, Arthur M, Bishai WR, Lamichhane G (2010) The *Mycobacterium tuberculosis* protein Ldt Mt2 is a nonclassical transpeptidase required for virulence and resistance to amoxicillin. Nat Med 16:466–469
45. Sauvage E, Kerff F, Terrak M, Ayala JA, Charlier P (2008) The penicillin-binding proteins: structure and role in peptidoglycan biosynthesis. FEMS Microbiol Rev 32:234–258
46. Bianchet MA, Pan YH, Basta LAB, Saavedra H, Lloyd EP, Kumar P, Mattoo R, Townsend CA, Lamichhane G (2017) Structural insight into the inactivation of *Mycobacterium tuberculosis* non-classical transpeptidase Ldt Mt2 by biapenem and tebipenem. BMC Biochem 18:8
47. Kim HS, Kim J, Im HN, Yoon JY, An DR, Yoon HJ, Kim JY, Min HK, Kim S-J, Lee JY (2013) Structural basis for the inhibition of *Mycobacterium tuberculosis* L,D-transpeptidase by meropenem, a drug effective against extensively drug-resistant strains. Acta Crystallogr D Biol Crystallogr 69:420–431

48. Chakraborti PK, Matange N, Nandicoori VK, Singh Y, Tyagi JS, Visweswariah SS (2011) Signalling mechanisms in Mycobacteria. Tuberculosis 91:432–440
49. Park HD, Guinn KM, Harrell MI, Liao R, Voskuil MI, Tompa M, Schoolnik GK, Sherman DR (2003) Rv3133c/dosR is a transcription factor that mediates the hypoxic response of *Mycobacterium tuberculosis*. Mol Microbiol 48:833–843
50. Shiloh MU, Manzanillo P, Cox JS (2008) Mycobacterium tuberculosis senses host-derived carbon monoxide during macrophage infection. Cell Host Microbe 3:323–330
51. Taneja NK, Dhingra S, Mittal A, Naresh M, Tyagi JS (2010) *Mycobacterium tuberculosis* transcriptional adaptation, growth arrest and dormancy phenotype development is triggered by vitamin C. PLoS ONE 5:e10860
52. Voskuil MI, Schnappinger D, Visconti KC, Harrell MI, Dolganov GM, Sherman DR, Schoolnik GK (2003) Inhibition of respiration by nitric oxide induces a *Mycobacterium tuberculosis* dormancy program. J Exp Med 198:705–713
53. Kaur K, Taneja NK, Dhingra S, Tyagi JS (2014) DevR (DosR) mimetic peptides impair transcriptional regulation and survival of *Mycobacterium tuberculosis* under hypoxia by inhibiting the autokinase activity of DevS sensor kinase. BMC Microbiol 14:195
54. Gupta RK, Thakur TS, Desiraju GR, Tyagi JS (2009) Structure-based design of DevR inhibitor active against nonreplicating *Mycobacterium tuberculosis*. J Med Chem 52:6324–6334
55. Fernandez P, Saint-Joanis B, Barilone N, Jackson M, Gicquel B, Cole ST, Alzari PM (2006) The Ser/Thr protein kinase PknB is essential for sustaining mycobacterial growth. J Bacteriol 188:7778–7784
56. Lougheed KE, Osborne SA, Saxty B, Whalley D, Chapman T, Bouloc N, Chugh J, Nott TJ, Patel D, Spivey VL (2011) Effective inhibitors of the essential kinase PknB and their potential as anti-mycobacterial agents. Tuberculosis 91:277–286
57. Singh N, Tiwari S, Srivastava KK, Siddiqi MI (2015) Identification of novel inhibitors of *Mycobacterium tuberculosis* PknG using pharmacophore based virtual screening, docking, molecular dynamics simulation, and their biological evaluation. J Chem Inf Model 55:1120–1129
58. Chiaradia LD, Mascarello A, Purificação M, Vernal J, Cordeiro MNS, Zenteno ME, Villarino A, Nunes RJ, Yunes RA, Terenzi H (2008) Synthetic chalcones as efficient inhibitors of *Mycobacterium tuberculosis* protein tyrosine phosphatase PtpA. Bioorg Med Chem Lett 18:6227–6230
59. Tan LP, Wu H, Yang P-Y, Kalesh KA, Zhang X, Hu M, Srinivasan R, Yao SQ (2009) High-throughput discovery of *Mycobacterium tuberculosis* protein tyrosine phosphatase B (MptpB) inhibitors using click chemistry. Org Lett 11:5102–5105
60. Wong D, Bach H, Sun J, Hmama Z, Av-Gay Y (2011) *Mycobacterium tuberculosis* protein tyrosine phosphatase (PtpA) excludes host vacuolar-H + –ATPase to inhibit phagosome acidification. Proc Natl Acad Sci 108:19371–19376
61. Sharma A, Khuller GK, Sharma S (2009) Peptide deformylase–a promising therapeutic target for tuberculosis and antibacterial drug discovery. Expert opinion on therapeutic targets 13:753–765
62. Serero A, Giglione C, Sardini A, Martinez-Sanz J, Meinnel T (2003) An unusual peptide deformylase features in the human mitochondrial N-terminal methionine excision pathway. J Biol Chem 278:52953–52963
63. Cynamon MH, Alvirez-Freites E, Yeo AE (2004) BB-3497, a peptide deformylase inhibitor, is active against *Mycobacterium tuberculosis*. J Antimicrob Chemother 53:403–405
64. Lofland D, Difuntorum S, Waller A, Clements JM, Weaver MK, Karlowsky JA, Johnson K (2004) In vitro antibacterial activity of the peptide deformylase inhibitor BB-83698. J Antimicrob Chemother 53:664–668
65. Sharma A, Sharma S, Khuller G, Kanwar A (2009) In vitro and ex vivo activity of peptide deformylase inhibitors against *Mycobacterium tuberculosis* H37Rv. Int J Antimicrob Agents 34:226–230

66. Gurcha SS, Usha V, Cox JA, Futterer K, Abrahams KA, Bhatt A, Alderwick LJ, Reynolds RC, Loman NJ, Nataraj V, Alemparte C, Barros D, Lloyd AJ, Ballell L, Hobrath JV, Besra GS (2014) Biochemical and structural characterization of mycobacterial aspartyl-tRNA synthetase AspS, a promising TB drug target. PLoS ONE 9:e113568
67. Soto R, Perez-Herran E, Rodriguez B, Duma BM, Cacho-Izquierdo M, Mendoza-Losana A, Lelievre J, Aguirre DB, Ballell L, Cox LR, Alderwick LJ, Besra GS (2018) Identification and characterization of aspartyl-tRNA synthetase inhibitors against *Mycobacterium tuberculosis* by an integrated whole-cell target-based approach. Sci Rep 8:12664
68. Palencia A, Li X, Bu W, Choi W, Ding CZ, Easom EE, Feng L, Hernandez V, Houston P, Liu L (2016) Discovery of novel oral protein synthesis inhibitors of *Mycobacterium tuberculosis* that target leucyl-tRNA synthetase. Antimicrob Agents Chemother 6271–6280
69. Rock FL, Mao W, Yaremchuk A, Tukalo M, Crépin T, Zhou H, Zhang Y-K, Hernandez V, Akama T, Baker SJ (2007) An antifungal agent inhibits an aminoacyl-tRNA synthetase by trapping tRNA in the editing site. Science 316:1759–1761
70. Abrahams KA, Cox JA, Spivey VL, Loman NJ, Pallen MJ, Constantinidou C, Fernandez R, Alemparte C, Remuinan MJ, Barros D (2012) Identification of novel imidazo [1, 2-a] pyridine inhibitors targeting *M. tuberculosis* QcrB. PloS one 7:e52951
71. Moraski GC, Seeger N, Miller PA, Oliver AG, Boshoff HI, Cho S, Mulugeta S, Anderson JR, Franzblau SG, Miller MJ (2016) Arrival of imidazo [2, 1-b] thiazole-5-carboxamides: potent anti-tuberculosis agents that target QcrB. ACS Infect Dis 2:393–398
72. Chandrasekera NS, Berube BJ, Shetye G, Chettiar S, O'Malley T, Manning A, Flint L, Awasthi D, Ioerger TR, Sacchettini J (2017) Improved phenoxyalkylbenzimidazoles with activity against *Mycobacterium tuberculosis* appear to target QcrB. ACS Infect Dis 3:898–916
73. McKinney JD, Zu Bentrup KH, Muñoz-Elías EJ, Miczak A, Chen B, Chan W-T, Swenson D, Sacchettini JC, Jacobs WR Jr, Russell DG (2000) Persistence of *Mycobacterium tuberculosis* in macrophages and mice requires the glyoxylate shunt enzyme isocitrate lyase. Nature 406:735–738
74. Muñoz-Elías EJ, McKinney JD (2005) *Mycobacterium tuberculosis* isocitrate lyases 1 and 2 are jointly required for in vivo growth and virulence. Nat Med 11:638–644
75. Huang H-L, Krieger IV, Parai MK, Gawandi VB, Sacchettini JC (2016) *Mycobacterium tuberculosis* malate synthase structures with fragments reveal a portal for substrate/product exchange. J Biol Chem 27421–27432
76. Krieger IV, Freundlich JS, Gawandi VB, Roberts JP, Gawandi VB, Sun Q, Owen JL, Fraile MT, Huss SI, Lavandera J-L (2012) Structure-guided discovery of phenyl-diketo acids as potent inhibitors of *M. tuberculosis* malate synthase. Chem Biol 19:1556–1567
77. Andries K, Verhasselt P, Guillemont J, Göhlmann HW, Neefs J-M, Winkler H, Van Gestel J, Timmerman P, Zhu M, Lee E (2005) A diarylquinoline drug active on the ATP synthase of *Mycobacterium tuberculosis*. Science 307:223–227
78. Kundu S, Biukovic G, Grüber G, Dick T (2016) Bedaquiline targets the ϵ subunit of mycobacterial F-ATP synthase. Antimicrob Agents Chemother 6977–6979
79. Pethe K, Bifani P, Jang J, Kang S, Park S, Ahn S, Jiricek J, Jung J, Jeon HK, Cechetto J (2013) Discovery of Q203, a potent clinical candidate for the treatment of tuberculosis. Nat Med 19:1157–1160
80. Bryk R, Arango N, Venugopal A, Warren JD, Park Y-H, Patel MS, Lima CD, Nathan C (2010) Triazaspirodimethoxybenzoyls as selective inhibitors of mycobacterial lipoamide dehydrogenase. Biochemistry 49:1616–1627
81. Venugopal A, Bryk R, Shi S, Rhee K, Rath P, Schnappinger D, Ehrt S, Nathan C (2011) Virulence of *Mycobacterium tuberculosis* depends on lipoamide dehydrogenase, a member of three multienzyme complexes. Cell Host Microbe 9:21–31
82. Tullius MV, Harth G, Horwitz MA (2003) Glutamine synthetase GlnA1 is essential for growth of *Mycobacterium tuberculosis* in human THP-1 macrophages and guinea pigs. Infect Immun 71:3927–3936

83. Harth G, Horwitz MA (2003) Inhibition of *Mycobacterium tuberculosis* glutamine synthetase as a novel antibiotic strategy against tuberculosis: demonstration of efficacy in vivo. Infect Immun 71:456–464
84. Kasbekar M, Fischer G, Mott BT, Yasgar A, Hyvönen M, Boshoff HI, Abell C, Barry CE, Thomas CJ (2016) Selective small molecule inhibitor of the *Mycobacterium tuberculosis* fumarate hydratase reveals an allosteric regulatory site. Proc Natl Acad Sci 113:7503–7508
85. Li M, Huang Y-J, Tai PC, Wang B (2008) Discovery of the first SecA inhibitors using structure-based virtual screening. Biochem Biophys Res Commun 368:839–845
86. Chen W, Huang Y-J, Gundala SR, Yang H, Li M, Tai PC, Wang B (2010) The first low μM SecA inhibitors. Bioorg Med Chem 18:1617–1625
87. Via LE, Savic R, Weiner DM, Zimmerman MD, Prideaux B, Irwin SM, Lyon E, O'Brien P, Gopal P, Eum S (2015) Host-mediated bioactivation of pyrazinamide: implications for efficacy, resistance, and therapeutic alternatives. ACS Infect Dis 1:203–214
88. Zhang Y, Wade MM, Scorpio A, Zhang H, Sun Z (2003) Mode of action of pyrazinamide: disruption of *Mycobacterium tuberculosis* membrane transport and energetics by pyrazinoic acid. J Antimicrob Chemother 52:790–795
89. Boshoff HI, Mizrahi V, Barry CE (2002) Effects of pyrazinamide on fatty acid synthesis by whole mycobacterial cells and purified fatty acid synthase I. J Bacteriol 184:2167–2172
90. Mdluli K, Ma Z (2007) Mycobacterium tuberculosis DNA gyrase as a target for drug discovery. Infect Disorders-Drug Targets (Form Curr Drug Targets-Infect Disorders) 7:159–168
91. Onodera Y, Tanaka M, Sato K (2001) Inhibitory activity of quinolones against DNA gyrase of *Mycobacterium tuberculosis*. J Antimicrob Chemother 47:447–450
92. Blondeau JM (2004) Fluoroquinolones: mechanism of action, classification, and development of resistance. Surv Ophthalmol 49:S73–S78
93. Piton J, Petrella S, Delarue M, André-Leroux G, Jarlier V, Aubry A, Mayer C (2010) Structural insights into the quinolone resistance mechanism of *Mycobacterium tuberculosis* DNA gyrase. PLoS ONE 5:e12245
94. Barancokova M, Kikelj D, Ilas J (2018) Recent progress in the discovery and development of DNA gyrase B inhibitors. Fut Med Chem 10:1207–1227
95. Chopra S, Matsuyama K, Tran T, Malerich JP, Wan B, Franzblau SG, Lun S, Guo H, Maiga MC, Bishai WR (2011) Evaluation of gyrase B as a drug target in *Mycobacterium tuberculosis*. J Antimicrob Chemother 67:415–421
96. Shirude PS, Madhavapeddi P, Tucker JA, Murugan K, Patil V, Basavarajappa H, Raichurkar AV, Humnabadkar V, Hussein S, Sharma S, Ramya VK, Narayan CB, Balganesh TS, Sambandamurthy VK (2013) Aminopyrazinamides: novel and specific GyrB inhibitors that kill replicating and nonreplicating *Mycobacterium tuberculosis*. ACS Chem Biol 8:519–523
97. Pedgaonkar GS, Sridevi JP, Jeankumar VU, Saxena S, Devi PB, Renuka J, Yogeeswari P, Sriram D (2014) Development of benzo[d]oxazol-2(3H)-ones derivatives as novel inhibitors of *Mycobacterium tuberculosis* InhA. Bioorg Med Chem 22:6134–6145
98. Reddy KI, Srihari K, Renuka J, Sree KS, Chuppala A, Jeankumar VU, Sridevi JP, Babu KS, Yogeeswari P, Sriram D (2014) An efficient synthesis and biological screening of benzofuran and benzo[d]isothiazole derivatives for *Mycobacterium tuberculosis* DNA GyrB inhibition. Bioorg Med Chem 22:6552–6563
99. Renuka J, Reddy KI, Srihari K, Jeankumar VU, Shravan M, Sridevi JP, Yogeeswari P, Babu KS, Sriram D (2014) Design, synthesis, biological evaluation of substituted benzofurans as DNA gyraseB inhibitors of *Mycobacterium tuberculosis*. Bioorg Med Chem 22:4924–4934
100. Davis CK, Nasla K, Anjana A, Rajanikant G (2018) Taxifolin as dual inhibitor of Mtb DNA gyrase and isoleucyl-tRNA synthetase: in silico molecular docking, dynamics simulation and in vitro assays. Silico Pharmacol 6:8

101. Cavusoglu C, Hilmioglu S, Guneri S, Bilgic A (2002) Characterization of rpoB mutations in rifampin-resistant clinical isolates of *Mycobacterium tuberculosis* from Turkey by DNA sequencing and line probe assay. J Clin Microbiol 40:4435–4438
102. Lin W, Mandal S, Degen D, Liu Y, Ebright YW, Li S, Feng Y, Zhang Y, Mandal S, Jiang Y (2017) Structural basis of *Mycobacterium tuberculosis* transcription and transcription inhibition. Mol Cell 66:169–179
103. Wang Q, Xu Y, Gu Z, Liu N, Jin K, Li Y, Crabbe MJC, Zhong Y (2018) Identification of new antibacterial targets in RNA polymerase of *Mycobacterium tuberculosis* by detecting positive selection sites. Comput Biol Chem 73:25–30
104. Kurabachew M, Lu SH, Krastel P, Schmitt EK, Suresh BL, Goh A, Knox JE, Ma NL, Jiricek J, Beer D (2008) Lipiarmycin targets RNA polymerase and has good activity against multidrug-resistant strains of *Mycobacterium tuberculosis*. J Antimicrob Chemother 62:713–719
105. Scharf NT, Molodtsov V, Kontos A, Murakami KS, Garcia GA (2017) Novel chemical scaffolds for inhibition of rifamycin-resistant RNA polymerase discovered from high-throughput screening. SLAS Discov Adv Life Sci R&D 22:287–297
106. Hong W, Deng W, Xie J (2013) The structure, function, and regulation of mycobacterium FtsZ. Cell Biochem Biophys 65:97–105
107. Huang Q, Kirikae F, Kirikae T, Pepe A, Amin A, Respicio L, Slayden RA, Tonge PJ, Ojima I (2006) Targeting FtsZ for antituberculosis drug discovery: noncytotoxic taxanes as novel antituberculosis agents. J Med Chem 49:463–466
108. Das D, Borah M, Singh AK, Das R, Boruah HPD (2015) Molecular docking of phytochemical as Ftsz cell division protein inhibitor in *Mycobacterium tuberculosis*. Int J Pharm Sci Res 6:463–472
109. Brötz-Oesterhelt H, Sass P (2014) Bacterial caseinolytic proteases as novel targets for antibacterial treatment. Int J Med Microbiol 304:23–30
110. Raju RM, Unnikrishnan M, Rubin DH, Krishnamoorthy V, Kandror O, Akopian TN, Goldberg AL, Rubin EJ (2012) *Mycobacterium tuberculosis* ClpP1 and ClpP2 function together in protein degradation and are required for viability in vitro and during infection. PLoS Pathog 8:e1002511
111. Gavrish E, Sit CS, Cao S, Kandror O, Spoering A, Peoples A, Ling L, Fetterman A, Hughes D, Bissell A (2014) Lassomycin, a ribosomally synthesized cyclic peptide, kills *Mycobacterium tuberculosis* by targeting the ATP-dependent protease ClpC1P1P2. Chem Biol 21:509–518
112. Schmitz KR, Carney DW, Sello JK, Sauer RT (2014) Crystal structure of *Mycobacterium tuberculosis* ClpP1P2 suggests a model for peptidase activation by AAA+ partner binding and substrate delivery. Proc Natl Acad Sci 111:E4587–E4595
113. Darwin KH, Ehrt S, Gutierrez-Ramos J-C, Weich N, Nathan CF (2003) The proteasome of *Mycobacterium tuberculosis* is required for resistance to nitric oxide. Science 302: 1963–1966
114. Gandotra S, Schnappinger D, Monteleone M, Hillen W, Ehrt S (2007) In vivo gene silencing identifies the *Mycobacterium tuberculosis* proteasome as essential for the bacteria to persist in mice. Nat Med 13:1515–1520
115. Debnath J, Siricilla S, Wan B, Crick DC, Lenaerts AJ, Franzblau SG, Kurosu M (2012) Discovery of selective menaquinone biosynthesis inhibitors against *Mycobacterium tuberculosis*. J Med Chem 55:3739–3755
116. Dhiman RK, Mahapatra S, Slayden RA, Boyne ME, Lenaerts A, Hinshaw JC, Angala SK, Chatterjee D, Biswas K, Narayanasamy P (2009) Menaquinone synthesis is critical for maintaining mycobacterial viability during exponential growth and recovery from non-replicating persistence. Mol Microbiol 72:85–97
117. Cagan R (2016) Drug screening using model systems: some basics. Dis Models Mech 9:1241–1244
118. Maitra A, Bates S, Kolvekar T, Devarajan PV, Guzman JD, Bhakta S (2015) Repurposing-a ray of hope in tackling extensively drug resistance in tuberculosis. Int J Infect Dis (IJID) Off Publ Int Soc Infect Dis 32:50–55

119. Amantea D, Certo M, Bagetta G (2015) Drug repurposing and beyond: the fundamental role of pharmacology. Funct Neurol 30:79–81
120. Boguski MS, Mandl KD, Sukhatme VP (2009) Drug discovery. Repurposing with a difference. Science 324:1394–1395
121. Kato S, Moulder SL, Ueno NT, Wheler JJ, Meric-Bernstam F, Kurzrock R, Janku F (2015) Challenges and perspective of drug repurposing strategies in early phase clinical trials. Oncoscience 2:576–580
122. Napper AD, Mucke HA (2015) A special focus on drug repurposing, rescue, and repositioning. Assay Drug Dev Technol 13:293
123. Oprea TI, Mestres J (2012) Drug repurposing: far beyond new targets for old drugs. AAPS J 14:759–763
124. Strittmatter SM (2014) Overcoming drug development bottlenecks with repurposing: old drugs learn new tricks. Nat Med 20:590–591
125. Solapure S, Dinesh N, Shandil R, Ramachandran V, Sharma S, Bhattacharjee D, Ganguly S, Reddy J, Ahuja V, Panduga V, Parab M, Vishwas KG, Kumar N, Balganesh M, Balasubramanian V (2013) In vitro and in vivo efficacy of beta-lactams against replicating and slowly growing/nonreplicating *Mycobacterium tuberculosis*. Antimicrob Agents Chemother 57:2506–2510
126. Correale S, Ruggiero A, Capparelli R, Pedone E, Berisio R (2013) Structures of free and inhibited forms of the L,D-transpeptidase LdtMt1 from *Mycobacterium tuberculosis*. Acta Crystallographica Section D 69:1697–1706
127. Zheng H, Colvin CJ, Johnson BK, Kirchhoff PD, Wilson M, Jorgensen-Muga K, Larsen SD, Abramovitch RB (2017) Inhibitors of *Mycobacterium tuberculosis* DosRST signaling and persistence. Nat Chem Biol 13:218–225
128. Singhal A, Jie L, Kumar P, Hong GS, Leow MK, Paleja B, Tsenova L, Kurepina N, Chen J, Zolezzi F, Kreiswirth B, Poidinger M, Chee C, Kaplan G, Wang YT, De Libero G (2014) Metformin as adjunct antituberculosis therapy. Sci Transl Med 6:263ra159
129. Marupuru S, Senapati P, Pathadka S, Miraj SS, Unnikrishnan MK, Manu MK (2017) Protective effect of metformin against tuberculosis infections in diabetic patients: an observational study of south Indian tertiary healthcare facility. Braz J Infect Dis Off Publ Braz Soc Infect Dis 312–316
130. Tyagi S, Ammerman NC, Li SY, Adamson J, Converse PJ, Swanson RV, Almeida DV, Grosset JH (2015) Clofazimine shortens the duration of the first-line treatment regimen for experimental chemotherapy of tuberculosis. Proc Natl Acad Sci 112:869–874
131. Levy L, Randall HP (1970) A study of skin pigmentation by clofazimine. International journal of leprosy and other mycobacterial diseases: official organ of the International Leprosy Association 38:404–416
132. Murashov MD, LaLone V, Rzeczycki PM, Keswani RK, Yoon GS, Sud S, Rajeswaran W, Larsen S, Stringer KA, Rosania GR (2017) The physicochemical basis of clofazimine-induced skin pigmentation. J Invest Dermatol 697–703
133. Tanaka E, Kimoto T, Tsuyuguchi K, Watanabe I, Matsumoto H, Niimi A, Suzuki K, Murayama T, Amitani R, Kuze F (1999) Effect of clarithromycin regimen for *Mycobacterium avium* complex pulmonary disease. Am J Respir Crit Care Med 160:866–872
134. Lebel M (1993) Pharmacokinetic properties of clarithromycin: a comparison with erythromycin and azithromycin. Canad J infectious Dis (Journal canadien des maladies infectieuses) 4:148–152
135. Cavalieri SJ, Biehle JR, Sanders WE (1995) Synergistic activities of clarithromycin and antituberculous drugs against multidrug-resistant *Mycobacterium tuberculosis*. Antimicrob Agents Chemother 39:1542–1545
136. Falzon D, Schünemann HJ, Harausz E, González-Angulo L, Lienhardt C, Jaramillo E, Weyer K (2017) World Health Organization treatment guidelines for drug-resistant tuberculosis, 2016 update. Eur Respir J 49:1602308

137. Merle CS, Fielding K, Sow OB, Gninafon M, Lo MB, Mthiyane T, Odhiambo J, Amukoye E, Bah B, Kassa F (2014) A four-month gatifloxacin-containing regimen for treating tuberculosis. N Engl J Med 371:1588–1598
138. Tang S, Yao L, Hao X, Zhang X, Liu G, Liu X, Wu M, Zen L, Sun H, Liu Y (2014) Efficacy, safety and tolerability of linezolid for the treatment of XDR-TB: a study in China. Eur Respir J 25–29
139. Weinstein EA, Yano T, Li L-S, Avarbock D, Avarbock A, Helm D, McColm AA, Duncan K, Lonsdale JT, Rubin H (2005) Inhibitors of type II NADH: menaquinone oxidoreductase represent a class of antitubercular drugs. Proc Natl Acad Sci 102:4548–4553
140. Crowle AJ, Douvas GS, May MH (1992) Chlorpromazine: a drug potentially useful for treating mycobacterial infections. Chemotherapy 38:410–419
141. Hollister LE, Eikenberry DT, Raffel S (1960) Chlorpromazine in nonpsychotic patients with pulmonary tuberculosis. Am Rev Respir Dis 81:562–566
142. Anquetin G, Greiner J, Mahmoudi N, Santillana-Hayat M, Gozalbes R, Farhati K, Derouin F, Aubry A, Cambau E, Vierling P (2006) Design, synthesis and activity against *Toxoplasma gondii*, *Plasmodium* spp., and *Mycobacterium tuberculosis* of new 6-fluoroquinolones. Eur J Med Chem 41:1478–1493
143. Mehra R, Rani C, Mahajan P, Vishwakarma RA, Khan IA, Nargotra A (2016) Computationally guided identification of novel *Mycobacterium tuberculosis* GlmU inhibitory leads, their optimization, and in vitro validation. ACS combinatorial science 18:100–116
144. Janardhan S, John L, Prasanthi M, Poroikov V, Narahari Sastry G (2017) A QSAR and molecular modelling study towards new lead finding: Polypharmacological approach to *Mycobacterium tuberculosis*. SAR QSAR Environ Res 28:815–832
145. Choudhury C, Priyakumar UD, Sastry GN (2015) Dynamics based pharmacophore models for screening potential inhibitors of mycobacterial cyclopropane synthase. J Chem Inf Model 55:848–860
146. Nandi S, Ahmed S, Saxena A (2018) Combinatorial design and virtual screening of potent anti-tubercular fluoroquinolone and isothiazoloquinolone compounds utilizing QSAR and pharmacophore modelling. SAR QSAR Environ Res 29:151–170
147. Singh S, Mandal PK, Singh N, Misra AK, Singh S, Chaturvedi V, Sinha S, Saxena AK (2010) Substituted hydrazinecarbothioamide as potent antitubercular agents: synthesis and quantitative structure–activity relationship (QSAR). Bioorg Med Chem Lett 20:2597–2600
148. Karan S, Kashyap VK, Shafi S, Saxena AK (2017) Structural and inhibition analysis of novel sulfur-rich 2-mercaptobenzothiazole and 1, 2, 3-triazole ligands against *Mycobacterium tuberculosis* DprE1 enzyme. J Mol Model 23:241
149. Zhang G, Guo S, Cui H, Qi J (2018) Virtual screening of small molecular inhibitors against DprE1. Molecules 23:524
150. Kandasamy S, Hassan S, Gopalaswamy R, Narayanan S (2014) Homology modelling, docking, pharmacophore and site directed mutagenesis analysis to identify the critical amino acid residue of PknI from *Mycobacterium tuberculosis*. J Mol Graph Model 52:11–19
151. Appunni S, Rajisha P, Rubens M, Chandana S, Singh HN, Swarup V (2017) Targeting PknB, an eukaryotic-like serine/threonine protein kinase of *Mycobacterium tuberculosis* with phytomolecules. Comput Biol Chem 67:200–204
152. Dkhar HK, Gopalsamy A, Loharch S, Kaur A, Bhutani I, Saminathan K, Bhagyaraj E, Chandra V, Swaminathan K, Agrawal P (2014) Discovery of *Mycobacterium tuberculosis*-1, 4-glucan branching enzyme (GlgB) inhibitors by structure-and ligand-based virtual screening. J Biol Chem 2015:76–89
153. Muddassar M, Jang JW, Gon HS, Cho YS, Kim EE, Keum KC, Oh T, Cho S-N, Pae AN (2010) Identification of novel antitubercular compounds through hybrid virtual screening approach. Bioorg Med Chem 18:6914–6921
154. Naidu KM, Srinivasarao S, Agnieszka N, Ewa A-K, Kumar MMK, Sekhar KVGC (2016) Seeking potent anti-tubercular agents: design, synthesis, anti-tubercular activity and docking study of various ((triazoles/indole)-piperazin-1-yl/1, 4-diazepan-1-yl) benzo [d] isoxazole derivatives. Bioorg Med Chem Lett 26:2245–2250

155. Prakash O, Ghosh I (2006) Developing an antituberculosis compounds database and data mining in the search of a motif responsible for the activity of a diverse class of antituberculosis agents. J Chem Inf Model 46:17–23
156. Dalecki AG, Wolschendorf F (2016) Development of a web-based tool for automated processing and cataloging of a unique combinatorial drug screen. J Microbiol Methods 126:30–34
157. Sharma A, Dutta P, Sharma M, Rajput NK, Dodiya B, Georrge JJ, Kholia T, Bhardwaj A (2014) BioPhytMol: a drug discovery community resource on anti-mycobacterial phytomolecules and plant extracts. J Cheminform 6:46
158. Lin K, O'Brien KM, Trujillo C, Wang R, Wallach JB, Schnappinger D, Ehrt S (2016) *Mycobacterium tuberculosis* thioredoxin reductase is essential for thiol redox homeostasis but plays a minor role in antioxidant defense. PLoS Pathog 12:e1005675
159. Shigyo K, Ocheretina O, Merveille YM, Johnson WD, Pape JW, Nathan CF, Fitzgerald DW (2013) Efficacy of nitazoxanide against clinical isolates of *Mycobacterium tuberculosis*. Antimicrob Agents Chemother 2834–2837
160. Li X, Hernandez V, Rock FL, Choi W, Mak YS, Mohan M, Mao W, Zhou Y, Easom EE, Plattner JJ (2017) Discovery of a potent and specific *M. tuberculosis* Leucyl-tRNA synthetase inhibitor:(S)-3-(Aminomethyl)-4-chloro-7-(2-hydroxyethoxy) benzo [c][1, 2] oxaborol-1 (3 H)-ol (GSK656). J Med Chem 60:8011–8026
161. Choi Y, Lee SW, Kim A, Jang K, Nam H, Cho YL, Yu K-S, Jang I-J, Chung J-Y (2017) Safety, tolerability and pharmacokinetics of 21 day multiple oral administration of a new oxazolidinone antibiotic, LCB01-0371, in healthy male subjects. J Antimicrob Chemother 73:183–190
162. Shoen C, DeStefano M, Hafkin B, Cynamon M (2018) In vitro and in vivo activity of contezolid (MRX-I) against *Mycobacterium tuberculosis*. Antimicrob Agents Chemother 00493-18
163. Wallis RS, Jakubiec WM, Kumar V, Silvia AM, Paige D, Dimitrova D, Li X, Ladutko L, Campbell S, Friedland G (2010) Pharmacokinetics and whole-blood bactericidal activity against *Mycobacterium tuberculosis* of single doses of PNU-100480 in healthy volunteers. J Infect Dis 202:745–751
164. Aagaard C, Hoang T, Dietrich J, Cardona P-J, Izzo A, Dolganov G, Schoolnik GK, Cassidy JP, Billeskov R, Andersen P (2011) A multistage tuberculosis vaccine that confers efficient protection before and after exposure. Nat Med 17:189–194
165. Bleicher KH, Böhm H-J, Müller K, Alanine AI (2003) A guide to drug discovery: hit and lead generation: beyond high-throughput screening. Nat Rev Drug Discov 2:369–378
166. Agatonovic-Kustrin S, Beresford R (2000) Basic concepts of artificial neural network (ANN) modeling and its application in pharmaceutical research. J Pharm Biomed Anal 22:717–727
167. Cherkasov A, Hilpert K, Jenssen H, Fjell CD, Waldbrook M, Mullaly SC, Volkmer R, Hancock RE (2008) Use of artificial intelligence in the design of small peptide antibiotics effective against a broad spectrum of highly antibiotic-resistant superbugs. ACS Chem Biol 4:65–74
168. Duch W, Swaminathan K, Meller J (2007) Artificial intelligence approaches for rational drug design and discovery. Curr Pharm Des 13:1497–1508
169. Martínez-Romero M, Vázquez-Naya JM, Rabunal RJ, Pita-Fernández S, Macenlle R, Castro-Alvariño J, López-Roses L, Ulla LJ, Martinez-Calvo VA, Vázquez S (2010) Artificial intelligence techniques for colorectal cancer drug metabolism: ontologies and complex networks. Curr Drug Metab 11:347–368
170. Fujiwara T, Kamada M, Okuno Y (2018) Artificial intelligence in drug discovery. Gan to kagaku ryoho Cancer Chemother 45:593–596

Turbo Analytics: Applications of Big Data and HPC in Drug Discovery

Rajendra R. Joshi, Uddhavesh Sonavane, Vinod Jani, Amit Saxena, Shruti Koulgi, Mallikarjunachari Uppuladinne, Neeru Sharma, Sandeep Malviya, E. P. Ramakrishnan, Vivek Gavane, Avinash Bayaskar, Rashmi Mahajan and Sudhir Pandey

Abstract In this current age of data-driven science, perceptive research is being carried out in the areas of genomics, network and metabolic biology, human, animal, organ and tissue models of drug toxicity, witnessing or capturing key biological events or interactions for drug discovery. Drug designing and repurposing involves understanding of ligand orientations for proper binding to the target molecules. The crucial requirement of finding right pose of small molecule in ligand–protein complex is done using drug docking and simulation methods. The domains of biology like genomics, biomolecular structure dynamics, and drug discovery are capable of generating vast molecular data in range of terabytes to petabytes. The analysis and visualization of this data pose a great challenge to the researchers and needs to be addressed in an accelerated and efficient way. So there is continuous need to have advanced analytics platform and algorithms which can perform analysis of this data in a faster way. Big data technologies may help to provide solutions for these problems of molecular docking and simulations.

Keywords Drug discovery · Drug repurposing · Hadoop · Big data
Molecular dynamics simulations

Abbreviation

PCA	Principal component analysis
RMSD	Root-mean-square deviation
RMSF	Root-mean-square fluctuation
MR	MapReduce

R. R. Joshi (✉) · U. Sonavane · V. Jani · A. Saxena · S. Koulgi
M. Uppuladinne · N. Sharma · S. Malviya · E. P. Ramakrishnan
V. Gavane · A. Bayaskar · R. Mahajan · S. Pandey
High Performance Computing-Medical & Bioinformatics Applications Group,
Centre for Development of Advanced Computing (C-DAC),
Savitribai Phule Pune University Campus, Pune 411007, India
e-mail: rajendra@cdac.in

C. G. Mohan (ed.), *Structural Bioinformatics: Applications in Preclinical Drug Discovery Process*, Challenges and Advances in Computational Chemistry and Physics 27, https://doi.org/10.1007/978-3-030-05282-9_11

1 Introduction

This decade has been witnessing a major shift in technologies which have been used in various sectors ranging from social media, agriculture, services, to science and technology. In the current age, new advances are being made in the field of satellites, robotics, micro- and nanotechnologies as well as revolution in computing. The stream of science has been impacted by this revolution. All disciplines of science have been generating and building newer technologies and different approaches for scientifically accurate experimentation. All these developments in various scientific disciplines are also changing our social life, health, environment, etc. One of the major streams of science is life sciences, which has been strongly affected and accelerated due to all these advancements in techniques and technologies.

Various technologies like next-generation sequencing (NGS) in genomics, high-throughput assays, and supramolecular chemistry are revolutionizing the life sciences and applied areas of human health, agriculture, livestock, and many more [1–4]. The robotics-based automation is generating volumes of data from various experiments and characterization techniques. The next-generation biology has been driven heavily by wet laboratory experimentation as well as dry laboratory computation.

Technologies like next-generation sequencing (NGS) enable sequencing of genomes of thousands of species in plants and animals at an extremely rapid rate [5–7]. Today, many genome sequencing centers are producing data of about terabytes per week. This results in petabytes of data of sequencing information per year. The figure is expected to grow exponentially and very soon will be facing challenges of storage and analysis of exabytes of sequence data [5–7]. To extend this further, there is already a race to sequence the genomes of all living species on the planet including humans, plants, animals, microbes to name a few. It is expected that this gigantic exercise will result in zetabytes to yottabytes of sequence data. Such large volumes of sequence data will be the genomic ocean of tomorrow [7–9].

Similarly, structural database of biomolecules like protein, nucleic acids, lipids, and membranes is also growing rapidly (shown in Fig. 1) due to methods like cryo crystallization, high-frequency NMR, and other characterization techniques along with computational modeling techniques [10]. Computational modeling and simulation of biomolecules have been drastically improving due to the advancement in high-performance computing (HPC) [11] and development of advanced enhanced sampling methods [12, 13]. It has paved the way for mimicking long timescale events occurring in different biological systems more efficiently. Owing to the better computing paradigm, today structural data generation is no more the major challenge, but analyzing this huge data has become one. Computer simulations help to determine mechanism of action of biomolecules in a cell, thereby suggesting their implication in various diseases and discovering their potential use in therapeutics. Hence, the computational techniques generate biomolecular structural and dynamical data via very long time scale simulations. Likewise, detailed and

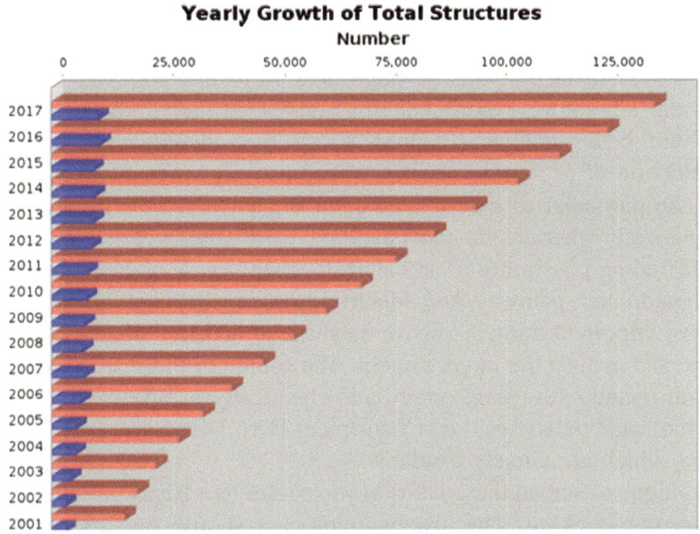

Fig. 1 Growth of structural data from 2001 onwards. *Source* https://www.rcsb.org/pdb/statistics

systematic analysis of data becomes an important part of any study, as it would further help to understand the entire mechanism of biomolecular action. Advances in crystallization, NMR, and computational methods are directly influencing and accelerating the drug discovery process.

2 Drug Discovery Process

Discovering a new drug is a very complex, time-consuming, expensive, and high-risk process for R&D and pharmaceutical laboratories [14–16]. It is also a multi-step process involving target identification, target validation, and screening of small molecules for validated targets. These steps need to be made easy, cost-effective, and fast. Computational method like computer aided drug discovery is one such process that involves identifying new ligand molecules for a particular target protein, which is an important step in drug discovery. Historically, the drug discovery process was involved extraction of chemical compounds from natural resources and testing them in the cell for disease treatment [17]. With the advancement of technology and ability to chemically synthesize small chemical moieties, various drug databases came into existence. The availability of vast structural resource of small molecules has made high-throughput screening of these databases against target protein a more feasible practice. Also, increasing affinity and reducing toxicity of already available ligand molecules needs to be addressed in drug discovery process.

Drug discovery process involves the following steps: (1) target identification, (2) validation of target protein, (3) creation of small molecule database,

(4) screening of small molecules against target protein, i.e., hit to lead identification, (5) lead optimization, (6) preclinical testing, and (7) clinical testing.

Almost all these steps generate huge data from experimental laboratory and computational laboratory experimentations and need better way of handling data with fast and better analytics approaches. Target identification and validation involve selection of protein molecules whose activity when blocked or enhanced can affect the particular disease-related cellular pathway. This involves a systems biology approach wherein an understanding of all the proteins involved in the pathway or finding possibility of any alternate pathway available, role of particular protein in particular pathway and identifying side effects of the target protein. Second most important thing is to have database of lakhs of small molecules which can be screened against the target protein. The source of these small molecules can be microbial metabolites, plant origin, and chemically synthesized. There are various drug molecule databases, i.e., Chemspider [18], DrugBank [19], ZINC [20] to name a few which are already available.

The technique to screen these lakhs of molecules to a target protein is performed using molecular docking. The screening process should be fast enough, which demands the use of and better computational or programming techniques. Each of these molecules tends to have conformational flexibility which in turn makes the docking process more time-consuming. Choice of efficient force field and scoring methodologies also plays an important role in screening of these molecules. In order to achieve this, high-throughput docking methods have been developed. Although, the analysis of these docked conformations to choose the best ligand becomes a big data analytics problem as it involves finding of various parameters and several interactions between the target protein and the docked ligand.

Docking or screening projects a static picture of the binding of ligand with the receptor [21]. However, the dynamic picture would be obtained from the molecular dynamics simulations which provide an understanding of the flexibility of protein and ligand. Molecular dynamics simulation gives an insight about various intermolecular interactions and binding affinities between protein-ligand complex, thereby ensuing binding efficiency [16]. Molecular docking followed by simulations generates huge molecular trajectories data. Thus, the management and fast analytics of this data have become the need of the hour.

The upcoming area of drug repurposing is again proving to be a bigger computational task, and it has the potential to deliver a drug molecule for a chosen disease [22, 23]. Various pharmaceuticals and R&D laboratories are working on drug repurposing which involves docking of already approved FDA drugs on new target protein. The involvement of FDA-approved drugs suggests that they have been already tested on humans for their toxicity and pharmacology. Hence, rejection of such drugs due to toxicity is ruled out, and entire duration required for the drug discovery process can be shortened by few years. HPC-based molecular docking and molecular dynamics simulations pose a challenging role in this area of drug repurposing.

In order to manage this rapidly increasing data and efficient analysis, there is need to develop tools with parallelization and thereby enhance the overall

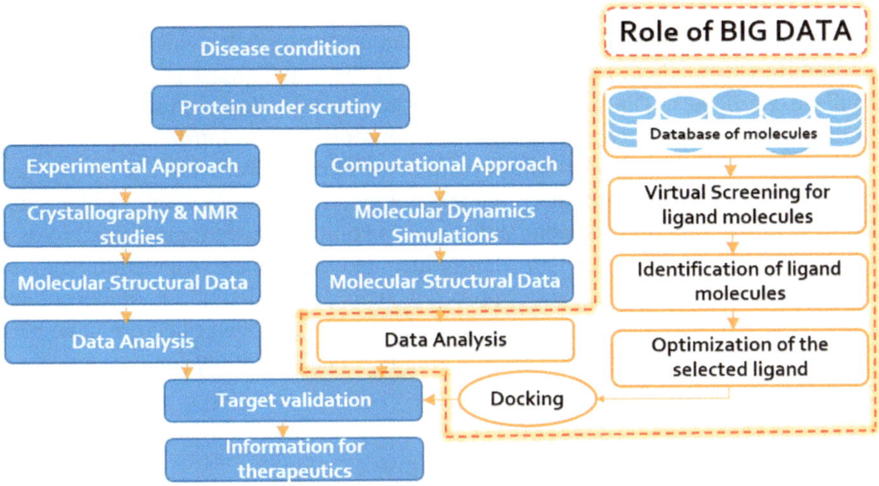

Fig. 2 Role of big data analytics in drug discovery

performance. This denotes a continuous need to have advanced analysis platform and algorithms which can perform analysis of the biological data in a faster way. Big data technologies may help to provide solutions for these problems of molecular docking and simulations (Fig. 2).

3 Big data Technologies: Challenges and Solutions

The context of big data is dependent on the problems and the existing technologies. Today's big data can be tomorrow's small data as the technologies and methods that are handling the data may become more advanced in the future. The big data is the data that cannot be handled using the existing traditional methods and requires specialized methods to solve the big data problem.

Big data is categorized by its three main properties, viz. volume, velocity, and variety [24]. Volume denotes the huge data that needs to be analyzed, velocity tells about the rate at which the data is generated of the data, and variety tells about the different types of data that can be generated by the various sources using different formats of data generations and exchange. Big data usually expands rapidly in the unstructured form and varies to such an extent that it becomes difficult to maintain the data in traditional databases. In such cases, specialized techniques like NoSQL [25] can be used to handle the problems of the unstructured data. Big data technologies are capable of managing huge data generated in different formats. Advancements in technologies like cloud computing offer a unified platform to store and retrieve the data. The Internet speed has increased to several manifolds, and the cloud technologies have effectively exploited the Internet capabilities to offer a scalable, multi-user platform for big data analytics in the field of Bioinformatics. The use of big data in the Bioinformatics is an emerging field which presents new opportunities to medical researchers and paves the way toward prediction of personalized medicines. The greatest challenge lies in designing a strategy to acquire the data followed by filtering it to meet the appropriate decision-making demands.

This can be achieved by bringing together experts from clinical medicines, computer science, bioinformatics, biotechnology, and statistics and address the challenge of the data management and analytics solutions toward precision biology. Hadoop [26]-based platform with MapReduce and spark-based algorithms may be useful to make all the analysis optimized with fast calculation. Hadoop- and MapReduce [27]-based algorithms implemented on scalable architecture have been discussed further along with drug repurposing big data case study for cancer protein.

4 Big data Technology Components

Hadoop

Apache Hadoop is an open-source software framework for storage and large-scale processing of datasets on clusters of commodity hardware. Hadoop has gained lots of popularity among the peer parallel data processing tools because of its simplicity, efficiency, cost, and reliability. Hadoop can be built on the commodity hardware. Hadoop has major three components. Hadoop Distributed File System (HDFS), YARN scheduler and resource negotiating framework and the MapReduce [27] programming framework. A typical framework of Hadoop test bed is shown in Fig. 3.

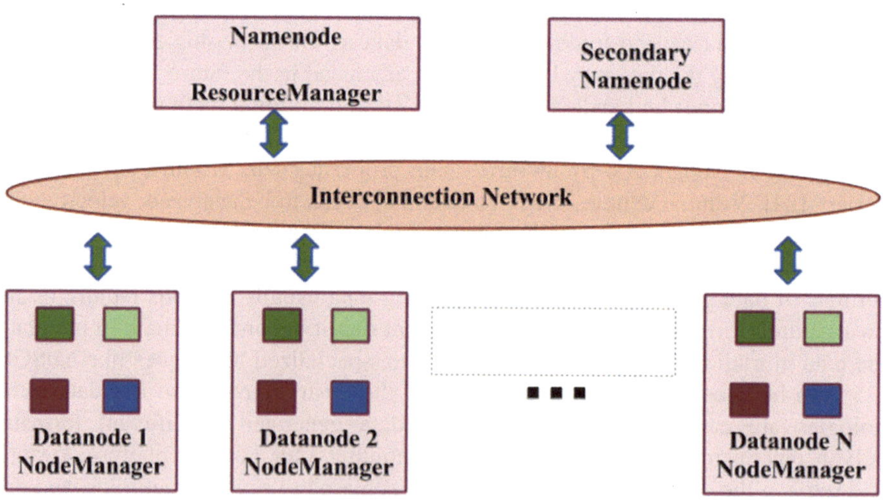

Fig. 3 Basic architecture diagram of hadoop test bed

I. *HDFS*

Hadoop Distributed File System (HDFS) is built to provide high-throughput, reliable, efficient, and fault-tolerant file system. It can provide streaming reads and writes for large files. The basic architecture diagram of HDFS is shown in Fig. 4. As shown in the figure, the HDFS has two main components, namenode, and datanode. HDFS is mainly designed for low-cost hardware, and hence, it can be built on cluster of commodity hardware. In HDFS, the file is divided into fixed size blocks or chunks of 128 MB each except the last chunk. The fixed size 128 MB can be configured with various needs. Namenode contains the metadata information of all the files. It stores information regarding the block of file stored on datanodes, while datanodes actually store the block of data. Each block is stored on three datanodes of the cluster. This policy provides reliability at the cost of redundancy. Generally, two copies of blocks are stored on two different datanodes of the same rack of cluster, while the third copy is stored on the datanodes of the different rack of the same cluster. These two racks are connected by a very high-speed network switch. This policy ensures the reliability of the HDFS file system. In case, if any two nodes fail, still the data can be accessed from the datanode having this third copy of the data. Datanodes periodically updates their state to the namenode so that namenode can be aware of the overall state of cluster. While scheduling MapReduce [27] job, the hadoop framework ensures with most possibility that the mapper task should run on the same datanode where the actual data is residing. This avoids significant network overhead. This policy of hadoop improves the performance of the overall cluster.

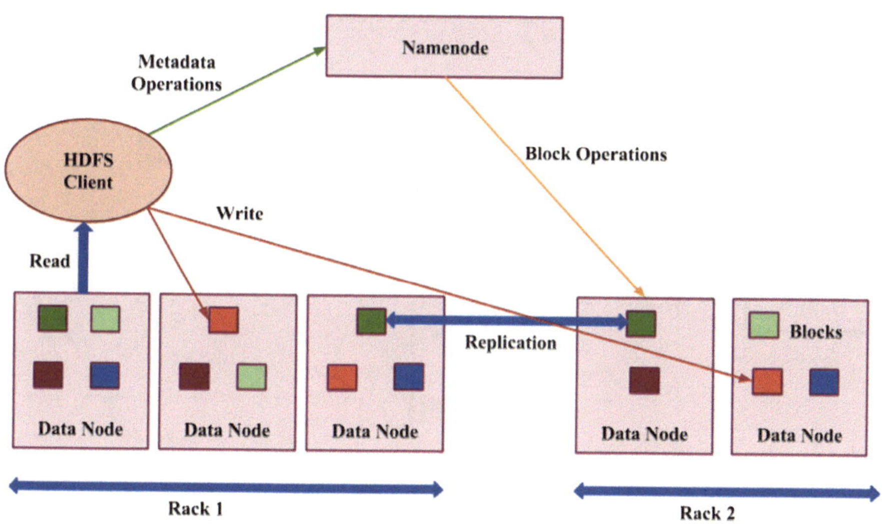

Fig. 4 Basic architecture diagram of Hadoop Distributed File System (HDFS)

HDFS major components:

(i) **Namenode**
Namenode stores the metadata about the file. It has the complete view of the distributed file system. It tracks which datanode is active and which are node. In case of any datanode failure, it initiates the operation regarding maintaining the replication factor by copying the data stored the failed nodes to the active datanodes. In case, namenode fails, the complete HDFS file system gets crashed.

(ii) **Datanode**
It stores the actual data. It performs the read and write operation once it receives the command from the namenode. It is responsible for block creation, deletion, and replication. It periodically sends the heartbeat signal to the namenode.

II. ***Map Reduce***

Hadoop MapReduce is the programming framework. It is one of the major parts of the Apache Hadoop project. It provides the programing model for data parallel application. The basic flow of MapReduce algorithm is shown in Fig. 5. MapReduce programming model makes use of HDFS and makes the application performance very efficient and fast. The MapReduce framework with the help of Hadoop framework places the mapper job on the datanode where the actual data resides. It improves the performance and removes the network bottleneck while processing huge amounts of data. The major phases of the MapReduce program are mapper, partitioner, combiner, shuffle and sort, and reducer.

The mapper reads the data from HDFS and processes it. This is followed by the partitioner ensuring that the processed data is sent to be the desired reducer. The

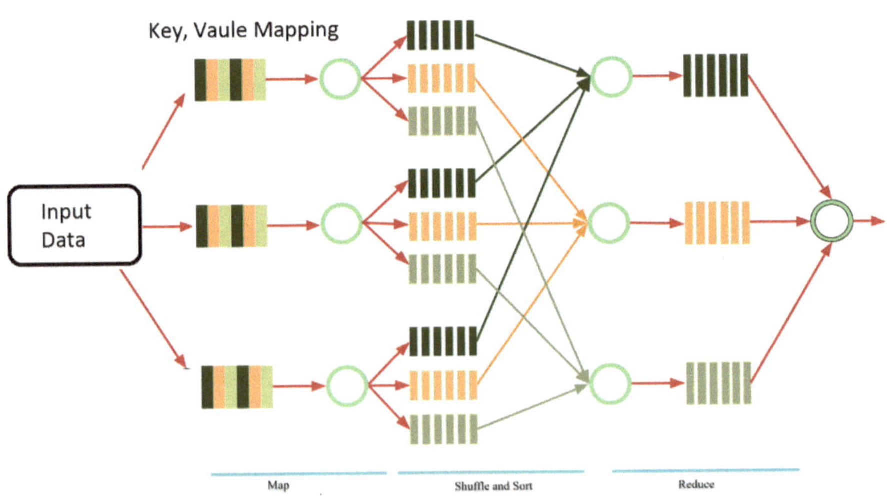

Fig. 5 Basic flow of MapReduce algorithm execution

data before being sent to the reducer is shuffled and sorted so that the reducer can easily process it. Finally, the reducer performs an operation of reduction or aggregation on the final data and this is followed by writing the final output to the HDFS. The combiner does a similar task as the reducer but at the mapper lever providing local lever aggregation or reduction.

III. ***YARN***

Apache YARN stands for Yet Another Resource Negotiator. Before Hadoop 2.x, the only framework which could run on Hadoop platform is MapReduce. The job scheduling and resource negotiation is integrated with the MapReduce framework and shared by Hadoop framework. The YARN provides the separate layer for job scheduling and resource negotiation. It provides the platform for other programming framework like spark and storm, and many can run on Hadoop framework. The basic architecture of YARN is shown in Fig. 6.

YARN has ResourceManager, NodeManager, Container, and ApplicationMaster. Each container on datanode is specified with amount of CPU and memory, and it is configurable. ResourceManager is run on namenode, and NodeManagers are run on datanodes. Whenever a job is submitted, one container is allocated by a ResourceManager on any datanode. This container process is called as ApplicationMaster. This ApplicationMaster is responsible for all job management and resource negotiation with ResourceManager. With the help of ResourceManager,

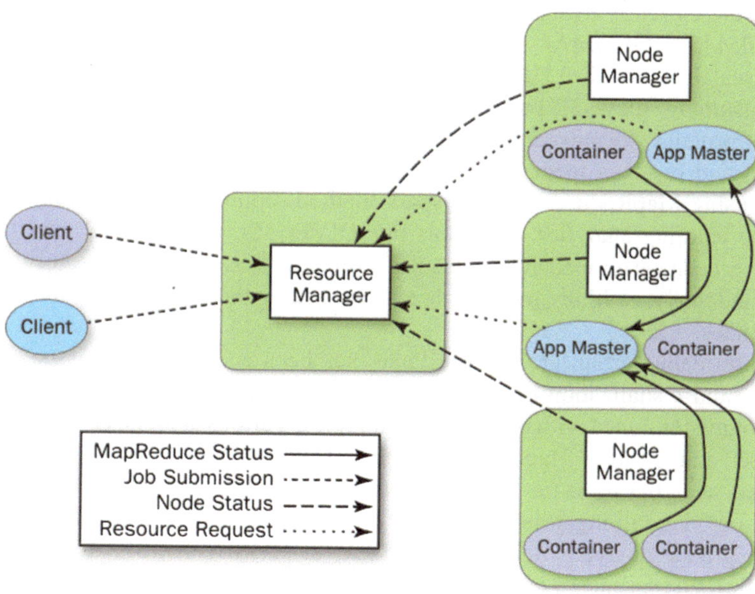

Fig. 6 Basic architecture of YARN showing various components

this ApplicationMaster allocates Containers from NodeManager for MapReduce task. This approach reduces the load on ResourceManager and distributes it across ApplicationMasters on the datanodes for each job. This way, using YARN the hadoop cluster can grow up to 10,000 nodes. Earlier benchmark without YARN on Hadoop 1.x was up to 4000 nodes. This way YARN provides scalability to the Hadoop cluster along with different programming platforms to be incorporated in hadoop framework.

5 Big data Tools Development for Drug Discovery

There have been efforts by various scientific groups to use HPC, grid technologies for drug discovery. Multiple docking tools like DOCK6 [28], Gold [29], Autodock Vina [30], and some others are already available in the parallel mode on HPC platform. Most of these tools are fast and robust; however, they have their own scoring functions based on molecular mechanics force fields and other geometrical descriptors. Although, improvements are still going on in enhancing the scoring function and guiding it further toward higher efficiency and accuracy. Docking with the concept of flexible ligand and protein still remains to be time-consuming calculation. Docking of multiple ligands to single protein or multiple ligands with multiple proteins may be some of the future challenges in docking area. Understanding the flexibility of both the proteins and ligands has been taken care by some of the currently available molecular simulation packages like AMBER [31], CHARMM [32], GROMACS [33], and NAMD [34]. All these packages are known to be scalable on the HPC platform. Although molecular simulations are time-consuming, they still prove to be the best in understanding the allowed flexibility of proteins, ligands, active sites, and other biomolecular entities. The advent of cloud and big data technologies promises to accelerate the drug development process using MapReduce [27] and spark methods coupled with machine learning and deep learning analytics. The tools like DIVE [35], HiMach [36], and HTMD [37] have been developed for molecular simulations as well as trajectory visualization and analysis. Many more tools may be getting developed using these newer technologies.

Bioinformatics group at C-DAC, Pune, has been addressing the issue on data analytics and visualization of trajectories in structural biology domain using HPC technologies combined with big data technologies. Various analytics tools have been developed and tested on Hadoop platform using MapReduce as shown in Fig. 7. At this stage, analytics tools for multiple molecular trajectories include hydrogen bond calculations, identifying water molecules and bridged water-mediated interactions. Other big data analytics tools for RMSD, 2DRMSD, RMSF, water density, WHAM-based free energy calculations are in the process of development. Few of the big data analytics tools which have been already developed proved to be useful in the process of drug discovery. These tools have described below.

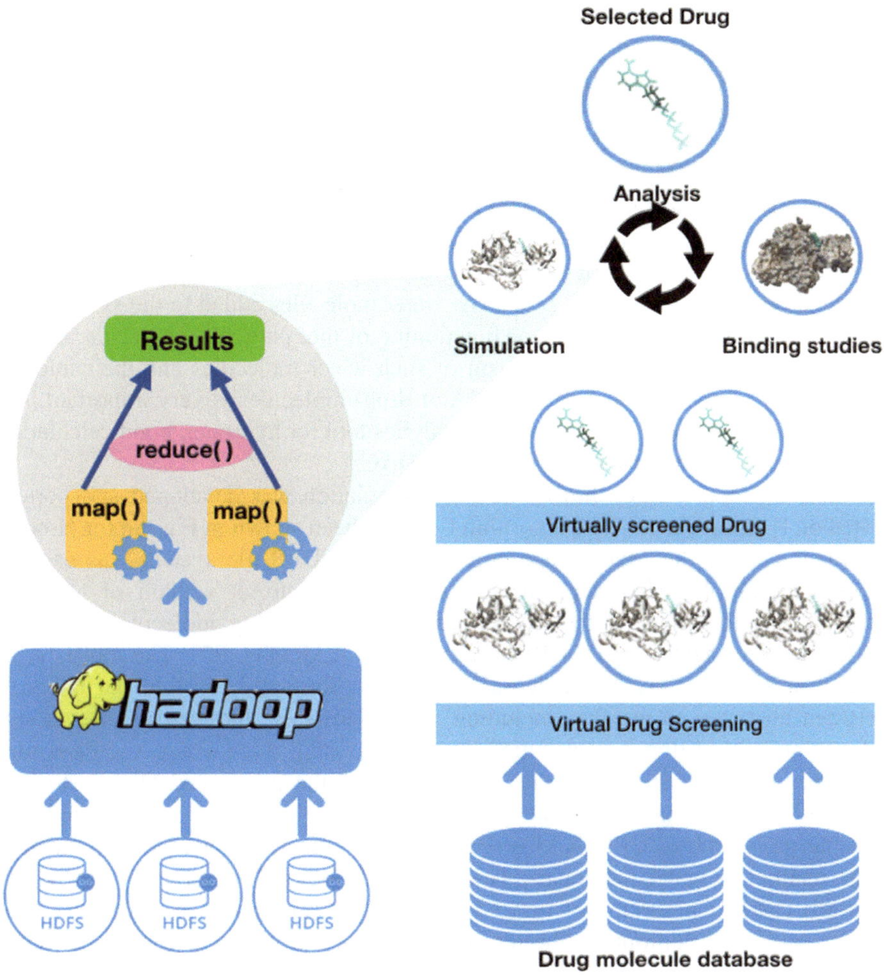

Fig. 7 Schematic representation of role of Hadoop and MapReduce paradigm in drug discovery process

5.1 *Hydrogen Bond Big data Analytics Tool (HBAT)*

The molecular dynamics (MD) simulations generate large trajectories which would be in the size of GBs to TBs depending on the size of the molecule and length of the simulation time. Many of the MD simulations use explicit solvation models in which water molecules are added explicitly to the solute to mimic the natural system. This increases the size of the system drastically in terms of number of atoms, and the analysis of such system becomes more compute intensive, iterative, and time-consuming. There are various analysis programs (ptraj, cpptraj [38], VMD [39], etc.) available corresponding to the different MD simulation packages.

All these programs have modules written for performing different analyses like RMSD, RMSF, radius of gyration, PCA [40], distance calculations, H-bond analysis, and MMGBSA [41] free energy calculations. However, many of these programs are either inefficient or very slow in calculating the H-bond interactions within solute and especially between the solute and the solvent (water molecules). These programs are highly time-consuming and also have constraint in dealing with the large size data for example 500 GB or beyond. This drawback of the existing tools suggests a strong need for the development of water-mediated H-bond analysis tool which is capable of handling a very large size of trajectories and also be executed parallel to reduce the time. The water molecules added to the system may play a crucial role in the activity or functioning of that particular molecule. Hence, understanding the role and mechanism of such water molecules and their interactions with the solute (protein/RNA/DNA or drug) molecules is very important [42, 43]. In order to achieve this, a big data analytics tool for hydrogen bond calculation was developed by Bioinformatics group C-DAC.

The MapReduce algorithm for H-bond calculation was developed and ported/ tested on Hadoop cluster. The algorithm flow has been shown in Fig. 8a for H-bond calculation using the MapReduce approach. The HDFS file system was used to store the multiple molecular trajectories data. The current version of tool can analyze trajectory data in the PDB format generated using molecular dynamics packages like AMBER [31], GROMACS [33], CHARMM [32]. The tool is scalable or portable on any distributed computing platform and can find out H-bonds between all types of residues including water. However, the tool requires a significant amount of time for executing the preprocessing stage where, the PDB files are generated from the trajectories and copied on the distributed HDFS storage. Despite this overhead, the overall performance of the tool is better than currently existing tools such as CPPTRAJ or PTRAJ [38], especially for trajectories with a large number of water molecules. The benchmarking of H-bond tool is shown in Fig. 8b. The benchmarking of up to 5.5 TB data is carried out, and it shows near linear scale up. Additionally, the tool can also help identify water-mediated interactions such as water bridges easily.

5.2 *Molecular Conformation Generation on Cloud (MOSAIC)*

Drug databases usually contain millions of ligands, and for each ligand, there can be billions of conformations [44, 45]. Such billions of conformations need to be docked on to a target which is a generally a protein molecule. Generation and optimization of such billions of ligand conformations is a huge computational problem, since it involves the use of advanced methods like molecular mechanics, semi-empirical and quantum techniques [46, 47]. The application of an embarrassingly parallel approach accompanied by virtualized resource scaling and an efficient structure optimization tool can handle billions of conformations with the help of cloud computing technologies.

The Bioinformatics group of C-DAC has developed a tool called MOSAIC, which stands for MOlecular Structure generator In the Cloud. MOSAIC is an OpenStack [48] cloud-based conformation search tool to explore potential energy

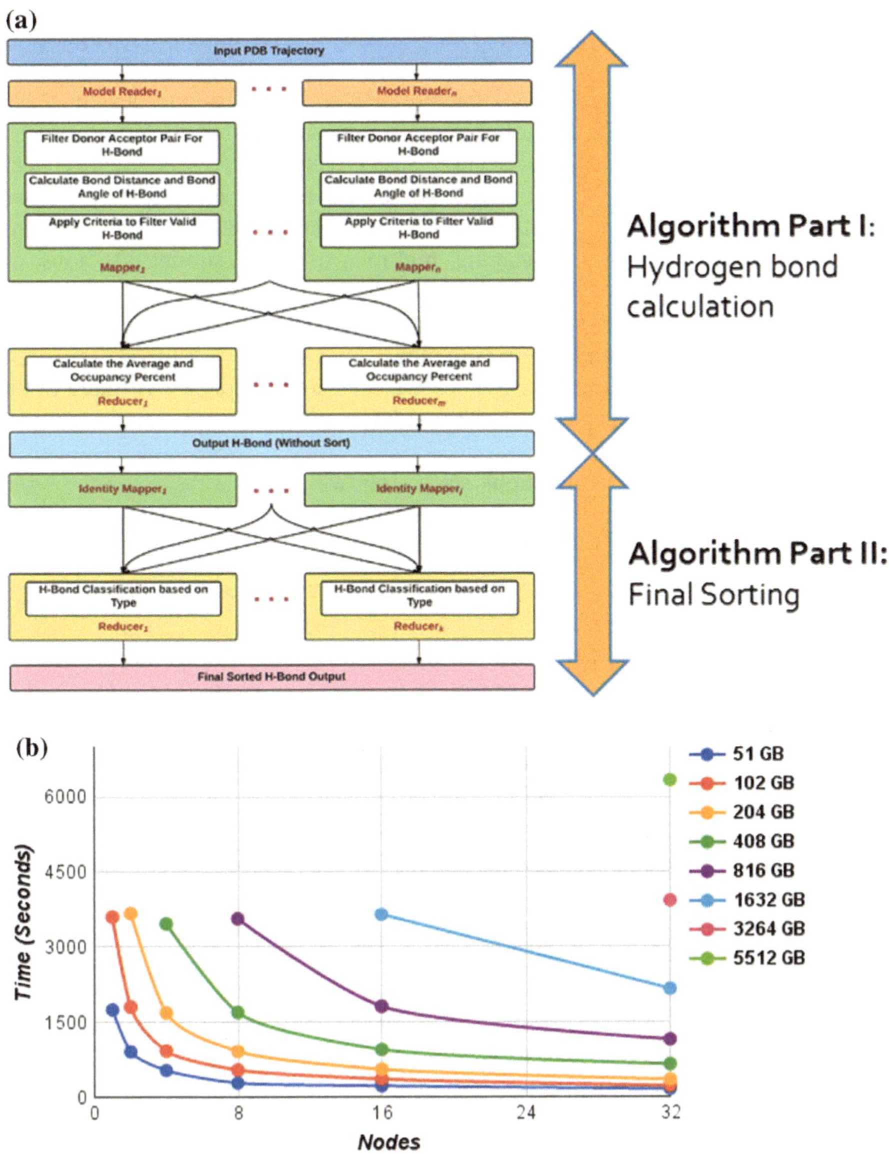

Fig. 8 **a** MapReduce algorithm for H-bond calculation implemented in MapReduce paradigm **b** Benchmarking of HBAT tool for data up to 5.5 TB

surface of biomolecules of interest in parallel mode using semi-empirical method. Molecular Orbital PACkage (MOPAC) is a general purpose semi-empirical molecular orbital package for the study of molecular structures and their energies [49]. The high-throughput energy calculations of the small molecules database can be done by MOPAC using hadoop and cloud technologies. Multiple instances of MOPAC are created for energy calculations of small molecules database. The tool can screen a database of millions of small drug-like molecules and understand their energetics and electrostatic behavior. The tool is useful for finding the target drug ligands. The torsion angle-driven conformational search method is useful in a range of chemical design applications [50], including drug discovery and design of targeted chemical hosts. MOSAIC has an easy-to-use interface for the bioinformatics community over Software as a Service (SaaS) platform. A user-friendly Web interface has been developed for MOPAC-based energy calculation of small molecule database. The Web interface has the capability of configuring any OpenStack-based cloud and managing multiple users to submit the jobs on dynamically created cloud VM. The Web interface has been developed using LAMP (Linux, Apache, Mysql, and PHP) framework [51]. The Web interface is shown in Fig. 9a, b. The application is deployed on OpenStack kilo version which provides platform for running the MOPAC with resources allocated virtually in the cloud. OpenStack cloud infrastructure provides scalable computational resources and scalable storage capacity.

The details of cloud configurations are as follows:

The cloud infrastructure is installed using multi-nodes architecture. The cloud test bed is deployed using following configurations:

- Controller node: 1 processor, 2 GB memory, and 5 GB storage and 2 NIC.
- Network node: 1 processor, 512 MB memory, and 5 GB storage and 3 NIC.
- Compute node: 1 processor, 2 GB memory, and 10 GB storage and 2 NIC.

To synchronize the clusters, there is a need to set up NTP server. The controller node acts as NTP server, and rest of the network along with compute nodes would be synchronize with this controller node. All the nodes in the cluster except controller node have mysql client service, and on controller mysql databases have been installed. Controller node also contains the messaging server for passing message across the nodes, and we have used the RabbitMQ [52] server. The configuration is depicted in Fig. 10.

MOSAIC is executed using underlying Open Stack-based cloud to distribute millions of molecules in .mop format across the cloud nodes. The cloud nodes can be dynamically scaled to accommodate the computing load. The drug database is in the sdf format having different conformations of the same molecule and containing millions of such molecules. The sdf is converted into the desirable input file, i.e., . mop format which is used by the code for semi-empirical optimizations. The output files generated are parsed based on the energy value, and a few best optimized ligand molecules are selected based on the energy profile. The best few optimized ligands may further be scrutinized for possible drug target. This tool may have

(a)

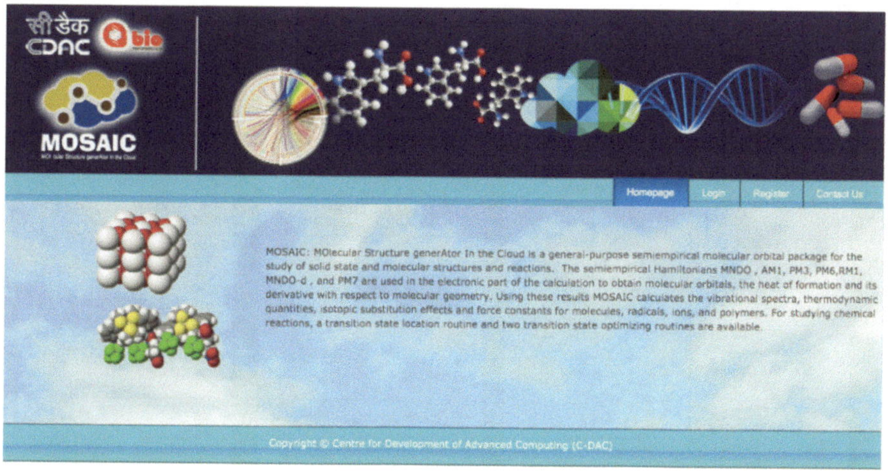

(b)

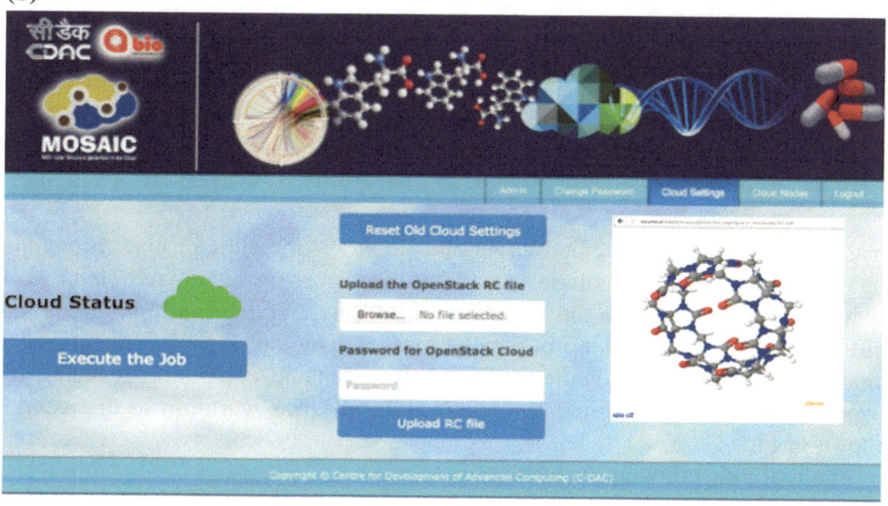

Fig. 9 **a** MOSAIC tool homepage **b** MOSAIC tool job submission page

tremendous potential in terms of ligand optimization, i.e., finding the best posture not just for one molecule but for ligand database. The tool can be easily deployable on any OpenStack-based cloud platform. MOSAIC has an easy-to-use interface for the scientific community as it abstracts the complexity of cloud-based job submission. It has a user-specific work area for managing secured private data and outputs. It has a configurable orchestration mechanism for virtual hardware configuration. The result is shared in the form of a few selected molecules favorable for drug target. It is anticipated that MOSAIC will accelerate the process of drug discovery by using high-throughput optimization of Ligand databases in parallel

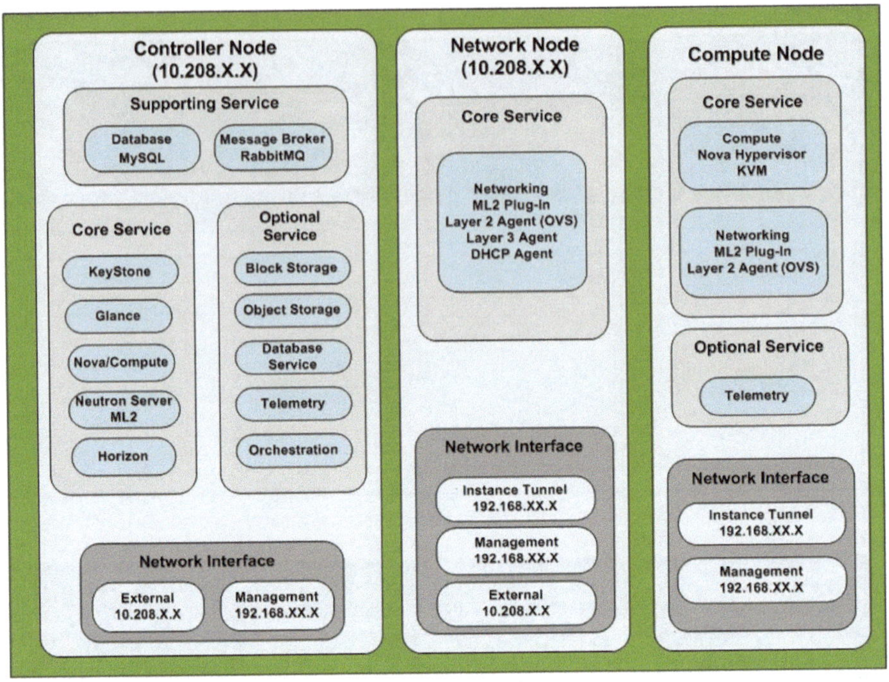

Fig. 10 Cloud configuration of MOSAIC tool

manner using distributed cloud environment. MOSAIC helps in high-throughput optimization of ligand database in parallel manner using distributed cloud environment. It will accelerate the scientific research by carrying out high-throughput virtual screening and docking in parallel manner. The tool uses the advantages of cloud computing like dynamic scaling and on-demand computing reducing the overall cost and helpful in finding optimized ligands. The workflow as discussed is shown in Fig. 11.

The tool has following features:

- Easy to use for the bioinformatics community which abstracts the complexity of cloud-based job execution.
- It is supported by a user-friendly interface with user-specific storage area with login time stamp features.
- Cloud-based high-throughput optimization of ligand database in parallel using distributed environment.
- Integrated browser-based visualization for optimized ligand molecules.
- OpenStack-based cloud environment facilitates users with on-demand scalable virtualized resources.
- Configurable orchestration mechanism for virtual hardware configuration.
- Generalized configurable solution for any OpenStack-based cloud using openrc script.

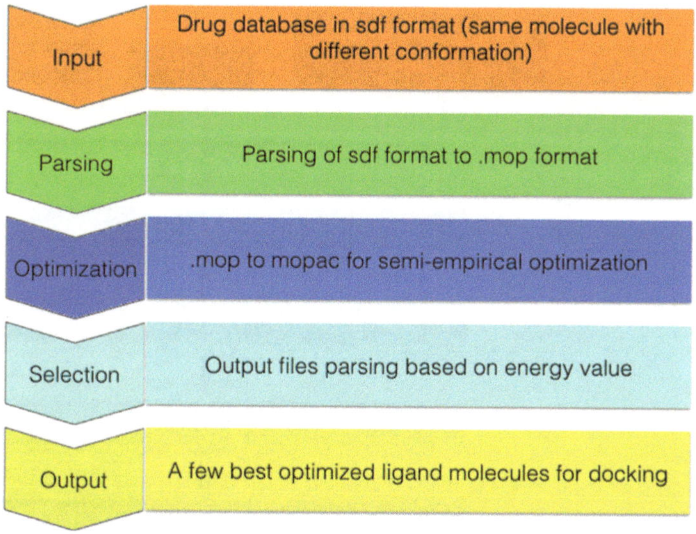

Fig. 11 MOSIAC tool workflow for cloud-based MOPAC implementation

5.3 *Embarrassingly Parallel Molecular Docking Pipeline*

Molecular docking or high-throughput screening has become increasingly important in the context of drug discovery [45]. High-throughput screening may be the only way to identify correct inhibitors of the specific target. However, high-throughput drug docking is cost-effective and very fast and could be very useful for pharmaceutical industry. An attempt has been made to develop a scalable workflow as shown in Fig. 12, for high-throughput conformational search and docking on the high-performance computing, Hadoop or cloud-based clusters. The workflow is divided into two sections. The first section performs conformational search, and the second section performs the molecular docking. The objective of the conformation search is to find the most stable conformation of the molecule along with alternative stable conformations. The semi-empirical program like MOPAC [49] is used for finding the stable structures as described in the previous section of MOSAIC. After getting the stable structures of the small molecule, docking is carried out in the parallel manner with protein of interest in the next part of the workflow. Docking of either multiple small molecules with one protein or multiple molecules with multiple proteins docking facility is available in the workflow. The testing of the workflow has been done for the drug repurposing strategy in the cancer. A test case/example of usage of this tool is given in the Sect. 6 below in the cancer K-Ras drug repurposing studies.

This tool is also deployable on any HPC, Hadoop, or cloud platform available worldwide. The current version is deployed on the computing resources of BRAF (Bioinformatics Resources & Applications Facility), C-DAC, Pune, India.

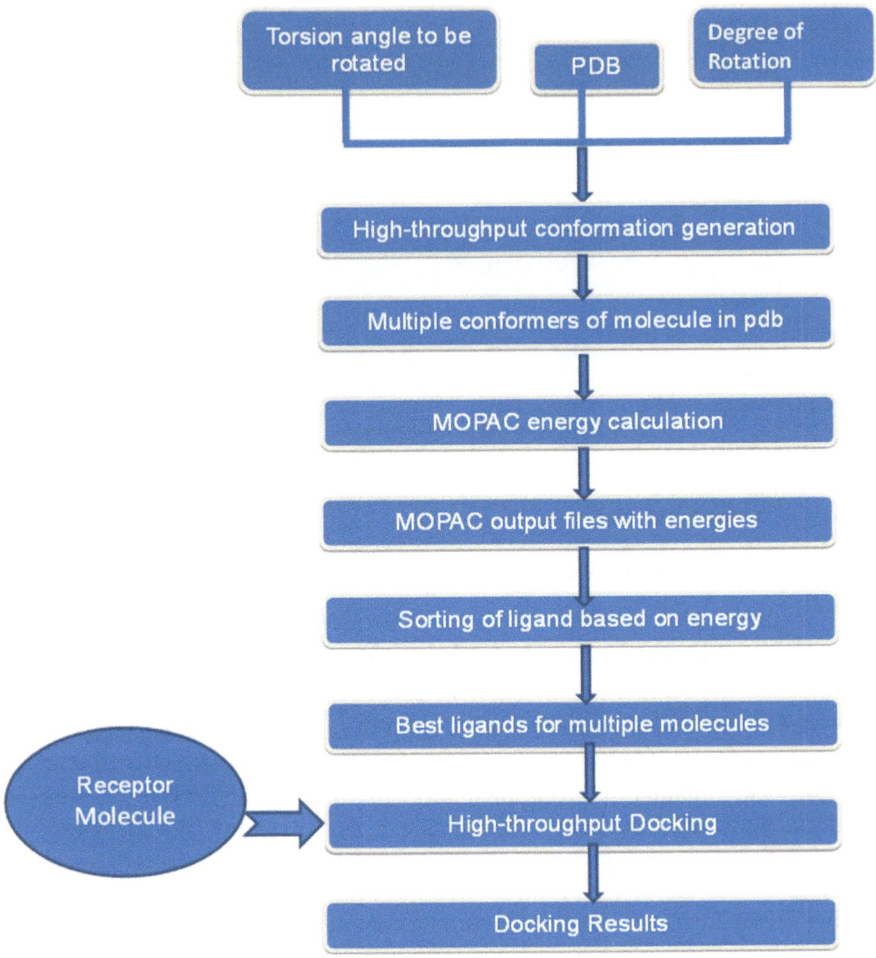

Fig. 12 High-throughput conformation generation and drug docking pipeline

5.4 Parallel Molecular Trajectories Visualization & Analytics (DPICT)

In any computational study of biomolecular systems, analysis and visualization play a pivotal role in understanding and interpretation. Molecular dynamics (MD) simulation studies of biomolecular systems, including proteins, nucleic acids, are no exceptions to this rule. The recent advances in MD techniques like REMD [53] generate multiple trajectory files whose size ranges in few gigabytes (GBs). The present-day tools often find it difficult to load a trajectory of a few GB size as it tends to occupy the entire CPU memory. The same problem is faced for loading multiple trajectories simultaneously, since most of the codes do not support parallel

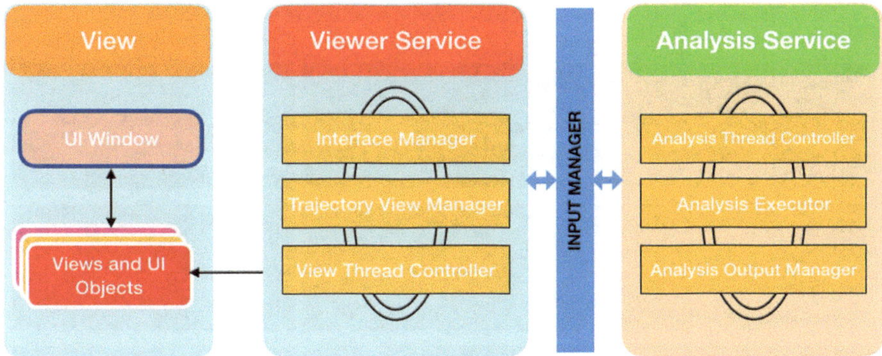

Fig. 13 Flowchart of the DPICT tool

architecture. Redundancy also occurs when the same set of calculations need to be carried out for all the trajectories individually. This often becomes a bottleneck in the research work, since recoding these programs to suit one's purpose is quite cumbersome. One often grapples around for an appropriate program/software, for analyzing and visualizing the multiple MD simulations data. And in the absence of a good program, one has to resort to writing codes and scripts. Also, loading trajectory files for visualization and analysis using the present tools often becomes extremely slow, since most of the codes are meant for serial processing and do not support multiple processors. VMD [39] tries to solve this issue by means of multi-threading, but the process becomes unresponsive when more than one trajectory is to be loaded at a time and visualized. The development of visualization and analysis tool capable of analyzing terascale and petascale data along with high-end visualization screens would accelerate the drug discovery process. Here, an attempt has been made to develop a new visualization and analysis tool capable of reading various file formats like AMBER [31], GROMACS [33] and doing most of the required analyses for a simulation in a parallel environment. The flowchart of the DPICT tool is shown in Fig. 13.

The tool has two distinct modules: one for visualization and rendering and the other for analysis of the MD simulations. The tool is an entirely GUI-based software meant to be run on Unix/Linux operating systems. The entire software tool is coded in C/C++ and OpenGL [ref] programming may be incorporated.

Features of DPICT:

- A tool to elucidate the visualization of huge molecular dynamics trajectories simultaneously for better understanding of the simulation data
- Supports visualization of nine molecules simultaneously
- Different rendering options for biomolecules like ribbon, cartoon, ball, and stick can be viewed
- Works in synchronous manner, where in nine trajectories may be handled simultaneously to perform certain operations

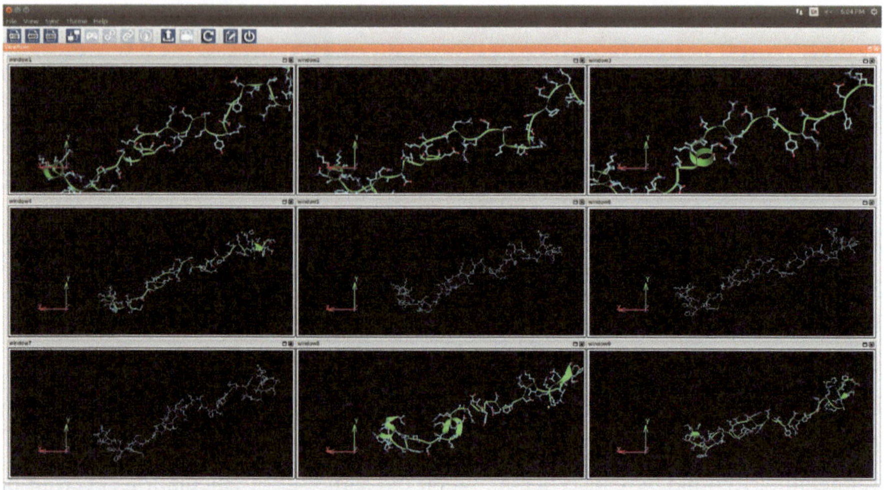

Fig. 14 DPICT tool showing simultaneous multiple trajectory visualization

- Widely used file formats of PDB, AMBER, and GROMACS are supported
- SSH feature enables the users to handle the transfer of large files from remote to local HPC clusters and vice versa.

DPICT tool in its current version is able to manage big data of multiple trajectories as shown in Fig. 14. However, future versions would be targeted to reach the goal of big data visualization.

Bioinformatics group at C-DAC has used the above tools on docking, simulations, and analytics for the drug repurposing studies for cancer protein. The details of it have been described below.

6 Drug Repurposing Study Using Big data Analytics

The drug repositioning or repurposing is a strategy to find new action mechanism of the FDA-approved drug for other disease protein than those for which it was originally intended. The repositioned drug need not go through complete drug development cycle of many years [54]. However, it can directly enter the preclinical testing and clinical trials, thereby reducing risk, time, and costs. One of the well-known examples of repurposed drug is sildenafil citrate (viagra), which was repositioned from a common hypertension drug to a therapy for erectile dysfunction [55, 56]. Similarly, use of off-label FDA-approved drugs for cancer medical practice is also known and accounts for 50–75% of drugs or biologic therapies for cancer in the USA [57, 58]. Owing to computational drug repurposing strategy, a large number of receptors can be tested with already FDA-approved drug, thereby increases the chance of identifying cure for disease within shortened time [59]. One of the proteins crucial Ras in a center pathway has been discussed as a case study.

RAt Sarcoma (RAS) protein is a crucial member of the protein family known as G-proteins. The protein Ras is encoded by one of the most common oncogene in humans. Ras belongs to GTPase class of the proteins, which possess an inherent property of GTP hydrolysis activity. Depending on its association with GDP/GTP, the protein is classified in two distinct conformations: GDP-bound inactive state and GTP-bound active state [60–62]. The malfunctioning of this protein is known to play a crucial role in human cancers, especially pancreatic cancer and various developmental disorders like Costello syndrome, Noonan syndrome [63–65]. The normal functioning of Ras plays pivotal role in the processes of cell proliferation, development, differentiation, and signal transduction [63]. The most common of the Ras mutations are found in pancreatic cancers. Most of the cancers causing mutations are reported to belong to the conserved switch (Sw I and Sw II) and GEF-binding regions of the protein. As these regions are involved in protein–protein interactions and other crucial features, and such mutations directly affect the Ras protein interaction with other proteins [66, 67]. Studies to understand the activation and deactivation Ras pathways and comparative studies of wild type and mutant have been carried out by various groups. A significant low-energy barrier in case of mutant counterparts of Ras is also well established by various experimental and computational studies. To further explore the crucial mutations and further comparison with the wild-type counterpart, computational studies are required to provide more insight about their dynamics and conformational features. Furthermore, for K-Ras which is inherently a less druggable molecule, the current trend of the drug discovery efforts is now directed toward the development of inhibitors of Ras downstream effectors. Related studies suggest that need of dual site inhibitors to effectively block oncogenic Ras signaling. Also, triple site inhibitors are also gaining more importance for improved cancer therapeutics. Considering this as a reference, simulations have been performed to explore and understand the dynamics of activation pathway of the reported hotspot mutants of Ras [68]. Similarly, the GTP hydrolysis-mediated inactivation pathways of the mutant Ras complexes have also been explored. This has helped to provide more information on the energetics of the mutant Ras complexes by calculating the energy barrier between the end states of the protein [69]. Molecular docking studies were carried out on Ras using the approach of drug repurposing with FDA-approved drug molecules database. The literature has suggested three active sites for Ras as shown in Fig. 15 where ligands can be docked [70]. The residues involved in three sites are (SITE1) residue 29–37, (SITE2) residue 68–74 and 49–57, (SITE3) residue 58–74 and 87–91. High-throughput docking has been done using the DOCK6 software employed in embarrassingly parallel molecular docking pipeline. Docking-based drug repurposing and simulation study is being carried out on four Ras systems, namely the wild type, Q61L, G12 V, and G12D mutants, each for 37 ligands. The multiple trajectories for these systems were visualized using parallel trajectory visualizer tool, DPICT. For understanding the ligand (drug candidate) properties, multiple conformations (Fig. 16) were generated using

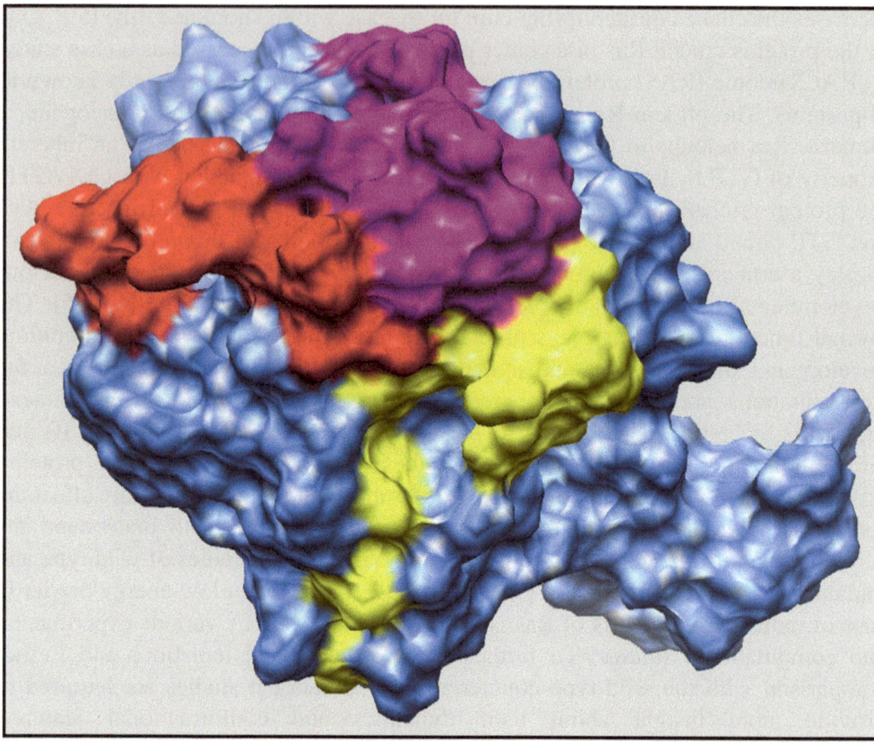

Fig. 15 KRas docking sites: SITE 1 (red): residue 29–37, SITE2 (yellow): 68–74 and 49–57, SITE3 (pink): 58–74 and 87–91

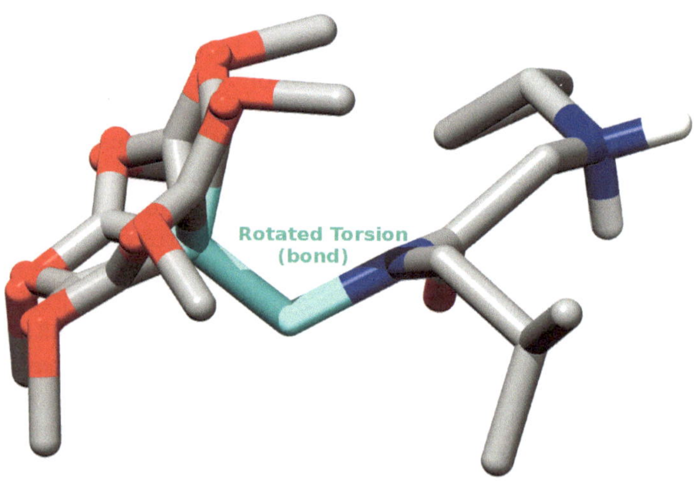

Fig. 16 Conformations generated for docking

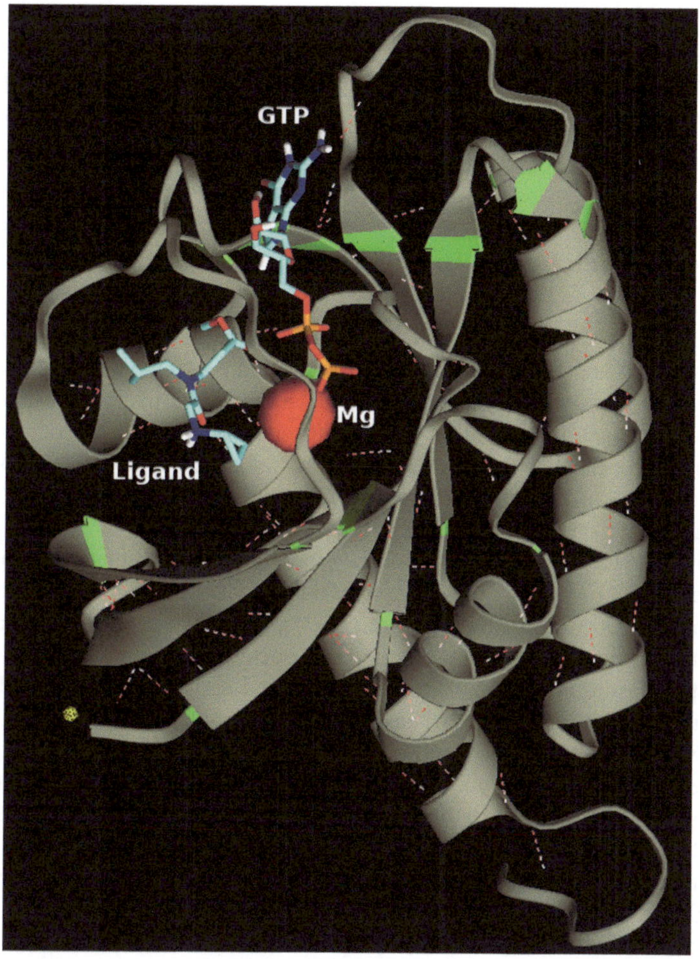

Fig. 17 KRas protein with ligand docked at SITE2

high-throughput conformation generator tool. Moreover, to study the protein–ligand complexes for the simulated systems, in-house developed tool was used. Docked pose of one of ligand is shown in Fig. 17. Preliminary analyses have been completed for the systems. The hydrogen bond and water density analyses have been performed using the in-house developed big data analytics tool, HBAT. MSM analyses are also being carried out for the same, and the results are being compared with the wild-type counterpart. Further, MD simulations were carried out for the best molecule per site in order to check the binding of the molecule with Ras (data unpublished). Classical simulations have been carried out using GROMACS software on Bioinformatics Resource and Applications Facility (BRAF). The standard protocol has been followed for minimization, heating, equilibration, and production run.

Various tools discussed earlier in this chapter have been used for parallel visualization and efficient and fast analysis of Ras docking and simulation trajectories data. In-house computational facility BRAF has been used where these tools are already deployed and tested. The results would help the experimentalist to select the better ligand for further steps of drug development.

7 Latest Development in Big data

Bioinformatics is a technology-driven science. There have been major technological shifts which are driving the data-driven science. With the ever-increasing data, the storage and analysis of huge data are becoming very tedious and most of the data remains unanalyzed. For example, the sequencing of genomes of various organisms is generating petabytes to zetabytes of data. Also, the development of new sequencing technology like nanopore is capable of producing long reads generating huge data [71]. The assembly of such genomes put out a huge challenge on the Big Data technologies. The Apache Hadoop has also enhanced to tackle such challenges like Yarn which allows different data processing engines including graph processing, stream processing as well as batch processing. The MapReduce framework provided by Apache Hadoop is good for batch processing. In case of iterative processing where the data need to be read many times, the MapReduce is not efficient. MapReduce relies heavily on disk input/output so it is slow. The Apache Spark addresses this limitation of Hadoop and provides in memory computing but reducing disk input/output. Spark supports in memory computing and optimizes disk performance by lazy loading and cache mechanism. Hence, spark is suitable for iterative computing.

Recent progressions have empowered the most precision analytics strategies at the "single cell" level. The sequencing of single cell brings about enormous volume and complexities of information and presents an extraordinary chance to comprehend the cell level heterogeneity. The latest developments highlight the inherent opportunities and challenges in Big Data analytics. The recently created technologies like erasure encoding mechanism [72] in Hadoop 3.x tend to resolve the difficulties postured by several big data problems like single cell transcriptome analysis in bioinformatics and present great opportunity to develop cutting-edge technologies for the future research problems. The HDFS uses redundancy for high availability of data. It provides great benefit at the cost of storage byte. Generally, with replication factor of 3, HDFS uses three times more storage data redundancy. So it is very costly in terms of storage. The erasure encoding mechanism in Hadoop 3.x provides same storage safety at the cost of 50% storage overhead. This is effective when data is more and its access frequency is less.

8 Conclusions

Future of medical science is to move toward personalized medicine for enhanced health care. The high-performance computing along with parallel and better algorithms would be generating volume of data from molecular docking and simulations. Advanced structural biology laboratories and techniques would also be generating different types of data. The only way which seems to be efficient in managing and analyzing such an extreme varied data may lie in the application of big data technologies. Similar kind of extreme data is being generated using advanced experimentation in life sciences in the area of agriculture for better crop production and reduced disease susceptibility and in the field of livestock to understand their genomics as well as protect them from various diseases. Data is also being generated in the field of microbes for genomics, drug discovery, vaccine ,and better environmental studies. The near future of biology/life sciences seems to be data-driven hypothesis rather than hypothesis-driven data generation, and newer computing paradigm of big data technologies may be very useful in this aspect.

References

1. Schmidt B, Hildebrandt A (2017) Next-generation sequencing: big data meets high performance computing. Drug Discov Today 22:712–717
2. Tripathi R et al (2016) Next-generation sequencing revolution through big data analytics. Front Life Sci 9(2):119–149
3. Taglang G, Jackson DB (2016) Use of "big data" in drug discovery and clinical trials. Gynecol Oncol 141(1):17–23
4. Leyens Lada et al (2017) Use of big data for drug development and for public and personal health and care. Genet Epidemiol 41(1):51–60
5. Richter BG, Sexton DP (2009) Managing and analyzing next-generation sequence data. PLoS Comput Biol 5(6):e1000369
6. Stephens ZD et al (2015) Big data: astronomical or genomical? PLoS Biol 13(7):e1002195
7. Zhao S et al (2017) Cloud computing for next-generation sequencing data analysis. In: Cloud computing-architecture and applications. InTech, London
8. Bhuvaneshwar K et al (2015) A case study for cloud based high throughput analysis of NGS data using the globus genomics system. Comput Struct Biotechnol J 13:64–74
9. da Fonseca RR et al (2016) Next-generation biology: sequencing and data analysis approaches for non-model organisms. Mar Genomics 30:3–13
10. https://www.rcsb.org/
11. Shaw DE et al (2008) Anton, a special-purpose machine for molecular dynamics simulation. Commun ACM 51(7):91–97
12. Bernardi RC, Melo MCR, Schulten K (2015) Enhanced sampling techniques in molecular dynamics simulations of biological systems. Biochimica et Biophysica Acta (BBA) 1850(5): 872–877
13. Sugita Y, Okamoto Y (1999) Replica-exchange molecular dynamics method for protein folding. Chem Phys Lett 314.1:141–151.APA
14. Swinney DC, Anthony J (2011) How were new medicines discovered? Nat Rev Drug Discov 10(7):507–519

15. Borhani DW, Shaw DE (2012) The future of molecular dynamics simulations in drug discovery. J Comput Aided Mol Des 26(1):15–26
16. Durrant JD, McCammon JA (2011) Molecular dynamics simulations and drug discovery. BMC Biol 9(1):71
17. Fabricant DS, Farnsworth NR (2001) The value of plants used in traditional medicine for drug discovery. Environ Health Perspect 109(Suppl 1):69
18. http://www.chemspider.com/
19. Wishart DS et al (2006) DrugBank: a comprehensive resource for in silico drug discovery and exploration. Nucleic Acids Res 34.suppl_1:D668–D672
20. Irwin JJ, Shoichet BK (2005) ZINC—a free database of commercially available compounds for virtual screening. J Chem Inf Model 45(1):177–182
21. Lengauer T, Rarey M (1996) Computational methods for biomolecular docking. Curr Opin Struct Biol 6(3):402–406
22. Sleigh Sara H, Barton Cheryl L (2010) Repurposing strategies for therapeutics. Pharm Med 24(3):151–159
23. Oprea TI, Mestres J (2012) Drug repurposing: far beyond new targets for old drugs. AAPS J 14(4):759–763
24. Sagiroglu, Seref, and Duygu Sinanc (2013) Big data: a review. In: International conference on collaboration technologies and systems (CTS). IEEE
25. Nayak A, Poriya A, Poojary D (2013) Type of NOSQL databases and its comparison with relational databases. Int J Appl Inf Syst 5(4):16–19
26. Hadoop A (2009) Hadoop. 2009-03-06. http://hadoop.apache.org
27. Zaharia M et al (2010) Spark: cluster computing with working sets. HotCloud 10(10-10):95
28. Allen WJ et al (2015) DOCK 6: impact of new features and current docking performance. J Comp Chem 36(15):1132–1156
29. Jones G et al (1997) Development and validation of a genetic algorithm for flexible docking. J Mol Biol 267(3):727–748
30. Trott Oleg, Olson AJ (2010) AutoDock Vina: improving the speed and accuracy of docking with a new scoring function, efficient optimization, and multithreading. J. Comput. Chem 31(2):455–461
31. Case DA et al (2005) The Amber biomolecular simulation programs. J Comput Chem 26.16:1668–1688
32. Brooks BR et al (2009) CHARMM: the biomolecular simulation program. J Comput Chem 30.10:1545–1614
33. Van Der Spoel D et al (2005) GROMACS: fast, flexible, and free. J Comput Chem 26(16): 1701–1718
34. Phillips JC et al (2005) Scalable molecular dynamics with NAMD. J Comput Chem 26(16): 1781–1802
35. Rysavy SJ, Bromley D, Daggett V (2014) DIVE: a graph-based visual-analytics framework for big data. IEEE Comput Graphics Appl 34(2):26–37
36. Doerr S et al (2016) HTMD: high-throughput molecular dynamics for molecular discovery. J Chem Theory Comput 12(4):1845–1852
37. Tu T et al (2008) A scalable parallel framework for analyzing terascale molecular dynamics simulation trajectories. In: International conference for high performance computing, networking, storage and analysis. SC 2008. IEEE
38. Roe DR, Cheatham TE III (2013) PTRAJ and CPPTRAJ: software for processing and analysis of molecular dynamics trajectory data. J Chem Theory Comput 9(7):3084–3095
39. Humphrey W, Dalke A, Schulten K (1996) VMD: visual molecular dynamics. J Mol Graph Model 14(1):33–38
40. Wold S, Esbensen K, Geladi P (1987) Principal component analysis. Chemometr Intell Lab Syst 2(1–3):37–52
41. Genheden S, Ryde U (2015) The MM/PBSA and MM/GBSA methods to estimate ligand-binding affinities. Expert Opin Drug Discov 10(5):449–461

42. Privalov PL, Crane-Robinson C (2017) Role of water in the formation of macromolecular structures. Eur Biophys J 46(3):203–224
43. Pace CN, Fu H, Lee Fryar K, Landua J, Trevino SR, Schell D, Thurlkill RL, Imura S, Scholtz JM, Gajiwala K, Sevcik J (2014) Contribution of hydrogen bonds to protein stability. Protein Sci 23(5):652–661
44. Irwin JJ, Shoichet BK (2005) ZINC—a free database of commercially available compounds for virtual screening. J Chem Inf Model 45(1):177–182
45. Yuriev E, Chalmers D, Capuano B (2009) Conformational analysis of drug molecules: a practical exercise in the medicinal chemistry course. J Chem Educ 86(4):477
46. Li J, Ehlers T, Sutter J, Varma-O'Brien S, Kirchmair J (2007) CAESAR: a new conformer generation algorithm based on recursive buildup and local rotational symmetry consideration. J Chem Inf Model 47(5):1923–1932
47. Lagorce D, Pencheva T, Villoutreix BO, Miteva MA (2009) DG-AMMOS: a new tool to generate 3D conformation of small molecules using distance geometry and automated molecular mechanics optimization for in silico screening. BMC Chem. Bio 9(1):6
48. Sefraoui O, Aissaoui M, Eleuldj M (2012) OpenStack: toward an open-source solution for cloud computing. Int J Comput Appl 55(3):38–42
49. Stewart JJP (1990) MOPAC: a semiempirical molecular orbital program. J Comput Aided Mol Des 4(1):1–103
50. Hawkins PC, Skillman AG, Warren GL, Ellingson BA, Stahl MT (2010) Conformer generation with OMEGA: algorithm and validation using high quality structures from the protein databank and cambridge structural database. J Chem Inf Model 50(4):572–584
51. Ware B (2002) Open source development with LAMP: using Linux, Apache, MySQL and PHP. Addison-Wesley Longman Publishing Co., Inc., Reading
52. https://www.rabbitmq.com/
53. Hukushima K, Nemoto K (1996) Exchange Monte Carlo method and application to spin glass simulations. J Phy Soc Jpn 65(6):1604–1608
54. Ashburn TT, Thor KB (2004) Drug repositioning: identifying and developing new uses for existing drugs. Nat Rev Drug Discov 3(8):673–683
55. Novac Natalia (2013) Challenges and opportunities of drug repositioning. Trends Pharmacol Sci 34(5):267–272
56. Smith Kelly M, Romanelli Frank (2005) Recreational use and misuse of phosphodiesterase 5 inhibitors. J Am Pharm Assoc 45(1):63–75
57. Pfister DG (2012) Off-label use of oncology drugs: the need for more data and then some. J Clin Oncol, 584–586
58. Jin G, Wong STC (2014) Toward better drug repositioning: prioritizing and integrating existing methods into efficient pipelines. Drug Discov Today 19(5):637–644
59. Neves SR, Ram PT, Iyengar R (2002) G protein pathways. Science 296(5573):1636–1639
60. Khrenova MG et al (2014) Modeling the role of G12 V and G13 V Ras mutations in the Ras-GAP-catalyzed hydrolysis reaction of guanosine triphosphate. Biochemistry 53(45): 7093–7099
61. Spoerner M et al (2010) Conformational states of human rat sarcoma (Ras) protein complexed with its natural ligand GTP and their role for effector interaction and GTP hydrolysis. J Biol Chem 285(51):39768–39778
62. Ma J, Karplus M (1997) Molecular switch in signal transduction: reaction paths of the conformational changes in ras p21. Proc Natl Acad Sci USA 94(22):11905–11910
63. White MA et al (1995) Multiple Ras functions can contribute to mammalian cell transformation. Cell 80(4):533–541
64. Schubbert S, Shannon K, Bollag G (2007) Hyperactive Ras in developmental disorders and cancer. Nat Rev Cancer 7(4):295
65. Gao C, Eriksson LA (2013) Impact of mutations on K-Ras-p 120GAP interaction. Comput Mol BioSci 3(02):9
66. Shurki A, Warshel A (2004) Why does the Ras switch "break" by oncogenic mutations? Proteins: Struct Funct Bioinf 55(1):1–10

67. Lu S et al (2016) Ras conformational ensembles, allostery, and signaling. Chem Rev 116(11): 6607–6665
68. Sharma N, Sonavane U, Joshi R (2017) Differentiating the pre-hydrolysis states of wild-type and A59G mutant HRas: an insight through MD simulations. Comput Biol Chem 69:96–109
69. Sharma N, Sonavane U, Joshi R (2014) Probing the wild-type HRas activation mechanism using steered molecular dynamics, understanding the energy barrier and role of water in the activation. Eur Biophys J 43(2-3):81–95
70. Wang W, Fang G, Rudolph J (2012) Ras inhibition via direct Ras binding—is there a path forward? Bioorg Med Chem Lett 22(18):5766–5776
71. Branton D, Deamer DW, Marziali A, Bayley H, Benner SA, Butler T, Di Ventra M, Garaj S, Hibbs A, Huang X, Jovanovich, SB (2010) The potential and challenges of nanopore sequencing. In: Nanoscience and technology: A collection of reviews from Nature Journals, pp 261–268
72. https://hadoop.apache.org/docs/r3.0.0/hadoop-project-dist/hadoop-hdfs/HDFSErasureCoding.html

Single-Particle cryo-EM as a Pipeline for Obtaining Atomic Resolution Structures of Druggable Targets in Preclinical Structure-Based Drug Design

Ramanathan Natesh

Abstract Single-particle cryo-electron microscopy (cryo-EM) and three-dimensional (3D) image processing have gained importance in the last few years to obtain atomic structures of drug targets. Obtaining atomic-resolution 3D structure better than ~2.5 Å is a standard approach in pharma companies to design and optimize therapeutic compounds against drug targets like proteins. Protein crystallography is the main technique in solving the structures of drug targets at atomic resolution. However, this technique requires protein crystals which in turn is a major bottleneck. It was not possible to obtain the structure of proteins better than 2.5 Å resolution by any other methods apart from protein crystallography until 2015. Recent advances in single-particle cryo-EM and 3D image processing have led to a resolution revolution in the field of structural biology that has led to high-resolution protein structures, thus breaking the cryo-EM resolution barriers to facilitate drug discovery. There are 24 structures solved by single-particle cryo-EM with resolution 2.5 Å or better in the EMDataBank (EMDB) till date. Among these, five cryo-EM 3D reconstructions of proteins in the EMDB have their associated coordinates deposited in Protein Data Bank (PDB), with bound inhibitor/ligand. Thus, for the first time, single-particle cryo-EM was included in the structure-based drug design (SBDD) pipeline for solving protein structures independently or where crystallography has failed to crystallize the protein. Further, this technique can be complementary and supplementary to protein crystallography field in solving 3D structures. Thus, single-particle cryo-EM can become a standard approach in pharmaceutical industry in the design, validation, and optimization of therapeutic compounds targeting therapeutically important protein molecules during preclinical drug discovery research. The present chapter will describe briefly the history and the principles of single-particle cryo-EM and 3D image processing to obtain atomic-resolution structure of proteins and their complex with their drug targets/ligands.

R. Natesh (✉)
School of Biology, Indian Institute of Science Education and Research Thiruvananthapuram (IISER-TVM), Maruthamala P.O.,Vithura, Trivandrum 695 551, Kerala, India
e-mail: natesh@iisertvm.ac.in

C. G. Mohan (ed.), *Structural Bioinformatics: Applications in Preclinical Drug Discovery Process*, Challenges and Advances in Computational Chemistry and Physics 27, https://doi.org/10.1007/978-3-030-05282-9_12

Keywords Single-particle cryo-EM · Drug development · Pharmacological targets Structural biology · High resolution

Abbreviations

3D	Three Dimension
CTF	Contrast Transfer Function
cryo-EM	Cryo-electron microscopy
CC	Cross-Correlation
DDD or DED	Direct Detection Device or Direct Electron Detector
ET	Electron Tomography
EMDB	Electron Microscopy Data Bank
EM	Electron Microscopy
FEG	Field Emission Gun
FSC	Fourier Shell Correlation
MSA	Multivariate Statistical Analysis
PDB	Protein Data Bank
PCA	Principle Component Analysis
SBDD	Structure-Based Drug Design
SNR	Signal-to-Noise Ratio
SSNR	Spectral SNR
TEM	Transmission Electron Microscopy

1 Introduction

The importance of structural biology in understanding the principles of molecular function of proteins, the workforce of cellular world, underpins its use in health science and pharma industries. Classically, protein crystallography was ruling the world of structure-based drug design (SBDD). This was mainly due to the capability of protein crystallography to solve high (better than 1.8 Å), atomic (better than 1.2 Å), and ultra-high (better than 0.95 Å)-resolution 3D structures, which give information of protein drug molecular interaction at various levels. Particularly, the positions of hydrogen atoms were located in many atomic and ultra-high-resolution protein structures. There were no other methods that could rival the versatility of obtaining 3D atomic-level macromolecular structures with which crystallography could achieve. Of the 131,108 protein structures in PDB (as on June 15, 2018), 90% of structures among them were solved by X-ray crystallography technique and 8% by NMR technique. The remaining 2% of structures by large were solved by electron microscopy, electron crystallography, hybrid, and other methods, which include neutron diffraction, solution scattering, fiber diffraction. Clearly, the PDB data suggests that the protein crystallography technique dominates till date. However, the protein crystallography method comes with a proviso. That is, we need diffractable protein crystals of reasonable 10–100s of micron size, in order to obtain a high-resolution X-ray crystallography protein structure. Also, as the unit cell parameter of the protein crystals increase, the resolution of diffraction data drops as the cube of unit cell parameter [1]. Moreover, many proteins, in particular membrane proteins and fibrous proteins, are recalcitrant to crystallization. An analysis of deposited protein structures in PDB by Kozma and co-workers in 2017 [2] showed that the majority of the solved structures (97.6%) are globular proteins and only ~2.4% of them are membrane protein structures. This is primarily because obtaining good diffraction quality 3D crystals for membrane proteins is challenging. As a result, single-particle cryo-EM has gained popularity nowadays for solving membrane protein structures as well along with globular proteins. Also, in cases where single-particle cryo-EM cannot give high-resolution maps, protein crystallography and cryo-EM can be used as hybrid method to visualize macromolecular assemblies at pseudo-atomic resolution as described in Natesh [3] and references cited therein.

SBDD is among one of the most important stages for drug discovery in industrial drug discovery pipelines [4]. It requires the best possible resolution protein structures, preferably better than 2.5 Å resolution. Until 2015, single-particle cryo-EM could not achieve the resolution comparable to resolution of structures in protein crystallography [5, 6]. Recently, Danev and co-workers have solved a structure of *Mus musculus* apo ferritin at 1.62 Å (EMD-9599). Others have solved the structures of proteins with bound ligands at resolution 2.5 Å or better [7–10], presented in Table 1. The foundation for this was laid 36 years ago in December 1981 when Jacques Dubochet (along with AW Mc Dowall) published the paper on vitrification (amorphous ice) of pure water for electron microscopy [11]. Jacques was excited about the prospects of making electron microscopy water friendly. Five years after that, they got the first cryo-EM virus structure at 35 Å resolution [12]. However, before that the first EM structure came from Henderson and Unwin [13] of purple membrane protein by electron crystallography, but however not using cryo, and hence, the resolution was bit low at 7 Å. This encouraged Joachim Frank to develop image processing algorithms for solving protein structures by building 3D reconstruction from fussy cryo-EM projection images of proteins [14–16]. These developments led to the first cryo-EM atomic model of the protein bacteriorhodopsin 15 years later in the year 1990 [17]. In recent years, other developments like field emission gun electron source, direct electron detectors, and movie-based cryo-EM imaging methods have led to an avalanche of high-resolution single-particle cryo-EM protein structures [5, 6, 18]. Thus, the full potential of cryo-EM in obtaining high-resolution structure of proteins was realized in 2015, which led to the Noble Prize in Chemistry in the year 2017 for "developing cryo-electron microscopy for the high-resolution structure determination of biomolecules in solution." The predictions made by Henderson in 1995 [19] that single-particle cryo-EM can be used for atomic-resolution structure determination of protein and protein complexes has become a reality today. Thus, single-particle cryo-EM technique can be used as a pipeline for obtaining atomic structures of druggable targets in preclinical SBDD.

Table 1 EMDataBank (EMDB) entries having single-particle cryo-EM 3D reconstruction with bound ligands at 2.5 Å or better resolution and their corresponding PDB codes

EMDB entry ID (deposition date)	Resolution (Å)	Fitted PDBs	Components	
			Protein	Ligand
EMD-2984 (April 26, 2015)	2.2	5a1a	*E.coli* beta-galactosidase (0.465 MDa)	Phenylethyl beta-D-thiogalactopyranoside (PETG)
EMD-3295 (January 12, 2016)	2.3	5ftj	*Homo sapiens* p97/VCP Transitional endoplasmic reticulum ATPase (0.54 MDa)	UPCDC30245 (an allosteric inhibitor of VCP)
EMD-7025 (September 9, 2017)	2.5	6az3	*Leishmania donovani* 91s ribosome LSU	Paromomycin
EMD-7770 (March 28, 2018)	1.9	6cvm	*E. coli* beta-galactosidase (0.465 MDa)	PETG
EMD-7638 (March 27, 2018)	2.43	6cvb	Enterovirus D68 (virus from *Homo sapiens*) vp1 (0.0330 MDa), vp3 (0.0272 MDa), vp2 (0.0276 MDa, vp4 (0.00734 MDa)	Glycan. 6′-sialyl-N-acetyllactosamine
EMD-7599 (March 20, 2018)	2.17	6csg	Enterovirus D68 vp1 (0.0329 MDa), vp3 (0.0271 MDa), vp2 (0.0276 MDa), vp4 (0.00734 MDa)	No bound inhibitor
EMD-8194 (May 17, 2016)	1.8	5k12	*Bos taurus* Glutamate dehydrogenase (0.334 MDa, 0.0616 MDa)	No bound inhibitor
EMD-8762 (June 8, 2017)	2.26	5w3m	Human rhinovirus B14 C5 antibody variable heavy domain (0.0120 MDa), C5 antibody variable light domain (0.0109 MDa), vp1 (0.0326 MDa), vp3 (0.0262 MDa), vp2 (0.0285 MDa), vp4 (0.00718 MDa)	No bound inhibitor
EMD-9012 (July 31, 2018)	1.86	6e9d	Adeno-associated virus - 2 (3.9 MDa), empty virus from Homo sapiens VP1 (0.0820 MDa)	No bound inhibitor

2 The Single-Particle Cryo-EM at High Resolution

The single-particle cryo-EM method for high-resolution structure determination of proteins and protein complexes involves four major steps, viz. (i) the sample preparation, (ii) specimen preparation, (iii) data collection, and (iv) image processing and 3D reconstruction (i.e., structure determination, which includes model building and refinement of the protein/ligand coordinates in the EM map). Sample preparation involves protein purification either from the source or expressed recombinantly in a heterologous host system. The amount of sample required for cryo-EM is very less (~ 1 μM) in comparison with protein crystallography or NMR spectroscopy techniques, where typically ~ 200 μM sample is required.

For single-particle electron microscopy (EM), there are two main ways of specimen preparation: (a) negative stain specimen preparation and (b) solution-state "vitrification" for cryo-EM. The former is used for quick characterization of macromolecules and their complexes. However, this type of specimen preparation involves inherent drawbacks (e.g., artifacts and visualizing stain rather than actual protein), which limits the resolution of EM reconstruction map from 30 to 20 Å at its best. Single-particle cryo-EM, the focus of this chapter, on the other hand is synonymous to solution-state structure, and the specimen preparation does not induce artifacts over the protein sample being studied. The vitrified specimen preserves the resolution of the protein structure that is being studied.

Single-particle cryo-EM technique has the capability to solve protein structures to better than 4 Å resolution nowadays. It is to be noted that, there is a consensus in the EM community that better then 4 Å depicts high-resolution structures, while, in the X-ray crystallography community, high resolution corresponds to better than 1.8 Å resolution, as described in the beginning of this chapter. Prior to the resolution revolution in the year 2015, most of the cryo-EM structures with resolution 4 Å or better were virus structures [20–22]. This was possible due to their large size and high symmetry (e.g., icosahedron symmetry). Most of these data were collected on photographic film (KODAK SO-163 FILM). However, the asymmetric particles (i.e., particles without higher-order symmetry) were limited to sub-nanometer (around 6–10 Å) resolution. Only 1/10th of the total number of structures in EMDB were with resolution 4 Å or better before the resolution revolution. This has significantly increased to 1/6th of the total number of single-particle cryo-EM structures in EMDB as on July 29, 2018, clearly indicating that, currently, there are more structures solved with resolution better than 4 Å in the database. These were possible due to the advancement in the hardware and software and the way the projection images are captured and processed during cryo-EM data collection and processing. Main steps involved in single-particle cryo-EM for obtaining high-resolution protein structure are presented in three subsections. First, we will begin with the details of the specimen preparation in Sect. 2.1, followed by data collection in Sect. 2.2, and finally image processing and 3D reconstruction in Sect. 2.3, respectively.

2.1 Specimen Preparation for Single-Particle Cryo-EM

The cryo-EM specimen preparation is the most challenging and a crucial step for high-resolution data collection. Most of the time spent on single-particle cryo-EM pipeline is in preparing the best protein specimen (which involves optimizing both the biochemistry and vitrification of the sample) for high-resolution cryo-EM data collection. Hence, it is worth to spend some time to get the best specimen out from the purified sample, which will save time and money later. The first step in specimen preparation is the purified sample (e.g., protein or protein complexes) typically 3–3.5 μL is applied on a pre-glow discharged holey carbon grid (in some special cases, continuous carbon grids laid over the holey grid are used). For the best specimen preparation, the quality of the protein sample, pre-treatment of holey grids, and the choice of the type of grid are important. After applying the sample on the holey carbon grid, the excess protein is blotted using a filter paper (usually Whatman filter paper 1) to leave a very thin layer of sample and immediately the grid is plunged into a pre-prepared liquid ethane well, surrounded and maintained at cryogenic temperature by a bath (surrounding the ethane well) with liquid nitrogen as shown in Fig. 1. Jacques Dubochet and co-workers standardized the vitrification

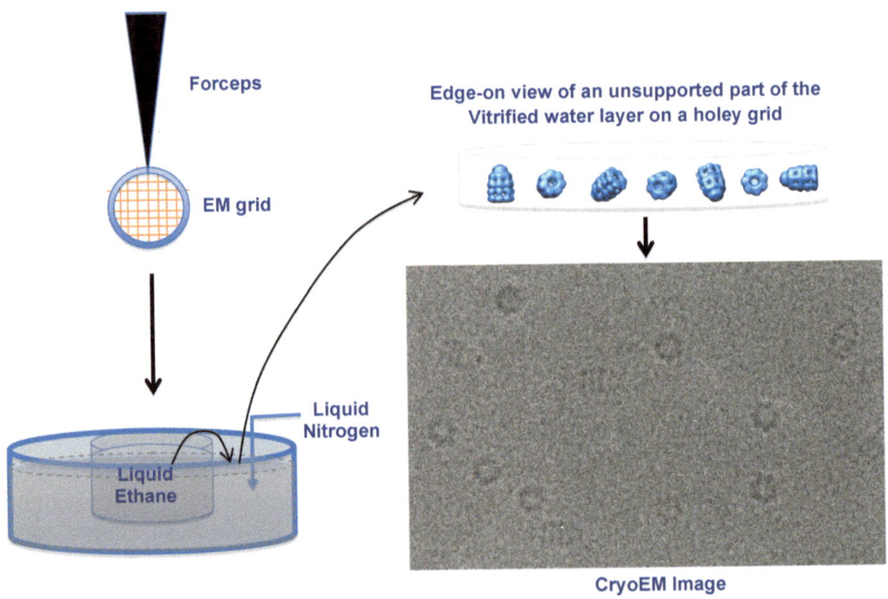

Fig. 1 Vitrification of cryo-EM specimens. A cryo-EM grid with a thin film of solution (<2000 Å of thickness) is plunged into liquid ethane for vitrification. The frozen specimen is transferred into liquid nitrogen before it is imaged at liquid nitrogen temperatures on a TEM. To the right top is a schematic edge-on view of a part of frozen water layer, with macromolecular complexes trapped in different orientations. Bottom right, part of a cryo-EM image showing weak and noisy views of the complexes. Figure reproduced from Natesh, 2014 [3], by permission of publisher—Indian Academy of Science, Bengaluru

process in the late 1970s and published the work in the year 1981 [11]. They showed that sample in buffer/water must be cooled in less than a millisecond to avoid the ice crystal formation and to get amorphous ice (i.e., vitrified). They also showed that if the temperature of specimen is kept sufficiently low below −160 °C, the vitrified state could be maintained for long time [11, 23]. This seminal discovery enabled proteins to be visualized in its native state under the vacuum of transmission electron microscope (TEM). For this discovery, Dubochet received one-third of the Noble Prize in Chemistry in the year 2017. The solution-state protein sample is frozen in time and space, maintaining the integrity of the protein's structural state in the vitrified water. The vitrification can be carried out with a homemade manual plunger or using a commercially available vitrification robot. A perfect vitrified specimen is one in which the thickness of the ice over the holes of the grid is such that there is one single layer of particles distributed, the particles are uniformly distributed (with distance between each particles at least 1.5 times the particle size), and the particles adopt as many different orientations as possible. The vitrified specimen grid is then placed in a cryo grid storage box that is preserved in liquid nitrogen storage Dewar, until the data collection is carried on a high-resolution cryo-TEM. An extensive description of the specimen preparation is given in Passmore and Russo [24].

2.2 *Data Collection*

Data collection is carried out on a cryo-TEM equipped with a 200 or 300 kV field emission gun (FEG) necessary to obtain a high-resolution single-particle data. The stored grids are transferred from the cryo grid storage box to a single tilt cryo-transfer holder pre-cooled on a cryo-workstation (Fig. 2b). In this case, only one grid can be inserted into the TEM by manually loading the holder into the cryo-TEM (Fig. 2a) and analyzed before the holder is taken out of the microscope at the end of data collection. Alternatively, each one of the stored grids can be transferred one by one to a cartridge, which is then placed on multiple grid holder cassette (which holds up to 12 grids). This cassette is then placed into the capsule, which is loaded into the cryo-TEM (Fig. 2c) through an autoloader robot that is built in the microscope. Thermo Fisher Scientific Talos Arctica/Glacios, Thermo Fisher Scientific Krios, and JEOL Cryo ARM 200/300 are microscopes with such autoloader capabilities. The robotic grid loader then can load one by one to the stage using inbuilt robot, which can load or unload the grid on the stage controlled by software. In case of high-end TEM analysis, grid atlas can be created to choose the square of right thickness from all the loaded grids. It is very important to keep the grid always under liquid nitrogen in order to avoid any ice crystal formation and contamination on the grid. Hence, all the processes described in Fig. 2, which involve handling of frozen specimen grid, are carried out under liquid nitrogen. Ice crystals destroy the view of particles by dark contrast, and hence, it is critical to avoid any exposure of plunge-frozen grid to the air.

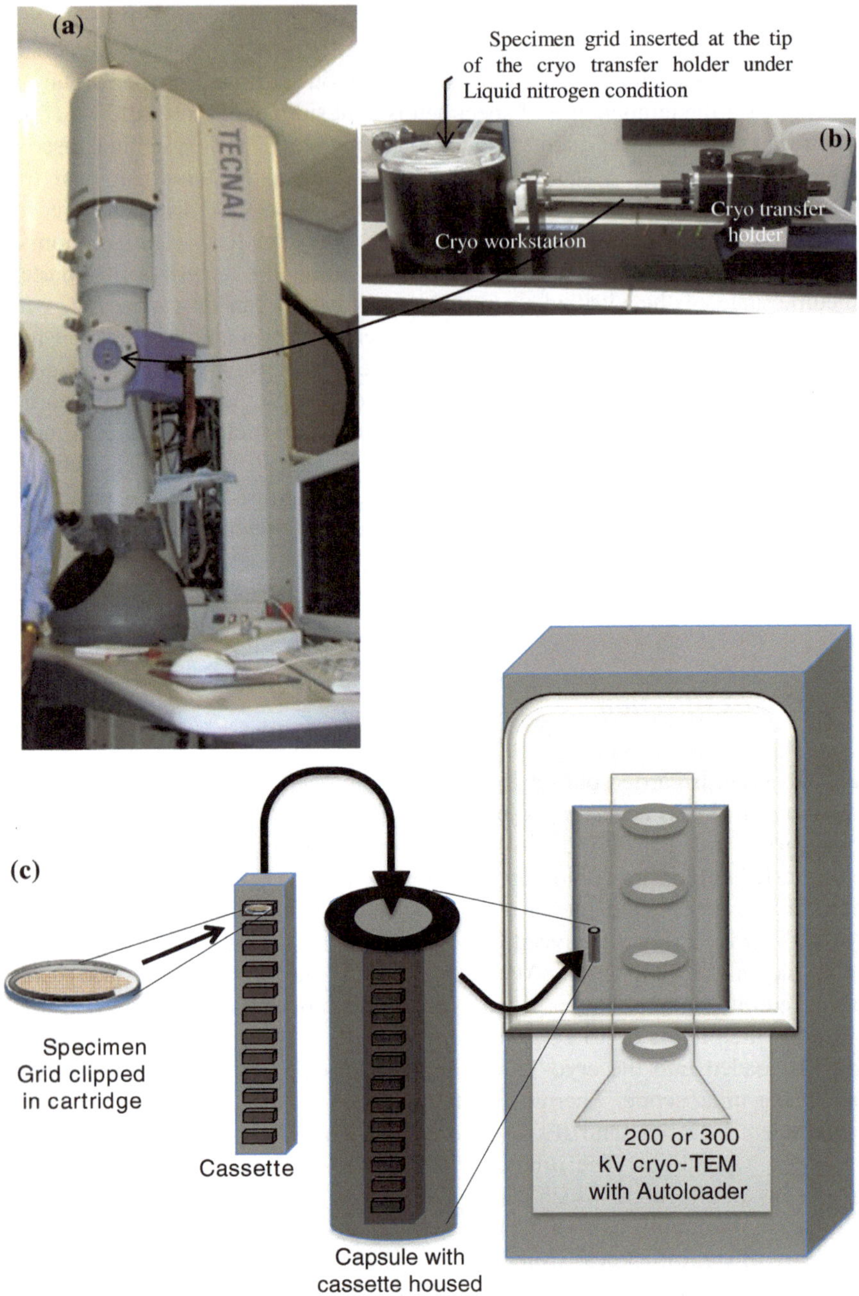
(a)
TECNAI
Specimen grid inserted at the tip of the cryo transfer holder under Liquid nitrogen condition
(b)
Cryo workstation
Cryo transfer holder
(c)
Specimen Grid clipped in cartridge
Cassette
Capsule with cassette housed
200 or 300 kV cryo-TEM with Autoloader

◀**Fig. 2** **a** 200 kV transmission electron microscope equipped with field emission gun (FEG). **b** Gatan CT3500 single tilt liquid nitrogen cryo-transfer holder docked onto cryo-workstation. After inserting the specimen grid onto the cryo-holder (not in scale to microscope), it is carefully transferred to the microscope as shown by the arrow mark. **c** A maximum of 12 grids can be loaded via cassette, housed in a capsule as described in the text. Each grid can be imaged one by one using an autoloader robot housed in a 200- or 300-kV cryo-TEM. (Fig. 2a, b was reproduced from Natesh [3], by permission of publisher—Indian Academy of Science, Bengaluru)

Once the grid is on the stage of the TEM, the data is collected on a highly sensitive direct detection device (DDD) also called as direct electron detector (DED) under low electron dose (typically <15 $e^{-}/Å^2$). Low electron dose is necessary since high dose (>15–20 $e^{-}/Å^2$) will cause radiation damage. However, high dose $\sim$1000 $e^{-}/Å^2$ is required for atomic-resolution reconstruction [25]. This problem can be overcome by averaging similar looking particles as described in image processing Sect. 2.3 below. DDD is more sensitive (technically, this feature is called improved detective quantum efficiency (DQE)) and can detect lower doses more effectively with low noise as compared to the conventional photographic film or the CCD (charged coupled device) detectors. Data collection at the focus gives the best resolution, but however the phase contrast is lost in the image (i.e., you cannot clearly visualize the particles). In order to visualize the particles, the images are captured at a defocus that restores the phase contrast in the image, which enables us to visualize particles. Hence, data is collected at a range of defocus between $\sim$4 μm (lower resolution) and $\sim$1 μm (higher resolution). Modern-day advancements in hardware have led to the use of phase plates and energy filters that can restore contrast in the images collected closer to focus. Thus, preserving high-resolution information in the images and at the same time preserving the image phase/amplitude contract as a result alleviate the need for contrast transfer function (CTF) modulation correction at image processing stage.

For high-resolution structure determination, the data is collected on DED as movie frames, which is actually a dose fractionated image stack. The movie frames collected can be corrected for loss of resolution due to stage drifts, charging, and beam-induced motion. The individual movie frames or subset of movie frames in batches are then aligned with respect to each other in order to restore the high-resolution information [26]. Relatively, high exposures up to 20 $e^{-}/Å^2$ can be used for movie mode while DEDs can also be used in electron counting mode where dose rate must be kept below 10 e^{-}/pixel/s [26, 27]. Movie corrections are applied immediately on the micrographs after the data collection using programs like MotionCor2 [28], optical flow algorithm as implemented in Xmipp [29, 30], Unblur/Summovie [31, 32]. In addition, improved stability of specimen can be provided by the use of grids with graphene and gold support [25, 33, 34]. Hence, in the last six years there has been many breakthroughs in detector, imaging, and image processing technology that has led to high-resolution data collection for even smaller proteins like hemoglobin with mass 64 kDa using Volta phase plate (VPP) [18], thus leading to resolution revolution with structures determination to better than 2.5 Å. Another aspect of data collection is the automation. Not all proteins give homogenous samples for atomic-resolution reconstruction. The fact that proteins are dynamic leads to heterogeneity and underlies the need for large amount of data collection (in a hope to group particles into homogenous groups), which is tedious to be done manually. In recent years, many software packages have been developed to interface with the advanced electron microscopes for automatic data acquisition. Some examples of such software that can be used for fully automated data collection on a well-calibrated cryo-TEM are Leginon [35], SerialEM [36], UCSFImage4 [37], FEI-EPU, JEOL-JADAS [38], GATAN-Latitude S. Most of the software is used for automated data collection for both single-particle cryo-EM and electron tomography (ET) work. Some programs like Appion [39] extend the automated data collection through a pipeline from automated data collection all the way through automated particle picking to image processing (CTF estimation, classification, and 3D reconstruction).

2.3 Image Processing and Three-Dimensional Reconstruction

Cryo-EM is different from X-ray crystallography because it uses "images" as primary data, rather than the diffraction patterns. Translated into Fourier lingo, the availability of images means that the "phase problem" known in X-ray crystallography (described in Sects. 2.3 and 2.4 of Natesh [3]) does not exist in EM. The electron microscope, in Hoppe's words, is a "phase-measuring diffractometer" [40]. Hence, extreme care has to be taken in image processing. Image processing involves preprocessing the collected data, particle picking, centering the particles in their selected boxes, 2D classification and determining their relative orientations and/or 3D classification and 3D reconstruction. An example of image processing and 3D reconstruction is shown in Figs. 3 and 4. The preprocessing step involves CTF correction and image normalization [41]. As mentioned in the data collection section, the data is collected at various defocus positions. As one gradually increases the defocus (i.e., under focus), the contrast of the image proportionally improves. Improvement in contrast comes at a cost, a loss in the higher spatial frequencies (i.e., high-resolution information is lost) in the image, and in addition, it introduces CTF modulation in the spatial frequencies of the image. Hence, the first step in image processing is to calculate the lens defocus and astigmatism, which is needed to correct the measured data for the CTF of the microscope [42, 43]. Software CTFFIND, ACE2, Gctf, or e2ctf.py [44–47] can be used to estimate the CTF that is used for CTF corrections.

After CTF correction, the images are normalized to set the mean density of the particles to zero and same standard deviation [41]. The particles are then manually or auto-picked into boxes of 1.5–2.5x, the size of the largest axis of the particle using suitable software. A guide for choosing the right box size is given at the online documentation http://blake.bcm.edu/emanwiki/EMAN2/BoxSize. Number of softwares are available for manual and automatic picking of particles and subsequent image processing. Examples of such programs are FindEM [48] (only for automated particle picking), EMAN (e2boxer.py) [49, 50], IMAGIC [51], Ximdisp [52] (only for interactive display, analyses, and particle picking; now a part of CCP-EM package [53, 54]), Xmipp [30], RELION-autopick [55], cryoSPARC [56], APPLE Picker [57] (completely automatic particle picking, a part of ASPIRE Suite [58]), gEMpicker [59] (only for template-based particle picking), SIGNATURE [60] (only for particle picking and data analysis), etc. Most of the auto-picking software employ initial manual picking routine (except APPLE picker), where a couple of thousands of particles are manually picked from a subset of available micrographs and use the best class averages generated from them (having as many different representative orientations) as templates to auto-pick particles from rest of the micrographs. This is the preferred method. Alternatively, the auto-picking programs can use low-pass-filtered EM maps as templates for particle picking (less preferred, but useful in protein drug complex where you have the apo-protein structure already). Using maps from PDB (Protein Data Bank) coordinates, as reference model is not preferred at this stage in order to avoid "Einstein-from-noise" effect [61], i.e., to avoid any 2D model bias. CTF corrections can also be performed on picked particle images as compared to whole micrographs in some software, e.g., EMAN [44].

After particle picking, the next stage is to get the 3D reconstruction of the biological macromolecules using the different but identifiable 2D projections of particles. The first 3D reconstruction from a 2D projection was carried out on negative stained tail of bacteriophage T4 by De Rosier and Klug [62]. However, the 2D projections of particle images cutout from the motion-corrected micrographs have still low signal-to-noise ratio (SNR) due to low electron dose data collection as described in data collection section. Hence, in order to improve the SNR of the particles, many identical looking particle images are aligned and summed (clustering) thus effectively increasing the SNR and dose without increasing the damage [62]. There are three main advantages of reference-free (unsupervised) 2D classification: (i) to select few 2D classes from which we can make starting 3D map, which can be projected as references for refinement. (ii) We can identify the fraction of bad classes (which may contain artifacts, invalid particles, or simply empty), and thus, those images with anomalies can be deleted from the data set in the beginning itself. (iii) It also helps in identifying the conformational and compositional variability in the data set [50]. Two-dimensional (2D) and 3D classifications are carried out by using various statistical analysis software suite IMAGIC [51], Spider [63], EMAN [44], RELION-3 [64], FREALIGN [65], Appion [39], cryoSPARC [56], ASPIRE Suite [58], Xmipp [30], SPHIRE (sphire.mpg.de), etc., or a combination of more than one of these suites. Several of these software packages are integrated into one processing framework, for example, as in Scipion [66]. An exhaustive list of EM software programs is available at EMDataBank (EMDB, http://www.emdatabank.org/emsoftware.html).

Spider [63, 67] and IMAGIC [51] were among the first programs to be developed for single-particle reconstruction in the year 1996 followed by FREALIGN

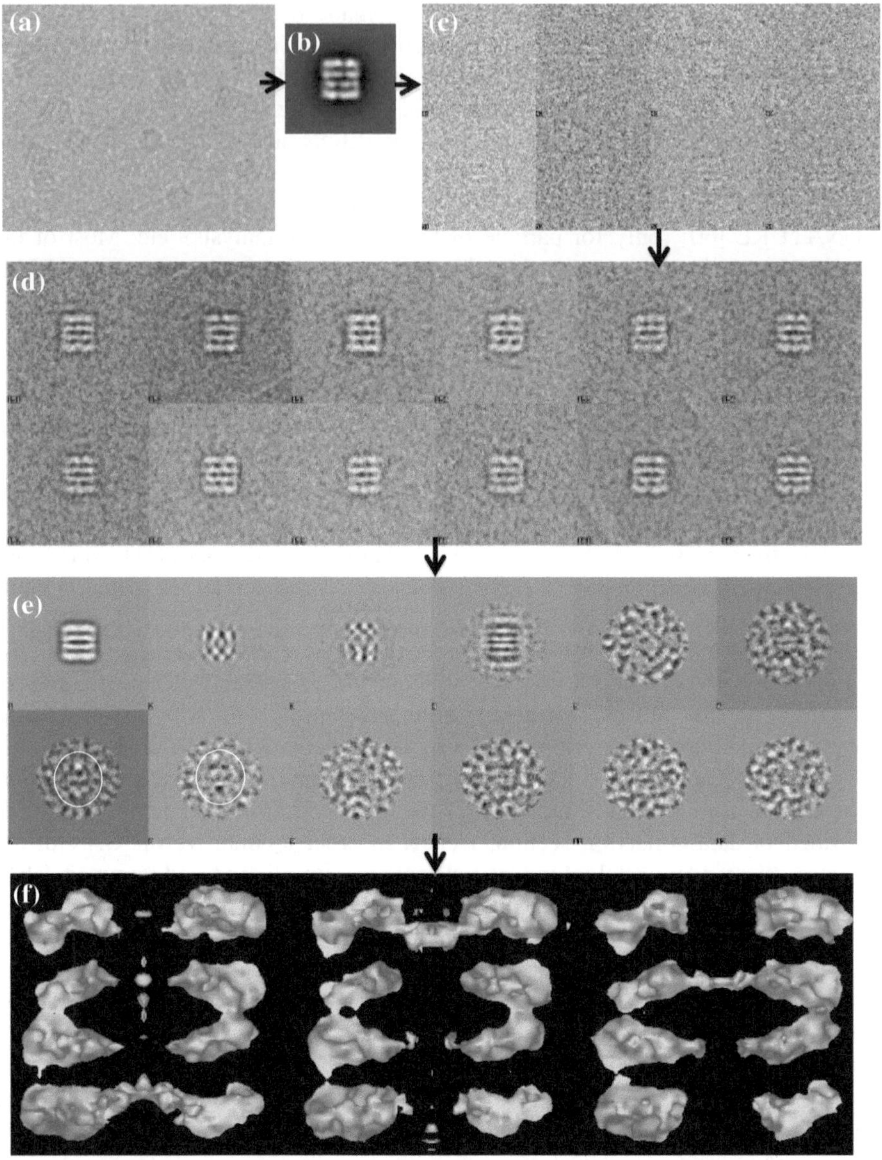

Fig. 3 Image processing and 3D reconstruction of GroEL and non-native protein RuBisCO complex [74]. **a** Raw micrograph (this image is not motion corrected, but at this stage if dose fractionated image stacks are collected on a DED, they are motion corrected) and **b** 30-Å-filtered brick view reference from empty GroEL cryo-EM map. (**c**) Particles from cryo-EM images like in micrograph (**a**) are extracted into boxes, CTF corrected, filtered, normalized, and aligned to reference to bring them to the same center. **d** Orientation separation by class averages of images using MSA shows significant improvement in signal-to-noise ratio. **e** Eigen image information (circled) was used to classify images into homogenous classes. **f** MSA classification into three homogenous groups; 3D reconstruction of three classes using projection matching is shown in Fig. 4

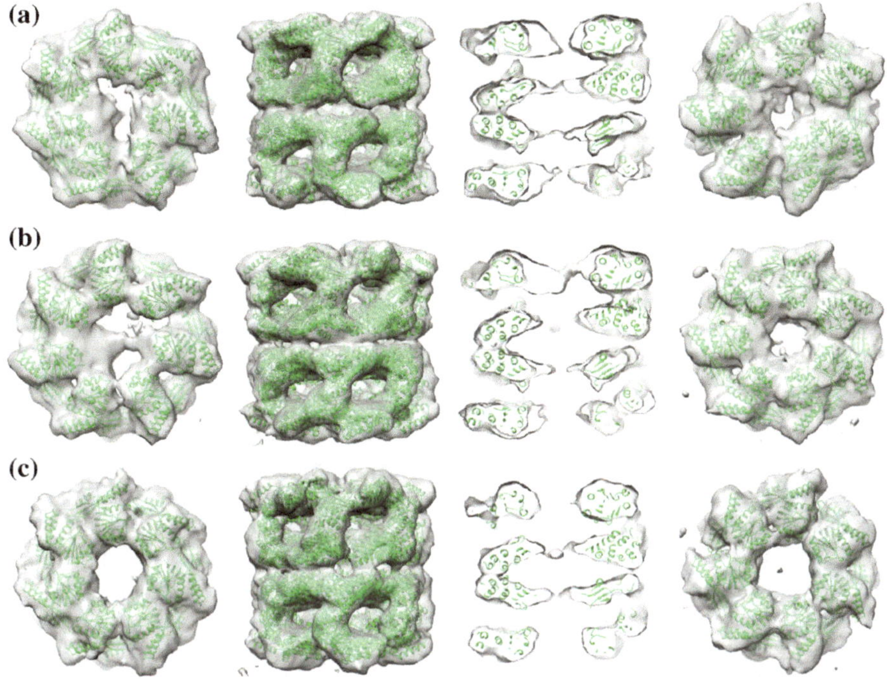

Fig. 4 Asymmetric (C1) 3D reconstructions of the three classes (structures). Each class (class 1 (**a**); class 2 (**b**); and class 3 (**c**)) is shown as a top view (top ring only), a side view, a central section through the side view, and a bottom view (bottom ring only). The fitted GroEL crystal structure is shown in green. The additional density in the upper rings of (**a**) and (**b**) is attributable to bound non-native RuBisCO substrate. All maps were contoured at the 1σ level without filtering. Figure generated with Chimera [85]. Figure produced by the author in https://doi.org/10.1016/j.ijbiomac.2018.06.120 [74] and reproduced here under Creative Commons Attribution License (CC BY)

[65] in the year 1998 and other program suites followed. The clustering of similar particle images was first introduced by van Heel and Frank [68] in the year 1981 using multivariate statistical analysis. Clustering in the currently available programs uses one of the following methods: multivariate statistical analysis (MSA)/principle component analysis (PCA), hierarchical clustering, k-means clustering, and the maximum-likelihood methods [41] or by recently proposed empirical Bayesian approach [69]. Currently, the EMAN2 [49] and RELION-3 [64] are among the popular program that do reference-free 2D class-averaging (references are generated from within the data set) and 3D reconstruction. EMAN2 uses iterative MSA-based reference-free 2D classification. The latest one, the RELION, uses empirical Bayesian likelihood approach for 2D classification [55].

Next step is to get the 3D reconstruction from selected good class averages. High-resolution 3D reconstructions require an initial 3D model that can be iteratively refined to obtain the best possible resolution for the data set. The first starting

3D model is obtained using experimental methods or by finding the relative orientations of 2D projection averages (and hence the particles) by computational methods. Assigning orientations by programs involves finding the location and Euler angles of the particles in the boxed region. The earliest one among them was the popular angular reconstitution method [70] by Marin van Heel, which uses real-space implementation of "common lines" principle to get relative orientations of the class averages as implemented in the program IMAGIC [51]. Thus, the Euler angles assigned 2D class averages can be used to get the starting 3D model. This method does not require reference for assigning relative orientation, while another program Spider by Joachim Frank and co-workers uses projection matching and cross-correlation approach [63, 71]. This method requires a starting 3D model which is generated from *ab initio* random conical tilt method [72] from EM images taken at a pair of know angles. Most of the present-day programs generate the starting 3D model by using statistical approach and comparison with back-projections to assign the Euler angles to a subset of manually selected good class averages. For example, EMAN2 uses a Monte Carlo method, RELION uses Bayesian methods, and VIPER [73] a module in SPHIRE suite (http://sphire.mpg.de/) uses a stochastic hill-climbing algorithm. Iterative rounds of projection matching with the references generated from starting 3D model (called as 3D projection matching procedure) followed by subsequent 3D reconstruction (using various algorithms) are used until the resolution of the reconstruction during subsequent refinement cycles does not further improve. This will lead to the final 3D reconstruction with the best possible resolution.

Figure 4 shows an asymmetric (C1 symmetry) 3D reconstruction carried out using IMAGIC and Spider. The non-native RuBisCO bound to GroEL is shown [74]. Figure 5 is another example of 1.9 Å high-resolution cryo-EM reconstruction with inhibitor phenylethyl β-D-thiogalactopyranoside (PETG) bound to β-galactosidase enzyme [75]. The quality of the final 3D reconstruction not only depends on the quality of the projection images and implementation of the clever

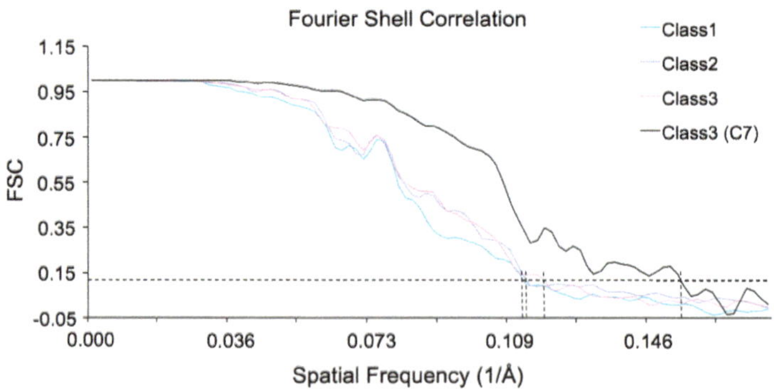

Fig. 5 Fourier shell correlation (FSC) curve for class 1, class 2, and class 3 asymmetric reconstruction and class 3 (C7 symmetry reconstruction) shown in Fig. 4. Vertical dashed lines show the spatial frequency for 0.143 "gold-standard" FSC which estimates classes 1, 2, and 3 resolution to be ∼9.0 Å and class 3 (C7 symmetry) as ∼7.6 Å

algorithms, but also on the angular distribution of the particles. Hence, in order to get the best resolution reconstruction, it is necessary for the particle (and thus its projections) to be distributed well in the Euler sphere [41]. By re-projecting the 3D reconstruction at the Euler angles of the class averages, we can assess the reliability of the 3D reconstruction. For a consistent reliable reconstruction, the re-projected image and the actual class average must match.

3 Resolution, Model Building, and Validation

3.1 Resolution

Resolution estimation of the EM maps is still subjective, with differences among various groups still not settled [76]. Resolution of 3D EM map is calculated from a plot of Fourier shell correlation (FSC) [77] as a function of spatial frequency (the resolution estimation of 3D reconstructions in Fig. 4 is shown in Fig. 5). FSC is the cross-correlation (CC) calculated between two 3D reconstruction maps, where each map is calculated from half the data images. The resolution that is reported in publication essentially as a single number is the value of maximum spatial frequency up to which the EM map is reliable. The identification of resolution is subjective as it is arbitrary what one considers as reliable. The procedure for resolution assessment is described in detail by Penczek [76]. There are several suggestions for identifying the cutoff: (i) the 3-sigma criteria where the spectral SNR (SSNR) = 0 in which case FSC = 0; (ii) point at which power of signal is equal to the power of noise, i.e., SSNR = 1 or FSC = 0.33; (iii) the classic midpoint of FSC curve, i.e., FSC = 0.5 [78] where SSNR = 2, which means signal dominates noise; and finally (iv) point where FSC = 0.143, derived by Rosenthal and Henderson [79] in comparison with X-ray crystallography. Hence, which cutoff is chosen is a matter of present-day debate. Recently, in order to reduce further any possible reference bias, "gold-standard FSC" was suggested with FSC calculated between two completely independent refinements and 3D reconstruction [80].

There are other computational ways to improve the resolution nominally without improving the image alignment, e.g., masking/threshold flattening. In any case, the resolution estimations have their own limitations and hence reported EM resolution should be treated as only broad guideline rather than a definitive number and cannot be used as validation. Nonetheless, it is an important parameter to be reported with each EM map deposition at the EMDB. Resolution anisotropy is common in cryo-EM structures, and it is a common practice to document it as color ramping from low to high resolution on the cryo-EM 3D reconstruction map using programs *ResMap* [81] and *blocres* [82]. The results can be visualized independently or with chimera (e.g., *blocres* with Local FSC plug-in for chimera).

With the booming medium- and high-resolution cryo-EM 3D structures, it is necessary to have consistency between crystallography and cryo-EM terms currently used for defining what is an atomic- or high-resolution structure. While it is very common to use the term "atomic resolution" for cryo-EM resolutions better than 3.5 Å, the crystallography definition of the term "***atomic resolution***" means the resolution is 1.2 Å or better [1] and ***ultra-high resolution*** means 0.95 Å or better ([83] and references cited therein). Similarly, 1.8 Å or better is called ***high resolution*** [84], 3.0 Å or better up to 1.9 Å is treated as ***medium resolution*** while ***low resolution*** is between 4 and 3.1 Å. Resolution below 4 Å is considered as ***poor resolution*** in protein crystallography. While the method of estimation of resolution is quite different between crystallography and cryo-EM techniques, the conventions for using the terms should be consistent, irrespective of the method. Hence, the author would like to suggest that it is necessary for the cryo-EM field to maintain consistency in the future, while using the terms ultra-high, atomic, high, medium, and low resolution.

3.2 Model Building

If the resolution of the 3D reconstruction (i.e., the electron potential map) is sufficiently high, e.g., better than 3 or 4 Å, it is often possible to build *ab initio* atomic model and do refinement with the EM map using known chemical constraints/restraints. If X-ray crystallography coordinates of the segment or its homologues are available, one can rigid body fit the segment coordinates into the cryo-EM map using programs like UCSF Chimera [85]. Where the resolution of the 3D reconstruction map is limited to worse than 4 Å, combining crystallography and cryo-EM as a hybrid method is a powerful tool to obtain a pseudo-atomic model (s). Iterative rounds of model building using programs like Coot [86], O [87] and refinement using programs like Coot, refmac [88], or PHENIX real-space-refinement [89] are carried out. De novo backbone tracing and model building can be carried out using programs like Pathwalking and Gorgon [90]; it can also build macromolecular assemblies at non-atomic resolution [90]. When the cryo-EM map shows variation by domain movements or flexibility to the available protein coordinates, programs like FlexEM [91, 92] and MDFF [93] with its graphical user interface VMD [94] can be used to flexibly fit the coordinates in the EM map. The fitted model and EM map can be visualized in programs like PyMOL [95]/Chimera [85] to generate publication quality figures.

3.3 Validation

Validation in cryo-EM reconstruction is important to avoid errors in particle alignment, reference bias, over-fitting of atomic coordinates, and over-estimation of resolution. Validation tools for cryo-EM similar to the free *R* value (R_{free}) in X-ray crystallography [96] have been introduced in 2003 by Joachim and co-workers [97] and recently by Chen and co-workers [98]. There are some general guidelines [80, 99] suggested, and they are actively evolving. As described in the resolution paragraph of this section, the reporting of resolution with one number cannot be used as validation; however, it is an important parameter to be reported during EMDB deposition. Further, FSC may fail when the particles are significantly misaligned. So one has to estimate the resolution properly [76] and use the reported single number resolution with caution. It is suggested that gold-standard FSC provides a realistic estimate of the true signal [100], and this will lead ultimately to a better map. In recent days, reporting local resolution has also formed a common practice in publication and thesis [81, 82]. Also, the local resolution will be helpful in avoiding over-interpretation of poor regions in the cryo-EM map. If the 3D map is of sufficient resolution (better than 4 Å), it can resolve the secondary structural features. A good validation would be especially if you can see a right-handed alpha helix or even the side chain residues, especially the bulky residues like tryptophan, phenylalanine, or tyrosine in the high-resolution cryo-EM map. Even the comparison of the new EM structure with the available EM structure will be one way of validating the newly reconstructed 3D EM map [101]. Further, programs TEMPy [102] and refmac [88] can be used to assess the validity of the fitted coordinates to the EM map. In Coot [86] program, one can use the Ramachandran plot (Validate →Ramachandran plot) and geometrical quality (Validate →Geometry analysis) to validate the quality of the refined model. One more way to validate is to compare the 3D reconstruction results from different techniques, e.g., projection matching and the angular reconstitution. For low-resolution maps (worse than 4 or 10 Å), measure of confidence can be provided by a priori random conical tilt experiments [103].

4 Heterogeneity

Though cryo-EM can handle heterogeneous particles, we need homogenous particles, which are equally dispersed in the vitrified ice in order to achieve atomic resolution. Ideally speaking, all data sets are heterogeneous! The question is how much one is willing to tolerate [104]. Further, during cryo-EM specimen preparation, non-physiological structural heterogeneity is often introduced [105]. While structural heterogeneity is a problem to obtain high resolution, it also provides a unique opportunity to study the conformational flexibility/dynamics of the macromolecular assemblies. Ideally, homogenous samples have to be biochemically standardized and prepared before the vitrification process. However, this is not possible with all protein samples due to the inherent protein flexibility which is necessary for its function, for example, rotation of 30S subunit of ribosome [106] or rotational states in case of eukaryotic V-ATPase [107]. In such cases, the heterogeneous sample data images can be classified computationally to classes containing homogenous particles (an example of such classification can be seen in Fig. 4).

Three main techniques are currently in use to identify and sort the macromolecular structural conformational variability or heterogeneity [41]. The first approach depends on classifying the 2D images based on the eigen images/ eigenvectors [74, 108–111] without any starting model. First, classify using MSA to obtain orientation classes, and then, the major variation among the picked particles in each orientation classes can be identified in the low-order eigen images by MSA, and using these, information particles can be classified into homogenous classes, leading to preliminary 3D reconstruction from a class containing majority of homogenous particles as shown in Fig. 3. The preliminary 3D reconstructions can be projected as references for competitive alignment. Further, the quality of 3D reconstruction can be iteratively improved until the eigen images show no major variations within the class and the particles stabilize from jumping to another class during competitive projection matching. In this manner, three class reconstructions were obtained as shown in Fig. 4. The second method depends on detection in 2D variations using starting model [41]. The third method also needs initial starting model and uses a statistical approach to obtain 3D classification. In this case, large number of 3D maps are calculated from randomly selected subset of particles (with previously assigned orientation based on initial 3D map). Determination of the 3D variance can be used to assess the heterogeneity, and estimation of covariance enables one to carry out the 3D classification according to variable regions. Alternatively, the molecular states can be separated using maximum likelihood classification [104, 112] or by the latest multi-body refinement method [113].

5 Single-Particle Cryo-EM Applications in SBDD

There are five 3D reconstructions in the EMDB with bound inhibitors or ligands at resolution better than 2.5 Å as shown in Table 1. We have focused at this resolution since this is at the center of medium (3.0 Å) and high resolution (1.8 Å), which is desired resolution for SBDD studies. Although we have highlighted reconstructions better than 2.5 Å, ligands have been visualized in the 3D reconstructions better than 4.0 Å. While there are only 10s of EM reconstructions with bound inhibitors at 2.5 Å or better, there are several 100s of structures in the EMDB at resolution between 2.5 and 4 Å with bound inhibitors or ligands. Here are couple of examples of 3D reconstructions with resolution better than 2.5 Å: In the Sect. 2.3 we have already come across the example of 3D reconstruction (by the Subramaniam group [75]) of the inhibitor PETG bound to beta-galactosidase at 1.9 Å resolution. His group used a similar approach to solve the cryo-EM structure of human p97ATPase, an important target for cancer, in complex with its allosteric inhibitor UPCDC30245 [10] as shown in Fig. 7c. Although they could not see a part of the inhibitor in the EM electron potential map, they could see at 2.3 Å resolution the other part where the inhibitor snugly fits into the protein pocket and proposed how the allosteric inhibitor UPCDC30245 inhibits the conformational changes necessary for the function of p97. They further could see three coexisting functional states of p97 in the presence and absence of ATPγS. Here are couple of examples of 3D reconstructions of proteins bound to inhibitors, with 3D reconstruction resolution

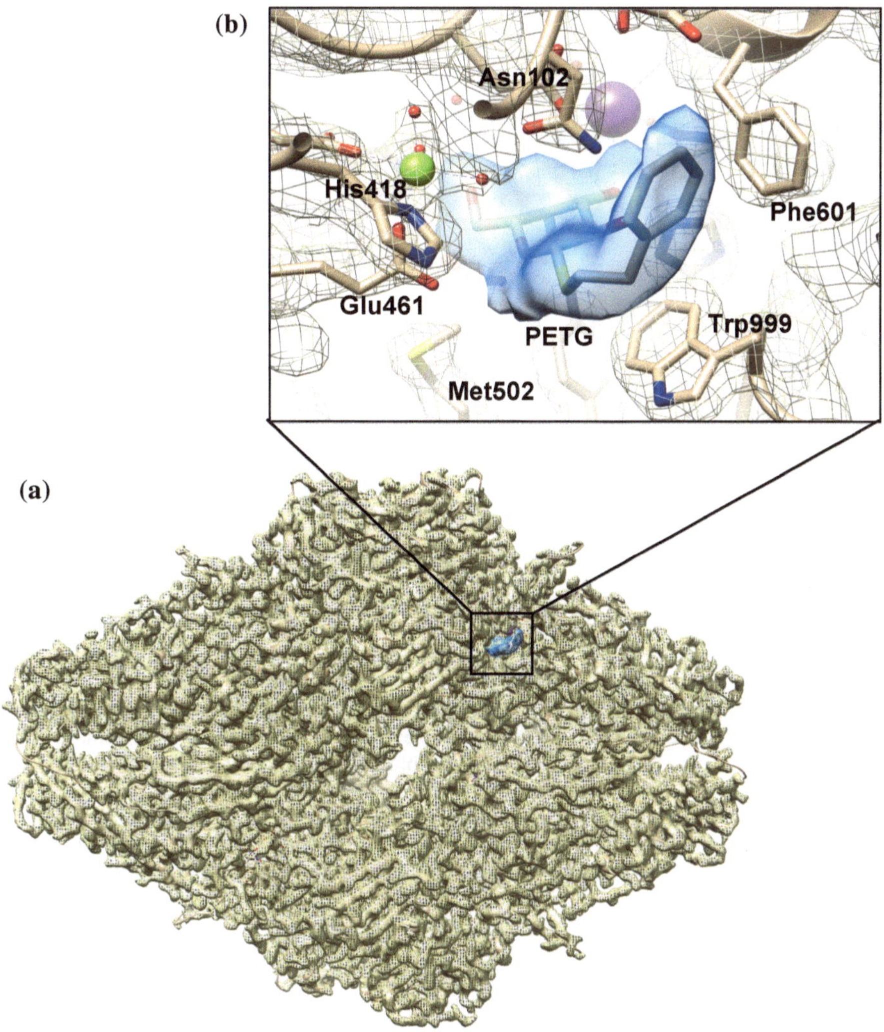

Fig. 6 **a** Inhibitor phenylethyl β-D-thiogalactopyranoside (PETG) (blue surface)-bound cryo-EM structure of β-galactosidase enzyme at 1.9 Å resolution [75]. **b** Zoom-in view of the squared area with bound inhibitor PETG (blue surface). The EM map is shown in yellow mesh. Sodium (magenta) and Mg^{2+} (green) ions and water molecules (red) can be seen in the pocket

poorer than 2.5 Å. For example, Paula and Ed solved the structure of 20S proteasome in complex with the inhibitor (EMD-3231) at ~3.6 Å resolution [114] as shown in Fig. 7a. Another example is the structure of 70S ribosome from *Escherichia coli* at 2.9 Å resolution in complex with elongation factor Tu, aminoacyl-tRNA, and the antibiotic kirromycin [115] as shown in Fig. 7b. With these examples, it is very clear that, in the future, the single-particle cryo-EM will play a very important role in the preclinical SBDD studies.

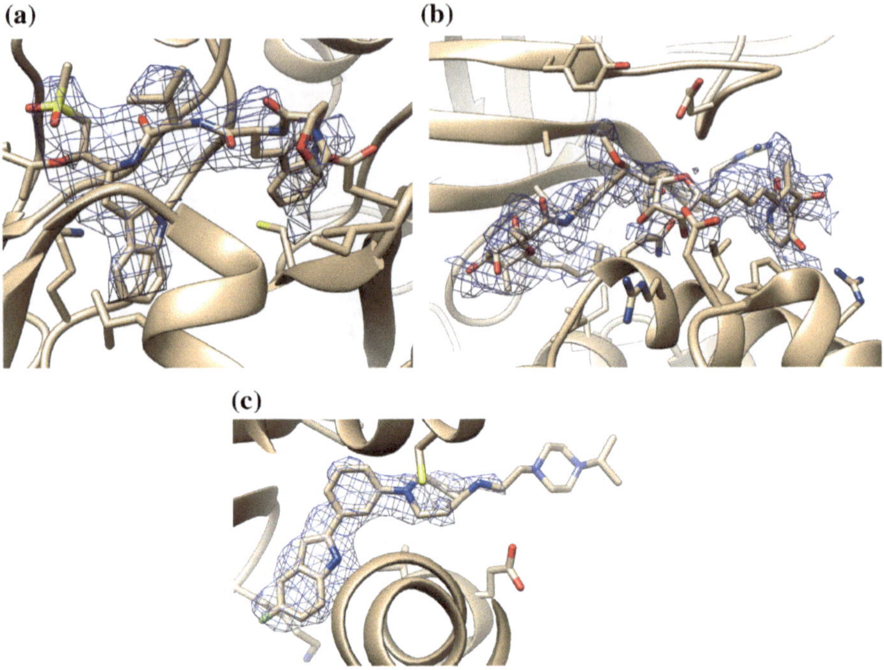

Fig. 7 Pharmacologically important target proteins (beige color) (modeled using the cryo-EM electron potential 3D reconstruction map) in complex with inhibitor (shown as stick model). The cryo-EM map of the inhibitors are shown as blue mesh. **a** 3.6 Å reconstruction of 20S *Plasmodium falciparum* proteasome [114] with bound inhibitor Mor-WLW vinyl sulfone (EMD3231) (PDB ID: 5fmg). **b** 2.9 Å cryo-EM reconstruction of complete 70S *Escherichia coli* ribosome with bound antibiotic kirromycin [115] (EMD 2847, PDB ID: 5afi). **c** 2.3 Å resolution 3D reconstruction of anticancer drug target human p97 with bound allosteric inhibitor UPCDC30245 [10]

6 Conclusions and Future Prospective

Recent advances in cryo-EM have enabled us to use single-particle cryo-EM as a method of choice to resolve solution-state 3D structures of proteins and protein complexes at atomic resolution, thus breaking the cryo-EM resolution barrier to facilitate SBDD [6]. In recent years, many pharmaceutical companies like Bayer, Merck Research Laboratories, Sonafi, AstraZeneca, Regeneron Pharmaceuticals, NovAliX, Genentech etc. have realized the importance of this method and started hiring experts in single-particle cryo-EM to get involved in their SBDD pipeline. Table 1 lists the protein structures with bound ligands solved by single particle cryo-EM at resolution 2.5 Å or better; i.e., five of the structures have bound inhibitors/glycans, which underscore the importance of single-particle cryo-EM in SBDD. Apart from these, there are many more structures with bound ligands in the EMDB at resolutions below 2.5 Å. The future of this technique will be in obtaining the sub-nanometer resolution and perhaps atomic-resolution structures of proteins and protein complexes in vivo. This methodology called the cellular tomography, although not the scope of this chapter, is a promising future technology for atomic-resolution structures of proteins and protein complexes in its native environment "the cell." With the advent of phase plates, energy filters, and automation in cryo-EM data collection, promising efforts are being made to achieve that goal and the realization of that goal may not be far away which would in turn potentially further accelerate the SBDD program.

Acknowledgements RN was supported by Ramalingaswamy Fellowship from DBT. RN would like to thank his laboratory members and colleagues for their constant support, valuable scientific and technical discussion. Last but not least, RN would like to thank IISER-TVM past Director Prof. ED Jemmis and present Director Prof. V. Ramakrishnan for their unstinted support.

References

1. Dauter Z, Lamzin VS, Wilson KS (1995) Proteins at atomic resolution. Curr Opin Struct Biol 5:784–790
2. Kozma D, Simon I, Tusnady GE (2013) PDBTM: protein data bank of transmembrane proteins after 8 years. Nucleic Acids Res 41:D524–D529
3. Natesh R (2014) Crystallography beyond crystals: PX and SP cryoEM. Resonance 19: 1177–1196
4. Mountain V (2003) Astex, structural genomix, and syrrx. i can see clearly now: structural biology and drug discovery. Chem Biol 10:95–98
5. Bartesaghi A, Merk A, Banerjee S, Matthies D, Wu X, Milne JL, Subramaniam S (2015) A resolution cryo-EM structure of beta-galactosidase in complex with a cell-permeant inhibitor. Science 348:1147–1151
6. Merk A, Bartesaghi A, Banerjee S, Falconieri V, Rao P, Davis MI, Pragani R, Boxer MB, Earl LA, Milne JLS, Subramaniam S (2016) Breaking cryo-EM resolution barriers to facilitate drug discovery. Cell 165:1698–1707
7. Shalev-Benami M, Zhang Y, Rozenberg H, Nobe Y, Taoka M, Matzov D, Zimmerman E, Bashan A, Isobe T, Jaffe CL, Yonath A, Skiniotis G (2017) Atomic resolution snapshot of Leishmania ribosome inhibition by the aminoglycoside paromomycin. Nat Commun 8:1589
8. Dong Y, Liu Y, Jiang W, Smith TJ, Xu Z, Rossmann MG (2017) Antibody-induced uncoating of human rhinovirus B14. In: Proceedings of the national academy of sciences of the United States of America, vol 114, pp 8017–8022
9. Danev R, Tegunov D, Baumeister W (2017) Using the Volta phase plate with defocus for cryo-EM single particle analysis. Elife 6:1–9
10. Banerjee S, Bartesaghi A, Merk A, Rao P, Bulfer SL, Yan Y, Green N, Mroczkowski B, Neitz RJ, Wipf P, Falconieri V, Deshaies RJ, Milne JL, Huryn D, Arkin M, Subramaniam S (2016) A resolution cryo-EM structure of human p97 and mechanism of allosteric inhibition. Science 351:871–875
11. Dubochet J, Mcdowall AW (1981) Vitrification of pure water for electron-microscopy. J Microsc Oxford 124:Rp3–Rp4
12. Vogel RH, Provencher SW, von Bonsdorff CH, Adrian M, Dubochet J (1986) Envelope structure of Semliki Forest virus reconstructed from cryo-electron micrographs. Nature 320:533–535
13. Henderson R, Unwin PN (1975) Three-dimensional model of purple membrane obtained by electron microscopy. Nature 257:28–32

14. Frank J (1975) Averaging of low exposure electron micrographs of non-periodic objects. Ultramicroscopy 1:159–162
15. Frank J, Al-Ali L (1975) Signal-to-noise ratio of electron micrographs obtained by cross correlation. Nature 256:376–379
16. Saxton WO, Frank J (1977) Motif detection in quantum noise-limited electron micrographs by cross-correlation. Ultramicroscopy 2:219–227
17. Henderson R, Baldwin JM, Ceska TA, Zemlin F, Beckmann E, Downing KH (1990) Model for the structure of bacteriorhodopsin based on high-resolution electron cryo-microscopy. J Mol Biol 213:899–929
18. Khoshouei M, Radjainia M, Baumeister W, Danev R (2017) Cryo-EM structure of haemoglobin at 3.2 A determined with the Volta phase plate. Nat Commun 8:16099
19. Henderson R (1995) The potential and limitations of neutrons, electrons and X-rays for atomic resolution microscopy of unstained biological molecules. Q Rev Biophys 28: 171–193
20. Yu X, Jin L, Zhou ZH (2008) 3.88 Å structure of cytoplasmic polyhedrosis virus by cryo-electron microscopy. Nature 453:415–419
21. Zhang X, Sun S, Xiang Y, Wong J, Klose T, Raoult D, Rossmann MG (2012) Structure of Sputnik, a virophage, at 3.5 Å resolution. Proc Nat Acad Sci USA 109:18431–18436
22. Chen DH, Baker ML, Hryc CF, DiMaio F, Jakana J, Wu W, Dougherty M, Haase-Pettingell C, Schmid MF, Jiang W, Baker D, King JA, Chiu W (2011) Structural basis for scaffolding-mediated assembly and maturation of a dsDNA virus. Proc Nat Acad Sci USA 108:1355–1360
23. Dubochet J, Adrian M, Chang JJ, Homo JC, Lepault J, McDowall AW, Schultz P (1988) Cryo-electron microscopy of vitrified specimens. Q Rev Biophys 21:129–228
24. Passmore LA, Russo CJ (2016) Specimen preparation for high-resolution cryo-EM. Methods Enzymol 579:51–86
25. Russo CJ, Passmore LA (2016) Progress towards an optimal specimen support for electron cryomicroscopy. Curr Opin Struct Biol 37:81–89
26. Li X, Mooney P, Zheng S, Booth CR, Braunfeld MB, Gubbens S, Agard DA, Cheng Y (2013) Electron counting and beam-induced motion correction enable near-atomic-resolution single-particle cryo-EM. Nat Methods 10:584–590
27. McMullan G, Clark AT, Turchetta R, Faruqi AR (2009) Enhanced imaging in low dose electron microscopy using electron counting. Ultramicroscopy 109:1411–1416
28. Zheng SQ, Palovcak E, Armache JP, Verba KA, Cheng Y, Agard DA (2017) MotionCor2: anisotropic correction of beam-induced motion for improved cryo-electron microscopy. Nat Methods 14:331–332
29. Abrishami V, Vargas J, Li X, Cheng Y, Marabini R, Sorzano CO, Carazo JM (2015) Alignment of direct detection device micrographs using a robust optical flow approach. J Struct Biol 189:163–176
30. de la Rosa-Trevin JM, Oton J, Marabini R, Zaldivar A, Vargas J, Carazo JM, Sorzano CO (2013) Xmipp 3.0: an improved software suite for image processing in electron microscopy. J Struct Biol 184:321–328
31. Brilot AF, Chen JZ, Cheng A, Pan J, Harrison SC, Potter CS, Carragher B, Henderson R, Grigorieff N (2012) Beam-induced motion of vitrified specimen on holey carbon film. J Struct Biol 177:630–637
32. Campbell MG, Cheng A, Brilot AF, Moeller A, Lyumkis D, Veesler D, Pan J, Harrison SC, Potter CS, Carragher B, Grigorieff N (2012) Movies of ice-embedded particles enhance resolution in electron cryo-microscopy. Structure 20:1823–1828
33. Russo CJ, Passmore LA (2014) Electron microscopy: ultrastable gold substrates for electron cryomicroscopy. Science 346:1377–1380
34. Russo CJ, Passmore LA (2014) Controlling protein adsorption on graphene for cryo-EM using low-energy hydrogen plasmas. Nat Methods 11:649–652

35. Suloway C, Pulokas J, Fellmann D, Cheng A, Guerra F, Quispe J, Stagg S, Potter CS, Carragher B (2005) Automated molecular microscopy: the new Leginon system. J Struct Biol 151:41–60
36. Mastronarde DN (2005) Automated electron microscope tomography using robust prediction of specimen movements. J Struct Biol 152:36–51
37. Li X, Zheng S, Agard DA, Cheng Y (2015) Asynchronous data acquisition and on-the-fly analysis of dose fractionated cryoEM images by UCSFImage. J Struct Biol 192:174–178
38. Zhang J, Nakamura N, Shimizu Y, Liang N, Liu X, Jakana J, Marsh MP, Booth CR, Shinkawa T, Nakata M, Chiu W (2009) JADAS: a customizable automated data acquisition system and its application to ice-embedded single particles. J Struct Biol 165:1–9
39. Lander GC, Stagg SM, Voss NR, Cheng A, Fellmann D, Pulokas J, Yoshioka C, Irving C, Mulder A, Lau PW, Lyumkis D, Potter CS, Carragher B (2009) Appion: an integrated, database-driven pipeline to facilitate EM image processing. J Struct Biol 166:95–102
40. Hoppe W (1983) Electron-diffraction with the transmission electron-microscope as a phase-determining diffractometer—from spatial-frequency filtering to the 3-Dimensional structure-analysis of ribosomes. Angew Chem Int Edit 22:456–485
41. Orlova EV, Saibil HR (2011) Structural analysis of macromolecular assemblies by electron microscopy. Chem Rev 111:7710–7748
42. Erickson HP, Klug A (1971) Measurement and compensation of defocusing and aberrations by fourier processing of electron micrographs. Philos T Roy Soc B. 261:105–118
43. Wade RH (1992) A brief look at imaging and contrast transfer. Ultramicroscopy 46:145–156
44. Bell JM, Chen M, Baldwin PR, Ludtke SJ (2016) High resolution single particle refinement in EMAN2.1. Methods 100:25–34
45. Rohou A, Grigorieff N (2015) CTFFIND4: fast and accurate defocus estimation from electron micrographs. J Struct Biol 192:216–221
46. Zhang K (2016) Gctf: real-time CTF determination and correction. J Struct Biol 193:1–12
47. Mallick SP, Carragher B, Potter CS, Kriegman DJ (2005) ACE: automated CTF estimation. Ultramicroscopy 104:8–29
48. Roseman AM (2004) FindEM–a fast, efficient program for automatic selection of particles from electron micrographs. J Struct Biol 145:91–99
49. Tang G, Peng L, Baldwin PR, Mann DS, Jiang W, Rees I, Ludtke SJ (2007) EMAN2: an extensible image processing suite for electron microscopy. J Struct Biol 157:38–46
50. Ludtke SJ, Bell JM, Chen M, Baldwin PR, Ludtke SJ (2016) Single-particle refinement and variability analysis in EMAN2.1, high resolution single particle refinement in EMAN2.1. Methods Enzymol 579:159–189
51. van Heel M, Harauz G, Orlova EV, Schmidt R, Schatz M (1996) A new generation of the IMAGIC image processing system. J Struct Biol 116:17–24
52. Smith JM (1999) Ximdisp–a visualization tool to aid structure determination from electron microscope images. J Struct Biol 125:223–228
53. Wood C, Burnley T, Patwardhan A, Scheres S, Topf M, Roseman A, Winn M (2015) Collaborative computational project for electron cryo-microscopy. Acta Crystallogr D Biol Crystallogr 71:123–126
54. Burnley T, Palmer CM, Winn M (2017) Recent developments in the CCP-EM software suite. Acta Crystallogr Sect D Struct Biol 73:469–477
55. Scheres SH (2012) RELION: implementation of a Bayesian approach to cryo-EM structure determination. J Struct Biol 180:519–530
56. Punjani A, Rubinstein JL, Fleet DJ, Brubaker MA (2017) cryoSPARC: algorithms for rapid unsupervised cryo-EM structure determination. Nat Methods 14:290–296
57. Heimowitz A, Andén J, Singer A (2018) APPLE picker: automatic particle picking, a low-effort Cryo-EM framework. J Struct Biol 204(2):215–227
58. Singer A et al. (2010) Algorithms for Single Particle Reconstruction (ASPIRE), http://spr.math.princeton.edu/

59. Hoang TV, Cavin X, Schultz P, Ritchie DW (2013) gEMpicker: a highly parallel GPU-accelerated particle picking tool for cryo-electron microscopy. BMC Struct Biol 13:25
60. Chen JZ, Grigorieff N (2007) SIGNATURE: a single-particle selection system for molecular electron microscopy. J Struct Biol 157:168–173
61. Henderson R (2013) Avoiding the pitfalls of single particle cryo-electron microscopy: Einstein from noise. Proc Nat Acad Sci USA 110:18037–18041
62. De Rosier DJ, Klug A (1968) Reconstruction of three dimensional structures from electron micrographs. Nature 217:130–134
63. Frank J, Radermacher M, Penczek P, Zhu J, Li Y, Ladjadj M, Leith A (1996) SPIDER and WEB: processing and visualization of images in 3D electron microscopy and related fields. J Struct Biol 116:190–199
64. Zivanov J, Nakane T, Scheres S (2018) A Bayesian approach to beam-induced motion correction in cryo-EM single-particle analysis, *bioRxiv*
65. Lyumkis D, Brilot AF, Theobald DL, Grigorieff N (2013) Likelihood-based classification of cryo-EM images using FREALIGN. J Struct Biol 183:377–388
66. de la Rosa-Trevin JM, Quintana A, Del Cano L, Zaldivar A, Foche I, Gutierrez J, Gomez-Blanco J, Burguet-Castell J, Cuenca-Alba J, Abrishami V, Vargas J, Oton J, Sharov G, Vilas JL, Navas J, Conesa P, Kazemi M, Marabini R, Sorzano CO, Carazo JM (2016) Scipion: a software framework toward integration, reproducibility and validation in 3D electron microscopy. J Struct Biol 195:93–99
67. Frank J, Shimkin B, Dowse H (1981) Spider—a modular software system for electron image processing. Ultramicroscopy 6:343–357
68. van Heel M, Frank J (1981) Use of multivariate statistics in analysing the images of biological macromolecules. Ultramicroscopy 6:187–194
69. Scheres SH (2012) A Bayesian view on cryo-EM structure determination. J Mol Biol 415:406–418
70. Van Heel M (1987) Angular reconstitution: a posteriori assignment of projection directions for 3D reconstruction. Ultramicroscopy 21:111–123
71. Penczek PA, Grassucci RA, Frank J (1994) The ribosome at improved resolution: new techniques for merging and orientation refinement in 3D cryo-electron microscopy of biological particles. Ultramicroscopy 53:251–270
72. Radermacher M, Wagenknecht T, Verschoor A, Frank J (1987) Three-dimensional reconstruction from a single-exposure, random conical tilt series applied to the 50S ribosomal subunit of Escherichia coli. J Microsc 146:113–136
73. Penczek PA (2014).http://sparx-em.org/sparxwiki/sxviper
74. Natesh R, Clare DK, Farr GW, Horwich AL, Saibil HR (2018) A two-domain folding intermediate of RuBisCO in complex with the GroEL chaperonin. Int J Biol Macromol 118:671–675
75. Bartesaghi A, Aguerrebere C, Falconieri V, Banerjee S, Earl LA, Zhu X, Grigorieff N, Milne JLS, Sapiro G, Wu X, Subramaniam S (2018) Atomic resolution cryo-EM structure of beta-galactosidase. Structure 26(848–856):e3
76. Penczek PA (2010) Resolution measures in molecular electron microscopy. Methods Enzymol 482:73–100
77. Harauz G, van Heel M (1986) Exact filters for general geometry three-dimensional reconstruction. Optik 73:146–156
78. van Heel M, Schatz M (2005) Fourier shell correlation threshold criteria. J Struct Biol 151:250–262
79. Rosenthal PB, Henderson R (2003) Optimal determination of particle orientation, absolute hand, and contrast loss in single-particle electron cryomicroscopy. J Mol Biol 333:721–745
80. Henderson R, Sali A, Baker ML, Carragher B, Devkota B, Downing KH, Egelman EH, Feng Z, Frank J, Grigorieff N, Jiang W, Ludtke SJ, Medalia O, Penczek PA, Rosenthal PB, Rossmann MG, Schmid MF, Schroder GF, Steven AC, Stokes DL, Westbrook JD, Wriggers W, Yang H, Young J, Berman HM, Chiu W, Kleywegt GJ, Lawson CL (2012) Outcome of the first electron microscopy validation task force meeting. Structure 20:205–214

81. Kucukelbir A, Sigworth FJ, Tagare HD (2014) Quantifying the local resolution of cryo-EM density maps. Nat Methods 11:63–65
82. Cardone G, Heymann JB, Steven AC (2013) One number does not fit all: mapping local variations in resolution in cryo-EM reconstructions. J Struct Biol 184:226–236
83. Natesh R, Manikandan K, Bhanumoorthy P, Viswamitra MA, Ramakumar S (2003) Thermostable xylanase from Thermoascus aurantiacus at ultrahigh resolution (0.89 A) at 100 K and atomic resolution (1.11 A) at 293 K refined anisotropically to small-molecule accuracy. Acta Crystallogr D Biol Crystallogr 59:105–117
84. Natesh R, Bhanumoorthy P, Vithayathil PJ, Sekar K, Ramakumar S, Viswamitra MA (1999) Crystal structure at 1.8 A resolution and proposed amino acid sequence of a thermostable xylanase from Thermoascus aurantiacus. J Mol Biol 288:999–1012
85. Pettersen EF, Goddard TD, Huang CC, Couch GS, Greenblatt DM, Meng EC, Ferrin TE (2004) UCSF Chimera–a visualization system for exploratory research and analysis. J Comput Chem 25:1605–1612
86. Emsley P, Cowtan K (2004) Coot: model-building tools for molecular graphics. Acta Crystallogr D Biol Crystallogr 60:2126–2132
87. Jones TA (2004) Interactive electron-density map interpretation: from INTER to O. Acta Crystallogr D Biol Crystallogr 60:2115–2125
88. Brown A, Long F, Nicholls RA, Toots J, Emsley P, Murshudov G (2016) Tools for macromolecular model building and refinement into electron cryo-microscopy reconstructions. Acta Crystallogr D Biol Crystallogr 71:136–153
89. Echols N, Moriarty NW, Klei HE, Afonine PV, Bunkoczi G, Headd JJ, McCoy AJ, Oeffner RD, Read RJ, Terwilliger TC, Adams PD (2014) Automating crystallographic structure solution and refinement of protein-ligand complexes. Acta Crystallogr D Biol Crystallogr 70:144–154
90. Baker ML, Baker MR, Hryc CF, Ju T, Chiu W (2012) Gorgon and pathwalking: macromolecular modeling tools for subnanometer resolution density maps. Biopolymers 97:655–668
91. Topf M, Lasker K, Webb B, Wolfson H, Chiu W, Sali A (2008) Protein structure fitting and refinement guided by cryo-EM density. Structure 16:295–307
92. Joseph AP, Malhotra S, Burnley T, Wood C, Clare DK, Winn M, Topf M (2016) Refinement of atomic models in high resolution EM reconstructions using Flex-EM and local assessment. Methods 100:42–49
93. Trabuco LG, Villa E, Schreiner E, Harrison CB, Schulten K (2009) Molecular dynamics flexible fitting: a practical guide to combine cryo-electron microscopy and X-ray crystallography. Methods 49:174–180
94. Humphrey W, Dalke A, Schulten K (1996) VMD: visual molecular dynamics. J Mol Graph 14(33–8):27–28
95. Schrodinger LLC (2015) The PyMOL molecular graphics system, Version 1.8 in
96. Brunger AT (1992) Free R value: a novel statistical quantity for assessing the accuracy of crystal structures. Nature 355:472–475
97. Shaikh TR, Hegerl R, Frank J (2003) An approach to examining model dependence in EM reconstructions using cross-validation. J Struct Biol 142:301–310
98. Chen S, McMullan G, Faruqi AR, Murshudov GN, Short JM, Scheres SH, Henderson R (2013) High-resolution noise substitution to measure overfitting and validate resolution in 3D structure determination by single particle electron cryomicroscopy. Ultramicroscopy 135:24–35
99. Rosenthal PB, Rubinstein JL (2015) Validating maps from single particle electron cryomicroscopy. Curr Opin Struct Biol 34:135–144
100. Scheres SH, Chen S (2012) Prevention of overfitting in cryo-EM structure determination. Nat Methods 9:853–854
101. Murray SC, Flanagan J, Popova OB, Chiu W, Ludtke SJ, Serysheva II (2013) Validation of cryo-EM structure of IP(3)R1 channel. Structure 21:900–909

102. Farabella I, Vasishtan D, Joseph AP, Pandurangan AP, Sahota H, Topf M (2015) TEMPy: a Python library for assessment of three-dimensional electron microscopy density fits. J Appl Crystallogr 48:1314–1323
103. Henderson R, Chen S, Chen JZ, Grigorieff N, Passmore LA, Ciccarelli L, Rubinstein JL, Crowther RA, Stewart PL, Rosenthal PB (2011) Tilt-pair analysis of images from a range of different specimens in single-particle electron cryomicroscopy. J Mol Biol 413:1028–1046
104. Scheres SH (2010) Classification of structural heterogeneity by maximum-likelihood methods. Methods Enzymol 482:295–320
105. Elmlund D, Le SN, Elmlund H (2017) High-resolution cryo-EM: the nuts and bolts. Curr Opin Struct Biol 46:1–6
106. Valle M, Zavialov A, Sengupta J, Rawat U, Ehrenberg M, Frank J (2003) Locking and unlocking of ribosomal motions. Cell 114:123–134
107. Zhao J, Benlekbir S, Rubinstein JL (2015) Electron cryomicroscopy observation of rotational states in a eukaryotic V-ATPase. Nature 521:241–245
108. White HE, Orlova EV, Chen S, Wang L, Ignatiou A, Gowen B, Stromer T, Franzmann TM, Haslbeck M, Buchner J, Saibil HR (2006) Multiple distinct assemblies reveal conformational flexibility in the small heat shock protein Hsp26. Structure. 14:1197–1204
109. White HE, Saibil HR, Ignatiou A, Orlova EV (2004) Recognition and separation of single particles with size variation by statistical analysis of their images. J Mol Biol 336:453–460
110. Elad N, Clare DK, Saibil HR, Orlova EV (2008) Detection and separation of heterogeneity in molecular complexes by statistical analysis of their two-dimensional projections. J Struct Biol 162:108–120
111. Elad N, Farr GW, Clare DK, Orlova EV, Horwich AL, Saibil HR (2007) Topologies of a substrate protein bound to the chaperonin GroEL. Mol Cell 26:415–426
112. Scheres SH, Gao H, Valle M, Herman GT, Eggermont PP, Frank J, Carazo JM (2007) Disentangling conformational states of macromolecules in 3D-EM through likelihood optimization. Nat Methods 4:27–29
113. Nakane T, Kimanius D, Lindahl E, Scheres SH (2018) Characterisation of molecular motions in cryo-EM single-particle data by multi-body refinement in RELION. Elife. 7:1–18
114. da Fonseca PC, Morris EP (2015) Cryo-EM reveals the conformation of a substrate analogue in the human 20S proteasome core. Nat Commun 6:7573
115. Fischer N, Neumann P, Konevega AL, Bock LV, Ficner R, Rodnina MV, Stark H (2015) Structure of the E. coli ribosome-EF-Tu complex at <3 Å resolution by Cs-corrected cryo-EM. Nature 520:567–570

Index

C. G. Mohan (ed.), *Structural Bioinformatics: Applications in Preclinical Drug Discovery Process*, Challenges and Advances in Computational Chemistry and Physics 27, https://doi.org/10.1007/978-3-030-05282-9